AF675965

DEEP-SEA TANAIDACEA (PERACARIDA) FROM THE GULF OF MEXICO

CRUSTACEANA MONOGRAPHS constitutes a series of books on carcinology in its widest sense. Contributions are handled by the Editor-in-Chief and may be submitted through the office of KONINKLIJKE BRILL Academic Publishers N.V., P.O. Box 9000, NL-2300 PA Leiden, The Netherlands.

Editor-in-Chief: J. C. von Vaupel Klein, Bilthoven, Netherlands.

Editorial Committee: N. L. Bruce, Wellington, New Zealand; Mrs. M. Charmantier-Daures, Montpellier, France; D. L. Danielopol, Mondsee, Austria; Mrs. D. Defaye, Paris, France; J. Dorgelo, Amsterdam, Netherlands; J. Forest, Paris, France; C. H. J. M. Fransen, Leiden, Netherlands; R. G. Hartnoll, Port Erin, Isle of Man; L. B. Holthuis, Leiden, Netherlands; P. P. Jaros, Oldenburg, Germany; E. Macpherson, Blanes, Spain; P. K. L. Ng, Singapore, Rep. of Singapore; H.-K. Schminke, Oldenburg, Germany; F. R. Schram, Langley, WA, U.S.A.; S. F. Timofeev, Murmansk, Russia; G. van der Velde, Nijmegen, Netherlands; W. Vervoort, Leiden, Netherlands; H. P. Wagner, Leiden, Netherlands; D. I. Williamson, Port Erin, Isle of Man.

Published in this series:

CRM 001 - *Stephan G. Bullard*	Larvae of anomuran and brachyuran crabs of North Carolina
CRM 002 - *Spyros Sfenthourakis et al.*	The biology of terrestrial isopods, V
CRM 003 - *Tomislav Karanovic*	Subterranean Copepoda from arid Western Australia
CRM 004 - *Katsushi Sakai*	Callianassoidea of the world (Decapoda, Thalassinidea)
CRM 005 - *Kim Larsen*	Deep-sea Tanaidacea (Peracarida) from the Gulf of Mexico

In preparation (provisional titles):

CRM 00x - *Darren C. Yeo & Peter K. L. Ng*	Southeast Asian freshwater crabs
CRM 00y - *Frank D. Ferrari & Hans-U. Dahms*	Copepod development
CRM 00z - *Chang-tai Shih*	Marine Calanoida of the China seas
CRM 00p - *Katsushi Sakai*	Upogebiidae of the world (Decapoda, Thalassinidea)

Author's address:

Dr. K. Larsen, Department of Marine Biology, Texas A&M University Galveston, Fort Crockett Campus, Galveston, Texas 77551, U.S.A. *Current address for correspondence*: Department of Invertebrate Zoology, Zoological Museum, University of Copenhagen, Universitetsparken 15, DK-2100 Copenhagen Ø, Denmark; e-mail: tanaids@hotmail.com

Manuscript first received 3 June 2004; final version accepted 25 November 2004.

Cover: Filitanais filiformis sp. nov., see p. 162.

DEEP-SEA TANAIDACEA (PERACARIDA) FROM THE GULF OF MEXICO

BY

Kim Larsen

CRUSTACEANA MONOGRAPHS, 5

BRILL
LEIDEN · BOSTON

This book is printed on acid-free paper.

Library of Congress Cataloging-in-Publication Data

The Library of Congress Cataloging-in-Publication Data is available from the Publisher.

ISBN 90 04 14210 X

© Copyright 2005 by Koninklijke Brill NV, Leiden, The Netherlands

Koninklijke Brill NV incorporates the imprints Brill Academic Publishers, Martinus Nijhoff Publishers and VSP.

All rights reserved. No part of this publication may be reproduced, translated, stored in a retrieval system, or transmitted in any form or by any means, electronic, mechanical, photocopying, recording or otherwise, without written permission of the publisher.

Authorization to photocopy items for internal or personal use is granted by Brill provided that the appropriate fees are paid directly to Copyright Clearance Center, 222 Rosewood Drive, Suite 910, Danvers, MA 01923, USA. Fees are subject to change.

CONTENTS

Preface ... 1
Introduction ... 2
Materials and methods ... 4
Generalized tanaidacean morphology ... 5
The body ... 5
The eyes ... 6
The antennae ... 6
The mouthparts ... 8
The labrum ... 8
The mandibles ... 8
The labium ... 10
The maxillules ... 10
The maxillae ... 11
The maxillipeds ... 11
The epignaths ... 12
The chelipeds ... 12
The pereopods ... 14
The pleopods ... 15
The uropods ... 15
Physiology ... 16
Respiration ... 16
Ecdysis ... 17
Internal anatomy ... 18
The nervous system ... 18
The digestive system ... 18
The circulatory system ... 18
Excretion ... 19
Gonad structure ... 19
Development ... 21
Tanaidomorpha ... 21
Neotanaidomorpha ... 23
Apseudomorpha ... 23
Variations in development between the suborders ... 24
Sexual polymorphism ... 25
Ecology ... 27
Habitats ... 28
Behaviour and interspecific relationships ... 29
Feeding strategies ... 29
Predators ... 31
Parasites ... 31
Commensal relationships ... 33
Parasitism ... 33
Behaviour ... 34
Communication ... 34

Tube construction 35
Fighting behaviour 38
Reproduction 38
Courtship 39
Copulation 39
Brood care 40
Fecundity 41
Locomotion 42
Apseudomorpha and Neotanaidomorpha 42
Tanaidomorpha 42
Systematics 44
Order Tanaidacea Dana, 1849 44
Fossil Tanaidacea 45
Key to the suborders of Tanaidacea in the Gulf of Mexico 45
Suborder Apseudomorpha Sieg, 1980 46
Key to the deep-sea genera of Apseudomorpha in the Gulf of Mexico 46
Family Apseudidae Leach, 1814 46
Subfamily Apseudinae Leach, 1814 47
Genus *Apseudes* Leach, 1814 48
A. olimpiae Guțu, 1986 50
A. cf. *spinosus* (M. Sars, 1858) 51
Key to the species of *Apseudes*, both sexes (Gulf of Mexico) 51
Genus *Atlantapseudes* Băcescu, 1978 52
A. lindae Meyer & Heard, 1989 53
Key to the species of *Atlantapseudes*, both sexes (global) 53
Genus *Dactyloprion* Guțu, 2002 54
D. anakanthes Guțu, 2002 54
Genus *Pectinapseudes* Băcescu & Williams, 1988 55
P. magnus Guțu, 2002 55
P. mesteri Guțu, 2002 62
Key to the species of *Pectinapseudes*, both sexes (global) 62
Subfamily Leviapseudinae Sieg, 1983 63
Genus *Leviapseudes* Sieg, 1983 64
L. bipartitus sp. nov. 65
L. cf. *sibogae* (Nierstrasz, 1913) 76
Key to the species of *Leviapseudes* (Gulf of Mexico) 76
Family Sphyrapodidae Guțu, 1980 77
Subfamily Sphyrapodinae Guțu, 1980 78
Genus *Sphyrapoides* Guțu & Iliffe, 1998 78
S. bicornis Guțu & Iliffe, 1998 79
Key to the species of *Sphyrapoides*, both sexes (global) 79
Subfamily Pseudosphyrapodinae Guțu, 1980 79
Genus *Pseudosphyrapus* Guțu, 1980 80
P. sp. A 81
Genus *Kudinopasternakia* Guțu, 1991 91
K. bispinosa Guțu & Heard, 2002 91
K. siegi (Viskup & Heard, 1989) 92
Key to the species of *Kudinopasternakia*, both sexes (global) 92
Family Parapseudidae Guțu, 1981 92

Genus *Parapseudes* G. O. Sars, 1882 93
P. cf. *latifrons* (Grube, 1864) 94
Suborder Neotanaidomorpha Sieg, 1980 96
Family Neotanaidae Lang, 1956 96
Genus *Neotanais* Beddard, 1886 96
N. armiger Wolff, 1956 98
N. curvimanus Larsen & Hansknecht, 2003 99
N. minimus Larsen & Hansknecht, 2003 99
Suborder Tanaidomorpha Sieg, 1980 101
Superfamily Tanaoidea Dana, 1849 101
Superfamily Paratanaoidea Lang, 1949 101
Family Agathotanaidae Lang, 1971 102
Key to the genera of Agathotanaidae in the Gulf of Mexico 102
Genus *Agathotanais* Hansen, 1913 103
A. hanseni Lang, 1971 104
Genus *Paragathotanais* Lang, 1971 104
P. medius Larsen, 2002 108
P. spinosus sp. nov. 108
P. typicus Lang, 1971 113
P. sp. A 118
Key to the species of *Paragathotanais*, females only (global) 118
Genus *Paranarthrura* Hansen, 1913 120
P. bispinosa sp. nov. 121
P. cf. *insignis* Hansen, 1913 129
P. tenuimanus sp. nov. 130
P. cf. *undulata* Lang, 1968 138
Key to the species of *Paranarthrura*, females only (global) 138
Family Anarthruridae Lang, 1971 141
Genus *Anarthruropsis* Lang, 1968 141
A. edentula sp. nov. 142
Key to the species of *Anarthruropsis*, females only (global) 147
Family Colletteidae Larsen & Wilson, 2002 147
Key to the genera of Colletteidae in the Gulf of Mexico 149
Genus *Caudalonga* n. gen. 149
C. quatropleon sp. nov. 150
Genus *Cetiopyge* Larsen & Heard, 2002 155
C. mira Larsen & Heard, 2002 156
Genus *Collettea* Lang, 1973 156
C. elongata Larsen, 2002 157
C. sp. A 157
Genus *Filitanais* Kudinova-Pasternak, 1973 159
F. filiformis sp. nov. 160
Key to the species of *Filitanais*, females only (global) 165
Genus *Isopodidus* Larsen & Heard, 2002 165
I. janum Larsen & Heard, 2002 166
Genus *Leptognathiella* Hansen, 1913 166
L. occidentis sp. nov. 167
L. vermiformis sp. nov. 174
Key to the species of *Leptognathiella*, females only (global) 181

Genus *Tumidochelia* Knight, Larsen & Heard, 2003 ... 181
T. randyi Knight, Larsen & Heard, 2003 ... 182
Family Leptocheliidae Lang, 1973 ... 182
Key to the deep-sea genera of Leptocheliidae in the Gulf of Mexico ... 183
Subfamily Heterotanaidinae Larsen & Wilson, 2002 ... 183
Genus *Bathyleptochelia* Larsen, 2003 ... 183
B. oculata Larsen, 2003 ... 184
Genus *Mesotanais* Dollfus, 1897 ... 184
M. longisetosus Sieg & Heard, 1989 ... 185
M. vadicola Sieg & Heard, 1989 ... 186
Key to the species of *Mesotanais*, females only (global) ... 186
Subfamily Leptocheliinae Lang, 1973 ... 186
Family Leptognathiidae Sieg, 1973 ... 187
Genus *Leptognathia* G. O. Sars, 1882 ... 188
L. gyreae sp. nov. ... 189
L. longimanus sp. nov. ... 194
Key to the species of *Leptognathia* in the Gulf of Mexico, females ... 200
Family Nototanaidae Sieg, 1976 ... 201
Genus *Meromonakantha* Sieg, 1986 ... 202
M. gilrowei sp. nov. ... 203
Genus *Typhlotanais* G. O. Sars, 1882 ... 209
T. cf. *trispinosus* Hansen, 1913 ... 210
Family Pseudotanaidae Sieg, 1976 ... 211
Key to the deep-sea genera of Pseudotanaidae in the Gulf of Mexico ... 211
Subfamily Cryptocopinae Sieg, 1977 ... 212
Genus *Cryptocope* G. O. Sars, 1882 ... 212
Genus *Iungentitanais* Sieg, 1977 ... 213
I. primitivus (Sieg, 1973) ... 213
Subfamily Pseudotanainae Sieg, 1976 ... 214
Genus *Parapseudotanais* Bird & Holdich, 1989 ... 214
Genus *Pseudotanais* G. O. Sars, 1882 ... 215
P. mexikolpos Sieg & Heard, 1988 ... 216
P. mortenseni Sieg, 1977 ... 217
Family Tanaellidae Larsen & Wilson, 2002 ... 217
Key to the genera of the Tanaellidae in the Gulf of Mexico ... 218
Genus *Araphura* Bird & Holdich, 1984 ... 218
A. curticauda sp. nov. ... 220
A. extensa Larsen, 2003 ... 224
A. higginsi Sieg & Dojiri, 1989 ... 225
A. spinithenari sp. nov. ... 225
Key to the species of *Araphura*, females (global) ... 231
Genus *Araphuroides* Sieg, 1986 ... 231
A. bombus sp. nov. ... 232
Key to the species of *Araphuroides*, females (global) ... 239
Genus *Tanaella* Norman & Stebbing, 1886 ... 240
T. mclellandi Larsen & Heard, 2004 ... 241
T. prolixcauda Larsen & Heard, 2004 ... 241
Familae indeterminatae ... 242

Key to the genera currently placed as familae indeterminatae, that occur in the Gulf of Mexico ... 242
Genus *Armaturatanais* n. gen. ... 243
A. minipodus sp. nov. ... 244
Genus *Chauliopleona* Dojiri & Sieg, 1997 ... 251
C. faini sp. nov. ... 252
Genus *Coalecerotanais* Larsen, 2003 ... 257
C. inflatus Larsen, 2003 ... 258
Genus *Spinitanaopsis* nomen novum ... 258
S. insolituchelia (Larsen, 2003) ... 259
Genus *Expina* Lang, 1968 ... 259
E. typica Lang, 1968 ... 260
Genus *Insociabilitanais* n. gen. ... 260
I. ventrospinatus sp. nov. ... 262
Genus *Nototanoides* Sieg & Heard, 1985 ... 268
N. trifurcatus Sieg & Heard, 1985 ... 268
Genus *Parafilitanais* Kudinova-Pasternak, 1989 ... 269
P. mexicanus Larsen, 2002 ... 269
Genus *Pseudoarthrura* n. gen. ... 270
P. heardi sp. nov. ... 271
Genus *Robustochelia* Kudinova-Pasternak, 1983 ... 276
R. solida sp. nov. ... 276
Key to the species of *Robustochelia*, females only (global) ... 281
Genus *Stenotanais* Bird & Holdich, 1984 ... 281
S. macrodactylus sp. nov. ... 282
Key to the species of *Stenotanais*, females only (global) ... 288
Filiform species ... 288
Tanaidacean distribution ... 290
Species level distributions in shallow waters ... 291
Rafting or human intervention ... 292
High mobility or transport by near-shore currents ... 292
Species level distributions in the deep sea ... 292
High mobility or transport by oceanic currents ... 293
Biotic factors affecting dispersal ... 293
Inbreeding ... 293
Sediment type and seasonality ... 293
Predation and mortality ... 294
Abiotic factors affecting dispersal ... 294
Temperature and salinity ... 294
Upwelling and benthic storms ... 294
Pressure gradients ... 295
Explanations for wide species distributions ... 295
Misidentification ... 295
Pressure tolerance ... 296
Behavioural modifications ... 297
Explanation for wide distributions at the generic level ... 298
Biological success ... 298
Conclusions ... 299
Tanaidacean distribution patterns in the Gulf of Mexico ... 300

Maps 1–65 305
Charts 1–2 338
Acknowledgements 340
Literature cited 341
Appendix 1, Station list DGoMB, R/V "Gyre" 365
Taxonomic index 372

PREFACE

A review of all information available on the various aspects of tanaidacean morphology, physiology, ecology, and biology is presented in this study.

A large number of deep-sea tanaidacean taxa from the Gulf of Mexico, predominantly collected during the Deep Gulf of Mexico Benthos study (DGoMB) and North Gulf of Mexico Continental Shelf Study (NGoMCSS) are examined. The new genera, *Armaturatanais*, *Caudalonga*, *Insociabilitanais*, and *Pseudoarthrura* are erected and described. Twenty-one new species belonging to the new genera and to *Anarthruropsis*, *Araphura*, *Araphuroides*, *Chauliopleona*, *Filitanais*, *Leptognathia*, *Leptognathiella*, *Leviapseudes*, *Meromonakantha*, *Paragathotanais*, *Paranarthrura*, *Robustochelia*, and *Stenotanais* are described, in many cases by both sexes. The female of *Paragathotanais typicus* and the male of *Pectinapseudes magnus* are also described here, for the first time.

Information about distribution and bathymetric range is included. Keys are presented for species in the genera *Atlantapseudes*, *Pectinapseudes*, *Sphyrapoides*, *Kudinopasternakia*, *Paragathotanais*, *Paranarthrura*, *Anarthruropsis*, *Filitanais*, *Leptognathiella*, *Mesotanais*, *Araphura*, *Araphuroides*, *Robustochelia*, and *Stenotanais*.

The genus *Crurispina* is renamed as *Spinitanaopsis*, as the name *Crurispina* is preoccupied by a fossil brachiopod.

The global distribution patterns and dispersal mechanisms applying to the Tanaidacea are discussed. Most deep-sea tanaidacean species appear widely distributed and show remarkably wide depth ranges. Misidentification is suggested as the cause of many of the apparently wide geographical distribution ranges of tanaidacean species. Wide bathymetric distributions are also recorded for many species, and their apparent pressure tolerance may contribute to dispersal as well. The known distribution patterns in the Gulf of Mexico seem to merely reflect sampling effort.

KIM LARSEN

INTRODUCTION

The order Tanaidacea is a group of small crustaceans within the superorder Peracarida (Malacostraca). More than 800 species of tanaidaceans have been described, but the order is estimated to contain several thousands of species (Dojiri & Sieg, 1997). Tanaidaceans range from 1 mm to several centimeters in length, the longest reaching 75 mm (Gamo, 1984), but the majority of species measure around a few millimeters and belong to the macrofaunal community.

There are four suborders: the Anthracocaridomorpha (known from fossils only), Apseudomorpha, Neotanaidomorpha, and Tanaidomorpha (cf. Sieg, 1980b). Members of the three recent suborders all have characteristic life styles. Neotanaidomorphans are free-living surface dwellers (Gardiner, 1975b; Thistle et al., 1985), tanaidomorphans are tube dwellers (Johnson & Attramadal, 1982a, b; Mendoza, 1982; Holdich & Jones, 1983b), and apseudomorphans are mostly burrowers (Sieg, 1983b).

While the vast majority of tanaidaceans are marine, several species are found in brackish water (Shiino, 1965; Muus, 1967; Boltt, 1969; Băcescu & Guțu, 1974; Levings & Rafi, 1978; Peck, 1993; Owen & Forbes, 1997) or even in fresh water (Gardiner, 1975a; Kocatas, 1977; Peck, 1993; Hickey & Martin, 1995; Wiedenburg et al., 1997) and possibly a ground-water component is also present (Larsen & Hansknecht, 2004).

Tanaidacean populations can attain sizeable proportions, in particular in shallow water, of up to 140 000 individuals/m^2 (Błażewicz & Jażdżewski, 1996). However, tanaidaceans have their greatest ecological importance on the abyssal plain where they might even rival that of polychaetes (on the level of order). Despite this abundance, tanaidaceans have been neglected in most ecological surveys (see chapter on Ecology herein).

Identifying tanaidaceans is notoriously difficult, in particular pertaining to the small members of the suborder Tanaidomorpha, but even the larger taxa are troublesome. While this is partly due to their small size and reduced morphology, the identification of Tanaidomorpha and Neotanaidomorpha is especially confounded by substantial sexual and ontogenetic variation. This polymorphism (see chapter on Development herein) is a consequence of a complicated reproductive strategy, involving protogynous hermaphroditism (Bückle-Ramirez, 1965; Larsen, 2001). Since the majority of taxonomic studies have ignored these variations, the diagnostic characters of tanaidacean taxa as currently in use, are often misleading.

The first description of a tanaidacean species from the Gulf of Mexico was that of *Apseudes propinquus* Richardson, 1903 [now *Hoplomachus propinquus* (cf. Guțu, 2002b)], but subsequently the group received little attention, in particular in deeper waters, although Pequegnat et al. (1990) did record an unusually high diversity of the deep-sea Tanaidacea. Specific information on this topic was, in the last century, limited to a few reports (Gardiner, 1975b; Sieg & Heard, 1983, 1985, 1988, 1989; Meyer & Heard, 1988; Viskup & Heard, 1989), which dealt with only six new species from the Gulf. Documentation in this century has already been far greater, and 14 new species and seven new genera have been described, excluding those in this study (Larsen, 2002a, 2003b; Larsen & Heard, 2002, 2004a; Knight et al., 2003; Larsen & Hansknecht, 2003). This progress has been made possible due to recently obtained material from the various sampling projects in the programmes: the Deep Gulf of Mexico, the Deep Gulf of Mexico Benthos study (DGoMB), the Northern Gulf of Mexico Continental Shelf Study (NGoMCSS), and the Shelf Slope Experimental Taphonomy Initiative (SSETI), all conducted by Texas A&M University.

The present study reports four additional new genera and 21 new species. More than 100 species of Tanaidacea were recorded in total. The material includes a large number of new species that form the basis of this monograph. Unfortunately, as often in deep-sea sampling, specimens of many new species were found in only a few numbers of individuals, often only present as one specimen, and much of the material was badly damaged. It is the opinion of the present author that preferably a complete holotype, as well as a complete array of dissected appendages are necessary for a formal systematic description, and thus many new or potentially new species had to be excluded from this study.

MATERIALS AND METHODS

The material examined in this study was mainly collected by Prof. G. Rowe and his party on board the R/V "Gyre", during the "Deepwater Program: Northern Gulf of Mexico Continental Slope Habitats and Benthic Ecology" (DGoMB), of Texas A&M University through contract no. 1435-01-99-CT-30991 from the Minerals Management Service, Department of the Interior. Older material from the NGoMCSS program (LGL Ecological Research Associates, Inc., through contract nos. 14-12-0001–30046 and 14-12-0001–30212) was also included.

Samples were taken with a 0.2 m^2 Gray-O'Hara modified GOMEX box corer (Boland & Rowe, 1991). This corer subsampled in six subcores with two different configurations and with a macrofauna subcore sample area of 0.0569 m^2 or 0.0475 m^2. The subcores were used for studies of the meiofauna (two subcores), geology (one subcore), bacteriology (one subcore), and geochemistry and contaminants (one subcore). The macrofauna from these five subcores has not been examined. The sediment was washed through a 300 μm sieve in the field, fixed for 24 hours in 10% buffered formalin, and preserved in 80% ethanol.

Dissections were conducted with chemically sharpened tungsten wire needles. If possible, appendages from both sides of the body were dissected and compared for conformity. Illustrations of whole animals and dissected appendages, mounted in glycerine to retain the possibility of manipulation, were prepared using a camera lucida attached to an Olympus BX51 compound microscope with Olympus optics. After illustration, dissected appendages were transferred to glycerine jelly for permanent storage and are deposited in the National Museum of Natural History, Washington, D. C.

Descriptions of the body of the new species are based on holotype material and those of dissected appendages are taken from paratypes.

The morphological terminology follows Larsen (2003a). The comparative quantifiers (short, long, many, scattered, dense) when used in descriptions of body/appendage ornamentation, are used relative to the area being described.

Sample sites for the DGoMB project are shown on map 1 and those for the NGoMCSS project on map 2. A station list is given in appendix 1.

Horizontal distribution ranges of cosmopolitan and bipolar species (maps 64 and 65) and bathymetric ranges of cosmopolitan species and genera (charts 1 and 2) are based on both reported and observed data. Shallow-water and deep-water tanaidacean genera and their number of species found in the Gulf of Mexico and the western Caribbean Sea, are listed in tables VII and VIII.

GENERALIZED TANAIDACEAN MORPHOLOGY

The body

The head and the following two somites (thoracomers) of tanaidaceans are fused to form a cephalothorax that is covered dorsally by a heavily calcified carapace (fig. 1). In certain taxa, *Bathytanais* and *Paratanais*, the carapace consists of plates separated by articulation membranes (Larsen & Wilson, 1998; Larsen, 2001). The fused thoracomers of the cephalothorax are not considered to be pereonites. The six (or rarely five) free somites after the cephalothorax are collectively called the 'pereon' and individually 'pereonites'. The next five somites (sometimes less) are termed 'pleonites' and together with a sixth pleonite fused to the telson (pleotelson), this region is termed the 'pleon' (Gardiner, 1975b). Various degrees of fusion occur in the pleon of some tanaids, but such phenomena are only rarely seen in the pereon (Sieg, 1980a; Bird & Holdich, 1988; Larsen, 2003b).

The apseudomorphan body plan is usually more dorso-ventrally flattened than that of the other suborders. Only in Apseudomorpha a rostrum can be present. A few Apseudomorpha have a reduced pleon (as in *Curtipleon*) but this is unusual. Sexual dimorphism of the body is usually limited to the pleonites (Apseudidae) and pleotelson (*Leviapseudes*).

The neotanaidomorphan body plan is more cylindrical and the male body is more or less similar to that of the female. The most striking difference is the much more triangular cephalothorax of the male. Also, there are smaller differences, e.g., in the shape of the pereonites, and the male pleonites are wider than those of the female.

The body plan of the Tanaidomorpha is usually also cylindrical. Males can have a body plan more or less similar to that of females (Tanaoidea, certain Paratanaoidea) or display a difference so great that the layman would never guess conspecificity (Leptognathiidae, Paratanaidae, Leptocheliidae). In cases of highly divergent males, the pleon is always comparatively longer than that of the female, and the pereonites are correspondingly shorter, although this configuration is much more pronounced in the Paratanaidae/Leptognathiidae than in the Leptocheliidae. In the Leptocheliidae, however, the cephalothorax of the male is much longer than that of the female.

The eyes

The Tanaidacea were earlier considered to belong to the "Sessile Eyed Crustacea" (Bate & Westwood, 1866). That perception was followed by later workers up to recent times (e.g., Dojiri & Sieg, 1997, and others). Lilljeborg (1864: 3–4) mentioned that a significant difference between tanaids and isopods was the more or less mobile nature of the tanaid eye, in particular in the male. This was confirmed by Claus (1888), who used the term "movable sessile eyes". Lauterbach (1970) was of a different opinion and designated the tanaids as "...'Stalked-eyed' crustaceans with a very short stalk". Andersson et al. (1978) claimed that the 'stalk' structures described by Lauterbach (1970) could not justify the term 'stalked eyes' since the eyes are immobile and lack optic neuropiles. Sieg (1984) complicated the issue further by confirming mobility and a well defined eye-muscle in the Tanaidomorpha, while rejecting the presence of such in the Apseudomorpha. A discussion of 'stalked' versus 'sessile' is not the topic of this study; however, the very detailed study by Lauterbach (1970) shows functional levator muscles and indicates that the definitions of such terms are not completely clear. What this means for the phylogeny of the Peracarida, one can only speculate.

The ultrastructure of the eye of *Tanais dulongii* (Audouin, 1826) has been described in minute detail (under the name *Tanais cavolinii* H. Milne Edwards, 1840) by Andersson et al. (1978), who described the eye as rather poorly developed and with low image-forming capacity. Andersson et al. (1978) also concluded that the tanaid eye only shares a few similarities with those of other peracarids.

In the Neotanaidomorpha, eye-lobes are present but never with visual elements. In the other suborders, eyes or eye-lobes may be either present or absent. In those taxa with sexual polymorphism in the Tanaidomorpha, the eyes (if present) are much larger in the male than in the female.

The antennae

The antennule is always "biramous" in the Apseudomorpha (i.e., has two flagella on its single ramus), but "uniramous" (bears a single flagellum) in the other suborders (fig. 1). The antenna can be either uniramous or biramous. In a biramous antenna only a single exopod article (the squama) is present and this may be subject to various degrees of reduction.

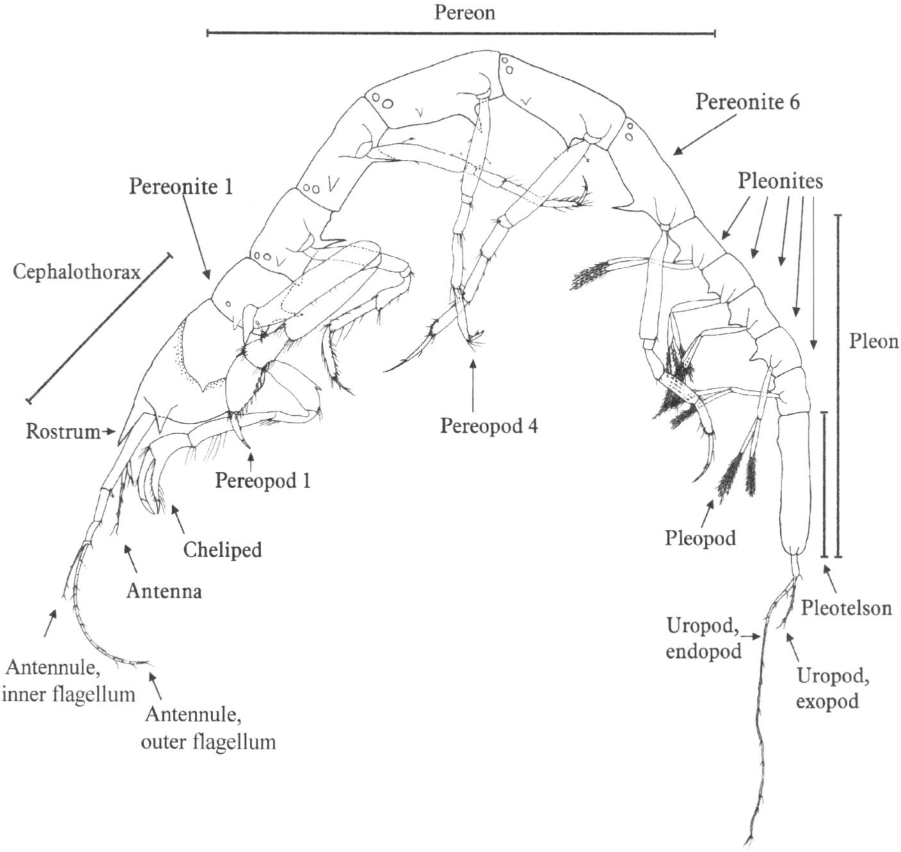

Fig. 1. General tanaidacean morphology, body.

In the Neotanaidomorpha the antennae are very conservative. They are always uniramous and equipped with the same number of articles: seven in the antennule and nine in the antenna in both sexes.

Within the Tanaidomorpha, a great deal of variation occurs in the antennae, i.e., between families, genera, species, and sexes. The number of antennular articles varies between three and five in females and between four and ten and possibly also higher, in males. The numbers of articles are always higher in the males than in females of the polymorphic species. The antenna may be partly or completely reduced in *Agathotanais* but in most families the number of articles varies between five and six and in some cases a subdivision (fusion line) of article 4 renders the antenna to appear seven-articulate, depending on how one chooses to define the fusion line.

The antennae and in particular the antennules, of the male are always equipped with more aesthetascs than those of the female, ranging from only one or just a couple more in the Colletteidae, to literally hundreds more in the Paratanaidae and Apseudidae.

The mouthparts

The mouthparts (fig. 2) of the Tanaidacea are arranged following the basic peracarid sequence, in anterior–posterior direction: labrum, paired mandibles, labium, maxillules, maxillae, and maxillipeds (although technically a thoracopod and not a mouthpart in the strict sense) with the epignaths. The mouthparts of the males of the Neotanaidomorpha and of several tanaidomorphan taxa are reduced and non-functional, usually retaining little more than a reduced maxilliped with slender palps.

The labrum

The labrum is an extension of the cephalothorax and not an appendage, although it is described in the same manner as a true appendage for taxonomic purposes. The labrum is usually featureless in most taxa (but frequently setose). Exceptions to this rule are found in many apseudomorphans, where the labral region is extended into an anteriorly directed spine termed epistome or hypostome. Exceptions are also seen in the tanaidomorphan genera *Anarthruropsis* and *Siphonolabrum*, in which the labrum is conspicuously enlarged for a purpose yet to be determined.

The mandibles

The mandibles (fig. 2A) are different for each suborder, family, and even some genera and species, so only a generalized overview is presented here. In the Apseudomorpha, the mandibles are almost always equipped with a palp (except in Sphyrapodinae), which is usually tri-articulate but may be reduced to one article. Incisor and setal rows are present on both mandibles. The lacinia mobilis is less developed on the right mandible (sometimes completely reduced). The molars are long and wide, with a distal array of spines or denticles.

In the Neotanaidomorpha the mandibles have no palp but otherwise have the same structure as in the Apseudomorpha. The incisors and molars are broad and very heavily calcified.

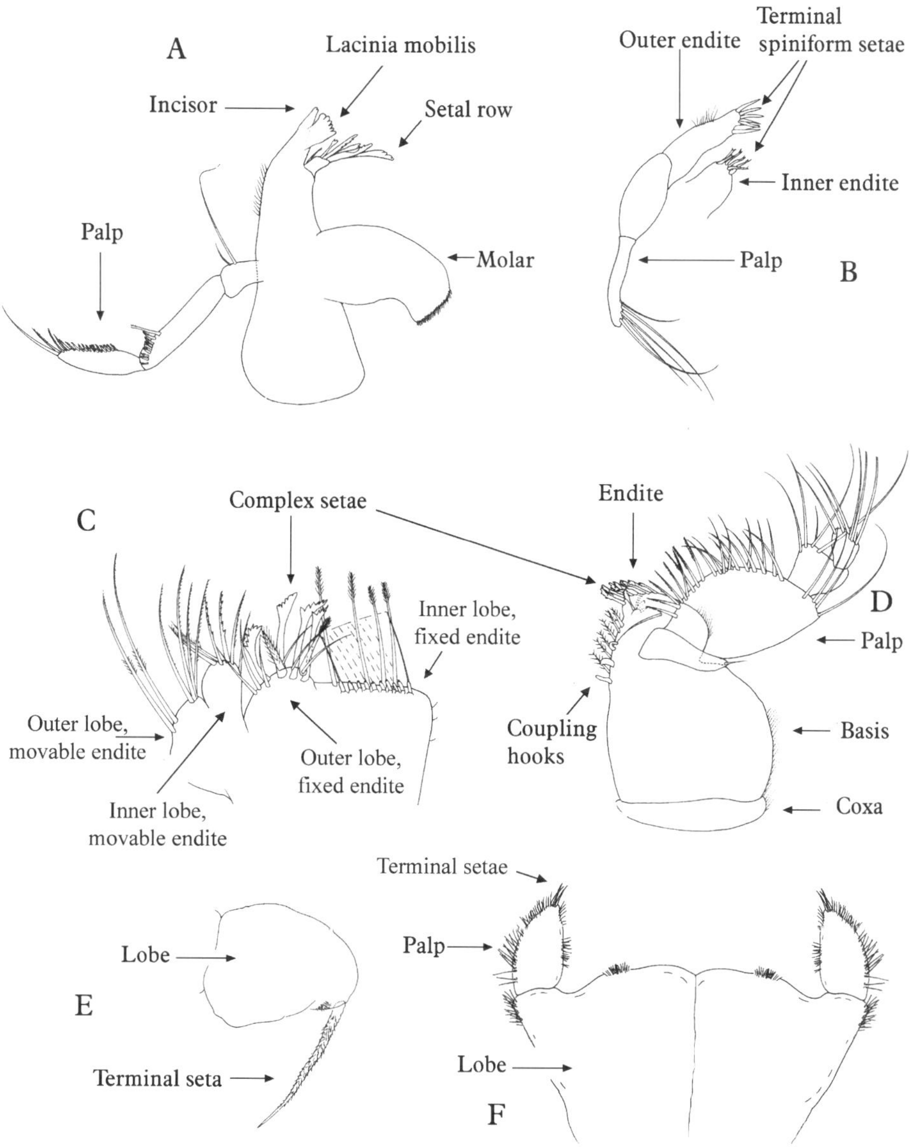

Fig. 2. General tanaidacean morphology, mouthparts: A, mandible; B, maxillule; C, maxilla; D, maxilliped; E, epignath; F, labium.

In the Tanaidomorpha, the mandibles vary considerably but are always without palp. In the superfamily Tanaoidea, the mandibles have a setal row and in some species a lacinia mobilis on both mandibles; the molars are usually quite broad (relative to those of other tanaidomorphans) and calcified. In the superfamily Paratanaoidea, there is no setal row, the lacinia mobilis is present only on the left mandible, and the molar comes in a multitude of shapes, ranging from completely reduced (*Coalecerotanais*, *Spinitanaopsis*, *Siphonolabrum*), to membranous (Agathotanaidae), pointed or styliform (Leptognathiidae and others), or broad and heavily calcified (Paratanaidae, Leptocheliidae).

The labium

The labium (fig. 2F) is equipped with a pair of basal lobes with a uni-articulate palp in the Apseudomorpha. Both the palp and the lobes are setulose and the palp is usually equipped with 2–3 complex setae at the apex.

In the Neotanaidomorpha, the palp may be more or less fused and two pairs of basal lobes will be present. Both lobes and palp are setose.

In the Tanaidomorpha, one or two pairs of lobes may be present. The palp is only retained in the Tanaoidea, though a process can sometimes be present on the lateral corners of the labium in some paratanaoideans. The lobes may be setose or not.

The maxillules

The apseudomorphan maxillule (fig. 2B) consists of an inner and an outer endite. The inner is much smaller than the outer and is equipped with about 3–6 setulated setae and a few simple setae. The outer endite is armed with several spiniform setae, distally. A bi-articulate palp with several long, terminal setae may be present or absent (Kalliapseudidae).

In the Neotanaidomorpha the maxillule is never equipped with a palp, but looks otherwise similar to that of the Apseudomorpha.

The maxillule of the Tanaidomorpha consists of only one endite and a palp. The terminal setae of the tanaidomorphan maxillule are longer and more slender than those in the other suborders and are usually more numerous. Those terminal setae are mostly simply spiniform, but may also be bifurcate or serrated or pinnate, and are always arranged in an oval ring.

In the superfamily Paratanaoidea the palp is always uni-articulate, with two long terminal setae. The palps of many paratanaoidean species have been omitted from their descriptions, and it is not yet clear if they have been lost during dissection, or are actually absent altogether. In the Tanaoidea, the palp is always present and often bi-articulate, and bears several terminal setae.

The maxillae

The maxilla is always well developed, and bears a multitude of different, specialized setae in the Apseudomorpha and Neotanaidomorpha (fig. 2C). The maxilla is divided into a movable (outer) and a fixed (inner) endite. Both endites are subdivided into inner and outer lobes (see Larsen, 2003a, for a comprehensive explanation of terms). The setation of the maxilla varies among families, but a basic pattern can be recognized almost everywhere. The outer lobe of the outer endite is always armed with the longest setae, which often have some kind of medial or distal setulation. The inner lobe of the outer endite is equipped with a more standard array of setae, consisting of rather thick setae, with different degrees of setulation. The outer lobe of the inner endite is characterized by a number of multifurcate, heavy, spiniform setae. These are hypothesized as having a scraping function, together forming a fixed comb. Setulated setae can be dragged through the comb and dislodge particles trapped in the setulation. The inner lobe of the inner endite usually has a number of sparsely setulated, stiff setae as well as a whole row of distally curved setae, arising from a wide proximal base that narrows abruptly. This row of setae may or may not be setulose and sometimes each seta ends in a tiny, bifurcate tip.

In the Tanaidomorpha, the maxilla is always reduced to an ovoid, featureless shape and is more or less naked, although setose in Tanaoidea. It is unlikely to have any function, and is here considered a vestigial structure only.

The maxillipeds

In all Tanaidacea, the maxilliped (fig. 2D) has a four-articulate palp, of which the three distal articles are armed with heavy and usually bipinnate setae. The palp may be attached to a pedestal on the basis, giving the appearance of having five articles, but the pedestal is really a part of the basis.

In the Apseudomorpha and Neotanaidomorpha the maxillipedal basis is divided into two separate halves and a proximal coxa is present. The distal endites are separate and contain several simple and specialized setae distally. In the Apseudomorpha these endites usually also have medial coupling hooks.

In the Tanaidomorpha, the maxillipedal basis is separate and often with a coxa in the Tanaoidea and in some taxa of the Paratanaoidea (Leptocheliidae, Pseudozeuxidae, and *Arthrura*), whereas the bases are fused with each other in all other taxa. The distal endites may be completely fused (certain species in the Pseudotanaidae), partially fused (certain species in Nototanaidae and Pseudotanaidae), or separate (all other taxa), and may have denticles, setae, or be naked.

Usually, there is not much variation between the maxilliped of a 'feeding' male and a female. Even non-feeding, terminal males will retain a reduced maxilliped, but in this case it has little resemblance to that of the female.

The epignaths

Associated with the maxilliped is the epignath (fig. 2E), the affinity of which has been under dispute. The epignath creates the ventilating current (Siewing, 1953; Johnson & Attramadal, 1982a, b; see chapter on Physiology, under Respiration). Some authors believe the epignath to be the original exopod of the maxilliped, owing to a connection between the musculature of the two appendages (Claus, 1884; Dennell, 1937; Sieg, 1984) but no articulation has ever been demonstrated. Lauterbach (1970) suggested that the epignath has a connection to the cheliped, but most authors consider the epignath an independent appendage (Blanc, 1884; Siewing, 1953, 1957).

In most of the Apseudomorpha the epignath is wide (kidney-shaped), strongly developed, and setose, and always has a long, terminal seta that is often setulose.

In the Neotanaidomorpha the epignath is also strongly developed, but it is more elongate than in the Apseudomorpha. The epignath is here divided into two almost equal-sized lobes and has a setulose terminal seta.

In the Tanaidomorpha the epignath is poorly developed and elongate, to the extent of being stick-shaped, with, or more often without, a terminal seta that may or may not be setulose. It is wider and better developed in Tanaoidea than in Paratanaoidea.

The chelipeds

Only in the Apseudomorpha can a true coxa of the cheliped be found (fig. 3A) but it is unclear how widespread this condition is, as this article is rarely described by workers on the suborder. The merus is well developed in the Apseudomorpha, in contrast to the other suborders. Sexual variation is usually restricted to processes of the carpus and of the propodal fixed finger and dactylus of the male, but in some taxa (*Synapseudes*, *Pagurapseudes*, *Macrolabrum*, *Curtipleon*) the male also has a greatly enlarged left cheliped (Shiino, 1951; Băcescu, 1976a, b; Guţu, 1997; Larsen, 2002b). In *Leviapseudes* there are no sexually determined chelipedal processes, but the species of this genus display sexual dimorphism in the antennules, mandibles, and pleotelson.

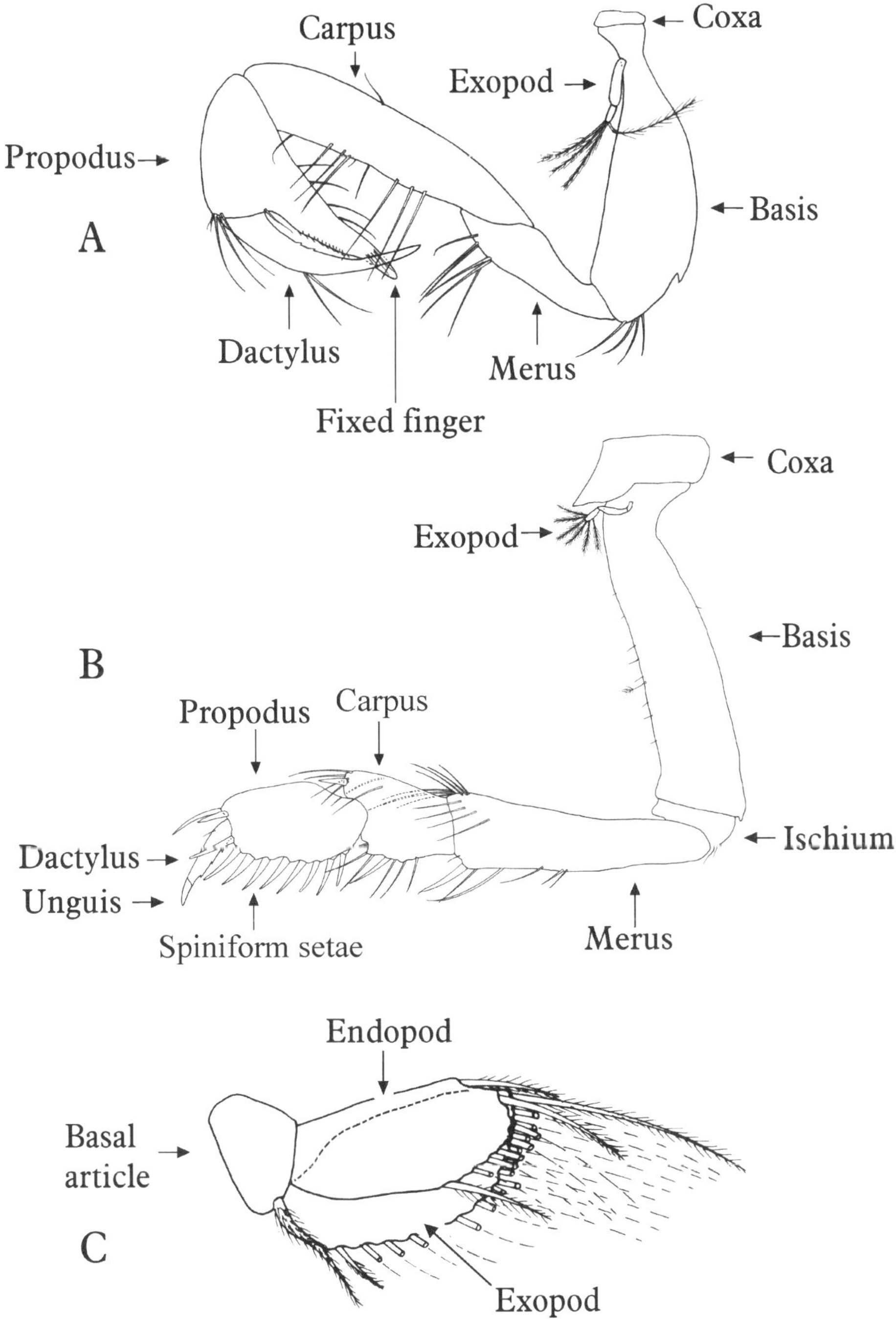

Fig. 3. General tanaidacean morphology, appendages: A, cheliped; B, pereopod; C, pleopod.

The cheliped of the Neotanaidomorpha is characterized by being attached to the cephalothorax through a large sclerite, frequently larger than the chelipedal basis. Also, the Neotanaidomorpha are the only group showing an identifiable ischium, shaped like an incomplete band (Larsen, 1999a). Lauterbach (1970) showed that this article persists in the Tanaidomorpha but in a state so reduced that it cannot be distinguished from the articulating membrane by the use of a compound microscope. Sexual dimorphism is more evident in the neotanaidomorphan chelae, where the carpus, the propodus including the fixed finger, and the dactylus are greatly enlarged in the male.

The tanaidomorphan cheliped is often (Paratanaidae, Colletteidae, Leptocheliidae, Leptognathiidae, Tanaellidae) attached through a sclerite apparently homologous to the one found in the Neotanaidomorpha (cf. Lauterbach, 1970), although much smaller. Sometimes the cheliped is attached directly to the ventral edge of the carapace (Agathotanaidae). In some cases (Anarthruridae, *Nesotanais*, *Paranarthrura*) the cheliped is attached via a pseudocoxa (Hansen, 1913; Lang, 1971b; Larsen, 2002a; Bird, 2004b) the origin of which is still uncertain (Larsen, 2003a). In males of polymorphic species, the cheliped can be smaller than that in the female (Paratanaidae, Leptognathiidae) or it can be larger (Nototanaidae, Leptocheliidae), and occasionally elongate, to the extent of being longer than the animal itself (Leptocheliidae).

The pereopods

Apseudomorphan pereopods are generally heavily armed with both simple and spiniform setae. The first pair (fig. 3B) is usually larger and wider than the following pereopods, while the dactylus is shorter. This specialization of the first pereopod is considered to be related to swimming (Lang, 1968), burrowing (Holdich & Jones, 1983a), or climbing (Guţu, 1996a). Pereopods 2 and 3 are generally similar, as is pereopod 4 except for a reduced dactylus in certain taxa (Gigantapseudidae, Parapseudidae, and *Leviapseudes*). Pereopod 6 has a specific setation but is otherwise of the same structure as pereopod 5.

In the Neotanaidomorpha all legs are more or less similar and of an unspecialized, stick-like 'walking' type.

In the Tanaidomorpha, the first pereopod is specialized for spinning mucus strands for tube construction (sometimes all three anterior pairs are equipped with spinning glands). The first leg is often longer than the following legs, in particular through the lengthy dactylus. The second and third pereopods are generally very similar to each other and the last three pairs are usually also similar to each other, but generally stouter than the anterior pairs. A special feature in many families is the lack of coxae and the fusion of the unguis/dactylus in the last three pairs of legs. Tanaidomorphan legs are usually much less armed than those of the other two suborders. The suborder Tanaoidea is unique by lacking the ischium in all pereopods. In males of polymorphic species, the pereopod armament is weaker than that in the females, possibly owing to the limited time these males spend in tubes (see chapter on Behaviour, under Tube construction).

The pleopods

Most apseudomorphans are equipped with pleopods (fig. 3C) and these are elongate, biramous, and armed with plumose setae. Some species of *Leviapseudes* lack pleopods in the female (Lang, 1968; Błażewicz-Paszkowycz & Larsen, 2004) and in *Curtipleon* both sexes lack pleopods (Larsen, 2002a).

Pleopods are present in both sexes of the Neotanaidomorpha. These are well developed, except in the rare deep-sea genus *Herpotanais,* where only the first pair of pleopods is retained and even these are strongly reduced and uniramous (Wolff, 1956a).

In the Tanaidomorpha all males have pleopods, although these may express all levels of reductions in both number and shape. Females may or may not have pleopods and these may or may not be reduced. Usually, the females of shallow-water taxa have well developed pleopods, whereas in most deep-sea species the female lacks pleopods.

The uropods

Apseudomorphan uropods may be uniramous (*Curtipleon*), but are generally biramous and both endites are multi-articulate (fig. 1A) except in *Macrolabrum* (cf. Guţu, 1997).

Neotanaidomorphan uropods are always biramous. The endopod is multi-articulate, while the exopod is bi-articulate, except in *N. gardineri*, where it is uni-articulate.

The uropods in Tanaidomorpha may be multi-articulate and uniramous (Tanaoidea), multi-articulate and biramous (Leptocheliidae), bi/uni-articulate and biramous (Colletteidae, Leptognathiidae, and others), or bi/uni-articulate and uniramous (Agathotanaidae, Tanaellidae). The uropods may even be uniramous, with the single ramus fused completely to the basal article (*Agathotanais*).

PHYSIOLOGY

Respiration

In tanaidaceans, oxygen is taken up through the respiratory inner surface of the posterior part of the carapace and possibly also through the surface of the epignath (Lauterbach, 1970). However, the ventilation procedure in the various suborders is quite different. In the Apseudomorpha and Neotanaidomorpha the ventilating water flow enters and exits ventrally. This system requires an efficient ventilation apparatus, which is reflected in the large and broad epignath (larger in the Apseudomorpha than in the Neotanaidomorpha). In both the Apseudomorpha and Neotanaidomorpha the ventilating current is filtered by the maxilla, which is well developed and heavily setose in these suborders. Also, the reduced exopod on the cheliped and pereopod 1 of some apseudomorphan taxa may contribute to ventilation (Dennell, 1937).

In the Tanaidae (*Tanais*) the ventilating current enters ventrally and exits postero-dorsally, between the carapace and the first pereonite. This unidirectional system is more efficient and thus the epignath is smaller and more elongate than in the other suborders. It is probable that the higher effectiveness of this system can explain the apparently higher tolerance to anoxic conditions of this taxon. Lauterbach (1970) considered this an adaptation to a tube-dwelling existence. While this appears logical, it is not supported by Lauterbach's (1970) own observation that the ventilating current of *Heterotanais* followed the 'apseudomorphan' direction, as *Heterotanais* also is a tube dweller.

In the Tanaidomorpha the ventilation current is not filtered and the maxilla is reduced and non-functional. The respiration of tube-dwelling tanaids has been described in detail by Gamble (1970) and Johnson & Attramadal (1982a). There is good evidence that some species use the pleopods to ventilate the tube (Johnson & Attramadal, 1982a). However, given the many taxa that lack pleopods, just as all mancas also lack pleopods, this pleopod function cannot be essential for respiration in the Tanaidacea as a group. Mancas probably utilize the ventilating current supplied by the female in the first phase of their life, but the manca III stage is found outside the parental tube and still has no developed pleopods. No explanation can be given for why almost all shallow-water species possess functional pleopods while many deep-water taxa do not. The only study of tanaid oxygen consumption (Gamble, 1970) shows no correlation between oxygen concentration and pleopod activity, but demonstrates a clear correlation between oxygen concentration and epignath activity. The mancas of the Kalliapseudidae retain exopods on pereopods 2–6 and recent observations (Drumm, 2005a) suggest that these are used to ventilate the tubes instead of the pleopods, which are not developed in the mancas.

The inner side of the respiratory chamber is cleaned by the maxillular palp in most taxa (Holdich & Jones, 1983a). The palp is missing in the Kalliapseudidae and Neotanaidomorpha, and the cleaning procedures in these taxa are unknown.

Ecdysis

The ecdysis of *Heterotanais oerstedii* (Krøyer, 1842) was examined in great detail by Bückle-Ramirez (1965) and described as follows: (1) About seven days before ecdysis, the animal begins to reduce food consumption and the stomach is completely empty by the time of ecdysis. This phase is defined as 'Proecdysis'. (2) Ecdysis proper can take place both within or outside the tube and can be initiated both by day and at night. The animal starts moving laboriously and slowly, making snapping movements with the chelipeds and soon falls motionless for a short period, the pleopods being the last appendages to move. The cuticle then rips right behind the cephalothorax and the animal quickly pulls the head, antennae, and chelipeds free, while the rest of the body follows later in small jerks. The whole process of actual extrication takes about three minutes when outside the tube, and 30 minutes when inside, possibly owing to the confinement of movements caused by the tube's wall. If inside the tube, the shed cuticle is often used as building material for tube construction. (3) The phase immediately after ecdysis is termed 'Akinesis' and is passed almost motionless, with only small pleopod movements. The animal is completely white during this phase, which lasts about 30 minutes. (4) A period of slow moving but not eating follows for about 90 minutes, while the cuticle remains soft. (5) In the next phase (3 hours) movements increase and the cuticle begins to harden. Eating is still not initiated. (6) A phase of 5 hours of full movement and limited feeding follows. The cuticle is still hardening. (7) After the end of phase 6, the cuticle has fully stiffened and the activity of the animal returns to normal.

INTERNAL ANATOMY

The nervous system

The nervous system is well developed and consists of a 'brain' (an aggregation of ganglia) in the cephalon, which is attached to an axon running parallel beneath the gut. This body axon runs through the length of the animal and connects to a pair of ganglia per body somite (Holdich & Jones, 1983a). Siewing (1953) reports all the ganglia in Apseudomorpha to be widely separated, while in Tanaidomorpha those in the cephalothoracic and pleonic regions are closer together than those in the pereon. This is possibly a reflection of the greater level of reduction in the pleon of Tanaidomorpha.

The digestive system

In the Tanaidomorpha, the alimentary tract is composed of an ectodermal foregut, a paired hepatopancreas, and a proportionally long ectodermal hindgut, leading to the anal opening on the ventral surface of the pleotelson between two large, movable, and heavily chitinous anal 'plates' (opercula). The foregut is divided into an oesophagus and a stomach region, and is located in the cephalothoracic section. The stomach is further divided into a filtering (pyloric) portion and a masticatory (cardiac) portion (Holdich & Jones, 1983a). The ultrastructure of the digestive system has been described in minute detail by Siewing (1953). In the Apseudomorpha and Neotanaidomorpha, the stomach is better developed. The hepatopancreas is located in the pereon and consists of a pair of diverticula opening into the masticatory region of the gut, between the pyloric stomach and the hindgut (Siewing, 1953).

The circulatory system

A tubular heart lies dorsally of the gut and extends into the pereonic region. It lies in a pericardial sinus, within a pericardial membrane. The heart is connected to anterior, posterior, and lateral arteries and blood sinuses. The heart is best developed in the Apseudomorpha, less in the Tanaoidea, and least in the Paratanaoidea (cf. Lauterbach, 1970). The pericardial sinus receives blood from the appendages and the ventral and posterior regions of the body. The anterior artery widens in the cephalothorax to form an accessory heart, which supplies the head and the cephalic appendages with blood. The blood is oxygenated in an extensive capillary network on the inner side of the carapace's respiratory chamber. The larger apseudomorphans and neotanaidomorphans have more lateral arteries than the smaller Tanaidomorpha (cf. Holdich & Jones, 1983a).

Excretion

Excretion is performed by maxillary glands and, in some primitive apseudomorphans, possibly also by rudimentary antennary glands (Holdich & Jones, 1983a; Sieg, 1984). Nephrocytes were reported by Siewing (1953) and supposed to be homologous with the rosette glands of isopods (Lauterbach, 1970). Lauterbach (1970) also reported glands of an unknown function in the mandibular region of *Tanais dulongii* (as *Tanais cavolinii*).

Spinning glands are located in the pereon of tanaidomorphans and these excrete mucus from the tip of the unguis of either the first pair or the first three pairs of pereopods, and from a yet unidentified source in the Kalliapseudidae (see chapter on Behaviour, under Tube construction).

Gonad structure

The gonads of female tanaids are composed of two longitudinal tubes that are divided into an anterior ovarian and a posterior testicular portion. Primary males lack the ovarian portion, but traces can be found in secondary males of some species (Gardiner, 1975b). Ovaries are attached to the pericardial septum at pereonite 3 and extend to pereonite 4. There, the oviducts lead to the gonopore on the inner side of the coxa of pereopod 4 or, in those species without coxae, on the proximal part of the basis (Gardiner, 1975b). The female gonopores are not visible under the compound microscope. Testes are found at each side of pereonite 6 (also in many females, albeit rudimentary). The vas deferens is formed ventrally and opens medially on pereonite 6, as paired gonopores in the Neotanaidomorpha and most Tanaidomorpha, or as a single gonopore in the Apseudomorpha. The vas deferens is better developed in the Tanaidomorpha than in the Neotanaidomorpha (cf. Blanc, 1884; Bückle-Ramirez, 1965). A small, clearly defined growth zone in the testes can be observed in the Neotanaidomorpha (cf. Gardiner, 1975b) but not in the Tanaidomorpha (cf. Bückle-Ramirez, 1965). Otherwise, the location and structure of the gonads are similar in the Neotanaidomorpha (*Neotanais micromopher* Gardiner, 1975b) and Tanaidomorpha (i.e., *Leptochelia* sp., cf. Roubault, 1937), and in *Heterotanais oerstedii* (cf. Blanc, 1884; Bückle-Ramirez, 1965).

Peracarid spermatozoa are remarkably uniform, and a 'tail', although immobile, is usually present (Siebold, 1836; Reger, 1966; Berreur-Bonnenfant, 1967; Steele, 1967; Kumé & Dan, 1968; Fain-Maurel et al., 1975; Cotelli et al., 1976; Cotelli & Donin, 1980; Jamieson, 1990). The ultrastructure of the peracaridean 'tail' reveals that it is not homologous to the spermatozoan flagellum of other taxa, and it is termed a pseudoflagellum (Jamieson, 1990). A pseudoflagellum appears to be present in most malacostracans observed thus far (Baccetti & Afzelius, 1976; Jamieson, 1990), except for some Tanaidacea. The pseudoflagellum is lacking in the spermatozoa of Tanaidomorpha (cf. Blanc, 1884; Roubault, 1937; Lang, 1953b; Bückle-Ramirez, 1965; Cotelli & Donin, 1980), but present in the Neotanaidomorpha (cf. Gardiner, 1975b). The situation is more complex in the Apseudomorpha, where the spermatozoa have been reported both with (Lang, 1953b) and without a pseudoflagellum (Cotelli & Donin, 1980). No explanation has yet been given for this variation. The ultrastructure of spermatozoa differs substantially within the Tanaidacea, and tanaidacean spermatozoa are considered highly advanced relative to those of other peracarids (Cotelli & Donin, 1980; Sieg, 1984). Several authors consider spermatozoan ultrastructure a good index of phylogenetic relationship (Mayr, 1971; Afzelius, 1972; Sieg, 1984; Jamieson, 1990). The presence of a pseudoflagellum is here perceived as plesiomorphic within the Peracarida. No correlation has been found between either gonad or gamete structure, and deep-sea adaptations (Eckelbarger, 1994).

DEVELOPMENT

Tanaidomorpha

Ontogenetic change is best studied from cultures where all stages can be recognized, described, and their sequence determined. Such cultures have been maintained for *Heterotanais oerstedii* (cf. Bückle-Ramirez, 1965), *Tanais dulongii* (cf. Johnson & Attramadal, 1982a), and to a lesser extent for *Pseudoleptochelia antarctica* (Lang, 1953) and *Leptochelia dubia* (Krøyer, 1842) [cf. Lang, 1953b; Highsmith, 1983]. Additional field studies have been carried out on *Heterotanais oerstedii* (cf. Jażdżewski, 1969), *Hargeria rapax* (Harger, 1879) [cf. Modlin & Harris, 1989], and *Leptochelia dubia* (cf. Masunari, 1983, [under the name *L. savignyi*]). There is much consensus between those studies. The development established in these has been suggested to be valid for all Paratanaoidea (cf. Lang, 1953b) and even for the whole order (Sieg, 1972, 1978). This, however, is an over-simplification. A tabular overview of the observations made by Bückle-Ramirez (1965) and Jażdżewski (1969) for *Heterotanais oerstedii*, is found in table I.

Larsen & Wilson (1998) and Larsen (2001, and in press) reported substantial ontogenetic variation in a number of important characters within *Bathytanais*, *Paratanais*, and *Protanais*. (1) The relative length of the pereonites does not increase proportionally to width with growth. (2) The number of ventral setae on the cheliped fixed finger is not constant. (3) The armament of the maxilliped does not completely develop until adulthood. (4) The relative length of the merus/carpus of pereopod 1 is ontogenetically dependent. (5) The armament of pereopods 4–6 does not fully develop until adulthood. (6) The number of plumose setae on the pleopods increases throughout life and the distance between the exopodal terminal seta and other setae decreases. (7) The number of terminal setae of the maxillule increases from manca to adult and may not even be stable in the adult. (8) Uropod articles are not fully separated until maturity and the exopod articles separate last or not at all, a situation also reported from the Leptocheliidae (cf. Masunari, 1983).

TABLE I

Developmental sequence of *Heterotanais oerstedii* (Krøyer, 1842) (cf. Bückle-Ramirez, 1965; Jażdżewski, 1969)

Manca I	Just hatched and only found in marsupium. Pereonites 1–5 visually separated. Pereonite 6 and pleon indistinct. Feeds on nutrients supplied with the egg and does not go through ecdysis when changing into a manca II.
Manca II	Found outside the marsupium. Lacks pereopod 6 and all pleopods. Other appendages more or less of adult appearance. Pereonite 6 visible and of same size as pleonites. Diet consisting of diatoms and detritus.
Manca III	Pereonite 6 of normal appearance. Traces of pereopod 6 and pleopods can be detected. Diet consisting of diatoms, detritus, parental faecal particles, and small invertebrates.
Neutrum I	Complete number of appendages, sexual characters indistinct. Neutrum I corresponds to Lang's (1953b) manca IV.
Neutrum II, Preparatory female	Traces of oostegites present.
Neutrum II, Preparatory male	Traces of penes present. Beginning of modifications in antennules and chelipeds.
Intermediate stage (?)	Lang (1953b) also suggested the presence of an additional stage, which he called the intermediate stage, but this has not been acknowledged by other authors.
Primary male	Penes present. Mouthparts reduced and non-feeding. Pleopods and chelipeds strongly developed. Smaller and with shorter carapace than secondary and tertiary males.
Mature female I	With marsupium.
Secondary male	Differs morphologically from primary male in a more elongate shape of the carapace, longer pereonites, and bigger eyes than primary male. Rudimentary ovaries can be traced. Larger than primary male but smaller than tertiary male.
Mature female II	Without marsupium, but with scars from shed oostegites.
Tertiary male	Differs morphologically from primary and secondary male in being even more elongate, almost syphon-shaped carapace, still longer pereonites and larger eyes. Rudimentary ovaries can be traced. Larger than primary and secondary males.
Mature female III	With marsupium. Considerably larger than other females.

TABLE II

Differences in development between *Neotanais micromopher* Gardiner, 1975(b) and *Heterotanais oerstedii* (Krøyer, 1842)

Feature	Species	
	N. micromopher	*H. oerstedii*
Number of female preparatory stages	2	1
Marsupium	Shed before moults of the preparatory female and the mature female	Retained between moults of the preparatory female and the mature female

Neotanaidomorpha

Bückle-Ramirez's (1965) study was carried out on animals in culture, but deep-sea species can rarely be maintained in culture. Gardiner (1975b) used the method of analysing large numbers of fixed specimens to determine developmental stages in the Neotanaidomorpha. Although many specimens were examined in Gardiner's (1975b) study, possible short-term developmental stages, as identified in *Pseudoleptochelia antarctica* (cf. Lang, 1953b, [under the name *Heterotanais antarcticus*]), may have been overlooked. Despite this shortcoming, Gardiner's (1975b) observations of *Neotanais* are comparable and to a large extent similar to those on *Heterotanais*. The major differences are given in table II.

Apseudomorpha

In most families of the Apseudomorpha, the sexes can be distinguished from as early as manca II (Gardiner, 1973b), whereas this is not the case for the Tanaidomorpha. No species of Apseudomorpha have been reported as having terminal, non-feeding males. Neither protandry nor protogyny have been conclusively demonstrated in the Apseudomorpha (cf. Gardiner, 1973b, 1975a, b), although protogyny has been suggested in *Apseudes hermaphroditicus* [synonymous with *A. spectabilis* Studer, 1883] (cf. Lang, 1953a), *A. intermedius* Hansen, 1895 (cf. Băcescu, 1961), *A. grossimanus* Norman & Stebbing, 1886 (cf. Lang, 1968), and *A. sarsi* [synonymous with *A. acutifrons* G. O. Sars, 1882] and *Parapseudes latifrons* (Grube, 1864) [cf. Lang, 1967], similarly, protandry may occur in *Leviapseudes galatheae* (Wolff, 1956a) and *L. gracillimus* (Hansen, 1913) [cf. Wolff, 1956a].

TABLE III

Differences in development of *Pagurapseudes largoensis* McSweeny, 1982 and *Heterotanais oerstedii* (Krøyer, 1842)

Feature	Taxon	
	P. largoensis	*H. oerstedii*
Number of instars between manca II and mature female I	2 or 3	1
Instars of preparatory female before first copulatory stage	2	1
Instars of preparatory female after first copulatory stage	2	1
Male development	Directly from manca II or from a neuter	Following one juvenile and one preparatory stage
Morphological variation within the instars	High	Low

The only species of Apseudomorpha maintained in culture, *Pagurapseudes largoensis* McSweeny, 1982, displays a development approximating that of the gonochoristic tanaidomorphans (Messing, 1983). The differences between *Pagurapseudes* and *Heterotanais* are given in table III.

Variations in development between the suborders

The location and number of oostegites vary significantly within the Tanaidacea. According to the definitions of Lang (1956b, 1960, 1968) four pairs of oostegites occur on pereopods 1–4. However, *Apseudes latreillii* (H. Milne Edwards, 1828) and *A. chilkensis* Chilton, 1924 (cf. Claus, 1887; Barnard, 1935), and *Cirratodactylus floridensis* Gardiner, 1973a have five pairs, located on the cheliped and on pereopods 1–4, the pair on the cheliped being reduced and non-functional. *A. spinosus* (M. Sars, 1858), *A. spectabilis* (under the name *A. nipponicus*), *Parapseudes latifrons* (under the name *P. pedispinis*) also have five pairs, but these are all functional (Moers-Messmer, 1936; Shiino, 1937; Menzies, 1953).

While neotanaidomorphans have four pairs of oostegites (on pereopods 1–4), the Tanaidomorpha display greater variation. In the family Tanaidae, the marsupium is formed by one pair of ovisacs, or in some case just one single sac, originating from the coxae of pereopod 4 (Sieg, 1980b). In the Paratanaoidea the marsupium is made up by oostegites that are structurally different from ovisacs. Within the family Pseudotanaidae the marsupium is formed by one pair of oostegites, also originating from pereopod 4. In the remaining families of Tanaidomorpha, the marsupium is formed by four pairs of oostegites originating from pereopods 1–4.

This kind of variation of the oostegites is also seen in other peracarids, including cumaceans (Forsman, 1938), isopods (Kjennerud, 1950; Kinne, 1954; Charniaux-Cotton, 1960; Wolff, 1962), and amphipods (Sexton, 1924).

In conclusion, the above-mentioned differences suggest that the scheme proposed by Bückle-Ramirez (1965) is limited to the family Leptocheliidae only and cannot, as postulated by Sieg (1972, 1978), be applied to the order Tanaidacea as a whole. In addition, the observed variation in development is further complicated by the differences in sexual variation, as discussed below.

Sexual polymorphism

Sexual polymorphism is, in general, not very pronounced in the Apseudomorpha. Males still display species-specific characters with the female, although they differ, sometimes radically, in the shape of the cheliped and the antennae, and in *Leviapseudes* also in the pleotelson and mandibles. Apseudomorphans are gonochoristic but a number of potential cases of hermaphroditism have been recorded (Lang, 1953a).

In some tanaidomorphan genera (*Paratanais*, *Leptognathia*) highly derived males are found and in the Leptocheliidae at least three separate male morphs are present. Male polymorphism is a result of a peculiar reproductive strategy, involving protogynous hermaphroditism (Bückle-Ramirez, 1965; Lang, 1973; Highsmith, 1983; Sieg, 1983b; Larsen, 2001). When males of several genera (*Sinelobus*, *Paratanais*, *Leptognathia*, *Leptochelia*) mature (or when females change sex), their mouthparts undergo different degrees of reduction, rendering the animal a non-feeding individual. This is considered a terminal stage, devoted solely to reproduction. However, when males of other taxa (Tanaidae) mature (or females change into males) they do not loose their mouthparts and retain feeding capacity for the duration of their lifespan (Lang, 1958; Sieg, 1980a).

Females may moult into males at several different instars. Thus, males can originate from: (1) a juvenile male (primary male), (2) a female having had the first brood (secondary male), (3) a female having had two broods (tertiary males), and (4) possibly also from later developmental female stages (Bückle-Ramirez, 1965; Sieg, 1978, 1983b; Highsmith, 1982, 1983; Stoner, 1986; Modlin & Harris, 1989). As mentioned above, primary male morphology is radically different from that of females but also from that of secondary and tertiary males, which also differ from each other (Bückle-Ramirez, 1965; Sieg, 1978, 1983b). Therefore, species assignments of males are, in many taxa, problematic and constitute a serious problem for the systematic study of the Tanaidacea.

Since females do not leave their tubes voluntarily (Mendoza, 1982), males locate females by roaming the substrate. This activity results in a high predation pressure, which has been observed in studies of fish gut contents (Mendoza, 1982). This, combined with the lifespan limitations resulting from a non-feeding life-style, may deplete the population of males. Bückle-Ramirez (1965) demonstrated that the presence of males or even of water in which males have been kept (i.e., released pheromones), prevent females from changing sex. In the absence of a male pheromone, some females will start developing into males but apparently not all females have this ability.

Older and larger individuals have a competitive advantage as they are subjected to less predation, can fertilize several females, and are stronger swimmers (Bamber, 1985). Also Highsmith (1983) suggested that sex reversal presents a reproductive advantage at the intra-specific level. Secondary males are almost twice as large as primary males and significantly superior in male-male fighting. A large female, having had one to several broods, can thus optimize its reproductive success by changing into a large, successful, fighting male (Highsmith, 1983). This is a double advantage in situations where the chances for reproduction as a female are seriously reduced because males are missing from the population.

The same situation is suggested for other families (Paratanaidae, Leptognathiidae, Nototanaidae), but in these only one type of male has been observed. The males in these families are also non-feeding and share no species-specific characters with the females, but it is unknown whether the females can moult into males.

In the majority of deep-sea taxa, however, sexual variation is not quite as radical. In the genera *Parafilitanais*, *Agathotanais*, *Collettea*, *Tanaella*, *Paragathotanais*, and *Paranarthrura*, the males retain functional mouthparts and are essentially similar to the female, except for a thicker antennule and the possession of pleopods with mostly simple setae (Larsen, 2002a). In *Agathotanais*, *Paragathotanais*, and *Paranarthrura* the pleon of the male is also somewhat thicker than that of the female.

In the Neotanaidomorpha polymorphism also occurs, but only two types of males have been identified: a primary male (without ovaries) and a secondary male (with rudimentary ovaries). The presence of testes in the female proves that females can change into males, but it is unknown if this holds true for all females (Gardiner, 1975b).

ECOLOGY

Tanaidaceans can occur in high densities. In shallow water, populations often exceed 10 000 individuals/m^2 (Woodin, 1974; Bailey-Brock, 1984; Modlin & Harris, 1989; Sheridan, 1997). Population densities of over 140 000 individuals/m^2 have been reported (Barnard, 1970; Delille et al., 1985; Krasnow, 1992; Błażewicz & Jażdżewski, 1996). The ecological importance of large, shallow water populations has been thoroughly illustrated by Kneib (1992). The large proportions of tanaidaceans in the diets of some prawns (Balasubramanian et al., 1979; and others) and juvenile fish (Monod, 1925; and others; see chapter on Behaviour, under Predators) further raise the appreciation of the ecological importance of tanaidaceans.

However, the relatively higher abundance of tanaidaceans in the deep sea suggests that they have their greatest ecological importance in the abyss. At least when considered at the level of the order, the Tanaidacea comprise one of the most diverse and abundant macrofaunal groups in the deep sea. The proportion of tanaids among the deep-sea macrofauna is high, and measurements range up to 19% (Wolff, 1977) and 22% (Sokolova, 1972) of the total deep-sea macrofauna in actual numbers of individuals. The proportion of tanaid abundance apparently rises in eutrophic environments (Cosson et al., 1997).

Tanaidacea are also reported as one of the most abundant crustaceans (in terms of numbers of species) on the abyssal plain by Gardiner (1977), Marshall (1979), Sieg (1982c), and Wilson (1987). Large numbers of tanaids also occur on the continental shelf and slope (Gardiner, 1977; Sieg, 1984; Sieg & Heard, 1985, 1989).

The abundance of tanaidaceans is strong evidence of their ecological importance, which is acknowledged by most workers (Segerstråle, 1962; Thorson, 1966; Sokolova, 1972; Woodin, 1976; Peterson, 1977; Wilson, 1980, 1981). Despite this, tanaidaceans have been neglected in most ecological surveys (Pires, 1980; Baldinger & Gable, 1996), and, when included, the Tanaidacea figure only at the level of the order (Sokolova, 1972; Pires, 1980; Baldinger & Gable, 1996).

Habitats

Tanaidaceans are found at all latitudes and in almost all marine habitats ranging from hard-bottom epifauna, soft-bottom infauna, coral reefs, and estuarine marshes, to mangrove swamps (Odum & Heald, 1972; Gage & Coghill, 1977; Livingston et al., 1977; Müller, 1992). Some are even considered fouling organisms (Hoese, 1972; El-Nassry, 1973; Ghobashy et al., 1980; Shoukr et al., 1991).

Neotanaidomorphans, although free-living, seem to be restricted to soft-bottom environments (Thistle et al., 1985), while tanaidomorphans and apseudomorphans have been found in all sediment conditions, including cold seep environments (Bird, 1999; Larsen, 2003b), hydrothermal vents (Larsen, in press), and interstitial habitats (Holdich & Jones, 1983b). Some families seem to be restricted to, or have preferences for, certain substrata (Sieg, 1986b; Leite, 1995); others seem to accept a wide variety of sediments. A few specialized tanaidaceans are found in marine caves (Guțu & Iliffe, 1989a, b, 1997, 2001) and large populations have even been found in sulfurous anoxic environments (Sieg & Heard, 1985).

The few brackish water species and the special conditions regarding the freshwater species (see below) indicate that tanaidaceans have not been successful in invading fresh water. Most freshwater species are more likely to have been cut off from the sea rather than to have invaded the freshwater habitats (Richardson, 1901; Miyadi, 1938; Gardiner, 1975a). Some brackish water tanaidaceans do, however, show tolerance to large variations in salinity (Sieg et al., 1982) and sometimes occur temporarily in fresh water, but appear unable to form stable colonies (Owen & Forbes, 1997). However, the diversity of the freshwater species is probably underestimated. *Sinelobus stanfordi* (Richardson, 1901) is reported to be found in both fresh water and in hyper-saline lakes on the Galapagos archipelago (Richardson, 1901; Gardiner, 1975a), in inland lakes of Japan (Miyadi, 1938), brackish water lakes on islands off Papua New Guinea, in river lagoons in the West Indies (Gardiner, 1975a), and in Florida marshlands (this study). Given the differences in habitat and in particular in the distance between populations, these reports are unlikely to deal with the same species. Also, the deep-water species *Paraleptognathia longiremis* (Lilljeborg, 1864) cannot possibly be the same species as the one found in the Balalaika Lake as suggested by Kudinova-Pasternak (1972). Relatively large, blind apseudomorphans have recently been discovered in freshwater springs in northern Australia (Larsen & Hansknecht, 2004) and since blind, shallow-water tanaidaceans are rarely seen outside the interstitial environment (where they are very small), a tanaidacean ground water component has also been suggested.

BEHAVIOUR AND INTERSPECIFIC RELATIONSHIPS

The behaviour and general biology of tanaidaceans are poorly known and the lack of reviews on the subject is remarkable. This study compiles information on all aspects of tanaidacean behaviour and interspecific relationships.

Feeding strategies

Shallow-water tanaidaceans are usually reported as being detritivores, scavengers, or browsers (Heard, 1982; Bierbaum & Wegner, 1993). Some species are raptorial carnivores, preying on a number of other invertebrates including echinoid larvae (Highsmith, 1982, 1983), polychaetes (Oliver & Slattery, 1985), and nematodes and harpacticoid copepods (Feller, 1978). However, most tanaidaceans are probably opportunists and feed on a diversity of small invertebrates, detritus, and algae. Only direct observations of feeding or the use of immunological techniques as those demonstrated by Feller (1978) can identify the specific prey items. Many shallow-water tanaids have been observed feeding on diatoms (Bückle-Ramirez, 1965; Johnson & Attramadal, 1982a; Kneib, 1992; Błażewicz-Paszkowycz & Ligowski, 2002). However, as elegantly illustrated by Highsmith (1983) and Oliver & Slattery (1985), even tanaids well known to subsist on detritus and diatoms, can play a significant role as opportunistic predators.

Most species of the Kalliapseudidae (Apseudomorpha) are filter-feeding and thus have mouthparts and chelipeds transformed into a filter-feeding apparatus with dense arrays of setulated setae (Menzies, 1953; Lang, 1956a). Dennell (1937) suggested that *Apseudes talpa* (Montagu, 1808) also employed a secondary filter-feeding strategy, using the heavily setose maxilla and maxilliped. Most Apseudomorpha do indeed filter the respiratory current with the maxilla (Holdich & Jones, 1983a; pers. obs.). It is the belief of the present author, however, that this is rather used to avoid clogging-up of the respiratory chamber and not a 'real' mode of feeding. The particles collected by the maxilliped setae are scraped off with the specialized setae of the maxillipedal endites and moved to the mouth via the maxillule (Dennell, 1937). This system is only found in the Apseudomorpha and to a lesser degree in the Neotanaidomorpha, never in the Tanaidomorpha.

Much is still unknown about the tanaidacean diet in the deep sea. While most deep-sea tanaidaceans undoubtedly are detritivores, additional trophic groups exist and many species are omnivores to some extent (Kudinova-Pasternak, 1991). Species that mainly feed on detritus are more likely to utilize the bacterial film enveloping detritus particles, than the particles themselves (Bird & Holdich, 1984). Dietary specialization may be reflected in the form of the mouthparts or appendages. Some species have slender, fragile molars, suggesting a diet different from that of species that have robust molars. It has been suggested that species with fragile mouthparts mainly feed on detritus (Kudinova-Pasternak, 1991) and that robust mouthparts reflect a carnivorous diet (Bird & Holdich, 1984). However, this is not supported by studies on other peracarids, where slender mouthparts are often associated with a predatory lifestyle (Costa, 1960). Furthermore, a diet consisting of diatoms or detritus particles would require mouthparts well developed for mastication, as the broad and solid molar found in the diatom-feeding species of the Paratanaidae and Leptocheliidae.

Neotanaidomorphans may feed primarily on detritus and small invertebrates and, as in *Apseudes talpa*, have been suggested to have an additional filter-feeding life-strategy owing to the setose maxilla (Gardiner, 1975b). However, as stated above and under 'Respiration', the present author considers the maxilla to filter the incoming respiratory current to avoid clogging rather than to comprise a feeding apparatus. Neotanaidomorphans, with their massive chelipeds and free-living nature, are much better suited for a predatory life-strategy. A likely prey for these relatively large tanaidaceans are Foraminifera, and the large biomass of Foraminifera found in the deep sea (Altenback & Sarnthein, 1989; Gooday et al., 1992) supports this hypothesis. Deep-sea asellote isopods have been reported to prey extensively on Foraminifera (cf. Svavarsson et al., 1993) and use their powerful mandibles to crush the tests. The neotanaidomorphans and several tanaidomorphan species could take advantage of their strong chelipeds to achieve the same goal (Bird & Holdich, 1984). The chelipeds are undoubtedly powerful enough, as tanaidomorphans can crush sand dollar tests with little effort (Highsmith, 1982). Schmidt & Brandt (2001b) also suggest a number of tanaidacean species (*Allotanais hirsutus* (Beddard, 1886), *Apseudes heroae* Sieg, 1986a) feed on Foraminifera, but these authors reach their conclusion (Schmidt & Brandt, 2001b: 224) on comparison of mouthpart morphology. However, the mouthparts of *A. hirsutus* and *A. heroae* do not distinguish themselves from those of other species in their respective families. Furthermore, it is doubtful if a tanaidomorphan, like *A. hirsutus*, would be able to break foraminiferan tests using its mouthparts, and it is more likely that both species employ the chelipeds for this purpose.

Predators

Shallow-water tanaidaceans have been observed to be preyed upon by a number of other organisms:

(1) Polychaetes (Oliver & Slattery, 1985; Dojiri & Sieg, 1997).

(2) Several decapod crustaceans, in particular prawns (Leber, 1983; Johnson & Attramadal, 1982a; Kneib, 1984; Mayer, 1985; Kneib, 1988). Tanaids are even reported to be the preferred prey of penaeid shrimps (Balasubramanian et al., 1979).

(3) Other peracarid crustaceans, such as isopods (Kneib, 1985) and amphipods (Guţu & Sieg, 1999).

(4) Tanaidaceans constitute a significant component of the diet of migratory birds on the mudflats of the northern hemisphere (Băcescu & Guţu, 1975).

(5) Tanaidaceans are also found in the stomach contents of a large number of fish species, in particular of larvae or juveniles (Monod, 1925; Valiela et al., 1977; Overstreet & Heard, 1978; Meredith & Lotrich, 1979; Hodson et al., 1981; Nelson, 1981; Mendoza, 1982; Ogle et al., 1982; Kneib, 1986; Hettler, 1989; Rozas & LaSalle, 1990; Gillanders, 1995; Hartney & Tumyan, 1998; Huh & Kwak, 1998; Platell et al., 1998; De Troch et al., 1998).

Little has been reported on predation of deep-sea Tanaidacea, although peracarid crustaceans are reported to constitute a major component of the diet of deep-sea fishes, and of those crustaceans the tanaids constituted about one-third (Bright, 1970).

Parasites

Several parasites have been identified in association with tanaidaceans. Various species of the genera *Typhlotanais*, *Leptognathia*, and *Haplocope* (cf. Greve, 1965), *Paranarthrura* (cf. Boxhall & Lincoln, 1987), and *Pseudobathytanais* (cf. Larsen & Heard, 2001), have been found with ectoparasitic tantulocaridean larvae attached (for reviews, see Grygier & Sieg, 1990; Huys, 1991). Specimens of *Leviapseudes*, *Paranarthrura*, *Leptognathiella*, and *Meromonakantha* were found during this study with several stages of tantulocarideans

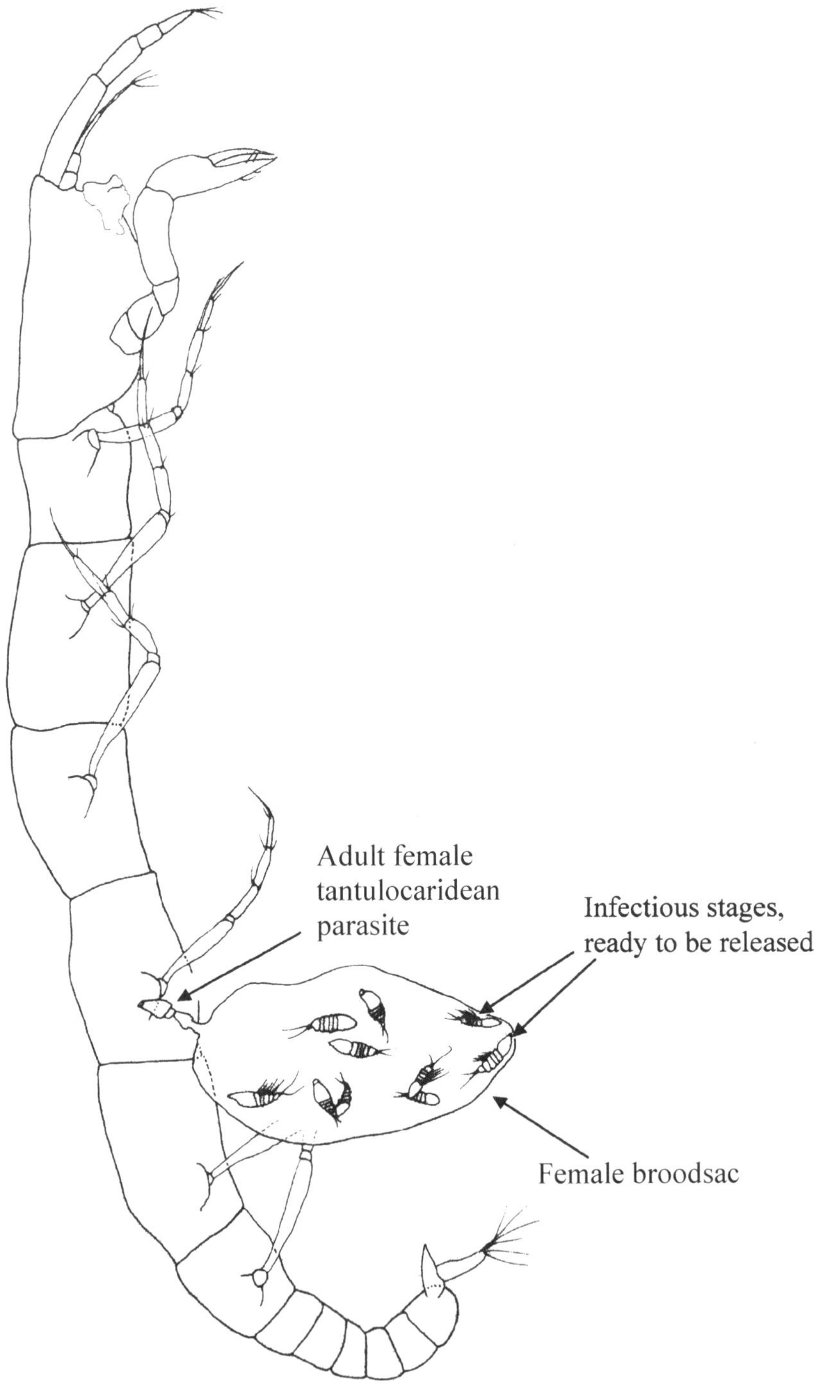

Fig. 4. Tantulocarida infestation on female *Paranarthrura* sp.

attached to their somites or appendages. One tantulocarid was found with a 'brood sack' containing ten completely developed infectious stages and extending from the body of *Paranarthrura* sp. (fig. 4). Tanaidaceans are also known as the intermediate host for acanthocephalan and nematode fish parasites (Shick et al., 1995; Escobar-Briones et al., 1999). Several species have been found with internal nematode parasites, sometimes filling out the entire exoskeleton (Gardiner, 1975b; this study).

Commensal relationships

Tanaidaceans are found associated with many other organisms in a variety of relationships:

(1) As epifauna on a number of organisms such as solitary corals (Sieg & Zibrowius, 1988), colonial corals and hydroids (Băcescu, 1981a; Preston & Doherty, 1994), or bryozoans (Larwood, 1954); on the shells of scallops (Browne & Berkman, 1992), oysters (Bamber, 1990), barnacles (Reimer, 1975; Bamber & Bird, 1997), or even sea turtles (Caine, 1986).

(2) As cleaning commensals on mobile bryozoan colonies (Larwood, 1954).

(3) Tanaidaceans have their own epifauna associates such as: stalked Protozoa (cf. Gardiner, 1975b; Larsen, in press), bivalves (Warén & Carrozza, 1994), or foraminiferans embedded in the cuticle (Larsen, 2000a).

(4) Sharing the shell together with gastropods (Howard, 1952) or tube-dwelling holothurians (Larsen, 2000a), in the canals of sponges (Stebbing, 1900; Brown, 1958; Greve, 1967; Sarma, 1979; Băcescu, 1981b; Reidenauer & Thistle, 1985; Thistle et al., 1985).

(5) Harpacticoid copepods, in turn, have been reported as a cleaning agent of *Tanais dulongii* (under the name *T. cavolinii*) inside its tube, and it has been suggested that the copepod may even reproduce inside the tanaid's tube (Johnson & Attramadal, 1982a). Ostracodes and mites are also seen inhabiting the tubes of Kalliapseudidae (cf. Drumm, 2005b).

Parasitism

Only indirect proof of parasitism has been recorded among tanaidaceans, but circumstantial evidence exists. One species, *Expina typica* Lang, 1968, was found living in the body cavity and intestines of holothurians, suggesting a parasitic way of life (Thurston et al., 1987). These authors proposed that the reduction of the mandibular molar and maxillule of *E. typica* support a parasitic life-style. The same is suggested for *Mirandotanais vorax* Kussakin & Tzareva, 1974, which has a greatly expanded pleon (also found in *E. typica*), also suggested as an adaptation associated with parasitism (Kussakin & Tzareva, 1974). Hassack & Holdich (1987) claim that the enlarged pleon of these species is a result of internal brooding, but this is not supported by the following evidence:

(1) Kudinova-Pasternak (1991) reported the expanded pleon of *M. vorax* to be filled with a lymphoid liquid, thus suggesting parasitism.

(2) During this study, a number of *M. vorax* was found with the pleon packed with blood, apparently outside the gut proper, also indicating parasitism, and with a specialized storage system of diverticula.

(3) Specimens of *M. vorax* were found breeding in a large cocoon with the mancas or eggs freely distributed and the female without a marsupium, thus excluding internal brooding (M. Błażewicz-Paszkowycz, pers. comm.).

(4) Furthermore, a large number of tanaidaceans possess similar, reduced mouthparts, apparently without having any parasitic affiliations.

Both *M. vorax* and *E. typica* have been sampled (Kussakin & Tzareva, 1974; Sieg, 1979, 1986a; Larsen, 2000a) from within tubes and/or cocoons in the sediment, clearly indicating that they are at least not obligate parasites, although the tube phase is probably related to breeding.

Behaviour

Communication

As with most other aspects of tanaidacean biology, the issue of intraspecific communication is largely unknown. It is known through the culture experiments of Bückle-Ramirez (1965) that tanaids can communicate through chemical cues (pheromones). This is supported by morphological evidence, such as a remarkable sexual variation in the presence of chemoreceptors as aesthetascs (Larsen, 2000a). Also, visual and tactile stimuli are involved in the courtship of *Heterotanais* (cf. Bückle-Ramirez, 1965), *Tanais* (cf. Johnson & Attramadal, 1982b), *Leptochelia* (cf. Stoner, 1986), and possibly *Curtipleon* (cf. Larsen, 2002b).

Menzies (1953) described a special structure in species of *Imitapseudes* (synonymous with *Apseudomorpha*) from Ecuador, i.e., a stridulating ridge or parallel series of raised cuticular ridges. Menzies (1953) thought this structure was a coupling device to keep the chelipeds together, since it is found on both chelipeds. However, no compelling reason can be raised for why a tanaid should attempt to keep its chelipeds together and Menzies also raised the possibility that this structure could be related to sound production. The tanaid could rub the chelipeds to produce sound in a similar way as crickets do. This suggestion has some merit, since this is only found in males and may be used for attracting females or warding off other males from the territory. A different form of stridulating ridge is formed on the outer cheliped propodal surface of the male *Nesotanais rugula* Bamber, Bird & Angsupanich, 2003. Bamber et al. (2003) also mention both the 'locking mechanism' and the 'sound production' possibilities while leaning towards the 'locking mechanism' hypothesis. The present author remains convinced that the sound production hypothesis is correct and the statements about the limitation for searching out females (reduced pleon and number of aesthetascs) in male *N. rugula*, given by Bamber et al. (2003), strengthen this conviction.

Tube construction

Gardiner (1975a, b) states that the Apseudomorpha and Neotanaidomorpha do not include tube builders, but several apseudomorphans (Kalliapseudidae) have been reported as tube dwellers (Morton, 1954; Delille et al., 1985). Spinning glands have not been reported from the Apseudomorpha. The glands reported by Hansknecht et al. (2002) in *Tanapseudes gutui* Hansknecht, Heard & Bamber, 2002 are not homologous with the spinning glands in the Tanaidomorpha. Since they are placed in pereonites 3–5, not pereonites 1–3 as in the Tanaidomorpha, and apparently have no outlet on the pereopods involved with tube building, they are unlikely to be involved in tube construction. There seem to be no obvious means for producing tube-mucus in most apseudomorphans, in particular because the first pereopods most often are modified for burrowing or climbing, and thereby are too robust and too heavily armed to be suited for tube construction. It is possible that certain burrows (closed off during brooding) can be mistaken for actually spun tubes, as some apseudomorphans burrow in connection with reproduction (Holdich & Jones, 1983a) or when hiding from predators. Some apseudomorphans can make deep burrows, i.e., about ten times their body length (Băcescu & Guțu, 1975).

During recent experiments of the feeding mechanisms of *Kalliapseudes*, D. T. Drumm (pers. comm.) noticed the building of tubes in the laboratory but, unfortunately, did not notice how the delivery of mucus occurred. However, *Kalliapseudes* must be considered a tube building apseudomorphan all the same.

Among the Tanaidomorpha, tube construction is a topic on which there seems to be little consensus. Sieg (1983a) claimed that among the Tanaidomorpha, tube building is common and may even be obligate. Holdich & Jones (1983b), however, report several tanaidomorphans as free-living. Lincoln (1979) and Lincoln et al. (1982) state that tube-dwelling occurs in only relatively few genera, and also Hassack & Holdich (1987) state that only a small proportion of tanaidomorphans builds tubes. Greve (1967) made a remark that tube building is very unusual, perhaps non-existing, among species of *Leptognathia*.

However, all female tanaidomorphans examined by the present author (including several species of *Leptognathia*) were equipped with spinning glands on (at least) the first pair of pereopods. A possible explanation for this discrepancy could be handling: when sampling tanaids, handling alone can contribute to expel tanaids from their tubes. Also the addition of fixing chemicals will cause the tanaids to attempt leaving their tube; this feature can be successfully used to extract live tanaids (or other tube dwellers) from coarse sediment samples (Larsen, 2002b; Larsen & Rayment, 2002).

When working on museum collections, it is not known whether a specimen has been removed from its tube before storing. The pores on the distal part of the unguis of either only the first pair, or all three anterior pairs of pereopods, are strongly indicative of tube building. These tiny pores, however, can be rather difficult to locate without the application of SEM observations.

The process of tube construction has been described in detail for members of the Leptocheliidae (cf. Bückle-Ramirez, 1965; Krasnow, 1992; Krasnow & Taghon, 1997) and Tanaidae (cf. Johnson & Attramadal, 1982a). In both families, tube construction appears to be related to feeding. The tanaid selects particles with its chelipeds, passes these on to the maxilliped and from there they are either eaten or passed on to the first pereopods. Sticky mucus, secreted from the tip of the dactylus of the first pereopods, glues the particles together in a non-random manner. The stickiness of the mucus is enhanced by microbiotic growth. The presence of tanaid tubes can act as a binding agent for the sediment (Krasnow & Taghon, 1997). The stickiness of the tube also prevents the tanaid from moving the tube but the female continues to extend the length of the tube throughout life, thus extending the individual feeding range (Krasnow, 1992; Krasnow & Taghon, 1997). The male, being of the non-feeding type in the Leptocheliidae, only makes short (body length) temporary tubes in *Leptochelia dubia* (cf. Mendoza, 1982; Highsmith, 1983) and *Heterotanais oerstedii* (cf. Bückle-Ramirez, 1965). Stoner (1986) reported that male *Lepto-*

chelia forresti (Stebbing, 1896a) are unable to build tubes, possibly owing to their extremely enlarged chelipeds. Males construct real tubes in the Tanaidae, albeit shorter and less well maintained than those of the female. Male *L. dubia* do build tubes, but significantly slower than the females (Mendoza, 1982). The general rate of tube construction is high, ranging from about 1.0 to 4.0 cm/day (Richards, 1969; Myers, 1977). Larger females build tubes faster than smaller ones (Krasnow & Taghon, 1997), whereas smaller males build faster than larger males (Mendoza, 1982). All kinds of materials are used for tube construction, such as: sand, silt, mud particles, sponge spicules, debris, and foraminiferan tests. Faecal particles and parts of shed cuticle are used as an inner lining of the tube (Bückle-Ramirez, 1965; Johnson & Attramadal, 1982a). Some species are selective in their choice of building material, whereas others seem to use indiscriminately whatever material is present (Hassack & Holdich, 1987). Large tanaidaceans appear to have a preference for relatively large (63 to 125 μm) particles, while small tanaidaceans are either indifferent or prefer small (32 to 63 μm) particles (Krasnow & Taghon, 1997).

Some tanaids invariably build tubes open at both ends (Holdich & Jones, 1983a), whereas in other species the female builds tubes only open at the front end, while the male keeps the tube open at both ends (Johnson & Attramadal, 1982a). The tubes are always sealed during brooding (Johnson & Attramadal, 1982b; Holdich & Jones, 1983a).

Little is known regarding deep-sea species. Most tubes of the deep-sea species examined by the present author consist of very fine detritus, sometimes with incorporated sponge spicules, and tend to be rather delicate.

While there are reports of several individuals of the same species sharing a single tube (Bird & Holdich, 1985), this phenomenon is rare, except when mancas are living in the parental tube.

Some tanaidaceans have been found in the open end of tubes of ammodiscaean foraminiferans (Gooday, 1984) and one brooding specimen of *Collettea cylindratoides* Larsen, 2000 was found using an abandoned tusk shell (Scaphopoda) as a tube (Larsen, 2000a). Some apseudomorphans (*Pagurapseudes*) also occupy abandoned bivalve (Howard, 1952) or gastropod shells, in a manner resembling that of hermit crabs (Messing, 1980, 1983; McSweeny, 1982).

A couple of species have been reported to spin a meshwork of strands extending from the front end of the tube. When these strands were mechanically affected, two responses were noted. A slight, gentle stimulus on the thread would provoke the tanaid to response by coming out in order to investigate; a more abrupt contact would induce the tanaid to immediately retreat down its tube (Bückle-Ramirez, 1965). These strands thus function as an analogue to a spider's web, both in alerting the tanaid to danger as well as informing it of potential prey. Tanaidomorphans are also reported to use a spun thread as an anchor line, thereby further extending the 'spider' analogue (Johnson & Attramadal, 1982a).

Fighting behaviour

The male chelipeds are used in male/male combat (Stoner, 1986) but apparently not for defence against predators (Balasurami et al., 1979). Bückle-Ramirez (1965) reported that males of *Heterotanais oerstedii* fight for the females with highly predictable results. Small males are certain losers and, even in cases where no other males are present, females fight with and expel smaller males from their tubes. In case of fighting between two equal-sized males, Bückle-Ramirez (1965) claims that the first to arrive is always successful, even though battles might last for up to an hour.

Highsmith (1983) reported a somewhat similar behaviour in *Leptochelia dubia* in great detail. Apparently, *L. dubia* fight almost like professional wrestlers, grappling and throwing each other. Larger males and defending males already inside a tube, were highly superior to smaller males or attacking males, but not exclusively so.

Agonistic behaviour is reported between male *Leptochelia dubia* and *L. forresti*, with *L. forresti* as by far the most successful combatant. This behaviour is suggested to be a result of the morphological similarity and sympatric distribution of these species. Agonistic behaviour of this sort is considered to be in some way mating-related, since females of both species are not subjected to aggression from males of any species and neither do the females fight among each other (Stoner, 1986).

Reproduction

The only extensive studies of tanaid reproduction are those on *Heterotanais oerstedii* (cf. Bückle-Ramirez, 1965) and *Tanais dulongii* (cf. Jones & Attramadal, 1982a). The results of these studies are generally considered valid for the Tanaidomorpha as a whole. As can be seen from the following review, however, the extrapolation of these results to other species may be hazardous.

Bückle-Ramirez (1965) claims, that if a large male enters a tube containing a small female, the female is expelled from the tube because copulation can only occur between equal-sized specimens. While this may occur in species displaying little sexual dimorphism in size, species of Leptognathiidae and Paratanaidae have highly modified males that are significantly smaller than the females (Sieg, 1976; Larsen, 2001). Consequently, similar size of the sexes cannot be a criterion for successful mating in these species, hence the universal application of this criterion seems flawed.

The male's chelipeds are used to gain entry to the female's tube. Even though the chelipeds are also used for grasping the female during copulation, males with amputated chelipeds can still mate, but have to enter the female tube from the entrance only and cannot fight any rival male.

Courtship

There is no 'pre-copulation' in Tanaidomorpha, but some form of courtship takes place instead. Both courtship and copulation only take place within the female's tube. Bückle-Ramirez (1965) described the mating procedures for *Heterotanais oerstedii* as follows: the male enters the tube of the female and initiates courtship. Courtship lasts for up to 11 hours, if not disturbed by another male. Several stereotypic movements are involved. As the male gains entry, the female first tries to defend the tube by attacking the male or attaining a diagonal posture with her carapace and chelipeds to block the access. The male touches the female with the antennae and beats a current towards the female with his pleopods. The male then places his chelipeds on the female's cephalothorax and initiates antennae-antennae and antenna-uropod contacts, sometimes for up to several minutes. Both male and female will then turn around to face away from each other. The male pushes his abdomen above the female's abdomen and moves his pleopods rapidly up and down on the female's pleonites. The female turns around again and grasps the male's uropods with her chelipeds.

Johnson & Attramadal (1982a) described the courtship in *Tanais dulongii* (as *T. cavolinii*) as almost similar. The female would initially defend the tube by blocking the entrance or trying to expel the male. After a while the agonistic behaviour subsides and the female will tolerate the male until copulation, which can only occur immediately after the female has moulted.

Copulation

In *Heterotanais oerstedii*, the male turns on its back and pushes with its chelipeds to obtain a position parallel under the female. Copulation itself lasts no longer than a few seconds. Sperms are delivered in the vicinity of the female's gonopore through the wide open oostegites. A few seconds later, the female starts to deposit the eggs into the marsupium. When the eggs begin to emerge into the marsupium, the female slides off the male. After one or two eggs have emerged, the female starts to attack the male's uropods and pleotelson and the male soon leaves the tube. The female then seals the tube and will attack any male trying to gain entrance (Bückle-Ramirez, 1965).

Tanais dulongii mates in much the same manner, but here the male grasps the female with his chelipeds. The male inserts the sperm through a slit in the female's ovisacs. The male then withdraws and the female immediately starts depositing the eggs into her ovisacs (Johnson & Attramadal, 1982a).

No observations have been made on apseudomorphans or on any deep-sea taxa; presumably, modifications may have evolved there. For example, in many deep-sea species the pleopods are rudimentary, immobile, or even absent, and the uropods are too small to be used as reported for *T. dulongii* and *H. oerstedii*.

Brood care

Bückle-Ramirez (1965) states that eggs of an intermediate size, when lost from the marsupium, are recovered by the female. Small eggs, however, are not recovered if lost but are left on the floor of the tube, while large eggs, if lost, are eaten. Recovering eggs is a complex process, involving both chelipeds and pereopods (Bückle-Ramirez, 1965: 734). Johnson & Attramadal (1982b) reported that *Tanais*, by releasing 'reserve' eggs, provides a source of food (egg yolk) for the developing brood-pouch mancas. Only if the original brood is lost will these reserve eggs be allowed to develop. This mechanism for providing nourishment for the mancas is not known in other peracarids. The reserve egg yolk, after ingestion, also seems to have a function in enabling the hatchlings to produce an enzyme dissolving the ovisacs, in order to release the mancas from the marsupium (Johnson & Attramadal, 1982b).

After leaving the marsupium, the mancas live for a while in the parental tube, as is also seen in several other peracarids (Lim & Alexander, 1986; Shillaker & Moore, 1987; Svavarsson & Davidsdottir, 1995; Thiel et al., 1997; Thiel, 1999). During this period the female will not allow males to enter the tube. The female may also construct barriers to prevent the mancas from leaving the tube prematurely and the females of some species are even reported to tie down 'unruly' mancas with secreted mucus strands (Johnson & Attramadal, 1982b). When sufficiently mature, the mancas leave the tube via the entrance or break through the tube walls and then start constructing their own tube, frequently in physical contact with the parental tube. After seven days, any remaining mancas will be physically removed from the tube by the female (Bückle-Ramirez, 1965). Mancas of deep-sea species probably do not construct their tubes in physical contact with the parental tube, because food availability in the deep sea would rarely allow settlement at a high density (see chapter on Distribution).

The rhythmic beating of the pleopods during brooding in *Tanais dulongii* and *Heterotanais oerstedii* seems not only to ventilate the tube, but also to create a current bringing in diatoms and other food particles on which the mancas will feed (Johnson & Attramadal, 1982b).

In at least one tanaid, *Typhlotanais magnificus* Kudinova-Pasternak, 1969, the oostegites are rudimentary and the eggs develop freely in the tube (Kudinova-Pasternak, 1969). Also in *Mirandotanais vorax* Kussakin & Tzareva, 1974, the females do not develop oostegites, possibly owing to the enlarged pleon and posterior pereon in this species. The mancas of *M. vorax* roam freely within a large cocoon, spun by the female.

Seasonal breeding has been reported from temperate environments: March to December in southern France (Salvat, 1967), June to July in northern Germany (Bückle-Ramirez, 1965). In warmer habitats, tanaids are reported to breed year-round (Masunari, 1983; Kneib, 1992).

Fecundity

Data on tanaidacean fecundity are limited. In *Pagurapseudes largoensis* McSweeny, 1982 the number of eggs, embryos, or unreleased mancas ranges between 4 and 17 (Messing, 1983), in *Tanais dulongii* up to 15 (Johnson & Attramadal, 1982b), and 2–15 in *Apseudes heroae* Sieg, 1986a (cf. Schmidt, 2001), in *Discapseudes surinamensis* Băcescu & Guțu, 1975 between 18 and 29 (Băcescu & Guțu, 1975), in *Allotanais hirsutus* (Beddard, 1886) 19–30 (Schmidt et al., 2002), and in *Apseudes latreillii* (H. Milne Edwards, 1828) up to 60 (Salvat, 1967). These variations appear to be dependent on female body size. *Typhlotanais magnificus* has been reported with over 80 eggs in its tube (Kudinova-Pasternak, 1969) but it is possible that some of those eggs were not fertilized and that they were intended as food for the mancas, as seen previously in *Tanais dulongii*. During this study, several tubes of an undescribed species of *Paranarthrura* contained up to 24 mancas II's which, with an unknown proportion of egg mortality, probably reflects a high fertility. Although the number of species and specimens with mancas examined during this study is limited, surprisingly the largest broods are found in species from greater depths (*T. magnificus* were found at 3620 m and *Paranarthrura* sp. at 1770 m). Since the deep-sea and other low-food environments normally dictate low fecundity (Smith & Hessler, 1987; Arntz et al., 1994), one should expect a lower fecundity in the deep sea, than in shallow-water tanaidaceans.

Locomotion

The locomotion of tanaids has been described by Hessler (1982) and Johnson & Attramadal (1982a). Hessler (1982) noted a marked difference in the movements of the legs, in particular in the body-coxa-basis movement, between Apseudomorpha (*Apseudes*) and Tanaidomorpha (*Leptochelia*). As the basic structure of the legs is the same in the Apseudomorpha and Neotanaidomorpha, these are treated as one, while the tube-dwelling tanaidomorphans are treated separately.

Besides walking, many shallow-water tanaids are capable of limited swimming. By a sudden flex of the body, the tanaid can bring itself into suspension, at which point a rapid beating of the pleopods allows the animal to swim short distances, but apparently without any sense of direction (pers. obs.).

Apseudomorpha and Neotanaidomorpha

The locomotion of most Apseudomorpha and Neotanaidomorpha does not distinguish itself from that of other benthic 'walking' Crustacea, as was thoroughly covered by Manton (1978).

The first pereopod of many apseudomorphans (= fossorial leg of some authors) is wider than the succeeding pereopods. This leg may be used for digging or climbing, but in general it does not participate in locomotion. An exception may be found in the family Sphyrapodidae. In this family, the first pereopod is not just wider than, but also almost twice as long as the other pereopods. Although no direct observation has been made of this, it is hard to view these enlarged pereopods as anything but large paddling appendages, designed for swimming.

The species of *Pagurapseudes* have pereopods greatly modified from those of the general apseudomorphans, and these are densely covered with suckers providing anchorage within the gastropod shells that these species reside in.

Tanaidomorpha

As *Leptochelia* is a tube dweller and has lost the coxa of the last three pereopods, it is not surprising that locomotion 'in the open' of the species in this taxon is reduced, compared to that of the free-living Apseudomorpha (cf.

Hessler, 1982). Within a tube, however, the adaptations of the tanaidomorphans (*Tanais*) are redeemed. The more posteriorly directed traction of their last three pereopods (Larsen, 2003a) is well suited for speedy retraction into the safety of a tube. Also, the last three tanaidomorphan pereopods are held in the shape of a 'V': if one imagines the basis as one leg of the 'V', the ischium as the tip and the more distal articles as the other leg of the 'V'. This 'V' shape allows the animal to gain traction from the tip of the 'V' (ischium region) as well as from the distal part of the leg around the dactylus/unguis (Johnson & Attramadal, 1982a). The usually heavy armament of the last three pereopods helps to provide traction and also presents a tactile sense organ when in contact with the tube wall (Johnson & Attramadal, 1982a).

The chelipeds are not used for locomotion inside the tube, but may be employed as a grasping tool to secure the animal when venturing outside the tube (Johnson & Attramadal, 1982a). The first pereopods are not used for locomotion but for spinning. They may, however, assist in securing the animal outside the tube, when spinning and attaching a safety line (Johnson & Attramadal, 1982a).

Almost all male tanaidomorphans have better developed pleopods than the female of the same species and often these are involved in swimming. In some taxa (*Leptognathia*, *Paratanais*) the male morphology reflects its swimming capacity to a degree where the animal appears almost amphipod-like. In most shallow-water species, the females have also retained the ability to swim, although not as good as the males.

SYSTEMATICS

Order TANAIDACEA Dana, 1849

Some confusion persists about the authority of the name 'Tanaidacea'. The family name Tanaidae was first used by Dana in 1849, at a time when tanaids were included in the Isopoda (Isopoda vagantia Thomson, 1880; Isopoda chelifera or 'Saksisopoder' G. O. Sars, 1882; Isopoda aberrantia Gerstaecker, 1883; Anisopoda Dana, 1852). Claus (1888) elevated the Anisopoda to the level of order and the name 'Tanaidacea' for that order was introduced by Hansen (1895). In the recent literature, the authority of the Tanaidacea has been given to either Dana (Sieg, 1983a; Bird & Bamber, 2000; and others) or Hansen (Bird & Holdich, 1984; Bamber, 1998; Sieg, 1986a; and others). From these examples it is clear that little consensus or consistency exist on this matter. There are no specific rules for supra-familial taxa in the International Code of Zoological Nomenclature. However, article 11 (family group names), article 12 (for names given before 1930), and article 34 (mandatory changes in spelling) can be applied to this situation. Following these articles, the first author to introduce a Latinized name to a family group will remain the authority as the name is raised or lowered in rank (International Code of Zoological Nomenclature, 1985). As the Tanaidae were first suggested a family by Dana (1849) and elevated to the rank of order by Claus (1888), Hansen (1895) merely applied an already used name, rather than introducing a new name. Therefore, Dana (1849) must be considered the authority of the name Tanaidacea.

The systematics and phylogeny of the Tanaidacea are incompletely understood (Guțu, 1981; Sieg, 1986a; Larsen & Wilson, 2002). Few attempts have been made to apply any phylogenetic approach to this order (Sieg, 1983b, 1984) and only one using computer assisted methods (Larsen & Wilson, 2002). While there is little consensus on the family and lower level systematics, there seems to be a general consensus in abandoning the old division of 'Monokonophora' and 'Dikonophora' for the three recent suborders, Apseudomorpha, Neotanaidomorpha, and Tanaidomorpha, as suggested by Sieg (1983b).

Fossil Tanaidacea

There is an extensive fossil record for tanaidaceans, dating back to the Lower Carboniferous (Schram et al., 1986). Most fossils are placed in the Apseudomorpha Sieg, 1980 (family Jurapseudidae Schram, Sieg & Malzahn, 1986; genus *Jurapseudes* Schram, Sieg & Malzahn, 1986; genus *Carlclausus* Schram, Sieg & Malzahn, 1986) or in the fossil suborder Anthracocaridomorpha Sieg, 1980 (family Anthracocarididae Schram, 1979; genus *Anthracocaris* Calman, 1933; genus *Ophthalmapseudes* Glaessner,1962; genus *Eucryptocaris* Schram, 1989).

Only one fossil has been reported in the literature as belonging to the Tanaidomorpha (cf. Schram et al., 1986) (family Cretitanaidae Schram, Sieg & Malzahn, 1986; genus *Cretitanais* Schram, Sieg & Malzahn, 1986). However, this fossil has little resemblance to any recent tanaidomorphan taxa and probably does not belong in the Tanaidomorpha (cf. Larsen, 2000a; Guțu, 2004). However, the author was recently presented with six specimens from the Cretaceous sealed in amber (R. Vonk., pers. comm.). These specimens undoubtedly do belong to the Tanaidomorpha and appear to represent a taxon with affinities that lie somewhere between the Tanaoidea and Paratanaoidea.

Recently, fossils of *Neotanais* dating back to the Middle Triassic have been recorded (Hannibal, 2003). This taxon was also attributed significant phylogenetic age by Sieg (1988).

Key to the Suborders of Tanaidacea in the Gulf of Mexico

1. Antennule biramous .... Apseudomorpha Sieg, 1980
– Antennule uniramous 2
2. Antennule with 7 articles. Antenna with 9 articles. Uropodal endopod with more than 5 articles .... Neotanaidomorpha Sieg, 1980 (*Neotanais* Beddard, 1886)
– Antennule with 5 or less articles (female). Antenna with 7 or less articles. Uropodal endopod with 5 or less articles .... Tanaidomorpha Sieg, 1980

SUBORDER APSEUDOMORPHA SIEG, 1980

KEY TO THE DEEP-SEA GENERA OF APSEUDOMORPHA IN THE GULF OF MEXICO

1. Mandible with palp 2
– Mandible without palp .... *Sphyrapoides* Guţu & Iliffe, 1998
2. Mandible body with spiniform setae near palp insertion ....
.... *Pectinapseudes* Băcescu & Williams, 1988
– Mandible body without spiniform setae near palp insertion 3
3. Pereopod 1 dactylus with membranous prolongation longer than unguis ....
.... *Dactyloprion* Guţu, 2002
– Pereopod 1 dactylus without membranous prolongation 4
4. Pereopod 1 without exopod .... *Atlantapseudes* Băcescu, 1978
– Pereopod 1 with exopod 5
5. Antenna without squama/pseudosquama .... *Pseudosphyrapus* Guţu, 1980
– Antenna with squama/pseudosquama 6
6. Antenna with pseudosquama medial on article 2 ....
.... *Kudinopasternakia* Guţu, 1991
– Antenna with squama distally on article 2 7
7. Pereonites without lateral spines. Pereopod 1 coxae without spiniform apophyses ....
.... *Parapseudes* G. O. Sars, 1882
– Pereonites with lateral spines. Pereopod 1 coxae with spiniform apophyses 8
8. Maxillipedal endite with modified seta on inner-distal margin. Pereopod 4 dactylus/unguis often modified. Male chelipeds without process on fixed finger and dactylus ....
.... *Leviapseudes* Sieg, 1983
– Maxillipedal endite without modified seta on inner-distal margin. Pereopod 4 dactylus/unguis not modified. Male chelipeds with process on fixed finger and dactylus ....
.... *Apseudes* Leach, 1814

Family APSEUDIDAE Leach, 1814

Apseudidae Leach, 1814, The Edinburgh Encyclopaedia, **7**: 386.
Apseudii Leach, 1814, The Edinburgh Encyclopaedia, **7**: 404.
Apseudiadae Samouelle, 1819, The entomologist's useful compendium: 109.
Apseudidæ G. O. Sars, 1882, Archiv for Matematik og Naturvidenskab, **7**: 6.
Apsendidæ Monod, 1923, Annales Société des Sciences Naturelles Charente Inférieure, **37**(4): 39.
Apseudides Monod, 1926, Bulletin de la Société des Sciences Naturelles du Maroc, **13**: 11.
Apseudidea Glaessner & Malzahn, 1962, Fortschritte in der Geologie von Rheinland und Westfalen, **6**: 262.
Whitelegiidae Sieg, 1983, Crustaceorum Catalogus, **6**: 162.

Diagnosis. — (Modified after G. O. Sars, 1896; Lang, 1968; Guțu, 1972, 1981, 1996a.) Female. Body more or less elongate, tapering distally; semi-cylindrical, dorso-ventrally flattened, or cylindrical. Eye-lobes present, often spiniform, with or without visual elements. Pereon with six free pereonites. Pleon with five free pleonites and pleotelson; pleotelson narrower than pereon. Pereonites 2–6 usually with lateral spines anterior to pereopod insertions. Antennule biramous and antenna with squama. Mandible with tri-articulate palp. Maxillule with bi-articulate palp. Epignath broad, kidney-shaped. Cheliped and pereopod 1 with or without exopod. Pereopod 1 longer and wider than other pereopods. Pereopod dactyli without aesthetascs. Pereopods 4–6 without sucker-type structures. Pleopods present, reduced, or absent. Uropods biramous, both rami multi-articulate.

Male. Antennule with higher number of (smaller, serially repeating) articles than in female, all with dense arrays of aesthetascs. Chelipeds with (Apseudinae, *Carpoapseudes*) or without (*Leviapseudes*) process on fixed finger and dactylus. Pleopods always present.

Type genus. — *Apseudes* Leach, 1814.

Remarks. — The heavy, spiniform setae on the ventral surface of the pereopod 1 propodus have previously been used to separate species or genera of the Apseudidae (cf. Lang, 1968) and the structure is claimed to constitute a very stable character (Guțu, 2002a). However, Lang (1966) regarded these spiniform setae as variable in *Parapseudes*. During this study, the number of these setae was observed to vary between the left and right side of individual *Leviapseudes bipartitus* sp. nov., and should thus, in the present authors opinion, not be given too much weight as a specific character.

Subfamily APSEUDINAE Leach, 1814

Apseudinae – Guțu, 1972, Revue Roumaine de Biologie, (Biologie Animale) **17**(5): 302.

Diagnosis. — Both sexes. Maxillipedal endopod without leaf-shaped setae subdistally. Cheliped and pereopod 1 with or without exopod. Pereopod 1 coxa with or without forward pointing spine. Pereopod 4 dactylus not reduced. Pereopod 1 dactylus and unguis not fused.

Type genus. — *Apseudes* Leach, 1814.

Genera included. — *Apseudes* Leach, 1814; *Atlantapseudes* Băcescu, 1978; *Bilobatus* Sieg, 1993; *Bunakenia* Guțu, 1995; *Dactyloprion* Guțu, 2002; *Fageapseudes* Băcescu, 1987; *Gollumudes* Bamber, 2000; *Gutuapseudes* Edgar, 1997; *Hainanius* Bamber, 1998; *Hoplomachus* Guţu, 2003; *Langapseudes* Băcescu, 1987; *Mendamanus* Bamber, 1998; *Paradoxapseudes* Guţu, 1991; *Pectinapseudes* Băcescu & Williams, 1988; *Spinosapseudes* Guţu, 1996; *Tuberapseudes* Băcescu & Guţu, 1971; *Typhlapseudes* Beddard, 1886.

Remarks. — Currently, there exist no unambiguous, valid diagnoses that can separate this subfamily from the Leviapseudinae (cf. Błażewicz-Paszkowycz & Larsen, 2004). Apseudinae and Leviapseudinae can be separated from the Pugiodactylinae by possessing an unfused dactylus and unguis of pereopod 1 (Guţu, 1995).

Genus **Apseudes** Leach, 1814

Apseudes Leach, 1814, The Edinburgh Encyclopaedia, **7**: 386.
Euphée Risso, 1816, Histoire naturelle des Crustacés des environs de Nice, **1**: 27.
Apseude Lamarck, 1818, Histoire naturelle des animaux sans vertébrés, **5**: 169.
Eupheus Desmarest, 1823, Dictionare Scientifique Naturelle, **28**: 369.
Rhoé H. Milne Edwards, 1828, Annales des Sciences Naturelles, Paris, **13**: 291.
Rhœa H. Milne Edwards, 1828, Annales des Sciences Naturelles, Paris, **13**: 291.
Rhoe Audouin & Milne-Edwards, 1829, Précis d'entomologie ou d'histoire naturelle des animaux articulés, **1**: 46.
Rhoea Latreille, 1831, Cours d'entomologie: 403.
Rhoés Westwood, 1832, Annales des Sciences Naturelles, Paris, **27**: 329.
Rhoë Rathke, 1843, Verhandlungen der Kaiserlichen Leopoldinisch-Carolinischen Deutschen Akademie der Naturförscher, **12**: 39.
Rhoëa G. O. Sars, 1886, Archiv for Mathematik og Naturvidenskab, **7**: 9.
Apseudopsis Norman, 1899, Annals and Magazine of Natural History, **3**: 330.
Apsendes Monod, 1923, Annales Société des Sciences Naturelles Charente Inférieure, **37**(4): 34.

Diagnosis. — (Modified after Guţu, 1996b.) Female. Body dorso-ventrally flattened or cylindrical. Carapace anteriorly not excavated, with conspicuous rostrum. Eye-lobes with spiniform apophysis, with or without visual elements. Pereonites and pleonites with lateral spines. Pleonites without paired ventral tubercles. Antennular article 2 short (less than 0.5 times as long as article 1). Mandible without spiniform setae near palp insertion. Cheliped and pereopod 1 with exopod. Pereopod 1 basis elongate (length/width ratio > 5). Pereopod 1 dactylus and unguis combined usually shorter than propodus; dactylus with weak serration and without membranous prolongation. Pereopod 1 carpus and propodus with heavy spiniform setae and few (< 10) simple setae. Pleopods present, biramous and well developed.

Male. Chelipeds with process on fixed finger and dactylus.

Type species. — *Cancer* (*Gammarus*) *talpa* Montagu, 1808.

Gender of generic name. — Masculine.

Remarks. — Currently, there is a broad consensus about the need for a rediagnosis of *Apseudes* (cf. Lang, 1968; Guţu, 1998a, 2002; Błażewicz-Paszkowycz & Larsen, 2004). However, there is also little to indicate that this task will be successfully accomplished in the near future. Guţu (2002) suggested a number of novel characters and character combinations to contribute towards a stable diagnosis, but some of those are hard to decipher (see below). Owing to the systematic confusion of the genus *Apseudes* and the numbers or species within this genus, a key is presented here to only the deep-sea species in the Gulf of Mexico.

The following characters have been suggested to have the potential to move several species, currently lodged within *Apseudes*, to other genera (Guţu, 2002a): (1) body length, (2) shape and chaetotaxy of the pereonites, (3) configuration of hyposphenians, (4) number of spiniform setae on pereopod 1 propodus, (5) appendage chaetotaxy, and (6) number of articles of the uropodal endopod.

Examination of these characters may indeed help resolve the confusion within *Apseudes*, but some features should be treated with caution. For example, Guţu (2002a: 21) suggested that the presence, size, and number of hyposphenians (including the genital cone) in the ovigerous female be considered a set of stable characters. The hyposphenians may be an underused character, but this is confounded by Guţu's (2002a) statement about a genital cone in the female, as this is only seen in males. As for the number of spiniform setae on the propodus of pereopod 1, this character is not valid (see remarks under Apseudidae).

Species and subspecies currently assigned to this genus. — *A. acutifrons* G. O. Sars, 1882; *A. africanus* Tattersall, 1925; *A. agulhensis* Barnard, 1920; *A. annabensis* Guţu, 2002; *A. apocryphus* Guţu, 2002; *A. arguinensis* Guţu, 2002; *A. armatus* Richardson, 1911; *A. babelmandebensis* Băcescu, 1978; *A. bacescui* Guţu, 2002; *A. bermudensis* Băcescu, 1980; *A. bowmani* Guţu & Iliffe, 1989; *A. browni* Băcescu, 1980; *A. bruneinigma* Bamber, 1998; *A. cedroensis* Menzies, 1953; *A. chilkensis* Chilton, 1924; *A. cooperi* Brown, 1954; *A. crozetensis* Shiino, 1978; *A. elisae* Băcescu, 1961; *A. espinosus* Moore, 1901; *A. galapagensis* Richardson, 1912; *A. galeroni* Băcescu, 1987; *A. graciloides* Stephensen, 1915; *A. graziellae* Băcescu, 1981; *A. grossimanus* Norman & Stebbing, 1886; *A. hastifrons* Norman & Stebbing, 1886; *A. heroae* Sieg, 1986; *A. holthuisi* Băcescu, 1961; *A. intermedius* Hansen, 1895; *A. lagenirostris* Lang, 1968; *A. latreillii* (H. Milne Edwards, 1828); *A. latus* Chilton,

1883; *A. levis* Kudinova-Pasternak, 1985; *A. mediterraneus* Băcescu, 1961; *A. meridionalis* Richardson, 1912; *A. minimus* Guţu, 2002; *A. minutus* Claus, 1888; *A. misarai* Băcescu, 1981; *A. mussauensis* Shiino, 1965; *A. nagae* Shiino, 1963; *A. nhatrangensis* Shiino, 1963; *A. olimpiae* Guţu, 1986; *A. orghidani* Guţu & Iliffe, 1989; *A. ostroumovi* (Băcescu & Cărăuşu, 1947); *A. paragracilis* Kudinova-Pasternak, 1975; *A. portnovensis* Balasubrahmanyan, 1962; *A. rikiteanus* Nobili, 1906; *A. robustus robustus* G. O. Sars, 1882; *A. robustus erythraeicus* Băcescu, 1984; *A. rotundifrons* Băcescu, 1981; *A. setiferus* Băcescu, 1981; *A. seurati* Nobili, 1906; *A. sicilianus* Băcescu, 1981; *A. siegi* Kudinova-Pasternak, 1985; *A. spectabilis* Studer, 1883; *A. spinosus spinosus* (M. Sars, 1858); *A. spinosus albidus* Băcescu, 1984; *A. talpa* (Montagu, 1808); *A. tenuimanus* G. O. Sars, 1882; *A. tenuipes* Băcescu, 1961; *A. tenuis* Hansen, 1913; *A. tridens* Guţu, 2002; *A. tropicalis* Miller, 1940; *A. tropicus* White, 1847; *A. uncidigitatus* Norman & Stebbing, 1886; *A. unicus* Kudinova-Pasternak & Pasternak, 1981; *A. vicinus* Hansen, 1913; *A. vitjazi* Kudinova-Pasternak, 1970.

Apseudes olimpiae Guţu, 1986

Apseudes olimpiae Guţu, 1986, Travaux du Muséum National d'Histoire Naturelle "Grigore Antipa", **28**: 37–48.

Diagnosis. — (Modified after Hansknecht & Heard, 2001.) Female. Body robust and heavily calcified; length about 5–6 mm. Carapace, including acute rostrum, as long as the first two pereonites combined. Pereonites 3–6 with anteriorly directed, hook-like lateral spiniform prolongation. Pereon and pleon with mid-ventral, curved spinous process on all somites. Cheliped and pereopod 1 with exopod. Pereopods 2–6 nearly cylindrical, slender, with few long setae and 0–3 spiniform setae on ventral edges of each merus, carpus, and propodus. Pleopods biramous and present on all pleonites.

Male. Chelipeds differ from those of female by having a bilobed ventral expansion on carpus; a very large propodus; and a tooth on inner margin of fixed finger and dactylus, respectively.

Distribution. — Reported from the shelf off Alabama and Florida. Bathymetric range: 19 to 47 m (Guţu, 1986; Hansknecht & Heard, 2001). See map 3.

Remarks. — This species is not a deep-sea species but there seems to be some uncertainty about the type locality (Hansknecht & Heard, 2001) as being "presumed off Bahamas" (Guţu, 1986). Thus, there are no records of depth for the type locality and it is possible that this species can reach greater depths, especially on the Florida escarpment, which is a habitat similar as that around the Bahamas, and which has not currently been surveyed.

Apseudes cf. **spinosus** (M. Sars, 1858)

Rhœa spinosa M. Sars, 1858, Forhandlinger i Videnskabsselskabet i Kristiania, **1858**: 151.
Apseudes talpa – Lilljeborg, 1864, Inbjudingsskrift, University of Uppsala: 12.
Apseudes spinosus – Norman, 1870, Proceedings of the Royal Society of London, **18**: 157.
Apseudes spinosa – Norman, 1880, Report from the British Association for the Advancement of Science: 387.
Rhoea spinosa – G. O. Sars, 1896, An Account of the Crustacea of Norway, **2**: 7.

Diagnosis. — (Modified after G. O. Sars, 1896.) Female. Carapace, including acute rostrum, longer than the first two pereonites combined. Eyes with weakly developed visual elements. Pereonites 3–6 without anteriorly directed, hook-like lateral spines. Pereon and pleon without mid-ventral process. Cheliped and pereopod 1 with exopod. Pereopods 2–6 nearly cylindrical, slender, with a few long setae and 0–3 spiniform setae on the ventral edges of each merus, carpus, and propodus.

Male. Cheliped differs from that of female by having a large process on inner margin of fixed finger and two small processes on inner margin of dactylus.

Distribution. — Off the coast of Norway, off the south coast of Ireland. Bathymetric range: 55–1300 m (G. O. Sars, 1896). In the Gulf of Mexico, it is reported from the west coast of Florida. Bathymetric range: 40 m (Dawson, 1966). See map 4.

Remarks. — Although listed from the Gulf of Mexico, it is hard to believe that a deep-sea cold water species from the northern tip of Norway could suddenly occur in the shallow waters of the Gulf of Mexico. The original specimens identified by Dawson (1966) appear to have been lost but new material from the area has revealed a species morphologically close to, but not identical with, *A. spinosus*. This species is currently being described (T. Hansknecht, pers. comm.).

KEY TO THE SPECIES OF *APSEUDES*, BOTH SEXES (GULF OF MEXICO)

1. Pereonites 3–6 with anteriorly directed hook-like lateral spiniform prolongation. Pereon and pleon with mid-ventral, curved, spinous process on all somites .... *A. olimpiae* Guțu, 1986

– Pereonites 3–6 without anteriorly directed, hook-like lateral spines. Pereon and pleon without mid-ventral process .... *A. spinosus* (M. Sars, 1858)

Genus **Atlantapseudes** Băcescu, 1978

Atlantapseudes Băcescu, 1978, Cahiers de Biologie Marine, **19**: 317.

Diagnosis. — (After Băcescu, 1978; Meyer & Heard, 1989; Guțu, 1996c.) Female. Body cylindrical. Carapace anteriorly not excavated, with conspicuous rostrum. Eye-lobes with spiniform apophysis, lacking visual elements. Pereonites with lateral spines. Pleonites without spines or paired ventral tubercles. Antennular article 2 short (less than 0.5 times as long as article 1). Mandibles without spiniform setae near palp insertion. Cheliped with [*A. diversus* (Lang, 1968)] or without exopod. Pereopod 1 lacking exopod. Pereopod 1 basis elongate (length/width ratio > 5). Pereopod 1 carpus and propodus with heavy spiniform setae and few (< 10) simple setae. Pereopod 1 dactylus and unguis combined longer than propodus; dactylus with weak serration, without membranous prolongation. Pleopods lacking or reduced (*A. diversus*).

Male. Antennule outer rami with many more articles than in female, all flagellum articles with numerous aesthetascs. Cheliped fixed finger and dactylus each with a proximal process. Pleopods present.

Type species. — *Atlantapseudes nigrichela* Băcescu, 1978.

Gender of generic name. — Masculine.

Remarks. — Meyer & Heard (1989) report that the type specimen of the type species is a juvenile male and therefore does not display any sexually dimorphic characters. This argument seems well founded as the sexually dimorphic characters (of an adult male) mentioned by Meyer & Heard (1989) are in accordance with those displayed by all other members of the Apseudidae, whereas the specimen of Băcescu (1978) does not exhibit any typical sexual characters.

Guțu (1996c: 87) considers *A. diversus* wrongly placed in the genus *Apseudes* and probably belonging to a new genus allied to *Carpoapseudes*, and this position seems well founded given the similarities of the genera mentioned. However, Guțu (1996c: 88) then transferred *A. diversus* to *Atlantapseudes*, a position that seems less well founded considering the impact this has on the generic diagnosis of *Atlantapseudes* (see above) and the forced exclusion of body armament as characteristic for this genus. Guțu (1996c: 88) based this inclusion on the short antennal squama of *A. diversus*, which is also found in *A. lindae*, but ignored that the third species in *Atlantapseudes*, *A. nigrichela*, does not share this character, which thus excludes that character as a generic one. Of other supporting characters Guțu (1996c: 88) mentions the dense aesthetasc covering of the male antennule and the thin peduncle of the female antennule. These characters, however, lend little support for inclusion in *Atlantapseudes*, as the aesthetasc covering of the male antennule is found in many other species of the Apseudidae (and in those cases where it is not, the male is probably juvenile) and the thin peduncle of the female antennule is found in some species of the Leviapseudinae as well.

Species currently assigned to this genus. — *A. lindae* Meyer & Heard, 1989; *A. diversus* (Lang, 1968); *A. nigrichela* Băcescu, 1978.

Atlantapseudes lindae Meyer & Heard, 1989

Apseudidae sp. A. Galloway et al., 1988, Mineral Management Service Study, **870061**: 210.
Atlantapseudes lindae Meyer & Heard, 1989, Gulf Research Reports, **8**: 97.

Diagnosis. — (Modified after Meyer & Heard, 1989.) Female. Pereonites 1 and 2 lacking antero-lateral spiniform processes. Pereonites 3–6 with weakly developed antero-lateral spiniform processes. Pleonites without lateral epimera and without ventral keel. Antenna squama not longer than third article. Maxilliped with three coupling hooks. Cheliped basis with spiniform seta ventrally; carpus with two long, simple setae dorsally. Pleopods absent.

Male. Pereonites 3–6 without antero-lateral spiniform processes. Pleonites with small ventral keel. Pleopods present.

Distribution. — Reported from the shelf and upper slope of the northern part of the Gulf of Mexico. Bathymetric range: 44–850 m. See map 5.

Remarks. — Juvenile males can be recognized by the incompletely developed pleopods.

KEY TO THE SPECIES OF *ATLANTAPSEUDES*, BOTH SEXES (GLOBAL)

1. Antennal squama three times as long as article 3. Antero-lateral spines (apophyses) on pereonites 1 and 2 present .... *A. nigrichela* Băcescu, 1978
– Antennal squama shorter than or as long as article 3. Antero-lateral spines (apophyses) on pereonites 1 and 2 present or absent 2
2. Antennal squama as long as article 3. Antero-lateral spines (apophyses) on pereonites 1 and 2 absent. Cheliped lacking exopod .... *A. lindae* Meyer & Heard, 1989
– Antennal squama shorter than article 3. Antero-lateral spines (apophyses) on pereonites 1 and 2 present. Cheliped with exopod .... *A. diversus* (Lang, 1968)*

(*) As stated in the generic remarks, *A. diversus* probably does not belong in *Atlantapseudes*.

Genus **Dactyloprion** Guțu, 2002

Dactyloprion Guțu, 2002, Travaux du Muséum d'Histoire Naturelle "Grigore Antipa", **44**: 41.

Diagnosis. — (Modified after, Guțu, 2002b.) Female. Body dorso-ventrally flattened. Carapace with rostrum. Eye-lobes with spiniform apophysis, without visual elements. Pereonites 2–6 with forward pointing antero-lateral spines. Pleonites without paired ventral tubercles. Antennular article 2 long (more than 0.5 times as long as article 1). Mandibles without spiniform setae near palp insertion. Cheliped and pereopod 1 with exopod. Pereopod 1 basis not elongate (length/width ratio < 3). Marsupium (where known) with 5 pairs of oostegites. Pereopod 1 carpus and propodus with heavy spiniform and simple setae. Pereopod 1 dactylus and unguis combined longer than propodus; dactylus with pronounced serration and membranous prolongation longer than unguis. Pereopod 1 coxa with prominent spine, carpus and propodus with heavy spiniform setae and few (< 10) simple setae. Pereopods 2–6 without coxal spines. Pleopods present, biramous and well developed.

Male. Unknown.

Type species. — *Dactyloprion anakanthes* Guțu, 2002.

Gender of generic name. — Masculine [as explicitly stated by Guțu, 2002].

Remarks. — This genus is easily recognized by the special features of the pereopod 1 dactylus and the long antennular article 2.

Species currently assigned to this genus. — *D. anakanthes* Guțu, 2002.

Dactyloprion anakanthes Guțu, 2002

Dactyloprion anakanthes Guțu, 2002, Travaux du Muséum d'Histoire Naturelle "Grigore Antipa", **44**: 42.

Diagnosis. — Monotypic genus, no species diagnosis given.

Distribution. — Reported from the upper slope of the northern part of the Gulf of Mexico. Bathymetric range: only taken from one station at 636 m. depth. See map 6.

Remarks. — This species resembles *Apseudes bowmani* Guțu & Iliffe, 1989 by the absence of spiniform setae on pereopods 2–6.

Genus **Pectinapseudes** Băcescu & Williams, 1988

Pectinapseudes Băcescu & Williams, 1988, Revue Roumaine de Biologie, (Biologie Animale) **33**(1): 3.

Diagnosis. — (Modified after Băcescu & Williams, 1988; Guțu, 2002b.) Female. Body semi-cylindrical. Carapace with rostrum. Eye-lobes with small spiniform apophysis, without visual elements. Pereonites and pleonites without lateral spines. Pereonites often of a rectangular appearance. Pleonites without paired ventral tubercles. Antennular article 2 short (less than 0.5 times as long as article 1). Mandibles with spiniform setae near palp insertion. Cheliped and pereopod 1 with exopod. Marsupium (where known) with 5 pairs of oostegites. Pereopod 1 coxa with prominent spine. Pereopod 1 basis not elongate (length/width ratio around 3). Pereopod 1 carpus and propodus with heavy spiniform and simple setae. Pereopod 1 dactylus and unguis combined longer than propodus and with small spiniform setae ventrally but without serration or membranous prolongation. Pereopods 2–6 without coxal spines. Pleopods present, biramous and well developed. Uropods with long rami, endopod longer than body.

Male. Cheliped fixed finger with proximal depression; dactylus without process.

Type species. — *Pectinapseudes carolinensis* Băcescu & Williams, 1988.

Gender of generic name. — Masculine.

Remarks. *Pectinapseudes* have only been recorded from North American waters (Atlantic off North Carolina, Gulf of Mexico). The fact that the two species found in the Gulf of Mexico are found at stations close to each other and that only two or three specimens of each species are ever found, could raise the suspicion that these species are not well separated. However, the spiniform setae near the mandibular palp insertion and the long uropod are unique characters, and seem stable for this genus.

Species currently assigned to this genus. — *P. carolinensis* Băcescu & Williams, 1988; *P. magnus* Guțu, 2002; *P. mesteri* Guțu, 2002.

Pectinapseudes magnus Guțu, 2002
(figs. 5–7)

Pectinapseudes magnus Guțu, 2002, Travaux du Muséum d'Histoire Naturelle "Grigore Antipa", **44**: 50.

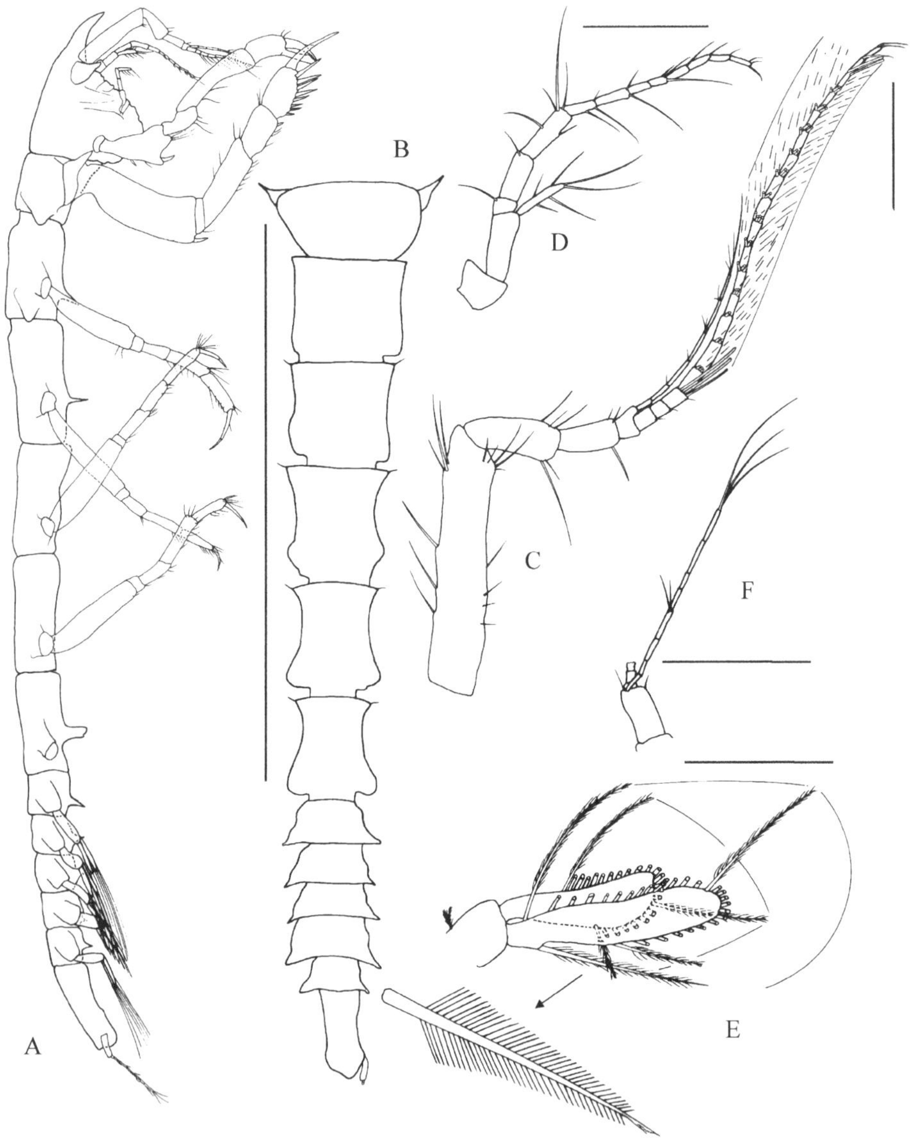

Fig. 5. *Pectinapseudes magnus* Guțu, 2002, male; A, lateral view; B, dorsal view of pereonites; C, antennule; D, antenna; E, pleopod; F, uropod. Scale bars equal: A-B, 10 mm; C-F, 0.5 mm.

Material examined. — 1 ♂, body length 1.82 cm, DGoMB Sta. S36-4.

Diagnosis. — (Modified after Guțu, 2002b.) Female. Mandible with several inconspicuous (much less robust than incisor) spiniform setae at palp insertion.

Mandibular palp article 1 with one inconspicuous (much less robust than incisor) spiniform seta. Pereopod 1 basis with one spiniform seta ventrally.

Male. Mandible with only two inconspicuous (much less robust than incisor) spiniform setae at mandibular palp insertion. Mandibular palp article 1 without spiniform setae. Cheliped fixed finger with proximal depression and propodal projection between fixed finger and dactylus.

Description. — Male. Body (fig. 5A, B) elongate (about 12 times as long as wide).

Carapace, including rostrum, shorter than pereonites 1 and 2 combined. Ocular spines well developed and without visual elements. No lateral spines. Rostrum reaching halfway first antennular article, almost triangular. Carapace not divided into indistinct lobes.

Pereonites almost rectangular, without lateral spine but with pointed anterior corners. Decreasing slightly in width distally. Pereonites 3–6 longer than wide, other pereonites wider than long. All pereonites with ventral hyposphenians.

Pleon less than 0.25 times as long as total body. Pleonites all wider than long, with pointed epimera and bearing pleopods. Pleotelson shorter than the last three pleonites combined.

Antennule (fig. 5C) 1.5 times as long as carapace. Peduncle article 1 three times as long as article 2, with several scattered simple setae on both sides; article 2 about 1.5 times as long as article 3, with simple setae distally; article 3 three times as long as first flagellum article, with simple setae distally. Common article naked. Outer flagellum 1.4 times as long as peduncle, consisting of about 19 serially repeating articles, all densely covered with aesthetascs. Inner flagellum almost 0.5 times as long as outer flagellum, consisting of six articles, all with simple setae distally.

Antenna (fig. 5D) longer than antennular peduncle, with 13 articles. Article 1 twice as wide as other articles, 0.5 times as long as article 2, naked. Article 2 longest, naked, squama inserted distally. Article 3 as long as wide. Remaining articles progressively smaller, all with setae distally. Squama reaching end of article 4, with several setae scattered on all edges.

Mouthparts. Labrum not recovered. Mandibles, molar process wider and longer than incisor, with numerous minute distal setules. Palp with three articles; article 1 with six long, equally sized setae distally; article 2 with medial row of pinnate, stout setae increasing in length in proximal direction; article 3 distal part armed with pinnate stout setae and 4–5 longer, simple setae. Mandibular body with two spiniform setae at palp insertion. Left mandible (fig. 6A) incisor blunt, with three denticles; lacinia mobilis narrower than incisor,

Fig. 6. *Pectinapseudes magnus* Guțu, 2002, male; A, left mandible; B, right mandible; C, labium; D, maxillule; E, maxilla; F, maxilliped; G, same, endite; H, epignath. Scale bars equal: A-F, 0.5 mm; G, 0.1 mm.

well calcified with four denticles on distal margin; setal row consisting of four multifurcate and two simple, spiniform setae arising from common peduncle. Right mandible (fig. 6B) incisor blunt to the point of being almost indistinct from the body of the mandible; setal row consisting of four multifurcate and one simple spiniform setae arising from a common peduncle. Labium (fig. 6C) lobes naked; palp with numerous simple setae on both margins; tipped by two multifurcate setae. Maxillule (fig. 6D) palp with two articles, distal articles tipped by eight simple setae; outer endite tipped by seven spiniform setae and outer row of setules; inner endite with six bipinnate, spiniform setae. Maxilla (fig. 6E) outer lobe of movable endite with three apparently simple setae; inner lobe of movable endite with five pinnate setae; outer lobe of fixed endite with three spiniform and three multifurcate spiniform setae; inner lobe of fixed endite with row of four distally setose setae and row of numerous distally curved and bifurcate setae arising from a peduncle. Maxilliped (fig. 6F) basis apparently naked; endite (fig. 6G) with row of five complex and five spiniform setae on inner distal margin: inner margin with row of 12 setulated setae and six coupling hooks; palp article 1 with one short seta on outer margin, article 2 with row of setae on inner margin and two setae on distal outer margin, article 3 with five setae on inner margin, article 4 tipped by four setae. Epignath (fig. 6H) lobe with proximal setules; terminal seta shorter than basal part and with sparsely scattered setules.

Cheliped (fig. 7A) basis shorter than carpus, with two medial spiniform setae ventrally and three ventro-distal setae. Merus 0.33 times as long as carpus, with several simple setae ventrally. Carpus with several setae ventrally, longer than propodus including fixed finger. Propodus with setose projection between fixed finger and dactylus. Fixed finger (fig. 7B) with ventral margin bearing six simple, inner margin (cutting surface) with several small blunt setae and weak serration; with proximal depression. Dactylus as long as fixed finger, without setae dorsally. Exopod consisting of two articles arising from a pedestal, first article naked, distal article with four plumose setae.

Pereopod 1 (fig. 7C) coxal spine conspicuous and reaching the cheliped insertion point (not illustrated). Basis less than 3 times as long as wide but hardly longer than the length of merus and carpus combined, with few scattered setules and conspicuous, spiniform seta ventro-distally. Ischium with two setae ventrally. Merus almost twice as long as carpus, with few simple setae ventro-distally and one spiniform seta ventro-distally. Carpus almost as wide as long, with one dorsal seta, two spiniform setae ventrally, and several scattered, simple setae. Propodus as long as carpus, with row of eight spiniform setae ventrally, two tiny accessory spiniform setae at dactylus insertion, and simple setae dorso-distally. Dactylus alone as long as propodus, with several tiny spiniform setae ventrally but without apparent serration or dorsal setae.

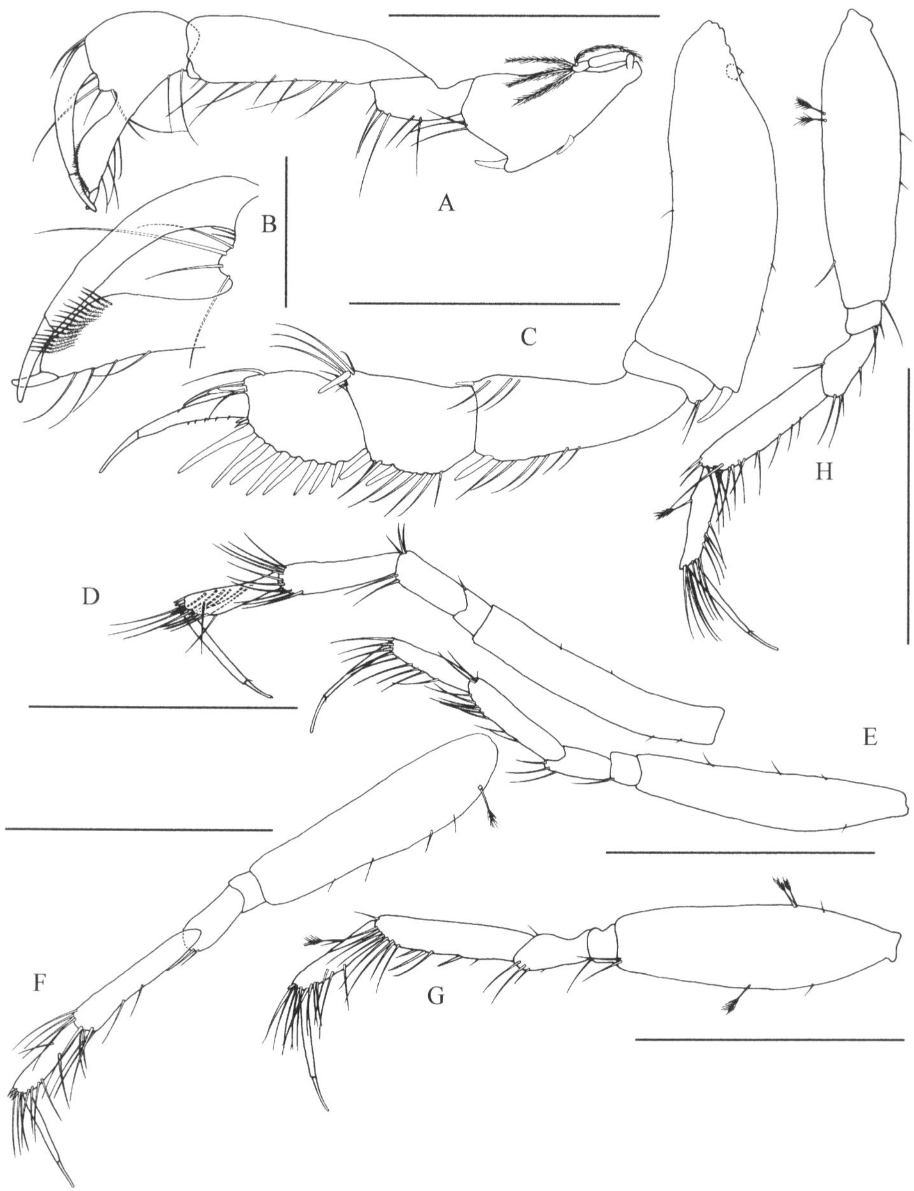

Fig. 7. *Pectinapseudes magnus* Guțu, 2002, male; A, cheliped; B, same, chela; C, pereopod 1; D, pereopod 2; E, pereopod 3; F, pereopod 4; G, pereopod 5; H, pereopod 6. Scale bars equal: A-G, 1 mm.

Unguis 0.33 times as long as dactylus. Exopod consisting of two articles arising from a pedestal, first article naked, distal article with four plumose setae.

Pereopod 2 (fig. 7D) basis as long as the three succeeding articles combined, with scattered minute setae. Ischium with one simple seta dorsally. Merus shorter than carpus, with simple setae distally. Carpus longer than propodus, with several simple setae on both distal margins. Propodus with several dorsal, simple setae distally and several small, pinnate spiniform setae distally. Dactylus 0.67 times as long as propodus, with one minute distal seta on ventral margin. Unguis 0.25 times as long as dactylus.

Pereopod 3 (fig. 7E) as pereopod 2 except: propodus without small pinnate spiniform setae.

Pereopod 4 (fig. 7F) as pereopod 2 except: basis with one setulose proximal seta. Ischium naked.

Pereopod 5 (fig. 7G) as pereopod 4 except: basis with one ventral and two setulated setae dorsally. Ischium with one seta ventrally. Propodus with one setulated dorsal seta, several simple setae dorsally and distally, and one small, pinnate spiniform seta distally. Dactylus 0.67 times as long as propodus, naked. Unguis 0.25 times as long as dactylus.

Pereopod 6 (fig. 7H) as pereopod 5 except: basis without setulose seta ventrally.

Pleopods (fig. 5E) individual pleopods shorter than pleotelson. Basal article longer than endites, with small plumose setae. Exopod with one article, with 30 plumose setae. Endopod shorter than exopod, with 29 plumose setae and one robust, specialized proximal seta.

Uropod (fig. 5F) peduncle three times as long as wide. Endopod lost. Exopod with eight articles. Most articles naked, others with one to three simple setae.

Distribution. — A very rare species as only two specimens are known. Bathymetric distribution: 1390–1827 m. See map 7.

Remarks. — The sexually dimorphic characters are: a somewhat more uneven lateral margin of the pereonites (in dorsal view), the presence of only two spiniform mandibular setae. Deeper proximal process on the cheliped fixed finger. Propodal projection between fixed finger and dactylus of the cheliped.

This species can be separated from *P. mesteri* Guţu, 2002 by having only one (distal) spiniform seta on the pereopod 1 basis and by not having spiniform setae on the first mandibular article. Also the sizes of the mandibular spines are far larger in *P. mesteri* than in *P. magnus*.

The specialized, proximal pleopod seta is reported from a large number of species including *Carpoapseudes kudinovae* Băcescu, 1981c; *Glabroapseudes larseni* Guerrero-Kommritz & Heard, 2003; *Carpoapseudes serratispinosus*

Lang, 1968; *C. longissimus* Lang, 1968; *C. oculicornutus* Lang, 1968; *Leviapseudes shiinoi* (Lang, 1968); *L. aberrans* (Lang, 1968); *L. wolffi* (Lang, 1968); *L. sibogae* (Nierstrasz, 1913); *L. weberi* (Nierstrasz, 1913); *Apseudella typica* Lang, 1968. The presence does not appear to be sexually dependent and even the subadult males of *L. conspicuus* (Lang, 1968), which otherwise have reduced pleopod setae, have this specialized seta (Lang, 1968). However, neither the female of *Pectinapseudes magnus* nor that of *P. mesteri* are described with this seta. Maybe this seta is sexually dependent in *Pectinapseudes*, but more likely this detail may have been overlooked when those species were first described (Guțu, 2002b, does indicate that this seta is somewhat thicker than the other pleopod setae).

Pectinapseudes mesteri Guțu, 2002

Pectinapseudes mesteri Guțu, 2002, Travaux du Muséum d'Histoire Naturelle "Grigore Antipa", **44**: 56.

Diagnosis. — (Modified after Guțu, 2002b.) Female. Mandibles with four conspicuous, strong (as robust as incisor) spiniform setae at mandibular palp insertion. Mandibular palp article 1 with one conspicuous, strong (as robust as incisor) spiniform seta. Pereopod 1 basis with three spiniform setae ventrally.

Male. Unknown.

Distribution. — A rare species, found only at two stations (three specimens) Bathymetric distribution: 357–595 m. See map 8.

Key to the Species of *Pectinapseudes*, both Sexes (global)

1. Mandible with inconspicuous (much less robust than incisor) spiniform setae at palp insertion ..... *P. magnus* Guțu, 2002
– Mandible with at least four conspicuous, strong (as robust as incisor) spiniform setae at palp insertion 2
2. Pereopod 1 with three spiniform setae ventrally. Mandibular palp article 1 with one conspicuous, strong (as robust as incisor) spiniform seta ..... *P. mesteri* Guțu, 2002
– Pereopod 1 with only one spiniform seta ventrally. Mandibular palp article 1 without spiniform setae ..... *P. carolinensis* Băcescu & Williams, 1988

Subfamily LEVIAPSEUDINAE Sieg, 1983

Leiopidae Lang, 1970a, Arkiv för Zoologi, **22**(16): 603.
Leiopinae – Guțu, 1972, Revue Roumaine de Biologie, (Biologie Animale) **17**(5): 302.
Leviapseudidae Sieg, 1980b, Zoologischer Anzeiger, **205**: 411, 412 [nomen nudum].
Leviapseudinae Sieg, 1983, Crustaceorum Catalogus, **6**: 163.

Diagnosis. — Female. Maxillipedal endopod with leaf-shaped setae subdistally. Cheliped and pereopod 1 with exopod. Pereopod 1 coxa with forward pointing spine. Pereopod 4 dactylus often reduced and then more so in the male. Pereopod 1 dactylus and unguis not fused. Pleopods present or absent.

Male. If reduced, pereopod 4 dactylus less so in the male than in the female. Pleopods always present in male.

Type genus. — *Leviapseudes* Sieg, 1983 [= *Leiopus* Beddard, 1886].

Genera included. — *Carpoapseudes* Lang, 1968; *Leviapseudes* Sieg, 1983.

Remarks. — The oldest name for this family group taxon is Leiopidae Lang, 1970. Guțu (1972) later demoted the family to a subfamily Leiopinae Lang, 1970. Both names, however, are invalid as they are based on the generic name *Leiopus* Beddard, 1886, which is a junior homonym of *Leiopus* Audinet-Serville, 1835 (Coleoptera, Cerambycidae). Sieg (1980b) realized this and proposed the new family name Leviapseudidae to replace the invalid Leiopidae Lang, 1970. Sieg's (1980b) action is perfectly correct under the rules of the ICZN. The fact as such that in establishing a new family name no type genus is expressly mentioned, does not invalidate that family name, as the inference of that generic name from the stem used in the new family name is considered sufficient (Art. 11.7.1.1 of the 1999 Code). Only for family names proposed after 1999 the indication of the name of the type genus is required (Code Art. 16.2). Sieg (1980b) did not assign a type genus to his Leviapseudidae either, but he did so in 1983 (cf. Sieg, 1983) by assigning *Leviapseudes* Sieg, 1983 (the renamed *Leiopus*) as the type genus, while concurrently demoting the family group name to Leviapseudinae. The rank of the taxon, i.e., as a family (Lang, 1970, 1973) or subfamily (Guțu, 1972, 1980a) apparently still is a matter of debate among taxonomists. Herein, I have retained the subfamily status for the time being: it is obvious that only cladistic analysis could provide a more definitive solution.

Genus **Leviapseudes** Sieg, 1983

Leiopus Beddard, 1886, Proceeding of the Scientific Meetings of the Zoological Society of London, **97**: 122.
Leviapseudes – Băcescu, 1981, Travaux du Muséum National d'Histoire Naturelle "Grigore Antipa", **23**: 37.
Leviapseudes Sieg, 1983, Crustaceorum Catalogus, **6**: 167.

Diagnosis. — (Modified after Błażewicz-Paszkowycz & Larsen, 2004.) Female. Pereonite 6 rectangular. Pleonites often much narrower than pereonite 6, sometimes as long as wide, with or without lateral epimera. Cheliped basis with or without ventral spine; carpus as long as or longer than fixed finger. Pereopod 1 merus longer than carpus. Pereopod 4 dactylus/unguis sometimes modified. Pleopods, if present, with bi-articulate exopod.

Male (see also remarks). Pereonite 6 with clearly identifiable genital cone. Antennule after terminal moult with numerous aesthetascs. Pereopod 4 dactylus/unguis, if modified, less so than in female. Pleonites always with lateral epimera, and epimera larger than in female.

Type species. — *Leiopus longidactylus* Beddard, 1886 [= *Leviapseudes*].

Gender of generic name. — Masculine.

Remarks. — The name *Leviapseudes* first appeared when Băcescu described *Leviapseudes zenkevitchioides* in 1981 and *Leviapseudes hanseni* in 1982. Băcescu did not assign a type species to this genus and apparently assumed that the genus was already established (L. B. Holthuis, pers. comm.). In Sieg's (1983) Crustaceorum Catalogus the generic name was formally changed from *Leiopus* to *Leviapseudes*, but it is apparent that this publication was in press already in 1980 (causing Băcescu to accept and use the name). Thus, in my view, Sieg (1983) is the correct authority for both Leviapseudinae and *Leviapseudes*, even though the name *Leviapseudes* appears already in 1981. However, even if Băcescu's generic name is not valid, according to the Code (Art. 13.3), as he did not assign a type species, his species names are valid (Art. 11.9.3.1) (L. B. Holthuis, pers. comm.).

Lang (1968: 88) reported sexual dimorphism in the pleotelson (larger distal processes in male) and mandibles (fewer denticles in the incisor and lacinia mobilis). With the material of *Leviapseudes bipartitus* sp. nov. (see below), for the first time sufficient material of both males and females was available to the present author for a comprehensive analysis. The dimorphic nature of the pleotelson and mandibles can be confirmed. The mandibular lacinia mobilis has less denticles in the male, as stated by Lang (1968), while this is less obvious on the incisor. However, both the incisor and in particular the lacinia mobilis are also much narrower in the male than in the female, and the setal row is less well developed in the male, in particular on the left mandible. Regarding the pleotelson, the most striking difference is the much larger size of the distal process on the male pleotelson, while the length of the pleotelson is proportionately shorter in the male than in the female.

Other noteworthy and apparently stable dimorphic characters include: pereopod 1 much shorter in the male than in the female and not significantly longer (although wider) than the other pereopods; unguis of male pereopod 4 narrower and more pointed (less modified) than that of the female.

In most apseudomorphan taxa, a typically dimorphic character is the fixed finger/dactylus of the male cheliped having a process. This is also seen in the Leviapseudinae in *Carpoapseudes* (cf. Lang, 1968; Larsen, 1999c) but this is **not** found in *Leviapseudes* in which the chelipeds seem almost identical in both sexes.

Species and subspecies currently assigned to this genus. — *L. aberrans* (Lang, 1968); *L. bipartitus* sp. nov.; *L. caecus* (Willemoes-Suhm, 1875); *L. conspicuus* (Lang, 1968); *L. demerarae* Băcescu, 1984; *L. desbruyeresi* Băcescu, 1987; *L. drachi* Băcescu, 1984; *L. galatheae* (Wolff, 1956); *L. gracilis* (Norman & Stebbing, 1886); *L. gracillimus* (Hansen, 1913); *L. hanseni* (Lang, 1968); *L. leptodactylus* (Beddard, 1886); *L. longispina* Băcescu, 1983; *L. preamazonica* Băcescu, 1984; *L. segonzaci segonzaci* Băcescu, 1981; *L. segonzaci gasconicus* Băcescu, 1984; *L. shiinoi* (Lang, 1968); *L. sibogae* (Nierstrasz, 1913); *L. thomassini* Băcescu, 1987; *L. weberi* (Nierstrasz, 1913); *L. wolffi* (Lang, 1968); *L. zenkevitchi* (Kudinova-Pasternak, 1966); *L. zenkevitchioides* Băcescu, 1981.

Leviapseudes bipartitus sp. nov.
(figs. 8–13)

Material examined. — Holotype, non-ovigerous ♀ (USNM 1068856), Paratypes: 1 non-ovigerous ♀ (USNM 1068857), body length 8.3 mm, DGoMB Sta. C7-4. 1 ♂ (USNM 1068858), body length 8.25 mm, DGoMB Sta. S36-2. 1 ♂ (USNM 1068859), NGoMCSS Sta. Cruise III S3-3, (C2-3) (dissected). 1 non-ovigerous ♀ (USNM 1068860), DGoMB Sta. S38-4 (dissected).

Additional material. — 1 non-ovigerous ♀, DGoMB Sta. S38-4. 4 mancas, DGoMB Sta. C7-4, 27. 1 non-ovigerous ♀, DGoMB Sta. NB3-4. 2 ♀♀, DGoMB Sta. RW4-3. 1 ♂, DGoMB Sta. NB2-3. 1 ♀, DGoMB Sta. RW2-1. 1 ♀, DGoMB Sta. NB5-4. 1 ♂, DGoMB Sta. MT4-3. 1 manca II, DGoMB Sta. S36-4. 1 ♂, DGoMB Sta. WC12-1. 2 ♀♀, 1 manca II, DGoMB Sta. RW3-1. 2 ♀♀, DGoMB Sta. NB2-4. 1 ♂, DGoMB Sta. RW4-1. 1 ♀, DGoMB Sta. NB3-2. 1 ♂, DGoMB Sta. RW3-3. 1 ♂, 1 manca II, DGoMB Sta. S37-3. 1 ♂, 2 ♀♀, DGoMB Sta. RW4-2. 1

Fig. 8. *Leviapseudes bipartitus* sp. nov., female; A, paratype, lateral view; B, antennule; C, antenna; D, uropod. Scale bars equal: A, 1.0 mm; B-D, 0.2 mm.

♀, DGoMB Sta. S36-5. 1 ♀, 1 manca III, DGoMB Sta. RW6-3. 1 ♀, 1 manca II, DGoMB Sta. RW5-1. 1 ♀, DGoMB Sta. NB2-2. 1 ♀, 1 manca II, DGoMB Sta. RW4-3. 1 ♂, 2 mancas II, DGoMB Sta. S42-3. 1 ♀, DGoMB Sta. C4-2. 1 ♀, DGoMB Sta. W4-1. 1 ♀, DGoMB Sta. S37-1. 1 ♂, DGoMB Sta. C4-1. 1 ♂, DGoMB Sta. S37-1. 1 ♂, 1 ♀, 3 mancas II, DGoMB Sta. RW2-2. 1 manca II, DGoMB Sta. MT3-2. 1 ♀, DGoMB Sta. RW5-3. 1 ♀, DGoMB Sta. NB4-1. 1 ♀, DGoMB Sta. S36-3. 1 manca II, DGoMB Sta. S42-2. 1 ♀, DGoMB Sta. W3-3. 1 ovigerous ♀, DGoMB Sta. RW2-3. 1 manca III, DGoMB Sta. W4-3.

Diagnosis. — Female. Pereonites 2–6 with paired tubercles. Pleonites wider than long and with blunt lateral epimera. Cheliped dactylus hardly longer than fixed finger. Pereopod 1 basis with conspicuous ventro-distal spine. Pereopod 4 with large gap between the dactylus process and unguis (gap wider than 0.5 times length of unguis); dactylus with extremely elongate, bifurcate ventral spine; unguis reduced to claw with two large spines. Pleopods present.

Male. Pleotelson with much larger distal expansions than in female. Antennule with many more serially repeating articles in outer flagellum than in female, all with numerous aesthetascs. Mandible incisor, lacinia mobilis, and setal row less well developed in male. Pereopod 1 shorter than in female and pereopod 4 dactylus less modified.

Etymology. — Named after the diagnostic shape of the pereopod 4 unguis (Latin, bipartitus = divided into two parts). The name thus is an adjective agreeing in gender with the (masculine) generic name.

Type locality. — DGoMB Sta. WC12-4, 27°19'1851"N 91°33'0650"W, depth 1169 m.

Description. — Female. Body, pleopods, and uropods from non-ovigerous female holotype, other appendages from dissected ovigerous specimen.

Body (fig. 8A) elongate, about ten times as long as wide.

Carapace, including rostrum, longer than pereonites 1 and 2 combined. Ocular spines well developed, without visual elements. Lateral spines small. Rostrum reaching halfway first antennular article, almost triangular. Carapace divided into three indistinct lobes, the posterior one, carrying the cheliped attachment, prominent, about 0.3 times total carapace length.

Pereon, all pereonites with a pair of anterior lateral spines. Decreasing in width distally. Pereonites 3–6 longer than wide, other pereonites wider than long. All pereonites with ventral hyposphenians. Pereonites 2–6 with paired rounded processes on antero-lateral margins. Pereonite 1 with only unpaired rounded processes.

Pleon less than 0.3 times as long as total body. Pleonites all wider than long, with blunt epimera and bearing pleopods. Pleotelson as long as the last three pleonites combined.

Antennule (fig. 8B) 1.5 times as long as carapace. Peduncle article 1 twice as long as the two succeeding articles combined, with three setulose setae ventrally and minute setules scattered on both sides; article 2 about 1.5 times as long as article 3, with four setulated setae distally; article 3 longer than outer flagellum article 1, naked. Common article naked. Outer flagellum 1.4 times as

Fig. 9. *Leviapseudes bipartitus* sp. nov., female; A, cheliped; B, same, chela; C, left mandible; D, right mandible; E, same, molar; F, same, palp setation; G, left mandible, incisor; H, right mandible, spine row; I, left mandible, spine row; J, labium; K, maxillule; L, maxilla; M, maxilliped; N, same, endite; O, epignath. Scale bars equal: A, B, 1.0 mm; C-O, 0.1 mm.

long as peduncle, consisting of about 17 articles. Inner flagellum 0.33 times as long as outer flagellum, consisting of four articles, all with setae distally.

Antenna (fig. 8C) article 1 three times as wide as other articles, 0.67 times as long as article 2, naked. Article 2 longest, with one proximal seta and elongate process at squama insertion. Article 3 as long as wide. Articles 4 and 5 each almost as long as article 2, with several setulose setae distally. Articles 5–9 each with setae distally. Squama arising from distal part of article 2, reaching past midlength of article 4, with several setae scattered on all edges.

Mouthparts. Labrum not recovered. Mandibular molar (fig. 9E) process wider and longer than incisor, bent distally and with numerous minute, distal setules. Mandibular palp with three articles; article 1 with two unequally sized setae distally; article 2 with distal row of pinnate, stout setae (fig. 9F); article 3 with distal part armed with pinnate, stout setae. Left mandible (fig. 9C) incisor pointed, with two denticles; lacinia mobilis (fig. 9G) wider than incisor, well calcified, with seven denticles on distal margin; setal row (fig. 9H) consisting of four multifurcate and one simple, spiniform setae arising from a common peduncle. Right mandible (fig. 9D) incisor with three denticles; setal row (fig. 9I) consisting of five multifurcate setae arising from a common peduncle. Labium (fig. 9J) lobes weakly setulose on inner medial edge; palp with numerous simple setae on both margins; tipped by three multifurcate setae. Maxillule (fig. 9K) palp with two articles, distal articles tipped by five simple setae; outer endite tipped by seven spiniform setae and outer row of setules; inner endite with six bifurcate spiniform setae. Maxilla (fig. 9L) outer lobe of movable endite with two medially bipinnate setae and three pinnate setae; inner lobe of movable endite with two pinnate and four serrated setae; outer lobe of fixed endite with five simple, one circumsetulose, one setulose, and four multifurcate spiniform setae; inner lobe of fixed endite with row of four distally setulose setae and row of numerous distally curved setae arising from a peduncle. Maxilliped (fig. 9M) basis setulose on proximal outer margin; endite with row of eight complex setae on inner distal margin, subdistal part with one leaf-shaped seta and three complex setae on other side; inner margin with row of six setulated setae and three coupling hooks; palp article 1 with one short seta on outer margin; article 2 with row of simple, bipinnate setae on inner margin and four long setae on distal outer margin; article 3 with five setae on inner margin; article 4 tipped by six long setae. Epignath (fig. 9O) basal part covered with setules; terminal seta shorter than basal part and apparently naked.

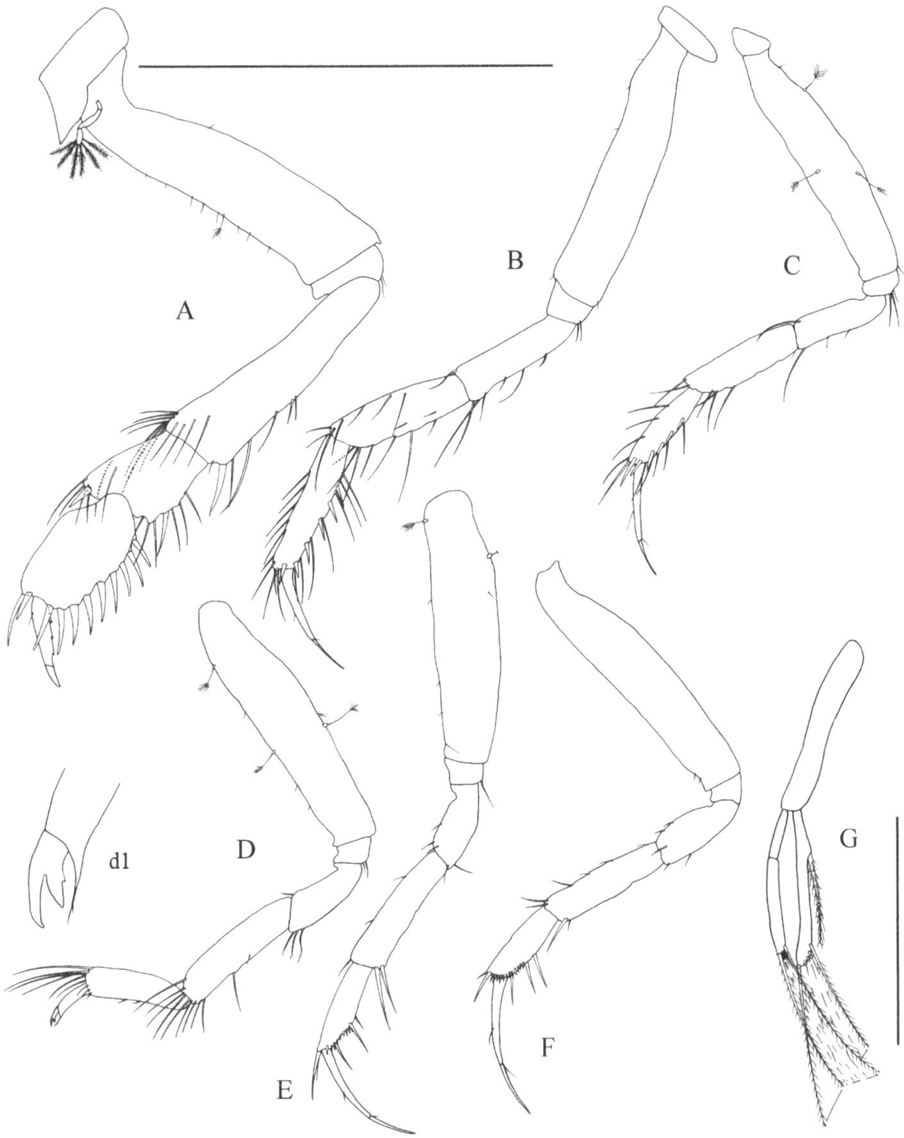

Fig. 10. *Leviapseudes bipartitus* sp. nov., female; A, pereopod 1; B, pereopod 2; C, pereopod 3; D, pereopod 4; d1, dactylus/unguis of same; E, pereopod 5; F, pereopod 6; G, pleopod. Scale bars equal: A-F, 0.1 mm; G, 0.5 mm.

Cheliped (fig. 9A) basis as long as carpus, with small spiniform seta ventrally and four simple setae ventrally. Merus 0.5 times as long as carpus with six simple setae ventrally. Carpus with numerous long setae ventrally and one seta at midlength dorsally. Propodus with tuft of simple setae dorsally, one simple seta at dactylus insertion, and row of simple setae on ventral margin extending onto fixed finger. Fixed finger (fig. 9B) with four simple setae on ventral margin and row of diagonal setae at unguis insertion; inner margin (cutting edge) with several setae and small spiniform setae. Dactylus hardly longer than fixed finger, with two simple setae at midlength dorsally; ventral margin with row of small spiniform setae. Exopod consisting of two articles, distal article with four plumose setae.

Pereopod 1 (fig. 10A) coxal spine moderately developed and no more than 0.5 times as wide as the coxa. Basis about 3.5 times as long as wide and shorter than merus and carpus combined, with scattered setules, one setulose seta on dorsal margin and one spiniform seta ventro-distally. Ischium with two setae ventrally. Merus more than twice as long as carpus, with several simple distal and medial setae and one spiniform seta ventro-distally. Carpus almost as wide as long, with one spiniform seta on each distal margin and several scattered, simple setae. Propodus longer than carpus but slightly narrower, with row of nine spiniform setae ventrally, two spiniform setae dorsally. Dactylus 0.75 times as long as propodus, with three weak serrations and one seta dorsally. Unguis 0.25 times as long as dactylus. Exopod consisting of two articles, distal article with six plumose setae.

Pereopod 2 (fig. 10B) basis as long as the three succeeding articles combined, with scattered minute setae dorsally. Ischium with two simple setae ventrally. Merus little shorter than carpus, with row of simple setae ventrally. Carpus as long as propodus, with several simple setae on both margins. Propodus with several simple setae on both margins and six spiniform setae ventrally. Dactylus 0.67 times as long as propodus, with one minute distal seta on each margin. Unguis 0.33 times as long as dactylus.

Pereopod 3 (fig. 10C) as pereopod 2 except: basis with several setulose setae. Carpus longer than propodus, with sparse setation only. Propodus as long as merus, with several simple and six spiniform setae ventro-distally.

Pereopod 4 (fig. 10D) basis longer than the three succeeding articles combined, with three setulose setae. Ischium with two simple setae ventrally. Merus and carpus with only sparsc sctation distally. Carpus longer than merus. Propodus with minute setae on ventral margin and several simple setae distally, shorter than carpus. Dactylus with extremely thin, elongate, bifurcate ventral spine; unguis (fig. 10d1) reduced to a bifurcate claw.

Pereopod 5 (fig. 10E) as pereopod 3 except: merus less than 0.33 times as long as carpus. Carpus with some minute setae medially and with two robust

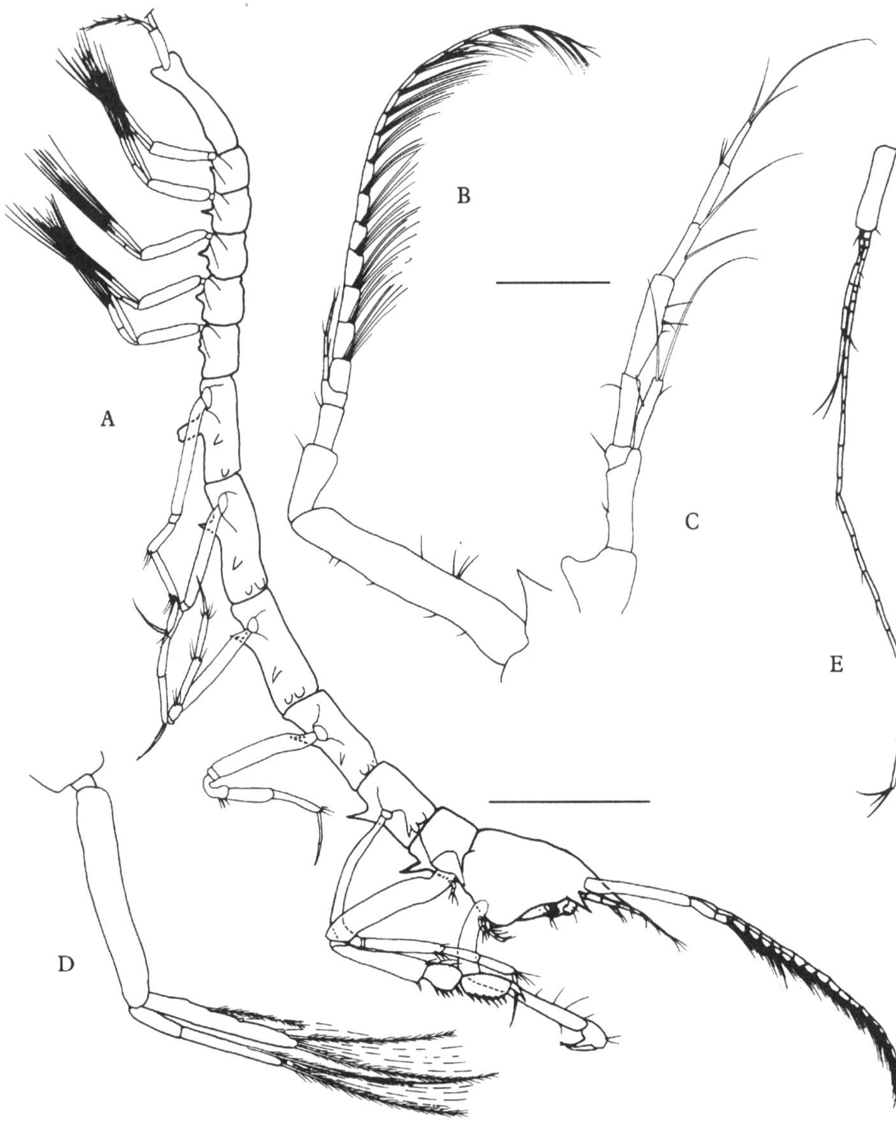

Fig. 11. *Leviapseudes bipartitus* sp. nov., male; A, paratype, lateral view; B, antennule; C, antenna; D, pleopod; E, uropod. Scale bars equal: A, 1.0 mm; B-E, 0.2 mm.

setae distally. Propodus 0.67 times as long as carpus, with row of short spiniform setae and two robust setae ventrally, dorsal margin with one spiniform and one simple setae distally. Dactylus with one minute seta dorsally and another at unguis insertion. Unguis 0.5 times as long as dactylus.

Pereopod 6 (fig. 10F) as pereopod 5 except: basis without setulose seta.

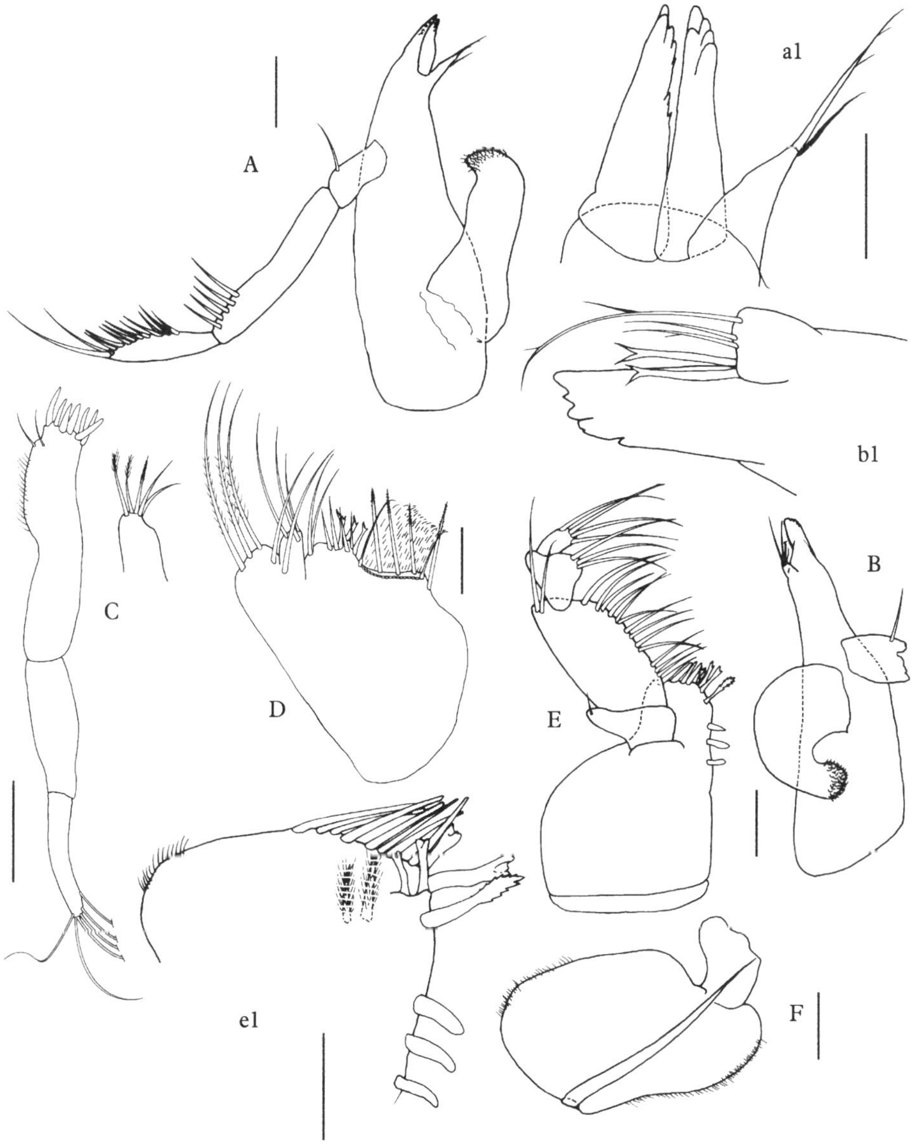

Fig. 12. *Leviapseudes bipartitus* sp. nov., male; A, left mandible; a1, same, close-up of incisor region; B, right mandible; b1, same, close-up of incisor region; C, maxillule; D, maxilla; E, maxilliped; e1, same, endite; F, epignath. Scale bars equal: 0.1 mm.

Pleopods (from holotype, fig. 10G), individual pleopods shorter than pleotelson. Basal article longer than endites, naked. Exopod with two articles, proximal article naked, distal article with six plumose setae distally. Endopod as long as exopod, with five plumose setae distally and one robust, circumplumose proximal seta.

Uropod (from holotype, fig. 8D) peduncle 3 times as long as wide. Endopod consisting of about 21 articles. Exopod of five articles. Articles with 0–3 simple setae distally.

Male, where differing from female. — Body from paratype, other appendages from dissected specimen.

Body (fig. 11A) elongate (about 11 times as long as wide).

Carapace as in female.

Pereonites with antero-lateral spines larger than in female, decreasing in width distally. Pereonites 2–6 longer than wide, pereonite 1 wider than long. All pereonites with ventral hyposphenians but smaller than in female. Pereonites 2–5 with paired rounded processes on antero-lateral margins. Pereonites 1 and 6 with only single rounded process.

Pleonite epimera more prominent than in female. Pleotelson shorter than the last three pleonites combined.

Antennule (fig. 11B) longer than in female, at least as long as carapace and the first three pereonites combined. Outer flagellum almost twice as long as peduncle, consisting of about 19 serially repeating articles, all with dense distal tuft of aesthetascs except most distal article. Inner flagellum about 0.2 times as long as outer flagellum and consisting of four articles.

Antenna (fig. 11C) as in female.

Mouthparts. Labrum not recovered. Mandibular molar process and palp as in female. Left mandible (fig. 12A) incisor pointed, with three or four denticles; lacinia mobilis (fig. 12a1) **not** wider than incisor, well calcified, with three denticles on distal margin; setal row consisting of only one multifurcate and one simple setae arising from common peduncle. Right mandible (fig. 12B) incisor with four to five denticles; setal row (fig. 12b1) consisting of five multifurcate or bifurcate setae arising from common peduncle; however, these are much thinner and less well developed than those in female. Labium not recovered. Maxillule (fig. 12C) as in female. Maxilla (fig. 12D) as in female. Maxilliped (fig. 12E) endite (fig. 12e1) with less setae and coupling hooks than in female, but otherwise as in female. Epignath (fig. 12F) identical to that of female.

Cheliped (fig. 13A) basis, carpus, and propodus appear more slender than in female. Dactylus longer than that of female.

Pereopod 1 (fig. 13B) conspicuously shorter than in female but otherwise as in female.

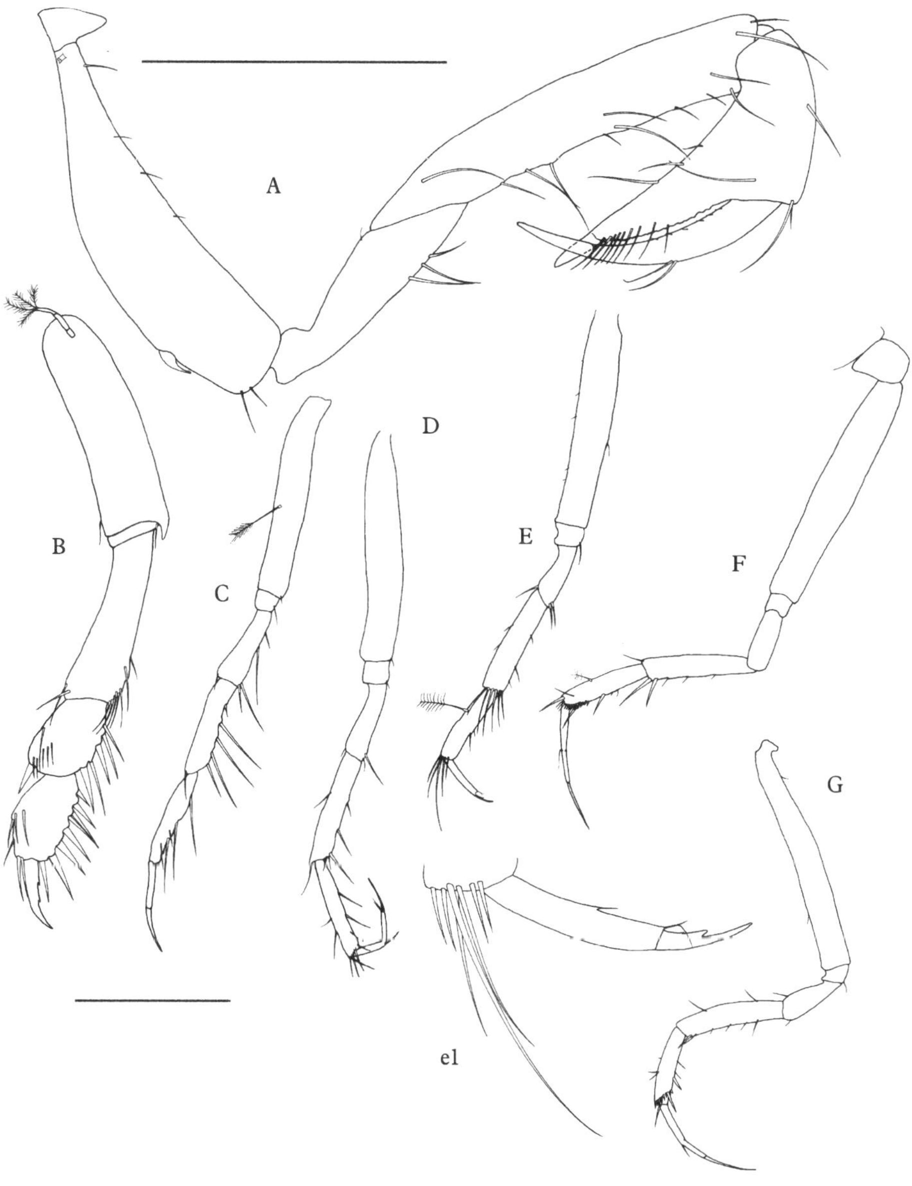

Fig. 13. *Leviapseudes bipartitus* sp. nov., male; A, cheliped; B, pereopod 1; C, pereopod 2; D, pereopod 3; E, pereopod 4; e1, same, dactylus/unguis; F, pereopod 5; G, pereopod 6. Scale bars equal: A, 0.5 mm; B-G, 0.5 mm.

Pereopods 2–4 (fig. 13C-E) less well armed with setae than in female. Dactylus of pereopod 4 (fig. 13e1) more slender and less modified, but otherwise as in female.

Pereopods 5–6 (fig. 13F-G) as in female.

Pleopods (fig. 11D), individual pleopods longer than pleotelson.

Uropod (fig. 11E) peduncle 3 times as long as wide. Endopod consisting of about 27 articles. Exopod of seven articles. Most articles naked, otherwise rarely with more than one simple seta.

Distribution. — A very common species found at almost all stations in the Gulf of Mexico. Bathymetric distribution: 636–3000 m. See map 9.

Remarks. — This species is the most commonly occurring apseudomorphan in the deep Gulf and was found in almost all samples. Tantulocaridean parasites were found on the bases of some male chelipeds. This is the first record of tantulocarideans on an apseudomorphan; only the "expanded trunk sac" (Boxhall & Lincoln, 1987) stages were present.

Leviapseudes cf. **sibogae** (Nierstrasz, 1913)

Apseudes sibogae Nierstrasz, 1913, Siboga Expeditie Mon., **32a**: 3.
Leiopus sibogae – Lang, 1968, Galathea Report, **9**: 88.

Diagnosis. — Female. Pleonites wider than long and with very acute lateral epimera. Cheliped dactylus as long as fixed finger. Pereopod 1 basis without ventro-distal spine. Pereopod 4 not modified. Pleopods present.

Male. Unknown. Nierstrasz (1913) mentioned that one male was found but did not describe it.

Distribution. — (No distribution map is presented for this species.) This species has not been recorded from the Gulf of Mexico, but was reported south of Cuba by Kudinova-Pasternak & Pasternak (1978). Otherwise only known from the type locality in the Banda Sea, Indonesia. Bathymetric distribution: 3060–4391 m.

Remarks. — It is unlikely that this Indo-Pacific species really occurs in the Caribbean as reported by Kudinova-Pasternak & Pasternak (1978). More likely the reported specimens belong to *L. bipartitus*.

Key to the Species of *Leviapseudes* (Gulf of Mexico)

1. Pleonites with blunt lateral epimera. Pereopod 1 basis with ventro-distal spine .... *L. bipartitus* sp. nov.
– Pleonites with very acute lateral epimera. Pereopod 1 basis without ventro-distal spine .... *L.* cf. *sibogae* (Nierstrasz, 1913)

Family SPHYRAPODIDAE Guțu, 1980

Sphyrapidae Guțu, 1980, Travaux du Muséum d'Histoire Naturelle "Grigore Antipa", **22**: 393.
Apseudidae – Sieg, 1983, Crustaceorum Catalogus, **6**: 26.

Diagnosis. — (Modified after Guțu, 1980b, 1991; Guțu & Iliffe, 1998.) Female. Body elongate, semi-cylindrical, dorso-ventrally flattened. Carapace without antero-lateral apophyses or visual elements. Pereon with five or six free pereonites, without spines. Pleon with five free pleonites. Antennule biramous, with short (1–3 articles) inner flagellum. Antennae with or without squama/pseudosquama. Mandibular palp absent, or present with one or three articles. Maxillule with bi-articulate palp. Epignath large, plate-shaped, with conspicuous terminal seta. Maxillipedal endites without leaf-shaped seta. Cheliped and pereopod 1 with exopod. Pereopod 1 strongly enlarged relative to other pereopods. Pereopod 1 dactyli without aesthetascs. Pereopods without sucker-like structures. Pleopods present. Uropods biramous, endopod multi-articulate, exopod with two or more articles.

Male. Only known in a few genera. No family diagnostic male characters assigned.

Type genus. — *Sphyrapus* G. O. Sars, 1882.

Remarks. — The characters of the enlarged pereopod 1, the lack of spiniform apophyses on the carapace, and the eye-lobes and pereopod coxa, are sound characters and thus make identification of this family simple.

I have herein emended Sphyrapidae Guțu, 1980 and Sphyrapinae Guțu, 1980, respectively, into Sphyrapodidae Guțu, 1980 and Sphyrapodinae Guțu, 1980, in order to conform to ICZN rules. G. O. Sars (1882), author of the name of the type genus *Sphyrapus*, indicated to have used this manuscript name of Norman. Norman himself published the name later with Stebbing (Norman & Stebbing, 1886, Trans. zool. Soc. London, **12**(4): 97) and the authors mentioned as derivatio nominis: "... from Greek "σφυρα", a hammer, and "πους", a foot". As the genitive of πους is "ποδος", the stem is ποδ-, whence according to the Code the family name has to read Sphyrapodidae. The incorrect spelling Sphyrapidae is definitely not "in prevailing usage" as Art. 29.5 of the Code (1999) demands, so the emendation can be executed by any qualified zoologist, as has been done herein. The same holds for family group names based on generic names derived from *Sphyrapus*, as in the case *Pseudosphyrapus* / Pseudosphyrapodinae.

Subfamily SPHYRAPODINAE Guțu, 1980

Sphyrapinae Guțu, 1980, Travaux du Muséum d'Histoire Naturelle "Grigore Antipa", **22**: 393.

Diagnosis. — (Modified after Guțu, 1980b, 1991; Guțu & Iliffe, 1998.) Female. Rostrum prominent. Pereon with five or six free pereonites, each wider than long. Pleon with five free pleonites all bearing biramous pleopods. Antennule with short inner flagellum (1–2 articles). Antenna with six articles, without squama/pseudosquama. Mandible without palp. Maxillule with bi-articulate, palp. Chelipeds with exopod. Pereopod 1 with or without exopod.

Male. Only known from *Sphyrapoides*. No subfamily diagnostic characters assigned.

Type genus. — *Sphyrapus* G. O. Sars, 1882.

Genera included. — *Anasphyrapus* Guțu, 2001; *Sphyrapus* G. O. Sars, 1882; *Sphyrapoides* Guțu & Iliffe, 1998.

Remarks. — The following characters: (1) absence of mandibular palp; (2) antennule inner flagellum with 1–2 articles; (3) antenna with 6 articles; are stable in this subfamily. The presence/absence and locality of the squama/pseudosquama are highly variable in this subfamily and useful for identification of the genus only.

Genus **Sphyrapoides** Guțu & Iliffe, 1998

Sphyrapoides Guțu & Iliffe, 1998, Travaux du Muséum d'Histoire Naturelle "Grigore Antipa", **40**: 211.

Diagnosis. — (Modified after Guțu & Iliffe, 1998.) Female. Eye-lobes large and visual elements present. Pereonite 1 fused with carapace. Pleon with five free pleonites all bearing biramous pleopods. Antennule inner flagellum bi-articulate. Antenna with squama. Mandible without palp. Chelipeds and pereopod 1 with exopod. Pereopod 1 longer than other pereopods. Uropods longer than pleon.

Male. Chela more robust than in female, dactylus with larger processes on inner margin.

Type species. — *Sphyrapoides bicornis* Guțu & Iliffe, 1998.

Gender of generic name. — Masculine.

Species currently assigned to this genus. — *S. bicornis* Guțu & Iliffe, 1998; *S. tuberculifrons* Guțu & Heard, 2002.

Sphyrapoides bicornis Guțu & Iliffe, 1998

Sphyrapoides bicornis Guțu & Iliffe, 1998, Travaux du Muséum d'Histoire Naturelle "Grigore Antipa", **40**: 211.

Diagnosis. — (After Guțu & Iliffe, 1998; Guțu & Heard, 2002.) Female. Rostrum divided into two processes.

Distribution. — (No distribution map presented.) Grand Exuma Island, Bahamas. Bathymetric range: only taken at type locality, at 140 m.

Remarks. — This species has not been reported from the Gulf of Mexico. Since it has been collected from crevices on the Bahamas Escarpment, a locality nearby and of strikingly similar habitat to the Florida Escarpment (which has yet to be studied), this species is likely to be present in the Gulf of Mexico as well.

The second species in this genus, *S. tuberculifrons* Guțu & Heard, 2002, is recorded from the Cayman Islands but is intertidal.

KEY TO THE SPECIES OF *SPHYRAPOIDES*, BOTH SEXES (GLOBAL)

1. Frontal margin of carapace (area between antennules) with several tubercles. Antennule inner rami uni-articulate, outer rami with three articles. Pereopod 1 propodus with two spiniform setae ventrally .... *S. tuberculifrons* Guțu & Heard, 2002
– Frontal margin of carapace with medial depression. Antennule inner rami bi-articulate, outer rami with five articles. Pereopod 1 propodus with three spiniform setae ventrally .... *S. bicornis* Guțu & Iliffe, 1998

Subfamily PSEUDOSPHYRAPODINAE Guțu, 1980

Pseudosphyrapinae Guțu, 1980, Travaux du Muséum d'Histoire Naturelle "Grigore Antipa", **22**: 393.

Diagnosis. — (Modified after Guțu, 1980b, 1991.) Both sexes. Rostrum prominent. Pereon with six free pereonites, each wider than long. Pleon with five free pleonites all bearing biramous pleopods. Antennule with short inner flagellum (2–3 articles). Antenna with eight articles, with or without squama/pseudosquama. Mandible with uni- or tri-articulate palp. Maxillule with bi-articulate palp. Chelipeds with exopod. Pereopod 1 with or without exopod.

Male. Only known from *Pseudosphyrapus*. No subfamily male diagnosis given.

Type genus. — *Pseudosphyrapus* Guțu, 1980.

Genera included. — *Francapseudes* Băcescu, 1981; *Kudinopasternakia* Guțu, 1991; *Pseudosphyrapus* Guțu, 1980.

Remarks. — The following characters are stable in this subfamily: (1) presence of mandibular palp; (2) antennule inner flagellum with two or three articles; (3) antenna with eight articles. The presence/absence and locality of the squama/pseudosquama are highly variable in this subfamily and useful for genus identification only.

Genus **Pseudosphyrapus** Guțu, 1980

Pseudosphyrapus Guțu, 1980, Travaux du Muséum d'Histoire Naturelle "Grigore Antipa", **22**: 393.

Diagnosis. — (Modified after Guțu, 1980b, 1991.) Antennule with bi- or tri-articulate inner ramus. Antenna without squama/pseudosquama. Mandible with tri-articulate palp.

Male. Antennule with dense rows of aesthetascs. Cheliped dactylus with inner process. Merus and carpus longer than those of female.

Type species. — *Sphyrapus serratus* G. O. Sars, 1882.

Gender of generic name. — Masculine.

Remarks. — The male of this genus is only known from the species mentioned below and from *P. anomalus* (G. O. Sars, 1869). The males seem to be characterized by: (1) having dense rows of aesthetascs on the inner flagellum of the antennule, (2) having at least one process on the serrated inner margin of the cheliped dactylus, while the dactylus is smooth in the female, (3) male chelipeds have a much longer merus and carpus, (4) male pereopod 1 is longer and thinner than in the female, (5) mouthparts fairly consistent, but a peculiar difference can be seen in the maxilliped endite, which has distal outer corners ornamented in males while smooth in females, (6) male maxillipedal endite setation is different from, in particular less setose than, that of female, (7) male pleopods with twice as many setae as in female (G. O. Sars, 1869; this study).

The character of the complete absence of a pseudosquama is stable in this genus. The presence of a tri-articulate mandibular palp is shared with *Kudinopasternakia*.

In view of the currently ongoing systematic revision of *Pseudosphyrapus* (cf. R. W. Heard, pers. comm.), no key to this genus is presented here.

Species currently assigned to this genus. — *P. anomalus* (G. O. Sars, 1869); *P. centobi* Băcescu, 1981; *P. gutui* Kudinova-Pasternak, 1985; *P. serratus* (G. O. Sars, 1882); *P. vladimiri* Guțu, 1989.

Pseudosphyrapus sp. A
(figs. 14–19)

Material examined. — 1 ♀ with rudimentary oostegites, body length 2.4 mm, 2 non-ovigerous ♀♀, 1 ♂ (badly damaged), DGoMB Sta. MT3-2. 1 non-ovigerous ♀, DGoMB Sta. W1-2.

Description. — Adult female. Body (fig. 14A, B) about 4.5 times as long as wide.

Carapace wider than any pereonite; including rostrum, longer than pereonites 1 and 2 combined. Without any spines. Rostrum (fig. 15C) widening anteriorly, with several small tubercles.

Pereonites all wider than long. Pereonites 1–2 without lateral demarcation. Pereonites 4–6 widening posteriorly.

All pleonites of similar size. Lateral edges without spiniform processes. Pleotelson about as long as last two pleonites combined, tapering to blunt apex.

Antennule (fig. 15A) peduncle article 1 as long as rest of antennule, with several simple setae. Article 2 less than 0.5 times as long as article 1, with one seta distally. Article 3 (common article), shorter than other articles, with one simple seta. Outer flagellum with three articles, bearing simple setae. Inner flagellum with five articles, all bearing simple setae.

Antenna (fig. 15B) almost as long as antennule. No rudiment of squama. Article 1 wider than following articles. Article 2 shorter than article 4 but longer than other articles, naked. Article 3 minute, naked. Article 4 longer than other articles, with six setulose and one simple setae. Articles 5–8 of diminishing size, all with setae.

Mouthparts. Labrum not recovered. Mandibular molar longer than incisor and tapering slightly; palp with three articles; article 1 naked, article 2 with one seta, article 3 with three setae. Left mandible (fig. 15D) lacinia mobilis bifurcate; incisor blunt; setal row with three apparently simple setae. Right mandible (fig. 15E) incisor with five denticles on flat apex; setal row minute. Labium (fig. 15F) with tuft of setae medially; palp small and narrow, with few setae distally. Maxillule (fig. 15G) outer endite with six spiniform setae; inner

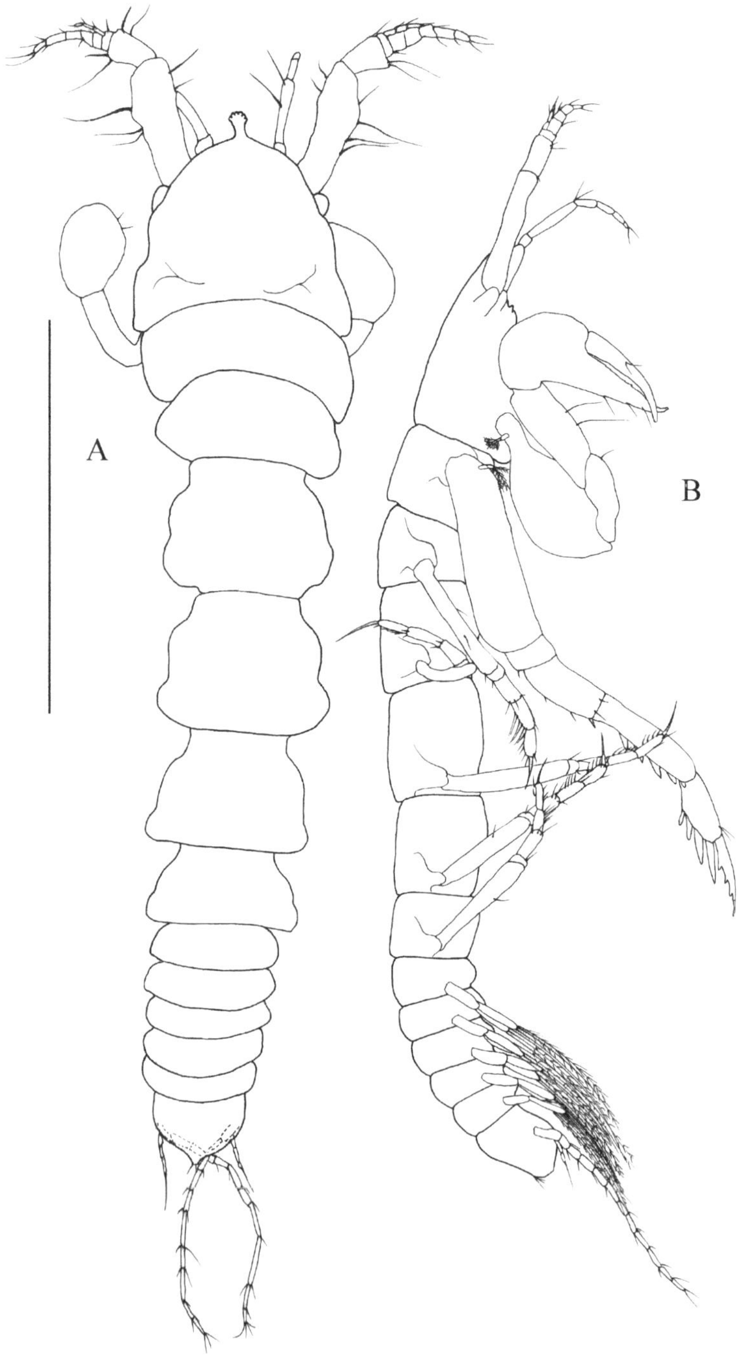

Fig. 14. *Pseudosphyrapus* sp. A, female; A, body, dorsal view; B, same, lateral view. Scale bar equals 1.0 mm.

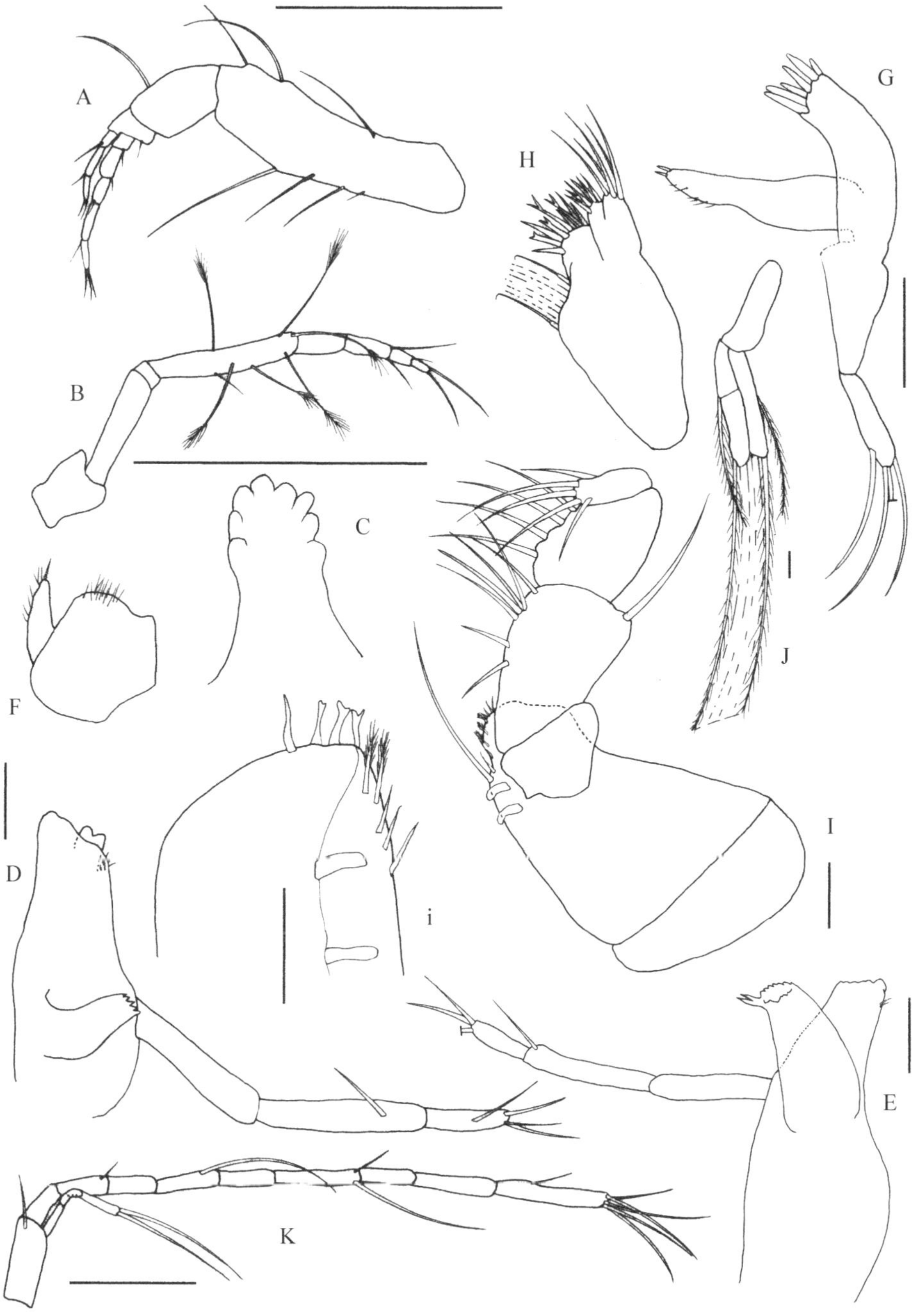

Fig. 15. *Pseudosphyrapus* sp. A, female; A, antennule; B, antenna; C, rostrum; D, left mandible; E, right mandible; F, labium; G, maxillule; H, maxilla; I, maxilliped; i, maxilliped endite; J, pleopod; K, uropod. Scale bars equal: A-B, K, 0.5 mm; C, 1.0 mm; D-J, 0.1 mm.

endite with two spiniform setae and setules; palp bi-articulate, distal article with four setae. Maxilla (fig. 15H) outer lobe of movable endite with five long simple setae; inner lobe of movable endite with six simple spiniform setae; outer lobe of fixed endite with five simple spiniform setae and four bifurcate spiniform setae; inner lobe of fixed endite with row of several distally curved setae. Maxilliped (fig. 15I) palp article 1 naked, article 2 with seven inner and one outer setae, article 3 with four inner setae, article 4 with four inner and one outer setae; endites (fig. 12i) outer corners smooth, distal edge with one simple and three bifurcate setae, inner corners with five spiniform setae of which two are pinnate, inner margin with two coupling hooks. Epignath not recovered.

Cheliped (fig. 19A) basis oval, longer than merus, with bi-articulate exopod bearing four plumose terminal setae. Merus clearly shorter than carpus, with one subdistal seta ventrally. Carpus shorter than basis, with three setae ventrally. Propodus (including fixed finger) longer than basis, with one seta at dactylus insertion. Fixed finger with four setae ventrally and three setae on inner margin; inner margin with weak serration. Dactylus as long as fixed finger, without processes or setae on inner margin.

Pereopod 1 (fig. 16A) 1.33 times as long as pereopods 2–6. Basis shorter than the three succeeding articles combined, with two proximal bipinnate setae and three simple setae distally. Ischium of irregular shape, naked. Merus almost as long as carpus, with three simple and one spiniform setae ventrally. Carpus 0.5 times as long as basis, with two spiniform setae ventrally, one spiniform seta dorsally, and several simple setae. Propodus shorter than merus, with four spiniform setae ventrally, two spiniform setae dorsally, and three simple setae dorsally. Dactylus and unguis combined as long as propodus and not fused; dactylus with marked serration.

Pereopod 2 (fig. 16B) only hardly longer than pereopods 3–6. Coxa naked. Basis as long as the three succeeding articles combined, with two bipinnate setae dorsally. Ischium with one seta ventrally. Merus as long as carpus, with two setae ventrally. Carpus longer than propodus, with five setae ventrally and three setae dorsally. Propodus as long as dactylus and unguis combined, with six setae ventrally and four setae dorsally. Dactylus and unguis smooth and with one seta at unguis insertion.

Pereopod 3 (fig. 16C) basis with one bipinnate seta. Ischium smooth. Merus with one seta ventrally. Carpus longer than propodus, with three setae ventrally and one seta dorsally. Propodus shorter than merus, with three setae ventrally and three setae dorso-distally. Dactylus and unguis combined longer than propodus, smooth and with one seta at unguis insertion.

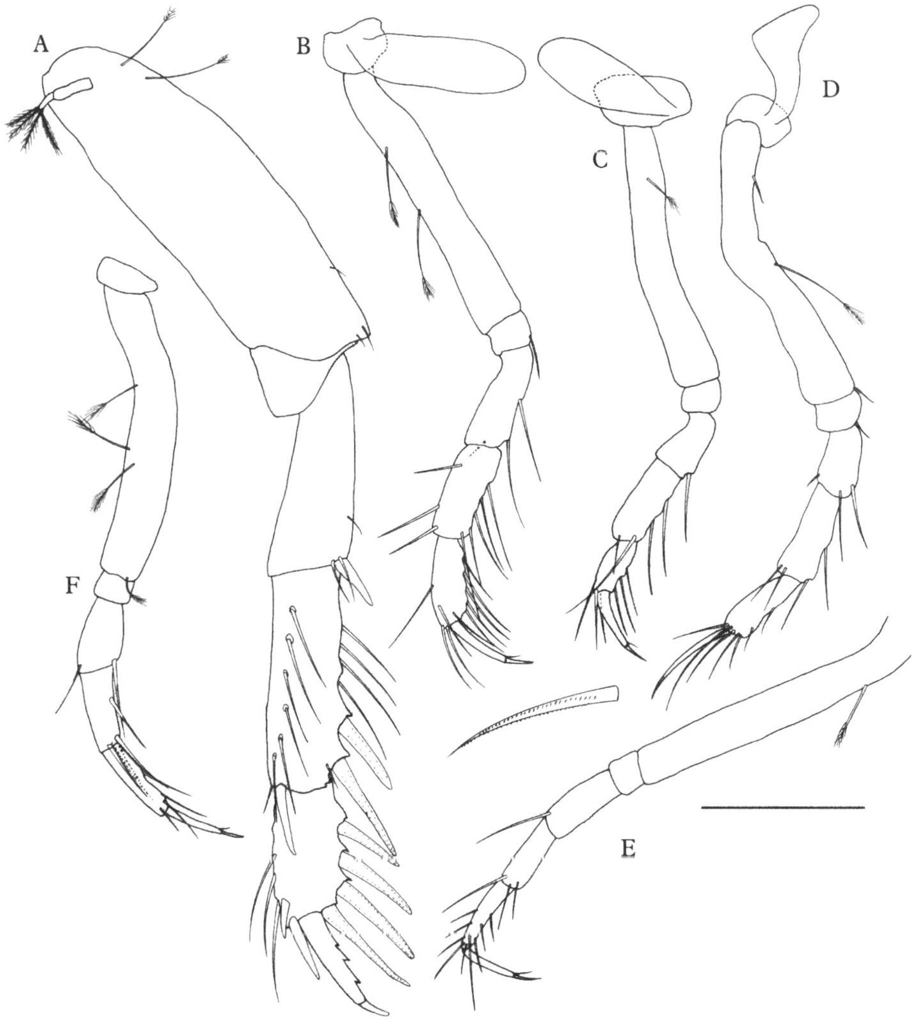

Fig. 16. *Pseudosphyrapus* sp. A, female; A, pereopod 1; B, pereopod 2; C, pereopod 3; D, pereopod 4; E, pereopod 5; F, pereopod 6. Scale bar equals 0.2 mm.

Pereopod 4 (fig. 16D) as pereopod 3 except: generally with fewer setae and dactylus without setae.

Pereopod 5 (fig. 16E) as pereopod 3 except: with fewer setae.

Pereopod 6 (fig. 16F) as pereopod 5 except: basis with three bipinnate setae and other articles with still fewer setae.

Pleopods (fig. 15J), basal article more than 0.5 times as long as endopod, naked. Exopod bi-articulate. Both exopod and endopod with six plumose setae.

Uropod (fig. 15K) about as long as pleon. Basal article with one seta. Endopod with nine articles of which most are setose, last article with incomplete sub-articulation (pseudo-articulation). Exopod with three articles, terminal article with two prominent setae distally.

Adult male. — Body too badly damaged to be described.

Antennule (fig. 17A) peduncle article 1 shorter than rest of antennule, with several simple setae. Article 2 less than 0.5 times as long as article 1, widening distally, with five setae distally. Article 3 (common article) shorter than other articles, with two setae at outer flagellum insertion. Outer flagellum with three articles bearing simple setae. Inner flagellum with seven articles, all bearing multiple aesthetascs (many more than illustrated).

Antenna (fig. 17B) almost as long as antennule. No rudiment of squama. Article 1 apparently not wider than following articles (this may be an artifact of dissection). Article 2 shorter than article 4 but longer than other articles, naked. Article 3 minute, naked. Article 4 longer than other articles, with seven setae. Articles 5–9 of subequal size, all with setae.

Mouthparts. Labrum (fig. 17C) naked and hood-shaped. Mandibular molar long and narrow. Left mandible (fig. 17D) lacinia mobilis and incisor bifurcate; setal row with apparently simple setae only. Right mandible (fig. 17E) incisor bifurcate; setal row with one multifurcate and two apparently simple setae; palp with three articles, article 1 naked, article 2 with only one seta, article 3 with only three setae. Labium (fig. 17F) with tuft of medial setae; palp small and narrow, with few setae distally. Maxillule (fig. 17G) outer endite with seven spiniform setae; inner endite with two spiniform setae and insertion points for several additional setae; palp bi-articulate, with three setae of unequal length. Maxilla (fig. 17H) outer lobe of movable endite with five long simple setae; inner lobe of movable endite with two simple spiniform setae, two bipinnate spiniform setae, and one bifurcate spiniform seta; outer lobe of fixed endite with five simple spiniform setae and two multifurcate spiniform setae; inner lobe of fixed endite with row of several distally curved setae. Maxilliped (fig. 17I) palp article 1 with one outer seta, article 2 with six inner and one outer setae, article 3 with four inner setae, article 4 with five inner and one outer setae; endites (fig. 17i) outer corners serrated, distal edge with three simple setae, inner corners with four spiniform setae, inner margin with two coupling hooks. Epignath not recovered.

Cheliped (fig. 19B, C) basis oval, longer than merus, with bi-articulate exopod bearing four plumose terminal setae. Merus almost as long as carpus,

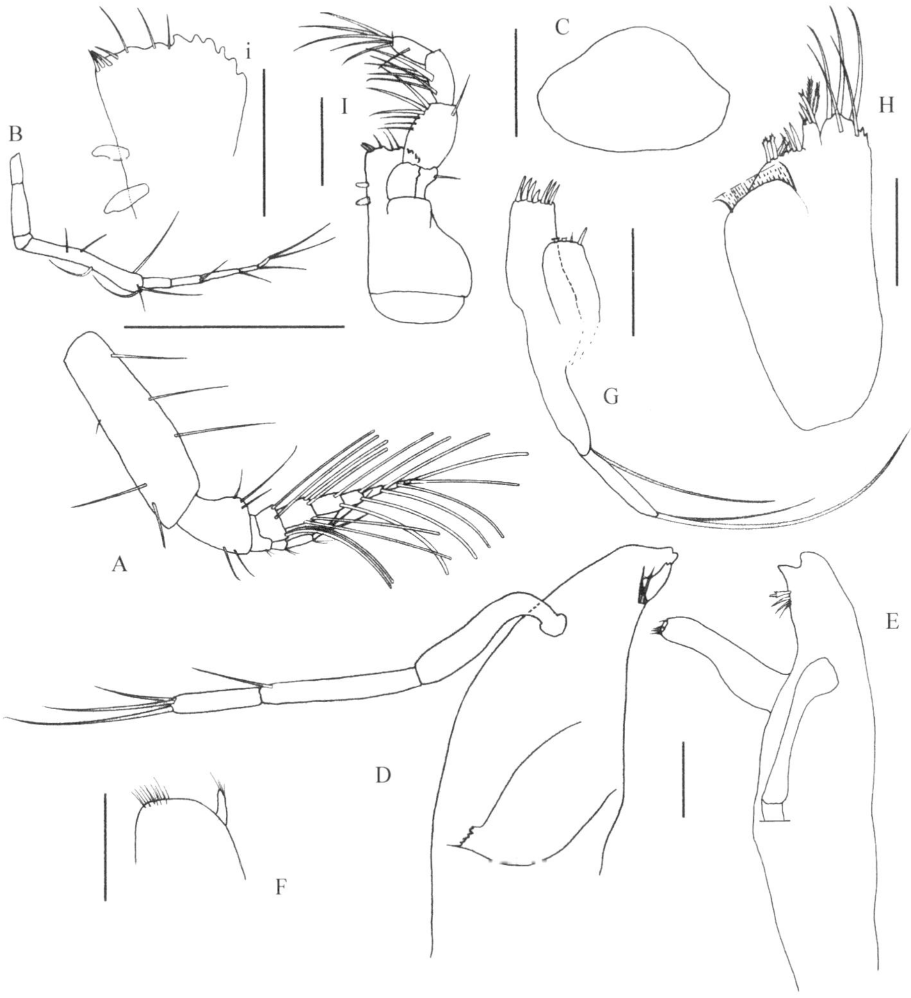

Fig. 17. *Pseudosphyrapus* sp. A, male; A, antennule; B, antenna; C, labrum; D, left mandible; E, right mandible; F, labium; G, maxillule; H, maxilla; I, maxilliped; i, maxilliped endite. Scale bars equal: A-B, I, 0.5 mm; C-E, 0.05 mm; F-H, i, 0.1 mm.

with one seta ventro-subdistally. Carpus shorter than basis, with three setae ventrally. Propodus (including fixed finger) shorter than basis, with one seta at dactylus insertion. Fixed finger with four setae ventrally and three setae on inner margin; inner margin with prominent serration and row of small setae originating from bottom of the serrations. Dactylus as long as fixed finger, with one proximal process and row of small spiniform setae on inner margin.

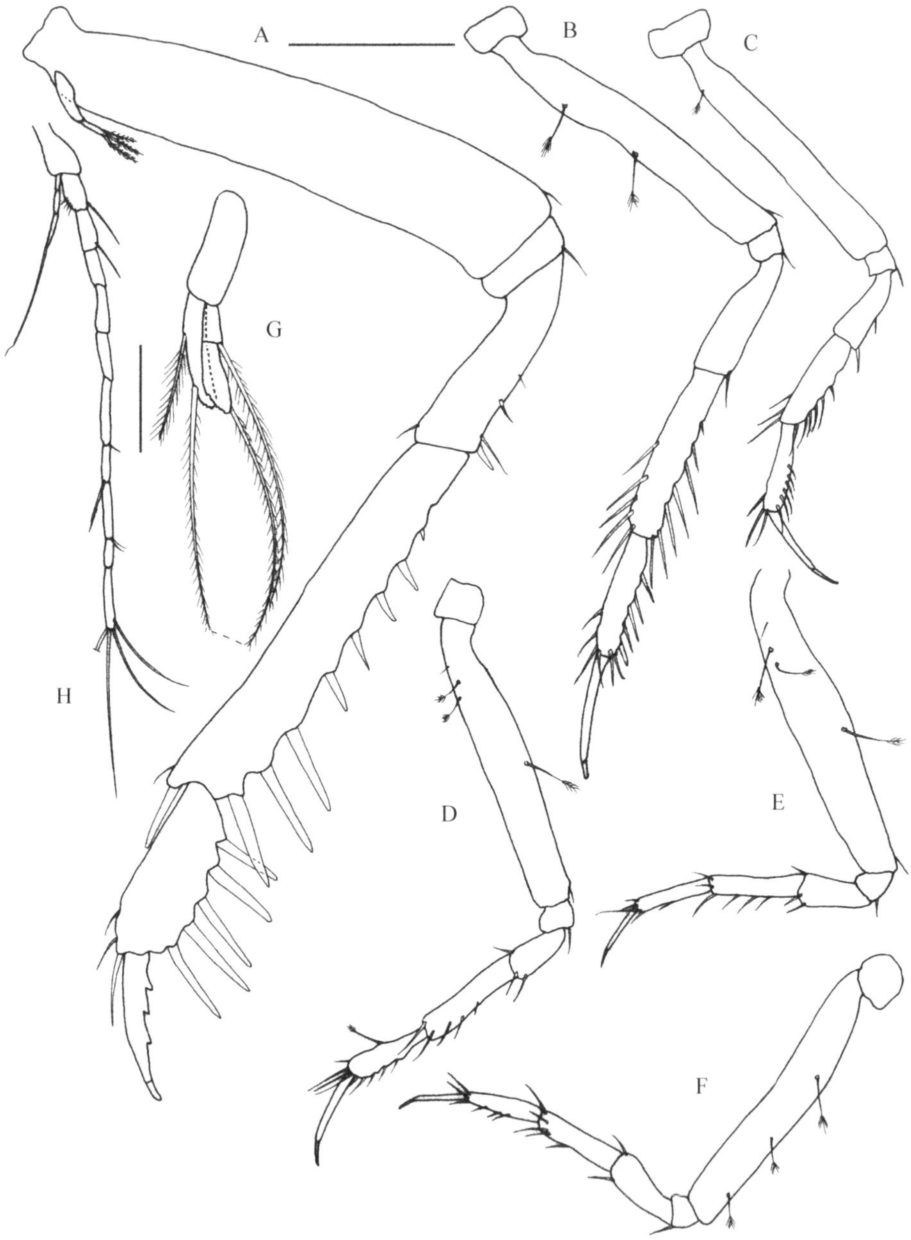

Fig. 18. *Pseudosphyrapus* sp. A, male; A, pereopod 1; B, pereopod 2; C, pereopod 3; D, pereopod 4; E, pereopod 5; F, pereopod 6; G, pleopod; H, uropod. Scale bar equals 0.2 mm.

Pereopod 1 (fig. 18A) almost twice as long as pereopods 3–6. Basis shorter than three succeeding articles combined, with one seta distally, exopod bi-articulate, bearing three plumose terminal setae. Ischium with one seta ventrally. Merus less than 0.5 times as long as carpus, with three simple and one spiniform setae ventrally. Carpus almost as long as basis, with eight spiniform setae ventrally, one spiniform seta dorsally, and one simple seta distally. Propodus shorter than merus, with five spiniform setae ventrally and three simple setae dorso-distally. Dactylus and unguis combined shorter than propodus and not fused; dactylus with marked serration.

Pereopod 2 (fig. 18B) 0.67 times as long as pereopod 1 but 1.33 times as long as pereopods 3–6. Coxa naked. Basis as long as the three succeeding articles combined, with two bipinnate setae dorsally and one simple seta ventro-distally. Ischium with one seta ventrally. Merus shorter than carpus, with one simple seta distally. Carpus longer than propodus, with seven setae ventrally and five setae dorsally. Propodus as long as merus, with seven setae ventrally and four dorsally. Dactylus and unguis combined as long as propodus, and smooth.

Pereopod 3 (fig. 18C) coxa naked. Basis longer than the three succeeding articles combined, with one bipinnate seta dorsally. Ischium with one seta ventrally. Merus with two setae ventrally. Carpus longer than propodus, with five setae ventrally and three setae dorsally. Propodus longer than merus, with six setae ventrally and three setae dorso-distally. Dactylus and unguis combined longer than propodus.

Pereopod 4 (fig. 18D) as pereopod 3 except: basis with three bipinnate and one simple setae. Merus with three setae distally. Carpus twice as long as merus, with four setae ventrally and one dorsally. Propodus longer than merus, with nine simple setae and one bipinnate seta dorsally.

Pereopod 5 (fig. 18E) as pereopod 4 except propodus with only one seta ventrally and four setae distally.

Pereopod 6 (fig. 18F) as pereopod 5 except propodus with three setae ventrally and three setae distally.

Pleopods (fig. 18G), basal article more than 0.5 times as long as endopod, naked. Exopod bi-articulate. Both exopod and endopod with 14 plumose setae.

Uropod (fig. 18H) basal article naked. Endopod with ten articles of which many are naked. Exopod with three articles, terminal article with one prominent seta distally.

Distribution. — This species was only found in a few samples in the northwestern part of the Gulf of Mexico. Bathymetric range 389–987 m. See map 10.

Remarks. — This species is not formally described since the genus is currently under revision (R. W. Heard, pers. comm.) and a suspicion exists that a

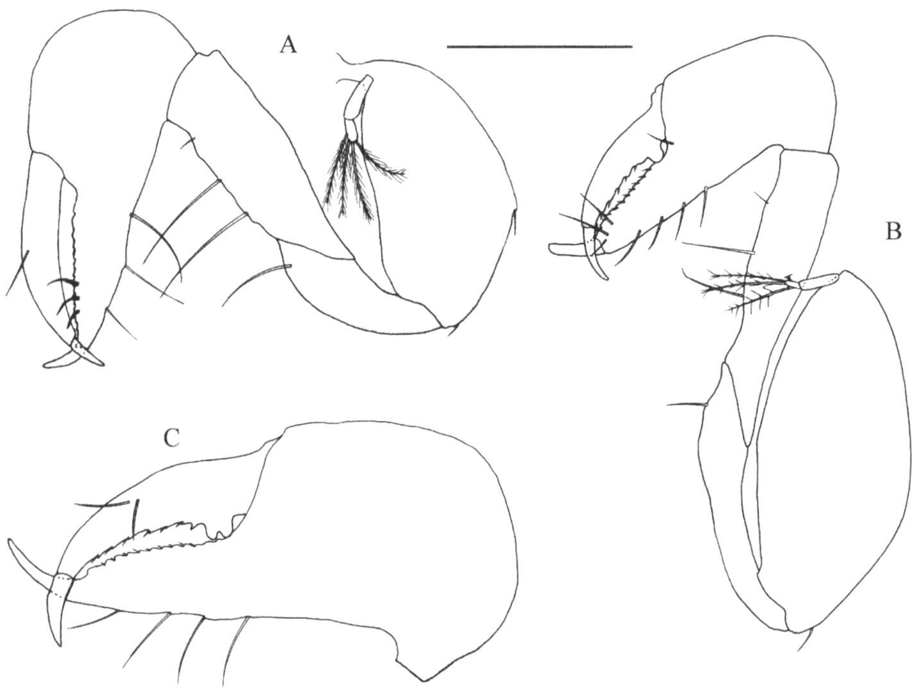

Fig. 19. *Pseudosphyrapus* sp. A, chelipeds; A, right cheliped, female; B, right cheliped, male; C, left cheliped propodus and dactylus enlarged, male. Scale bar equals 0.4 mm.

species complex of *Pseudosphyrapus*, all with a multi-tuberculate rostrum, is present in the Gulf of Mexico (R. W. Heard, pers. comm.).

The sexual dimorphism in this species is not extensive. The female cheliped dactylus is smooth while the male has a serration, a row of small spiniform setae, and one large process on the inner dactylus margin. The merus of the male cheliped is almost as long as the carpus, while conspicuously shorter than the carpus in the female. The shape of the dactylus process differs between the left and right cheliped. The process on the right cheliped is just one simple process, while the left one is divided into two processes. Male pereopod 1 longer than that of female.

The antennule of both male and female *Pseudosphyrapus* sp. A appears to have three articles in the peduncle, while the other species in the genus have four. A thorough examination of the apparent species complex in the Gulf of Mexico is needed to clarify if this fusion justifies the erection of a new genus.

Genus **Kudinopasternakia** Guţu, 1991

Pseudosphyrapus – Guţu, 1989 [p. p.], Travaux du Muséum d'Histoire Naturelle "Grigore Antipa", **30**: 135.
Pseudosphyrapus – Viskup & Heard, 1989, Gulf Research Reports, **8**(2): 107.
Kudinopasternakia Guţu, 1991, Travaux du Muséum d'Histoire Naturelle "Grigore Antipa", **31**: 341.

Diagnosis. — (Modified after Guţu, 1991a.) Both sexes. Antennule with bi-articulate inner flagellum. Antenna with pseudosquama located medially on article 2. Mandible with tri-articulate palp.

Type species. — *Pseudosphyrapus larisae* Guţu, 1989.

Gender of generic name. — Feminine.

Remarks. — The mancas of this genus have exopods on pereopods 4–5 (Guţu, 1989), a condition otherwise only known from the Kalliapseudidae.

The character of the pseudosquama (located halfway on antenna article 2) is stable in this genus.

Species currently assigned to this genus. — *K. bispinosa* Guţu & Heard, 2002; *K. dispar* (Lang, 1968); *K. larisae* (Guţu, 1989); *K. siegi* (Viskup & Heard, 1989).

Kudinopasternakia bispinosa Guţu & Heard, 2002

Kudinopasternakia bispinosa Guţu & Heard, 2002, Travaux du Muséum National d'Histoire Naturelle "Grigore Antipa", **44**: 69.

Diagnosis. — (Modified after Guţu & Heard, 2002.) Female. Rostrum triangular. Carapace lateral margins straight or slightly convex, laterally fused with pereonite 1. Cheliped carpus lacking distal process on anterior margin. Pereopod 1 propodus with only two spiniform setae ventrally. Pleonites with postero-lateral margins rounded in dorsal view.

Male. Unknown.

Distribution. — Only known from one specimen taken in the northern part of the Gulf of Mexico. Bathymetric range: 625 m. See map 11.

Remarks. — This species was described from one specimen, a manca II. Guţu & Heard (2002) mention that *Kudinopasternakia bispinosa* can be separated from *K. siegi* (Viskup & Heard, 1989) by having only two spiniform setae ventrally on the propodus of pereopod 1. However, as stated above (see remarks on *Leviapseudes bipartitus*), this character is not stable and likely to be ontogenetically dependent. Without this setal character, there is nothing to separate *K. bispinosa* from *K. siegi*.

Kudinopasternakia siegi (Viskup & Heard, 1989)

Pseudosphyrapus siegi Viskup & Heard, 1989, Gulf Research Reports, **8**(2): 107.

Diagnosis. — (Modified after Viskup & Heard, 1989.) Both sexes. Rostrum triangular. Carapace lateral margins straight or slightly convex, laterally fused with pereonite 1. Mandibular palp article 2 with seven setae distally. Maxillipedal basis with five short spiniform setae on lateral margin; palp article 4 with 9–10 setae. Maxillule distal palp article with 5–6 harpoon-shaped setae subdistally. Maxilla, outer lobe of fixed endite with two large, four-pronged spiniform setae, one bifurcate spiniform seta, four serrated spiniform setae, and one simple spiniform seta. Cheliped carpus lacking distal process on anterior margin. Pereopod 1 propodus with five spiniform setae ventrally. Pleonites with postero-lateral margins rounded in dorsal view.

Distribution. — This species has been found only in the northern part of the Gulf of Mexico so far. Bathymetric range 595–2400 m. See map 12.

Key to the Species of *Kudinopasternakia*, both Sexes (Global)

1. Pereopod 1 propodus with two spiniform setae ventrally .... *K. bispinosa* Guţu & Heard, 2002
– Pereopod 1 propodus with more than two spiniform setae ventrally 2
2. Cheliped carpus with spiniform process dorsally .... *K. dispar* (Lang, 1968)
– Cheliped carpus without spiniform process dorsally 3
3. Pleonite epimera rounded. Rostrum triangular .... *K. siegi* (Viskup & Heard, 1989)
– Pleonite epimera pointed. Rostrum lateral margins convex .... *K. larisae* (Guţu, 1989)

Family Parapseudidae Guţu, 1981

Parapseudidae Guţu, 1981, Travaux du Muséum d'Histoire Naturelle "Grigore Antipa", **23**: 82.
Apseudidae – Sieg, 1983, Crustaceorum Catalogus, **6**: 26.

Diagnosis. — (Modified after Guţu, 1981, 1996b.) Female. Body elongate, robust. Eye-lobes present, with or without visual elements. Pereon with six free pereonites. Pleon consisting of five free pleonites and pleotelson. Antennule biramous. Antenna with squama. Mandibles with tri-articulate palp. Maxillule with bi-articulate palp. Maxillipedal endite without leaf-shaped setae subdistally at inner corners. Epignath cup-shaped. Cheliped and pereopod 1 with exopod. Pereopod coxae without spiniform apophyses. Pereopod 1 stouter than other pereopods, carpus and propodus densely equipped with long setae. Pereopod dactylus without aesthetascs. Pereopods without specialized sucker-like structures. Pleopods biramous. Uropods filiform, biramous, both rami multi-articulate.

Male. Generally as female but with differently shaped chela.

Type genus. — *Parapseudes* G. O. Sars, 1882.

When describing the family Parapseudidae Guțu, 1981, that author indicated with an asterisk that the genus *Pseudoapseudes* Guțu, 1981 would be the type genus of that family. However, either Guțu himself, or the type-setter of his paper made a mistake by wrongly placing the asterisk, which should, of course, have been placed with the genus *Parapseudes* G. O. Sars, 1882. The rules (Code, 1999, Art. 29.1) say: "A family group name is formed by adding to the stem of the name of the type genus ... a suffix specified in Article 29.2." (which is: -idae or -inae). The asterisk thus was placed erroneously with *Pseudoapseudes* instead of with *Parapseudes*, which mistake must be corrected as has now been done herein.

Genera included. — *Ascumnella* Guțu & Heard, 2003; *Ctenapseudes* Bamber, Ariyananda & Silva, 1996; *Discapseudes* Băcescu & Guțu, 1975; *Halmyrapseudes* Băcescu & Guțu, 1974; *Longiflagrum* Guțu, 1995; *Pakistanapseudes* Băcescu, 1978; *Parapseudes* G. O. Sars, 1882; *Pseudoapseudes* Guțu, 1981; *Saltipedis* Guțu, 1995; *Trichapseudes* Barnard, 1920.

Genus **Parapseudes** G. O Sars, 1882

Parapseudes G. O. Sars, 1882, Archiv for Matematik og Naturvidenskab, **7**: 16.
Dalapseudes Boone, 1923, Proceedings of the Biological Society of Washington, **36**: 147.

Diagnosis. — (Modified after Guțu, 1998b.) Female. Body dorso-ventrally flattened. Carapace short and wide (no length/width ratio given). Pereon with six free pereonites, the first two pereonites conspicuously shorter than the others. Antennule with short peduncle and long flagellum. Antenna peduncle with five short articles, squama present. Maxillipedal endite with widely serrated seta subdistally at inner corners. Cheliped and pereopod 1 with exopod. Pereopods 4–6 with basis thickest medially. Pereopod 4 with reduced dactylus. Marsupium consisting of five pairs of oostegites, the pair on the cheliped reduced.

Male. As female except for chelipeds being better developed in males.

Type species. — *Rhoea latifrons* Grube, 1864.

Gender of generic name. — Masculine.

Species currently assigned to this genus. — *P. francispori* (Băcescu, 1980); *P. inermis* (Brum, 1973); *P. latifrons* (Grube, 1864); *P. trispinosus* Guțu, 1998.

Parapseudes cf. **latifrons** (Grube, 1864)

Rhoëa latifrons Grube, 1864, Die Insel Lussin und ihre Meeresfauna: 75.
Apseudes latifrons – Stossich, 1881, Bolletino della Società Adriatica di Scienze Naturali in Trieste, **6**: 219.
Parapseudes latifrons – G. O. Sars, 1882, Archiv for Matematik og Naturvidenskab, **7**: 17.
Parapseudes goodei Richardson, 1906, Transactions of the Connecticut Academy of Arts and Sciences, **11**: 283.
Parapseudes similis Vanhöffen, 1914, Deutsche Südpolar Expedition, **15**: 462.
Dalapseudes pedispinis Boone, 1923, Proceedings of the Biological Society of Washington, **36**: 147.
Parapseudes neglectus Miller, 1940, Occasional Papers from the Bernice P. Bishop Museum, Honolulu, **15**: 298.
Apseudes algicola Shiino, 1952, Publications of the Seto Marine Biological Laboratory, **2**(2): 53.
Parapseudes pedispinis – Menzies, 1953, Bulletin of the Museum of Comparative Zoology at Harvard College, **107**: 450.
Parapseudes spongicola Brown, 1958, Annals and Magazine of Natural History, **13**(1): 453.
Parapseudes algicola – Lang, 1966, Arkiv för Zoologi, **2**: 552.
Parapseudes latinfrons – Lang, 1970, Arkiv för Zoologi, **22**(16): 599.
Parapseudes grubei Băcescu, 1977, Revue Roumaine de Biologie, (Biologie Animale) **21**(1): 123.

Diagnosis. — (After Guțu, 1998b.) Female. Pereopod 1 basis with less than three spiniform setae dorso-proximally.

Male. As female except for better developed cheliped basis and carpus.

Distribution. — Generally recorded in the littoral zone but also taken on the upper shelf in the northern Gulf of Mexico (Ogle, 1977) and off the west coast of Cuba (Makkaveeva, 1968). Bathymetric range: 1–70 m. See map 13.

Remarks. — It has not been possible to track down the material mentioned by Makkaveeva (1968) or Ogle (1977). It is unlikely that the animal mentioned in these reports really is *P. latifrons*, which has its type locality off Tanzania. It is likely to be a misidentification of another, undescribed species of *Parapseu-*

des found in the Gulf of Mexico. This new species was found during the completion of a guidebook to the shallow-water Tanaidacea for the state of Florida (Heard et al., 2004) and is currently under investigation (T. Hansknecht, pers. comm.), so no key has been constructed.

As no precise locality data are available, the map locations are to indicate a general area only and should not be considered precise. Usually, this family consists of shallow-water species only (mostly intertidal) and the record from 70 m is unusual.

SUBORDER NEOTANAIDOMORPHA SIEG, 1980

Family NEOTANAIDAE Lang, 1956

Neotanaidae Lang, 1956, Arkiv för Zoologi, **2**(9)(21): 469.

Diagnosis. — (Modified after Lang, 1956c; Gardiner, 1975b; Larsen & Błażewicz-Paszkowycz, 2003.) Female. Eye-lobes present but never with visual elements. Always bearing pleopods. Antennule uniramous with 7–8 articles. Antenna without squama and with nine articles. Mandibles without palp, with strong and heavily calcified molars. Labium with two pairs of lobes and palp (more or less fused), both pairs of lobes and palp setose. Maxillule consisting of two endites with simple and specialized setae, palp missing. Maxilla well developed with multiple, specialized setae. Maxillipeds with unfused coxae and bases; endites always separate and bearing several simple and specialized setae. Epignath strongly developed, divided into two almost equal-sized lobes and with setulose terminal seta. Chelipeds attached to the cephalothorax by a large sclerite, frequently larger than the cheliped basis; small, incomplete ischium present. Pereopods all more or less similar and all of an unspecialized 'walking' type; coxae present on all pereopods; all pereopods without spinning glands. Pleopods biramous or uniramous (*Herpotanais*) and with plumose setae, endopod often bi-articulate. Uropods biramous, endopod multi-articulate, exopod with one or (normally) two articles.

Male. Sexual polymorphism considerable. Antennae slimmer than in female but with same number of articles; antennule with numerous aesthetascs in particular on article 4. Mouthparts reduced and non-functional. Cheliped carpus, dactylus, and fixed finger greatly enlarged in the male.

Type genus. — *Neotanais* Beddard, 1886.

Genera included. — *Carololangia* Gardiner, 1975; *Herpotanais* Wolff, 1956; *Neotanais* Beddard, 1886; *Venusticrus* Gardiner, 1975.

Genus **Neotanais** Beddard, 1886

Neotanais Beddard, 1886, Proceeding of the Scientific Meetings of the Zoological Society of London, **97**: 122.

Alotanais Norman & Stebbing, 1886, Transactions of the Zoological Society of London, **12**(4): 107.

Diagnosis. — (Modified after Larsen & Błażewicz-Paszkowycz, 2003.) Female. Pleotelson pentagonal (not acorn-shaped). Antennule with seven articles. Pereopod attachment on lateral surface of the body. Pleopods biramous and with plumose setae, endopod often bi-articulate. Other appendages as in family diagnosis.

Male. See family diagnosis.

Type species. — *Neotanais americanus* Beddard, 1886.

Gender of generic name. — Masculine.

Remarks. — Several characters used for systematic considerations in the Tanaidacea do not carry much information in *Neotanais*. (1) In contrast to most tanaidacean species the number of articles in the antennule (seven) and the antenna (nine) is constant for both sexes of all species known so far. These characters are thus not informative for species identification. However, the stoutness of the antennule and the relative length of the articles are often useful characters, in particular that of article 1 and that of article 4, relative to those of articles 5–7 combined. (2) The mouthpart characters, are, with a few exceptions (Gardiner, 1975b; Larsen, 1999a), of little value owing to great similarity among the species as well as some intraspecific variation. (3) Setal number on just about all appendages varies greatly with development and is thus not a useful character (Gardiner, 1975b). (4) The number of denticulations on the cheliped fixed finger and the setation formula of the pereopods, are characters that have traditionally been attributed much information throughout the literature (Wolff, 1956a, b; Lang, 1968; Gardiner, 1975b). The value of these characters is, however, greatly over-rated as they are also ontogenetically dependent.

There are five useful characters: (1) The presence and shape of the midventral keel on the pleonites. This is one of the first characters the identifier will look at in a *Neotanais*, but it has to be remembered that in some species (notably *N. americanus* and *N. micromopher*) this character is quite variable. (2) The relative length of the cheliped dactylus to that of the fixed finger. (3) The number of dorsal setae on the cheliped carpus. (4) The shape (in dorsal view) of the margins of the pleonites, although this is somewhat subjective. (5) Morphometric characters such as the relative length/width of the carapace, pereonites, pleotelson, cheliped propodus, antennae, and uropods. Such characters are less obvious, but hold the greatest potential for discriminating between the species of *Neotanais*.

Males usually display more features for separation than females, especially in the shape of the chelipeds, but since protogynic hermaphroditism exists in

Neotanais and since males are not as common as females, one cannot rely on male characters alone.

Owing to the conservative morphology of the genus and the large number of species, a key to the genus is only effective if combined with illustrations of the characters involved, and even so a number of character combinations have to be applied. An illustrated key to the genus was given by Larsen (1999a).

Species currently assigned to this genus. — *N. affinis* Wolff, 1956; *N. americanus* Beddard, 1886; *N. antarcticus* Kussakin, 1966; *N. armiger* Wolff, 1956; *N. bacescui* Lang, 1968; *N. barfoedi* Wolff, 1956; *N. calcarulus* Gardiner, 1975; *N. curvimanus* Larsen & Hansknecht, 2003; *N. dinotomer* Gardiner, 1975; *N. gardineri* Larsen, 1999; *N. giganteus* Hansen, 1913; *N. hadalis* Wolff, 1956; *N. hamatus* Kudinova-Pasternak, 1990; *N. hastiger* (Norman & Stebbing, 1886); *N. hessleri* Gardiner, 1975; *N. insignis* Kudinova-Pasternak, 1978; *N. krappschickelae* Larsen & Błażewicz-Paszkowycz, 2003; *N. kurchatovi* Kudinova-Pasternak, 1975; *N. laevispinosus* (Norman & Stebbing, 1886); *N. magnificus* Kudinova-Pasternak, 1972; *N. mesostenoceps* Gardiner, 1975; *N. micromopher* Gardiner, 1975; *N. minimus* Larsen & Hansknecht, 2003; *N. noelitaiti* Larsen & Hansknecht, 2003; *N. peculiaris* Lang, 1968; *N. persephone* Messing, 1977; *N. pfaffi* Wolff, 1956; *N. pfaffioides* Lang, 1968; *N. robustus* Wolff, 1956; *N. sandersi* Gardiner, 1975; *N. triangulocephalus* Kudinova-Pasternak, 1973; *N. tricarinatus* Gardiner, 1975; *N. tuberculatus* Kudinova-Pasternak, 1970; *N. vemae* Gardiner, 1975; *N. wolffi* Kudinova-Pasternak, 1966; *N. zenkevitchi* Kudinova-Pasternak, 1973.

Neotanais armiger Wolff, 1956

Neotanais armiger Wolff, 1956, Videnskabelige Meddelelser fra Dansk Naturhistorisk Forening i København, **118**: 47.

Diagnosis. — (Modified after Wolff, 1956b.) Female. Pereonite 4 wider than long. Pleonites with pronounced, posteriorly directed midventral keel. Pleonites with lateral margin pointed. Pleotelson with posterior margin and lateral margins straight, narrower than pleonites. Antennular article 1 long (length/width ratio about 5), article 4 longer than 2.0 times the lengths of articles 5–7 combined. Cheliped with dactylus and fixed finger of subequal length, fixed finger bearing a few small denticles on inner margin. Last pair of pleopods not reaching apex of pleotelson. Uropods longer than pleon. Uropodal exopod with two articles, shorter than endopod article 1. Uropodal endopod last article with distal setae longer than article itself. Uropod attachment midlength on pleotelson and at 45° angle.

Male. Antennular article 1 long (longer than carapace), pleonites with pronounced ventrally directed midventral keel.

Distribution. — A very common species on the western hemisphere, found along both the Pacific and Atlantic coasts of the Americas. Bathymetric range wide, 600–6100 m (Gardiner, 1975b). See map 14.

Neotanais curvimanus Larsen & Hansknecht, 2003

Neotanais curvimanus Larsen & Hansknecht, 2003, Journal of Natural History, London, **37**(23): 2794.

Diagnosis. — (After Larsen & Hansknecht, 2003.) Female. Pereonite 4 wider than long. Pereonite 6 only 0.5 times as long as wide. Pleonites lateral margins rounded. Pleonites with weakly defined midventral keel and several setulose setae laterally. Antennular article 1 stout (length/width ratio about 3), article 4 less than twice as long as the lengths of articles 5–7 combined. Cheliped with dactylus and fixed finger of subequal length, fixed finger bearing conspicuous denticles on inner margin. Last pair of pleopods not reaching apex of pleotelson. Uropods as long as or longer than pleon. Uropodal exopod with two articles, shorter than endopod article 1. Uropod attachment midlength on pleotelson and at 45° angle.

Male. Cheliped fixed finger only half as long as the dactylus; dactylus incurved from the fixed finger at about 40° angle.

Distribution. — A very rare species only known from three specimens from one station in the northern part of the Gulf. Bathymetric range: only collected at one station from 1000 m depth. See map 15.

Neotanais minimus Larsen & Hansknecht, 2003

Neotanais minimus Larsen & Hansknecht, 2003, Journal of Natural History, London, **37**(23): 2788.

Diagnosis. — (After Larsen & Hansknecht, 2003.) Female. Pereonite 4 wider than long. Pleonites with small, straight midventral keel. Pleonites lateral margins rounded. Pleotelson with straight posterior margin. Pleotelson with lateral margins expanded laterally, as wide as pleonites. Antennular article 1 long (length/width ratio > 5), article 4 less than twice as long as lengths of articles 5–7 combined. Cheliped with dactylus and fixed finger of subequal length, fixed finger bearing conspicuous denticles on inner margin. Last pair of pleopods reaching apex of pleotelson. Uropods longer than pleon. Uropodal exopod with two articles, 0.5 times as long as endopod article 1. Uropodal endopod last article with distal setae longer than article itself. Uropod attachment midlength on pleotelson and at 45° angle.

Male. Carapace more sharply divided into narrow anterior and broad posterior part than in female. Antennule longer than that of female, article 4 greatly expanded and densely packed with aesthetascs. Cheliped basis circular and smaller than that of female; fixed finger as long as dactylus; dactylus attached parallel to fixed finger.

Distribution. — A rare species found only in the northern part of the Gulf. Bathymetric range: 379–1825 m. See map 16.

SUBORDER TANAIDOMORPHA SIEG, 1980

Superfamily TANAOIDEA Dana, 1849

Of the two superfamilies in the Tanaidomorpha, the Tanaoidea are reasonably well understood owing to the monograph of Sieg (1980). Unfortunately, since 1980, only four new species belonging to this superfamily have been described. No higher level taxa have been erected and no revisionary work conducted, despite the fact that members of this superfamily are encountered in just about any shallow water habitat of the world (Sieg, 1980; Bamber & Bird, 1997; Heard et al., 2004) and even have representatives on the shelf (Sieg, 1983c) and in the deep sea (Kudinova-Pasternak, 1970; Larsen, in press). However, as no member of the Tanaoidea was found in the deep Gulf of Mexico, this superfamily is not treated further here.

Superfamily PARATANAOIDEA Lang, 1949

The superfamily Paratanaoidea is the most advanced taxon of the Tanaidacea, and it is here that we find the highest number of reductions and reversals (Larsen & Wilson, 2002). The recent re-classification of the Paratanaoidea recognizes 10 families, 98 genera (including those described herein), and more than half the total number of species in the Tanaidacea (cf. Larsen & Wilson, 2002). Unfortunately, the classification of Larsen & Wilson (2002) leaves a large number of genera without family designation. Seven genera had to be excluded from the analysis as they were either described too incompletely to allow a phylogenetic analysis, or had been described from males only. Seventeen more genera could not be assigned to a family on the basis of the analysis, due to reversal and lack of synapomorphies (Larsen & Wilson, 2002).

Since then, 12 new genera have been erected belonging to the Paratanaoidea, and five of these have not been given family designation. Clearly this situation is untenable, but it is not easily resolved, either. Larsen & Wilson (2002: 205, 215) state that their analysis is a 'first step' phylogeny, and that the support values for most families are low, thus indicating the phylogeny is still far from satisfactorily resolved.

Family AGATHOTANAIDAE Lang, 1971

Agathotanaidae Lang, 1971, Arkiv för Zoologi, **2**(23): 367.
Anarthruridae – Kudinova-Pasternak, 1986, Zoologicheskii Zhurnal, **65**: 66.
Agathotanaididae Larsen & Wilson, 2002, Journal of Zoological Systematics and Evolutionary Research, **40**(4): 206, 211, 216.

Diagnosis. — (Modified after Larsen & Wilson, 2002.) Female. Rigid body calcification. No plates in carapace. Eyes and eye-lobes absent. Pereon with 5–6 free pereonites. Pereonite 1 not reduced. Pleon with five free pleonites or all pleonites fused with telson and pereonite 6; narrower than pereon or as wide as pereon (*Metagathotanais*), no articulated setae on pleonites. Antennule with 3–4 articles. Antenna completely reduced or with up to six articles, article 3 without spiniform setae dorsally. Molar of mandible process pointed or reduced. Labium with one pair of lobes, medial spiniform setae present or absent. Maxillipedal bases fused, endites not fused, narrower than basis and without flat setae or denticles. Cheliped attachment ventral or via ventral pseudocoxa. Marsupium (where known) with four pairs of oostegites. Pereopod coxae present on pereopods 1–3 and present or absent on pereopods 4–6. Pereopods 4–6 with dactylus and unguis not fused. Pleopods absent. Uropodal endopod with 1–2 articles or fused with basal article, exopod absent, reduced to spiniform process or with two articles (*Paranarthrurella*).

Male (where known) antennule with 3–7 articles, without numerous aesthetascs. Functional mouthparts retained. Pleopods present but reduced, often fused into immobile cone and with simple setae only. Pleon as wide as pereon or wider.

Type genus. — *Agathotanais* Hansen, 1913.

Genera included. — *Agathotanais* Hansen, 1913; *Metagathotanais* Bird & Holdich, 1988; *Paragathotanais* Lang, 1971; *Paranarthrura* Hansen, 1913; *Paranarthrurella* Lang, 1971.

KEY TO THE GENERA OF AGATHOTANAIDAE IN THE GULF OF MEXICO

1. Antenna with three or less articles. Uropodal endopod fused with basal article .... *Agathotanais* Hansen, 1913
– Antenna with four or more articles. Uropodal endopod not fused with basal article 2
2. Uropodal exopod process shaped as an acute spine .... *Paranarthrura* Hansen, 1913
– Uropodal exopod not shaped as an acute spine .... *Paragathotanais* Lang, 1971

Genus **Agathotanais** Hansen, 1913

Agathotanais Hansen, 1913, Danish Ingolf Expedition, **3**(3): 8, 63.

Diagnosis. — (Modified after Larsen, 1999b.) Female. Cuticle heavily calcified and deeply pitted. Pereon with six free pereonites. Pleon with five free pleonites. All body somites and appendages finely setulose except for distal part of cheliped and pereopod dactyli. Antennule with three articles. Antenna completely reduced or with 1–3 articles. Mandibular molar process reduced. Labium usually with medial processes. Maxillipedal bases partially fused; endites not fused, almost circular, narrower than basis and without flat setae or denticles. Cheliped attachment ventral. Marsupium (where known) with four pairs of oostegites. Pereopod coxa present. Pereopods 4–6 with dactylus and unguis not fused. Pleopods absent. Uropodal endopod fused with basal article, exopod reduced.

Male. Sexual dimorphism weak. Antennule with three articles, without numerous aesthetascs. Functional mouthparts retained. Pleopods present but reduced, fused into immobile cone and with simple setae only. Pleon as wide as pereon.

Gender of generic name. — Masculine.

Type species. — *Agathotanais ingolfi* Hansen, 1913.

Remarks. — The diagnostic characters are considered stable in this genus. The antenna is usually reduced to just one article but both lesser (with three articles) or greater reductions (complete) may be found. For species identification, the following characters are useful: (1) The level of antenna reduction. (2) The length/width ratios of the pereonites. (3) The length of antennular article 1. (4) The spination of the propodus of pereopod 4.

Only rarely do the mouthparts or cheliped reveal useful characters. Both the shape of the anterior part of the lateral carapace margins (convex/concave) and the length/width ratio of the pereonites (Lang, 1971a) are problematic characters and should not be attributed taxonomic value uncritically.

The genus *Agathotanais* is recorded from every major ocean in the world and occurs commonly in deep-sea samples. A key to the species is given in Larsen (1999b).

Species currently assigned to this genus. — *A. ahyongi* Larsen, 1999; *A. brevis* Kudinova-Pasternak, 1990; *A. ghilarovi* Kudinova-Pasternak, 1989; *A. hanseni* Lang, 1971; *A. ingolfi* Hansen, 1913; *A. manganicus* Larsen, 1999; *A. spinipoda* Larsen, 1999; *A. splendidus* Kudinova-Pasternak, 1970.

Agathotanais hanseni Lang, 1971
(figs. 20, 21)

Agathotanais hanseni Lang, 1971, Crustaceana, **21**: 59.

Material examined. — 3 non-ovigerous ♀♀, 2 ♂♂, 1 manca, DGoMB Sta. MT3-4. 1 non-ovigerous ♀, DGoMB Sta. RW3-3. 1 juvenile, DGoMB Sta. NB4-4. 1 non-ovigerous ♀, DGoMB Sta. C4-4. 2 non-ovigerous ♀♀, DGoMB Sta. W4-4. 1 non-ovigerous ♀, DGoMB Sta. MT4-4. 1 manca, DGoMB Sta. W2-1.

Diagnosis. — (Modified after Lang, 1971a; Larsen, 1999b.) Female. Carapace antero-lateral margin convex. Pereonites 4–5 width/length ratio less than 1.

Male. Pereonites 5 and 6 of subequal length.

Distribution. — Found in the northern part of the Gulf of Mexico, southern Caribbean Sea, Gulf of Panama, and off Ecuador (and one dubious record from the Azores). Bathymetric range in the Gulf of Mexico: 625–2050 m. In the southern Caribbean Sea: 3944–3761 m; in the Gulf of Panama: 2875 m; off Ecuador: 2861–2864 m; (around the Azores: 4165 m). See map 17.

Remarks. — Since Lang (1971a) described the male only superficially and did not illustrate a whole body view of any specimens of *A. hanseni*, illustrations of the body of both sexes are presented here (figs. 17–18). Lang's (1971a) description is accurate but his diagnostic characters (length/width ratio of pereonites 4–5; shape of carapace) are unstable. Lang (1971a) did not illustrate these diagnostic characters in his description of this species. The holotype is lost and the paratypes are all in poor condition (dried out) so the character can not be ascertained (Larsen, 1999b). Also some *Agathotanais* (most likely *A. ingolfi*) specimens were found in the Atlantic with a mixture of 'hanseni' and 'ingolfi' characters, an indication that these characters are variable (Larsen, 1999a). The DGoMB material shows variation in the length of pereonites 4–5, although they are never shorter than wide.

Genus **Paragathotanais** Lang, 1971

Paragathotanais Lang, 1971, Crustaceana, **21**: 65.

Diagnosis. — (Modified after Lang, 1971a; Bird & Holdich, 1988; Larsen, 2002a.) Female. Cuticle heavily calcified but not pitted or setulated. Pereon with six free pereonites. Pleon with five free pleonites. Pleonites most often narrower than pereonites and pleotelson. Pleotelson with posterior dorsal plate.

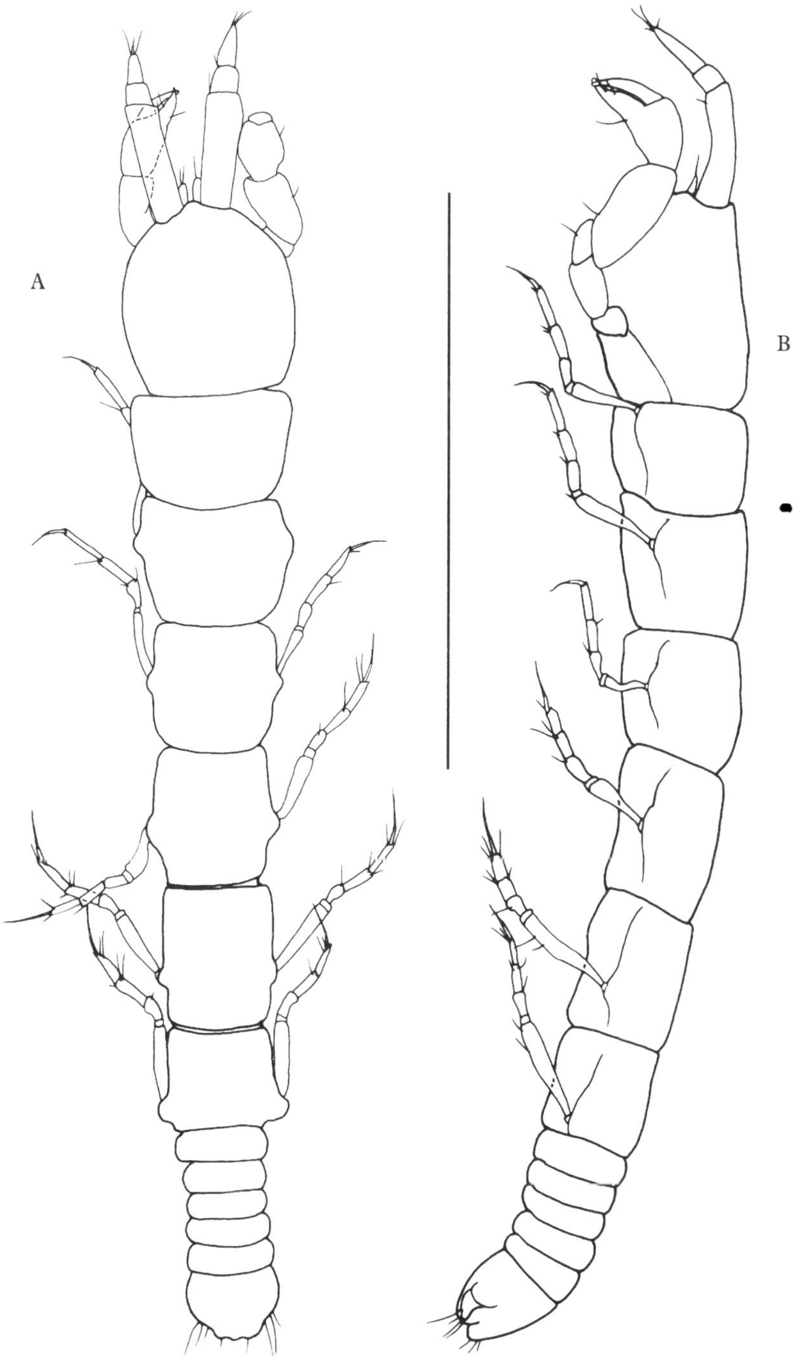

Fig. 20. *Agathotanais hanseni* Lang, 1971, female; A, dorsal view; B, lateral view. Scale bar equals 1.0 mm.

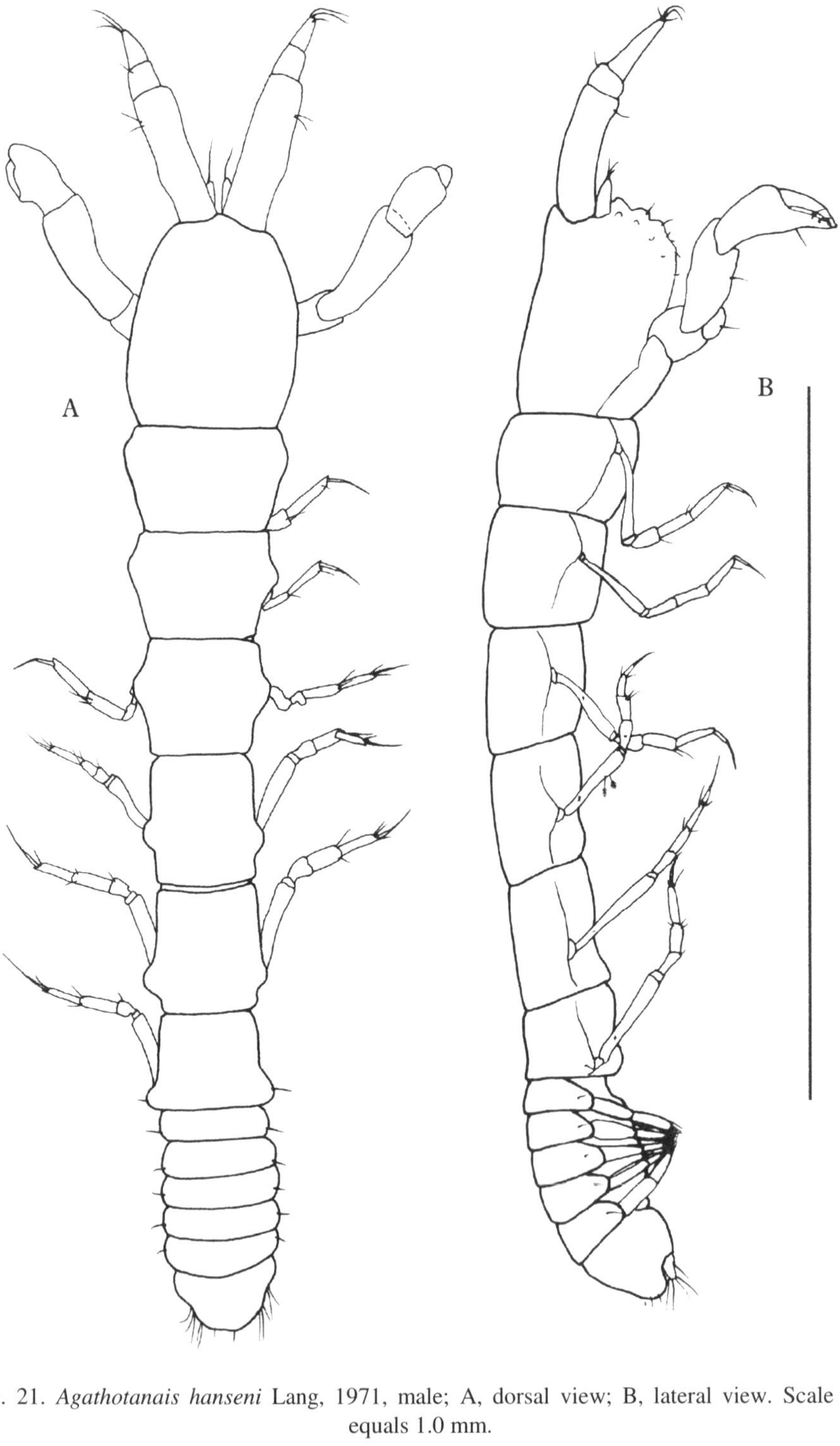

Fig. 21. *Agathotanais hanseni* Lang, 1971, male; A, dorsal view; B, lateral view. Scale bar equals 1.0 mm.

Appendages not setulose. Antennule with four articles. Antenna with 4–6 articles. Mandibles with posteriorly directed, pointed, or membranous molar; lacinia mobilis of left mandible tiny and spiniform. Labium with or without medial processes. Maxillule with 10–11 spiniform terminal setae. Maxillipedal bases fused, endites not fused, narrower than basis and without flat setae. Epignath without terminal seta or setules. Cheliped attached to ventral surface of cephalothorax, no pseudocoxa present. Pereopods with coxae. Marsupium (where known) with four pairs of oostegites. Pereopods 4–6 with basis not enlarged relative to bases on pereopods 1–3; dactylus and unguis not fused. Pleopods absent. Uropods small and not visible from dorsal side; basal article not fused with uni-articulate endopod; exopod reduced to a minute projection, hardly visible, not acute (not spiniform), giving rise to one or more setae.

Male. Pleonites as wide as pereonites and pleotelson. Antennule thicker than that of female, with extra, reduced fifth article, without numerous aesthetascs. Functional mouthparts retained, resembling, but not identical to, those of female; maxillule with fewer spiniform terminal setae. Labium with less pronounced distal processes. Pleopods present but reduced, with simple setae, fused to an immobile cone. Pleon always as wide as pereon or wider.

Gender of generic name. — Masculine.

Type species. — *Paragathotanais typicus* Lang, 1971.

Remarks. — *Paragathotanais* was initially described as having a 5-articulate antenna (Lang, 1971a) while Bird & Holdich (1988) claimed that *Paragathotanais* have a 6-articulate antenna. The number of antenna articles separates *Paragathotanais* from *Agathotanais*, which has a reduced antenna. Larsen (1999b) observed considerable variation in the reduction of the antenna in *Agathotanais* and also in *Paragathotanais* (cf. Larsen, 2002a), thus making this character subject to variation in both genera.

Reduction in width of the pleonites is usually a good diagnostic character of the females, but ontogenetic variations were observed during this study in the following respects: the larger the female, the less reduced pleonites, and in large females the pleonites are as wide as the pereonites and pleotelson. For the male, thc width of the pleonites seems constant, irrespective of ontogenetic stage.

Paragathotanais is a genus with a cosmopolitan distribution, many species, and conservative morphology. One character, the spination of the dactylus of pereopods 4–6, seems very promising and can be observed without dissection; a rare commodity indeed in tanaidacean systematics.

Species currently assigned to this genus. — *P. gracilis* Bird & Holdich, 1988; *P. macrocephalus* Kudinova-Pasternak, 1986; *P. medius* Larsen, 2002;

P. nanus Bird & Holdich, 1988; *P. robustus* Bird & Holdich, 1988; *P. spinosus* sp. nov.; *P. typicus* Lang, 1971.

Paragathotanais medius Larsen, 2002

Paragathotanais medius Larsen, 2002, Gulf and Caribbean Research, **14**: 61.

Diagnosis. — (Modified from Larsen, 2002a.) Female. Carapace shorter than the lengths of pereonites 1 and 2 combined. Pereonite shoulders inconspicuous. Pereonite 4 longer than wide. Pleon narrower than pereon and pleotelson. Antenna with four articles, article 3 with distal fusion line. Labium without lateral setae. Pereopods 4–6 with dactylus without spines.

Male. Antennule thicker than in female. Pleopods present. Pleon wider than in female (as wide as or wider than pereon and pleotelson).

Distribution. — A species only recorded from the northern part of the Gulf. Bathymetric range: 213–3007 m. See map 18.

Remarks. — Larsen (2002a: 66) remarked that the presence of a cheliped pseudocoxa is surprising and unexplainable. Re-examination of the type and other material has revealed that the cheliped 'pseudocoxa' is an artifact, caused by an attached sclerite from the carapace.

Paragathotanais spinosus sp. nov.
(figs. 22, 23)

Material examined. — Holotype, non-ovigerous ♀ (USNM 1068861), body length 2.5 mm, DGoMB Sta. S36-1. Paratypes: 1 non-ovigerous ♀ (USNM 1068862), body length 2.0 mm, DGoMB Sta. MT3-4. 1 non-ovigerous ♀ (USNM 1068863), DGoMB Sta. S36-1 (dissected).

Additional material. — 1 juvenile ♀, 1 manca III, DGoMB Sta. MT3-4. 1 manca III, DGoMB Sta. S37-4. 1 non-ovigerous ♀ (badly damaged), DGoMB Sta. S35-4.

Diagnosis. — Female. Carapace shorter than the lengths of pereonites 1–2 combined. Pereonite shoulders inconspicuous. Pereonite 4 longer than wide. Antennule with four articles but with abruptly narrowing apex (rudiment of fifth article). Antenna with five articles; article 3 without distal fusion line. Labium without lateral setae. Pereopods 4–6 with dactylus with one long distal spine (reaching beyond insertion of unguis) and two more proximal smaller spines (less than 0.5 times as long as distal spine). Pleon as wide as pereon and pleotelson.

Male. Unknown.

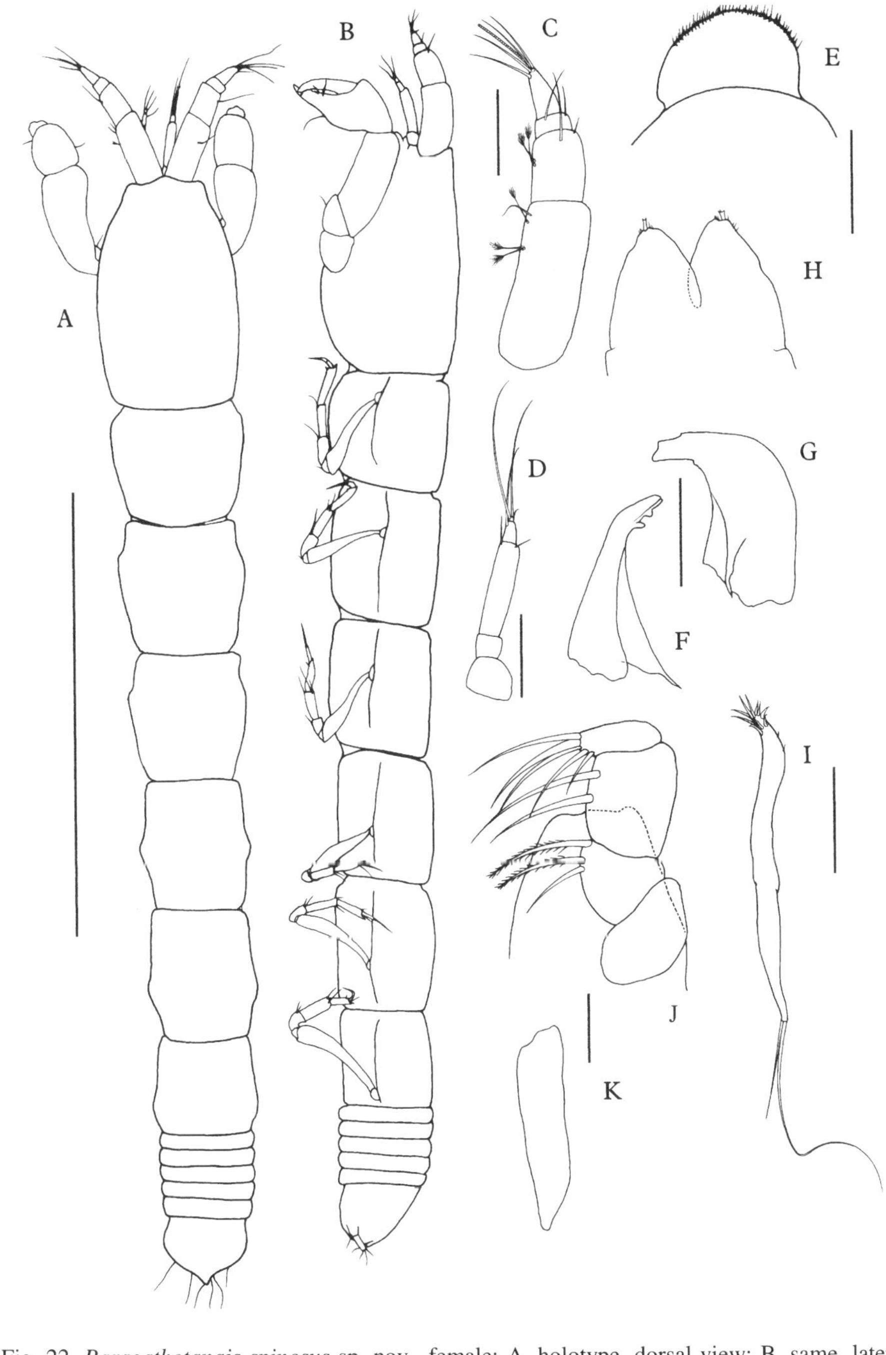

Fig. 22. *Paragathotanais spinosus* sp. nov., female; A, holotype, dorsal view; B, same, lateral view; C, antennule; D, antenna; E, labrum; F, left mandible; G, right mandible; H, labium; I, maxillule; J, maxilliped; K, epignath. Scale bars equal: A, B, 1.0 mm; C-D, 0.1 mm; E-K, 0.05 mm.

Etymology. — Named after the many pereopodal propodus and dactylus spines (Latin, spinosus = spiny). The name is an adjective agreeing in gender with the generic name.

Type locality. — DGoMB Sta. S36-1, 28°55'1647"N 87°40'2232"W, depth 1825 m.

Description. — Adult female. Body (fig. 23A, B) 7.7 times as long as wide.

Carapace hardly shorter than the lengths of pereonites 1 and 2 combined, almost rectangular.

Pereon with pereonite shoulders inconspicuous. Pereonites 1 and 6 wider than long. Pereonite 2 length subequal to width. Pereonites 3, 4, and 5 longer than wide.

Pleon with all pleonites subequal and as wide as pereonites and pleon. Pleotelson with posterior dorsal plate.

Antennule (fig. 22C) with four articles but with abruptly narrowing apex (rudiment of a fifth article). Length less than that of carapace. Article 1 longer than rest of antennule, with one simple and three setulate setae. Article 2 longer than article 4, with two simple and two setulate setae distally. Article 3 band-shaped and shorter than other articles, with two simple setae distally. Article 4 less than 3 times as long as article 1, with four simple setae and one aesthetasc distally.

Antenna (fig. 22D) with five articles. Hardly shorter than the lengths of antennular articles 1 and 2 combined. Article 1 twice as wide as other articles, naked. Article 2 square, naked. Article 3 longer than the lengths of other antenna articles combined, without fusion line, with two simple setae distally. Article 4 as long as article 2, tapering, and with one simple seta distally. Article 5 minute, with two long and one tiny setae distally.

Mouthparts. Labrum (fig. 22E) hood-shaped, with setose distal edge. Left mandible (fig. 22F) lacinia mobilis short and shaped as a blunt spine; incisor broad and bifurcate; molar apparently without denticles. Right mandible (fig. 22G) incisor broader than on left mandible, with three denticles. Labium (fig. 22H) tapering distally, with distal setules and one complex seta on each lobe. Maxillule (fig. 22I) endite with ten spiniform setae distally; palp as long as endite, with two setae of unequal length. Maxilla not recovered. Maxilliped (fig. 22J) endites with uneven distal margin, without setae; palp article 1 naked, article 2 with three setulose inner setae, article 3 with two inner setae, article 4 with four setae distally. Epignath (fig. 22K) without setules.

Cheliped (fig. 23G) basis shorter but wider than merus. Merus triangular, with one seta ventrally. Carpus longer than propodus including fixed finger,

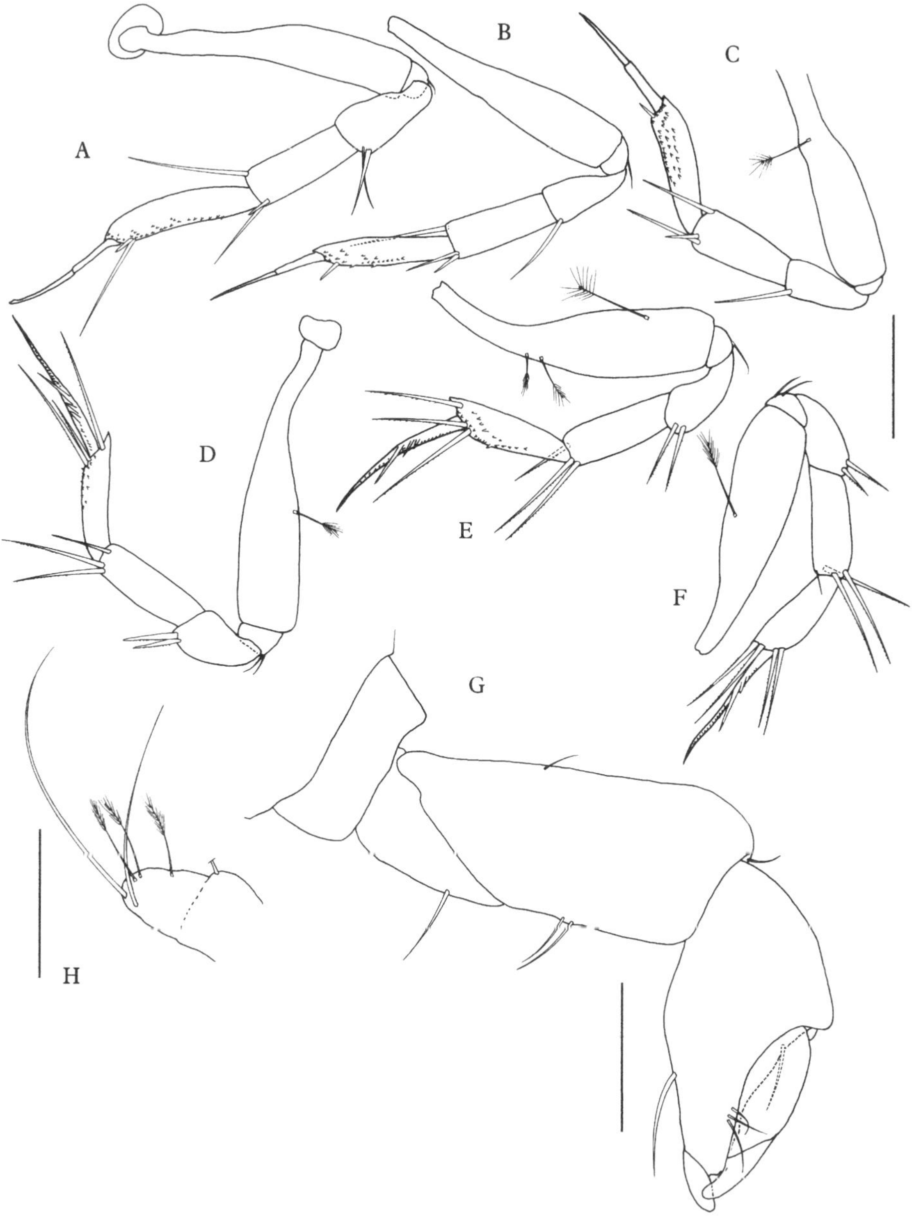

Fig. 23. *Paragathotanais spinosus* sp. nov., female; A, pereopod 1; B, pereopod 2; C, pereopod 3; D, pereopod 4; E, pereopod 5; F, pereopod 6; G, cheliped; H, uropod. Scale bars equal: 0.1 mm.

with one seta dorso-medially and one seta dorso-distally and with two setae ventrally. Propodus with one seta at dactylus insertion. Fixed finger with one seta ventrally, and three setae and one denticle distally on inner margin. Dactylus as long as fixed finger.

Pereopod 1 (fig. 23A) coxa naked. Basis hardly shorter than the lengths of the four succeeding articles combined, naked. Ischium with one seta ventrally. Merus shorter than carpus, widening distally, with two stiff, simple setae ventrally. Carpus shorter than propodus, rectangular, with two stiff simple and one spiniform setae distally. Propodus about 0.5 times as long as basis, with one stiff and one spiniform seta ventro-distally and one spine dorsally, ventral edge covered with small spines. Dactylus and unguis combined shorter than propodus, unguis with distal spatulate process.

Pereopod 2 (fig. 23B) as pereopod 1 except: merus with one stiff seta. Propodus with one spiniform simple seta ventrally and a spine dorsally, covered with small spines. Unguis without distal spatulate process.

Pereopod 3 (fig. 23C) as pereopod 1 except: basis with one setulate seta dorsally.

Pereopod 4 (fig. 23D) basis not wider than those of pereopods 1–3, with one setulate seta ventrally. Ischium with two setae ventrally. Merus with two spiniform pinnate setae, shorter than carpus. Carpus with three stiff pinnate setae distally. Propodus with three pinnate, long (more than 0.5 times as long as dactylus/unguis) setae distally and a spine dorsally, ventral margin covered with small spines. Dactylus and unguis combined longer than propodus and with ventral row of spinules, dactylus with two small proximal spines and one long spine (2 times as long as proximal spines).

Pereopod 5 (fig. 23E) as pereopod 4 except: basis with two simple dorsal and one setulate setae ventrally. Ischium with one seta ventrally.

Pereopod 6 (fig. 23F) as pereopod 4 except: propodus with four pinnate, long (more than 0.5 times as long as dactylus/unguis) setae distally and a spine dorsally, ventral margin covered with small spines.

Pleopods lacking in females.

Uropod (fig. 23H) basal article wider but shorter than endopod. Exopod process almost completely reduced, with one simple seta distally. Endopod with two long simple setae distally, one seta medially, and two setulate setae distally.

Distribution. — Only found in the northern part of the Gulf. Bathymetric range: 658–2382 m. See map 19.

Remarks. — The condition of pereopods 4–6 having a dactylus with one long distal spine and two more proximal smaller spines is unique for this species. The characters of the dactylus spination are easy to use and do not require dissection. These features are likely to be adopted for widespread use in species identifications of this otherwise difficult genus.

Paragathotanais typicus Lang, 1971
(figs. 24–26)

Paragathotanais typicus Lang, 1971, Crustaceana, **21**: 65.

Material examined. — 1 non-ovigerous ♀, DGoMB Sta. S42-4. 1 ovigerous ♀ (damaged), DGoMB Sta. RW1-4. 1 ♂ (immature), DGoMB Sta. C4-4.

Diagnosis. — Female. Carapace as long as pereonites 1 and 2 combined. Pereonite shoulders inconspicuous. Pereonite 4 square. Antenna with five articles; article 3 without distal fusion line; article 2 without spinules. Labium with lateral setae. Pereopods 4–6 with dactylus without spines. Pleon narrower than pereon and pleotelson.

Male. (Modified after Lang, 1971a.) Antenna with five articles; article 3 without fusion line.

Description. — Adult female. Body (fig. 24B) 6.5 times as long as wide.

Carapace as long as lengths of pereonites 1 and 2 combined, almost rectangular.

Pereon with pereonite shoulders present but not conspicuous. Pereonites 1, 2, and 6 wider than long. Pereonites 3 and 4 longer than wide. Pereonite 5 square.

Pleon with all pleonites subequal and hardly narrower than pereonites, but as wide as pleotelson. Pleotelson with posterior dorsal plate.

Antennule (fig. 24C) hardly more than 0.5 times as long as carapace, with four articles. Article 1 longer than rest of antennule, with one simple seta distally. Article 2 longer than article 4, with two simple setae distally. Article 3 band-shaped and shorter than other articles, with one simple seta distally. Article 4 less than three times article 1, with four simple setae distally.

Antenna (fig. 24D) shorter than antennular article 1, with five articles. Article 1 wider than other articles, with one seta dorsally. Article 2 square, with one seta dorsally. Article 3 as long as lengths of other antenna articles combined, without fusion line, naked. Article 4 as long as article 2, tapering, and with one simple seta distally. Article 5 minute, with two long and one tiny setae distally.

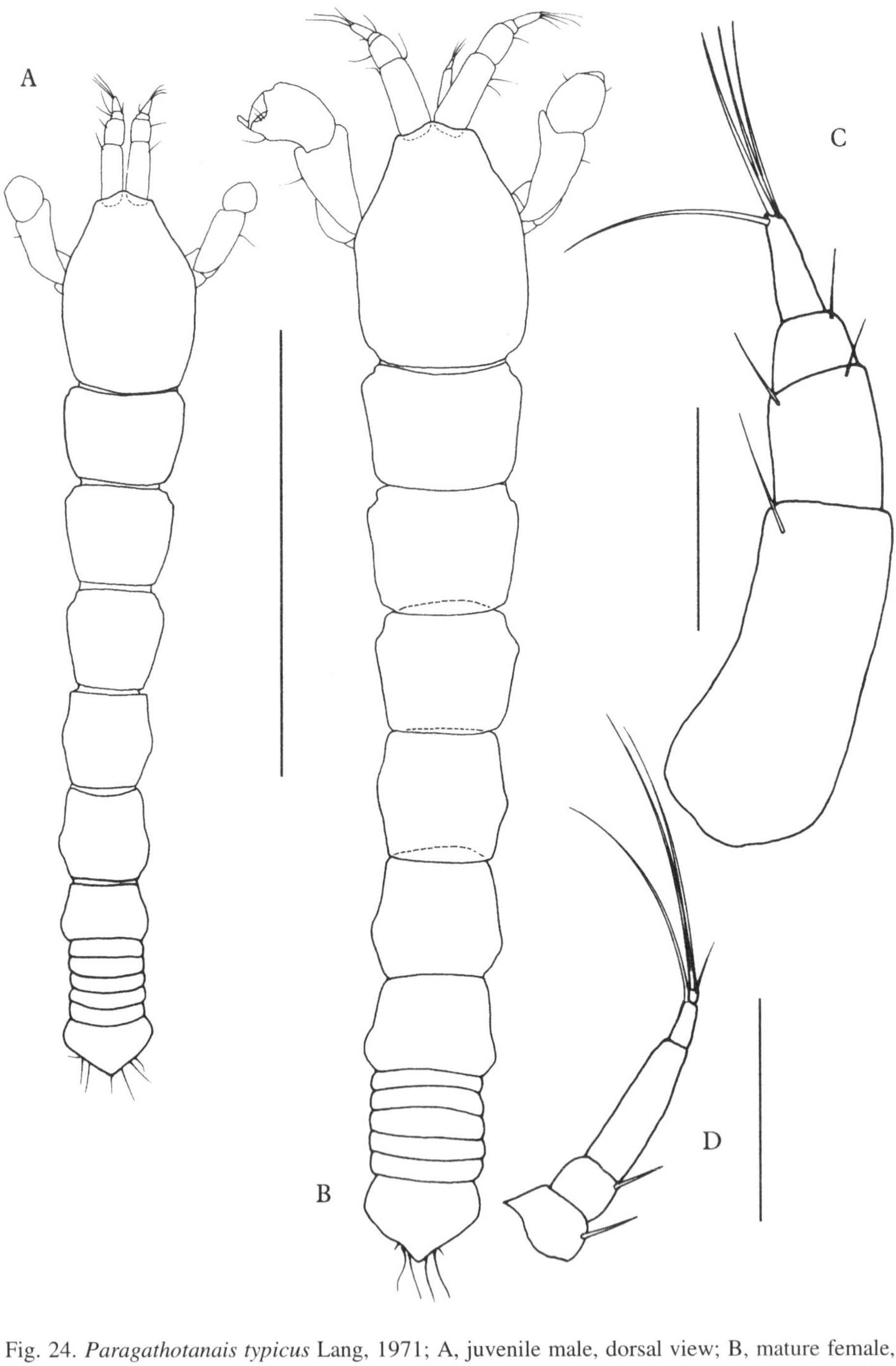

Fig. 24. *Paragathotanais typicus* Lang, 1971; A, juvenile male, dorsal view; B, mature female, dorsal view; C, antennule, female; D, antenna, female. Scale bars equal: A, B, 1.0 mm; C-D, 0.1 mm.

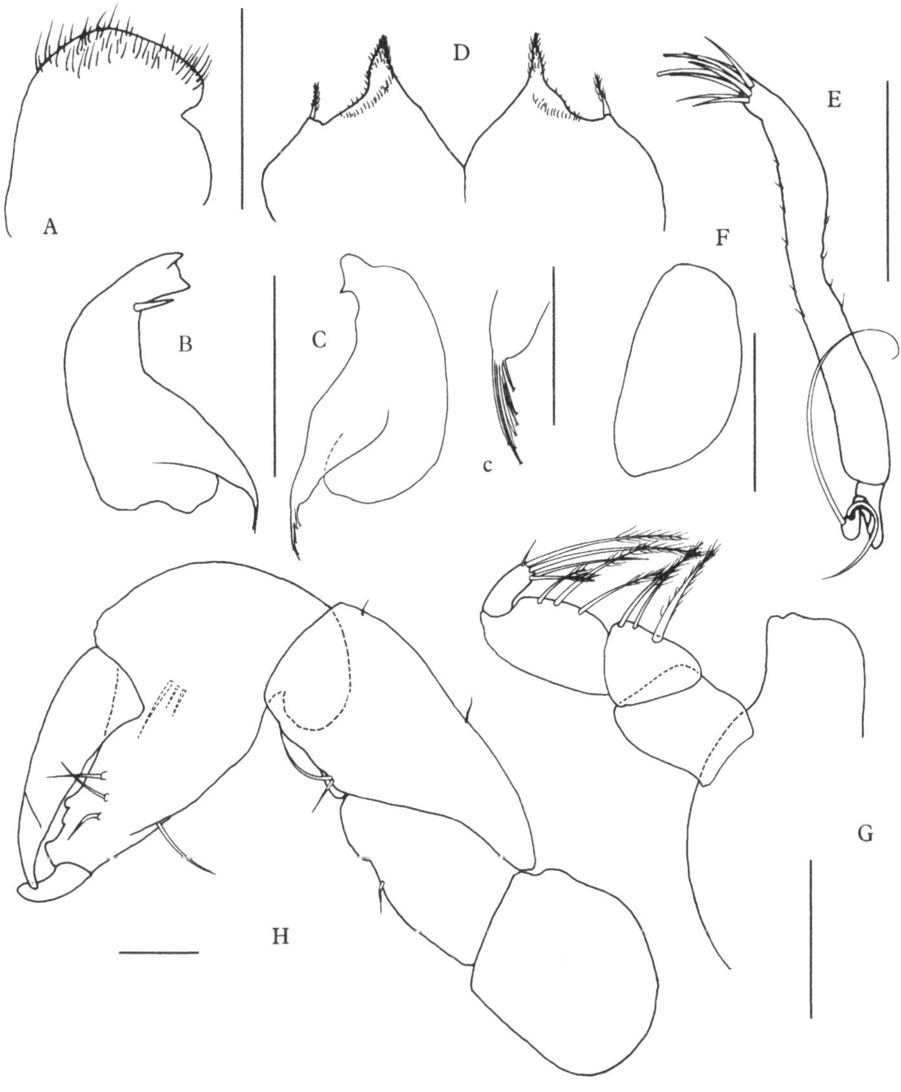

Fig. 25. *Paragathotanais typicus* Lang, 1971, female; A, labrum; B, left mandible; C, right mandible; c, same, distal part of molar; D, labium; E, maxillule; F, maxilla; G, maxilliped; H, chclipcd. Scalc bars equal: 0.05 mm.

Mouthparts. Labrum (fig. 25A) hood-shaped, with setose distal edge. Mandibular molar with peculiar array of bi- or trifurcate setae (fig. 25c). Left mandible (fig. 25B) lacinia mobilis short and shaped as a blunt spine; incisor broad and bifurcate. Right mandible (fig. 25C) incisor wide and with medial depression. Labium (fig. 25D) tapering distally into groove-shaped, setose, pointed apex, with one setulose lateral seta on each lobe. Maxillule (fig. 25E) endite with eight spiniform setae distally; palp less than 0.5 times as long as endite (curled on fig. 25E), with two terminal setae of unequal length. Maxilla (fig. 25E) ovoid and without any special features. Maxilliped (fig. 25G) endites with uneven distal margin, without setae; palp article 1 naked, article 2 with three setulose inner setae; article 3 with three inner setae; article 4 with five distal setae. Epignath not recovered.

Cheliped (fig. 25H) basis shorter but wider than merus. Merus triangular, with one seta ventrally. Carpus shorter than propodus including fixed finger, with two setae dorsally and two ventrally. Propodus with three setae at dactylus insertion. Fixed finger with one seta ventrally, three setae and three denticles distally on inner margin. Dactylus as long as fixed finger.

Pereopod 1 (fig. 26A) coxa with one seta. Basis longer than the lengths of the three succeeding articles combined, naked. Ischium with one seta ventrally. Merus shorter than carpus, widening distally, with one stiff pinnate and one simple setae ventrally. Carpus shorter than propodus, rectangular, with two stiff pinnate setae distally. Propodus about 0.5 times as long as basis, with one stiff pinnate and one spiniform seta ventro-distally and a spine dorsally. Dactylus and unguis combined shorter than propodus.

Pereopod 2 (fig. 26B) as pereopod 1 except: merus with one stiff pinnate seta. Carpus with three stiff pinnate setae distally. Propodus with one spiniform seta ventrally and one spine dorsally.

Pereopod 3 (fig. 26C) as pereopod 2 except: merus with one stiff pinnate and one simple seta. Propodus covered with spiniform scales.

Pereopod 4 (fig. 26D) basis with one setulate seta ventrally. Ischium with two setae ventrally. Merus with two stiff pinnate setae, shorter than carpus. Carpus with two stiff pinnate and one simple setae distally. Propodus with three pinnate setae, more than 0.5 times as long as dactylus/unguis, distally and a spine dorsally. Dactylus and unguis combined longer than propodus and with ventral row of spinules.

Pereopod 5 (fig. 26E) as pereopod 4 except: basis naked. Ischium with one seta ventrally. Carpus with three stiff pinnate setae distally.

Pereopod 6 (fig. 26F) as pereopod 4 except: basis naked. Propodus with four stiff pinnate setae distally and one spine dorsally.

Pleopods lacking in females.

Uropod (fig. 26G) basal article wider but shorter than endopod. Exopod process completely reduced, with one simple seta distally. Endopod with two long simple setae distally and three setae subdistally.

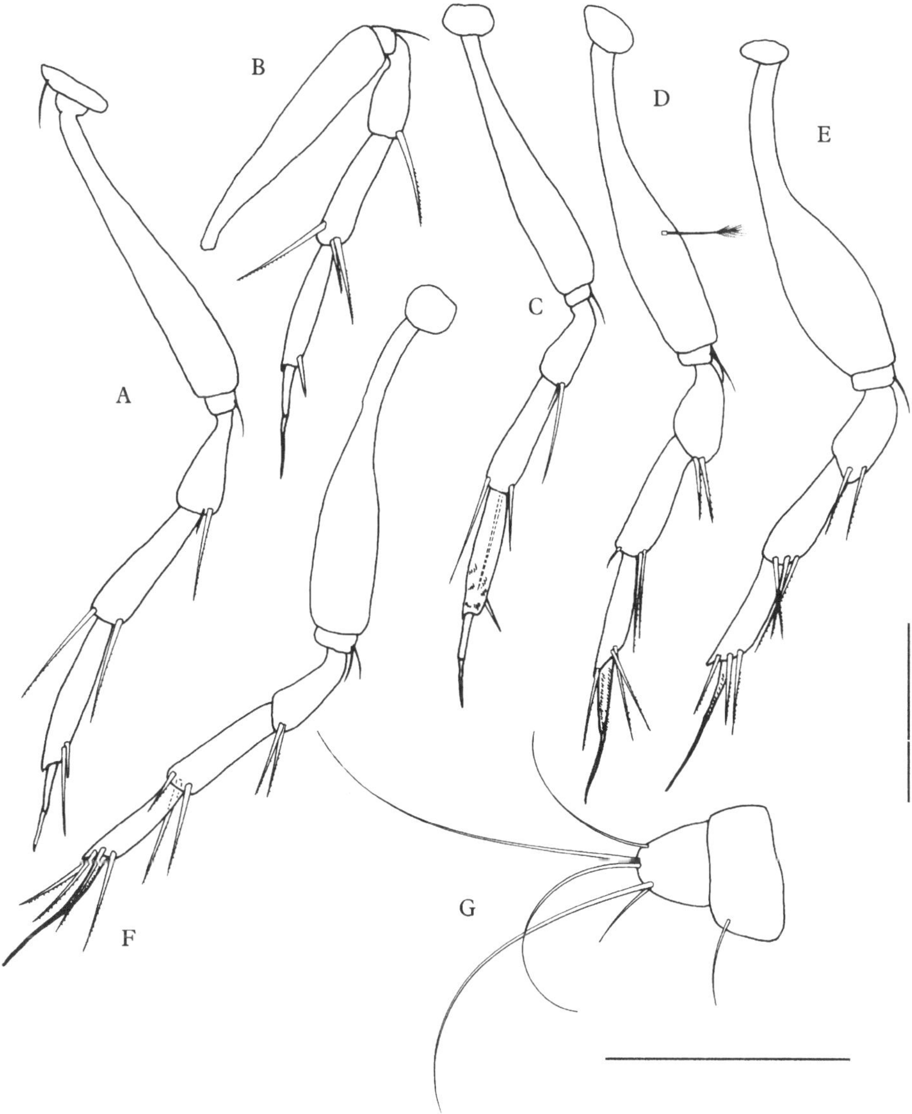

Fig. 26. *Paragathotanais typicus* Lang, 1971, female; A, pereopod 1; B, pereopod 2; C, pereopod 3; D, pereopod 4; E, pereopod 5; F, pereopod 6; G, uropod. Scale bars equal: 0.1 mm.

Distribution. — This species was originally described from the Caribbean Sea but was found in the Gulf during this project. *Paragathotanais typicus* was also reported from the NGoMCSS study. Bathymetric range: 212–3659 m. See map 20.

Remarks. — Since Lang (1971) never described the female or prepared a 'whole body' illustration of the male, a full description of the female and a body illustration of the juvenile male are presented here. A neotype has not been designated, since the specimens used in this study do not come from the actual type locality. Despite the large difference between the depth of the type locality and that of the collection site in the Gulf, everything else suggests that this is indeed *P. typicus*. *Paragathotanais typicus* was also recorded south of Cuba and off Panama (Caribbean side) by Kudinova-Pasternak & Pasternak (1978).

Paragathotanais sp. A
(fig. 27)

Material examined. — 1 ♂, DGoMB Sta. DGoMB Sta. C7-4.

Diagnosis. — Male. Pereonite shoulders inconspicuous. Pereopods 4–6 with dactylus with one long ventral spine and one smaller dorsal spine.

Female. Unknown.

Distribution. — Only found at one station in the northern part of the Gulf. Bathymetric range: 1079 m. See map 21.

Remarks. — This species is not described since the material consists of only one male specimen. This species is, however, easily recognizable by the dactylus of pereopods 4–6 having one long, ventral spine and one smaller dorsal spine, giving the animal the appearance of having a tri-spinose unguis. This may be the male of *P. spinosus*, but the habitus and the pereopodal dactylus spines are different and suggest that it is a new species.

KEY TO THE SPECIES OF *PARAGATHOTANAIS*, FEMALES ONLY (GLOBAL)

1. Pereopods 4–6 with dactylus with spines 2
– Pereopods 4–6 with dactylus without spines 5
2. Antenna article 2 with spinules .... *P. gracilis* Bird & Holdich, 1988
– Antenna article 2 without spinules 3
3. Carapace shorter than the lengths of first two pereonites combined. Pereopods 4–6 with dactylus with two or three spines 4
– Carapace longer than the lengths of first two pereonites combined. Pereopods 4–6 with dactylus with only one spine .... *P. macrocephalus* Kudinova-Pasternak, 1986
4. Pereopods 4–6 with dactylus with one long spine (twice as long as other spines) and two small ventral spines .... *P. spinosus* sp. nov.
– Pereopods 4–6 with dactylus with one long spine (twice as long as other spines), one small ventral, and one small dorsal spine .... *P.* sp. A
5. Cheliped stout (length/width ratio of carpus 1.5). Pereonite shoulders absent .... *P. nanus* Bird & Holdich, 1988

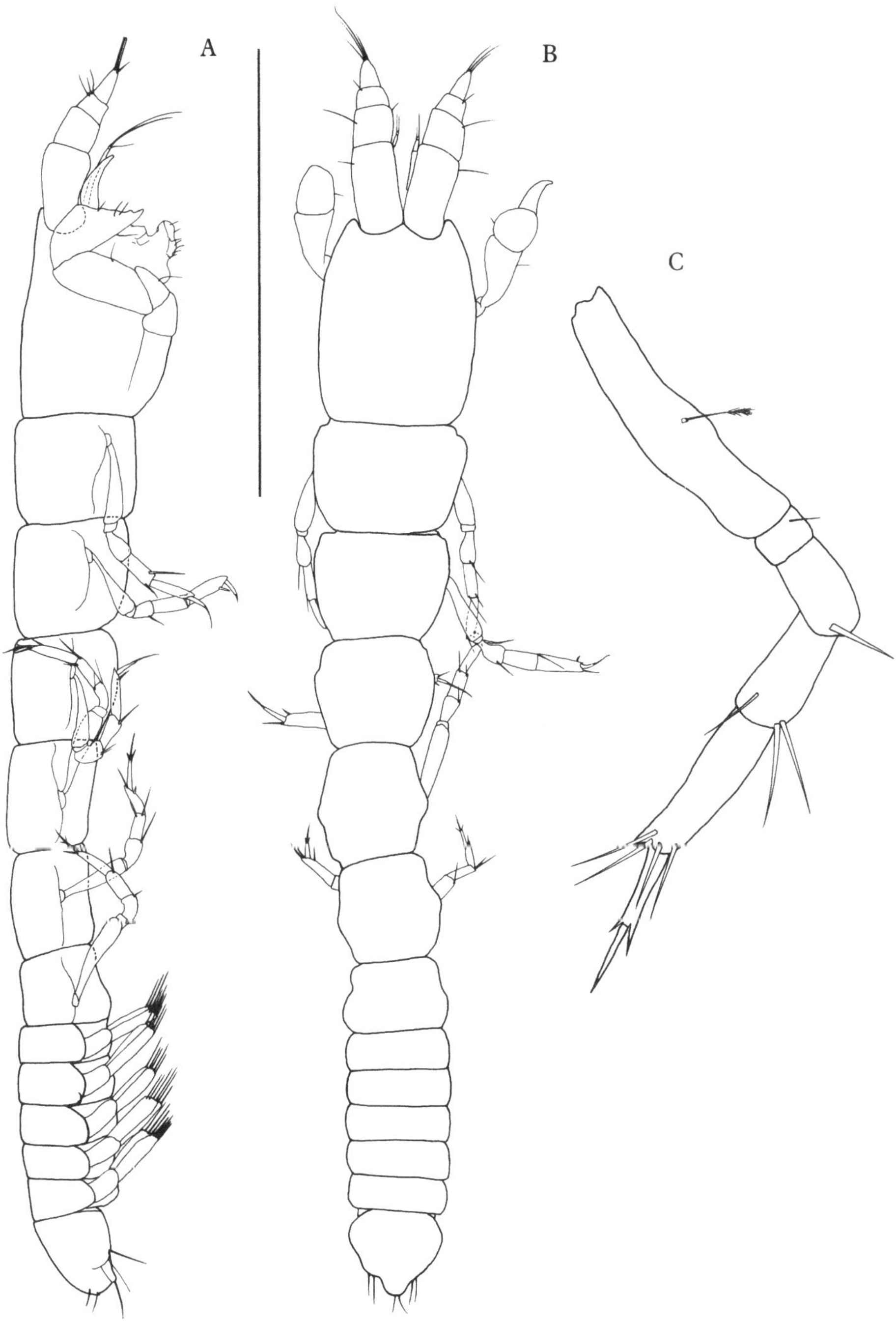

Fig. 27. *Paragathotanais* sp. A, male; A, body, dorsal view; B, same, lateral view; C, pereopod 5. Scale bar equals: for A, B, 1.0 mm; C, 0.1 mm.

– Cheliped slender (length/width ratio of carpus 2.3). Pereonite shoulders present 6
6. Pereonites 4 and 5 wider than long .... *P. robustus* Bird & Holdich, 1988
– Pereonites 4 and 5 square or longer than wide 7
7. Pereonite 4 longer than wide. Antenna with four articles. Labium without lateral setae .... *P. medius* Larsen, 2002
– Pereonite 4 square. Antenna with five or more articles. Labium with lateral setae. .... *P. typicus* Lang, 1971

Genus **Paranarthrura** Hansen, 1913

Paranarthrura Hansen, 1913, Danish Ingolf Expedition, **3**(3): 121.

Diagnosis. — (Modified after Bird & Holdich, 1989a.) Female. Cuticle heavily calcified, not setulose. Eyes and eye-lobes absent. Pereon with six free pereonites. Pleon with five free pleonites. Antennule with four articles. Antenna with 5–6 articles. Molar process of mandible reduced. Labium lobes pointed. Cheliped attachment via ventral pseudocoxa. Marsupium (where known) with four pairs of oostegites. Pereopod coxa present. Pereopods 4–6 with basis not thickened relative to bases of pereopods 1–3; dactylus and unguis not fused. Pleopods absent. Uropod about as long as pleotelson and clearly visible from dorsal side; endopod with 1–2 articles and not fused with basal article; exopod process clearly visible, acute (shaped as a spur), but fused with basal article.

Male. Functional mouthparts retained. Antennule thicker than that of female, without numerous aesthetascs. Pleopods present but coalesced into immobile cone and with simple setae only. Pleon always as wide as pereon or wider.

Gender of generic name. — Feminine.

Type species. — *Paranarthrura insignis* Hansen, 1913.

Remarks. — Most generic characters are stable and the genus is easily recognizable. However, the pleon is reported to be narrower than the pereon in females (Bird & Holdich, 1989a). This is not always easily identifiable and appears to vary even within one species. Within the genus, the shapes of the chelipeds, body, and uropods vary but, as a generic character, the process of the exopod of the uropod is decisive.

Species identification within this genus is difficult. Useful characters are: (1) uropod articulation, (2) uropodal exopod length, (3) uropod total length, and (4) cheliped shape.

The male of *Paranarthrura* shares the same typically male characters with the males of *Agathotanais*, *Paragathotanais*, *Collettea*, *Tanaella*, and *Parafilitanais* (cf. Larsen, 1999b, 2001; Larsen & Heard, 2004a).

In addition to the species described from the Gulf below, at least seven other new species were found in the DGoMB material. Some of these had quite interesting features, but unfortunately the number of the specimens was too low and their condition too poor, for full taxonomic treatment.

Species currently assigned to this genus. — *P. angolensis* Guerrero-Kommritz, Schmidt & Brandt, 2002; *P. bacescui* Kudinova-Pasternak, 1986; *P. bispinosa* sp. nov.; *P. borealis* Bird & Holdich,1989; *P. crassa* Bird & Holdich,1989; *P. fortispina* Sieg, 1986; *P. insignis* Hansen, 1913; *P. intermedia* Kudinova-Pasternak, 1982; *P. kurchatovi* Kudinova-Pasternak, 1975; *P. lusitanus* Bird & Holdich,1989; *P. meridionalis* Sieg, 1986; *P. similis* Lang, 1971; *P. subtilis* Hansen, 1913; *P. tenuimanus* sp. nov.; *P. tridens* Bird & Holdich, 1989; *P. undulata* Lang, 1968; *P. vitjazi* Kudinova-Pasternak, 1970; *P. zevinae* Kudinova-Pasternak, 1970.

Paranarthrura bispinosa sp. nov.
(figs. 28–32)

Material examined. — Holotype, non-ovigerous ♀ (USNM 1068864), body length 2.5 mm, DGoMB Sta. MT5-2. Paratypes: 1 non-ovigerous ♀ (USNM 1068865), body length 2.9 mm, same locality. 1 (immature) ♂ (USNM 1068866), body length 2.15 mm, DGoMB Sta. W5-2. 1 non-ovigerous ♀ (USNM 1068868), DGoMB Sta. MT5-2 (dissected). 1 ♂ (USNM 1068867), DGoMB Sta. AC1-3 (dissected).

Additional material. — 1 non-ovigerous ♀, 7 mancas, DGoMB Sta. MT5-1. 1 non-ovigerous ♀, 1 manca, DGoMB Sta. NB4-2. 1 non-ovigerous ♀, 1 ♂ (immature), 3 mancas, DGoMB Sta. S37-1, 1 non-ovigerous ♀, 1 manca, DGoMB Sta. AC1–3. 2 non-ovigerous ♀♀, DGoMB Sta. NB5-1. 1 non-ovigerous ♀, DGoMB Sta. MT6-2.

Diagnosis. — Female. Carapace in dorsal view with smooth lateral edges; longer than wide; shorter than pereonites 1 and 2 combined. Pereonite 1 wider than long. Antennular article 1 shorter than rest of antennule. Cheliped propodus with inferior margin 1.4 times as long as carpus; fixed finger and apex of dactylus not thickened. Pereopods 1–3 with unguis without dorsal process but with sharp reduction in width at midlength. Pereopod 4–6 propodi with one short, very stout distal spiniform seta dorsally next to dorsal spine; dactyli without setae. Pleon almost as wide as pereon. Pleotelson acorn-shaped with pointed apex. Uropods longer than pleotelson, endopod bi-articulate, longer than basal article. Uropodal exopod 0.25 times as long as endopod.

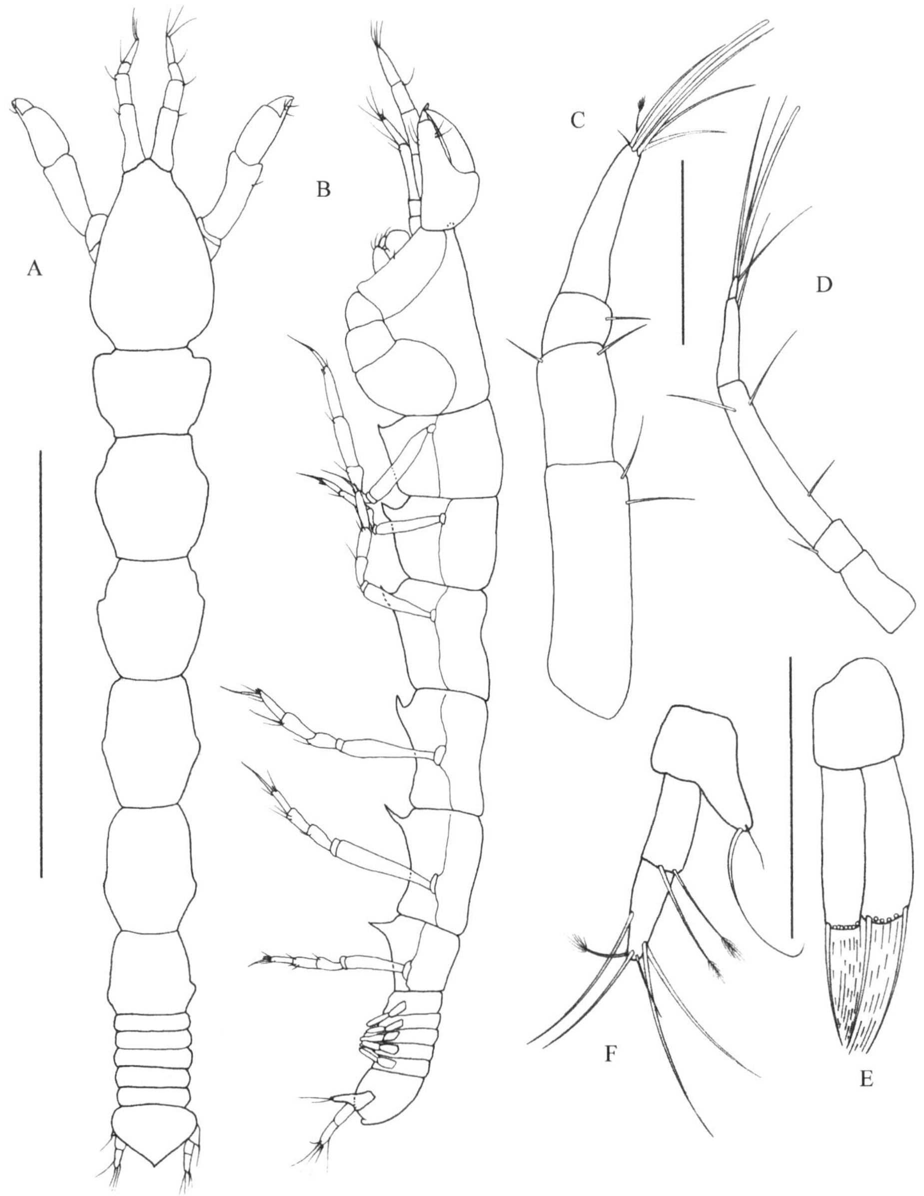

Fig. 28. *Paranarthrura bispinosa* sp. nov.; A, holotype, female, dorsal view; B, paratype, male, lateral view; C, antennule, male; D, antenna, male; E, pleopod, male; F, uropod, male. Scale bars equal: A, B, 1.0 mm; C-F, 0.1 mm.

Male. Essentially as female except for the presence of pleopods with simple setae.

Etymology. — Named after the presence of the short, stout spiniform seta next to the dorsal spine on the propodus of pereopods 4–6, giving the impression of a double spine: (Latin, bis = two, spinosus = with spines). The name is thus an adjective agreeing in gender with the (feminine) generic name.

Type locality. — DGoMB Sta. MT5-2, 27°19'5819"N 88°40'1733"W, depth 2290 m.

Description. — Adult female. Body (fig. 28A) 7.7 times as long as wide. Lateral shoulders weakly defined.

Carapace with smooth lateral edges in dorsal view. Longer than wide. Shorter than the lengths of pereonites 1 and 2 combined.

Pereonite 1 wider than long. Pereonites 2–5 longer than wide. Pereonite 6 as long as wide.

Pleon short (including pleotelson only 0.15 times total body length). All pleonites subequal. Pleotelson longer than the lengths of three free pleonites combined, acorn-shaped, apex pointed and covered by dorsal plate.

Antennule (fig. 29A) hardly shorter than carapace, with four articles. Article 1 as long as rest of antennule, with five setulated setae distally. Article 2 twice as long as article 3, with two setulated and one simple setae distally. Article 3 shorter than other articles, with one simple seta distally. Article 4 longer than article 2, with three simple setae and one aesthetasc distally.

Antenna (fig. 29B) with five articles, 0.8 times as long as antennule. Article 1 longer than article 2, with one simple seta dorsally. Article 2 is 0.5 times as long as article 4, with one seta dorso-distally. Article 3 longer than other articles, with one seta medially and two setulated and one simple setae distally. Article 4 shorter than article 1, with one setulated seta medially and one simple seta distally. Article 5 minute, with two long setae and one aesthetasc distally.

Mouthparts. Labrum (fig. 29C) smooth, hood-shaped. Mandibular molar indistinct and membranous. Left mandible (fig. 29D) lacinia mobilis longer than incisor; incisor broad with two denticles. Right mandible (fig. 29E) incisor blunt with three denticles. Labium (fig. 29F) lobes pointed. Maxillule (fig. 29G) endite with eight spiniform setae distally; palp not recovered. Maxilla not recovered. Maxilliped (fig. 29H) endites with medial process; palp article 1 naked, articles 2 and 3 with three setae on inner margin, article 4 only 0.5 times as wide as article 3, with five setae. Epignath not recovered.

Cheliped (fig. 30A) pseudocoxa more than 3 times as long as basis. Merus prominent, naked. Carpus only 0.6 times as long as propodus including fixed finger, with two setae dorsally and one simple seta ventrally. Propodus with

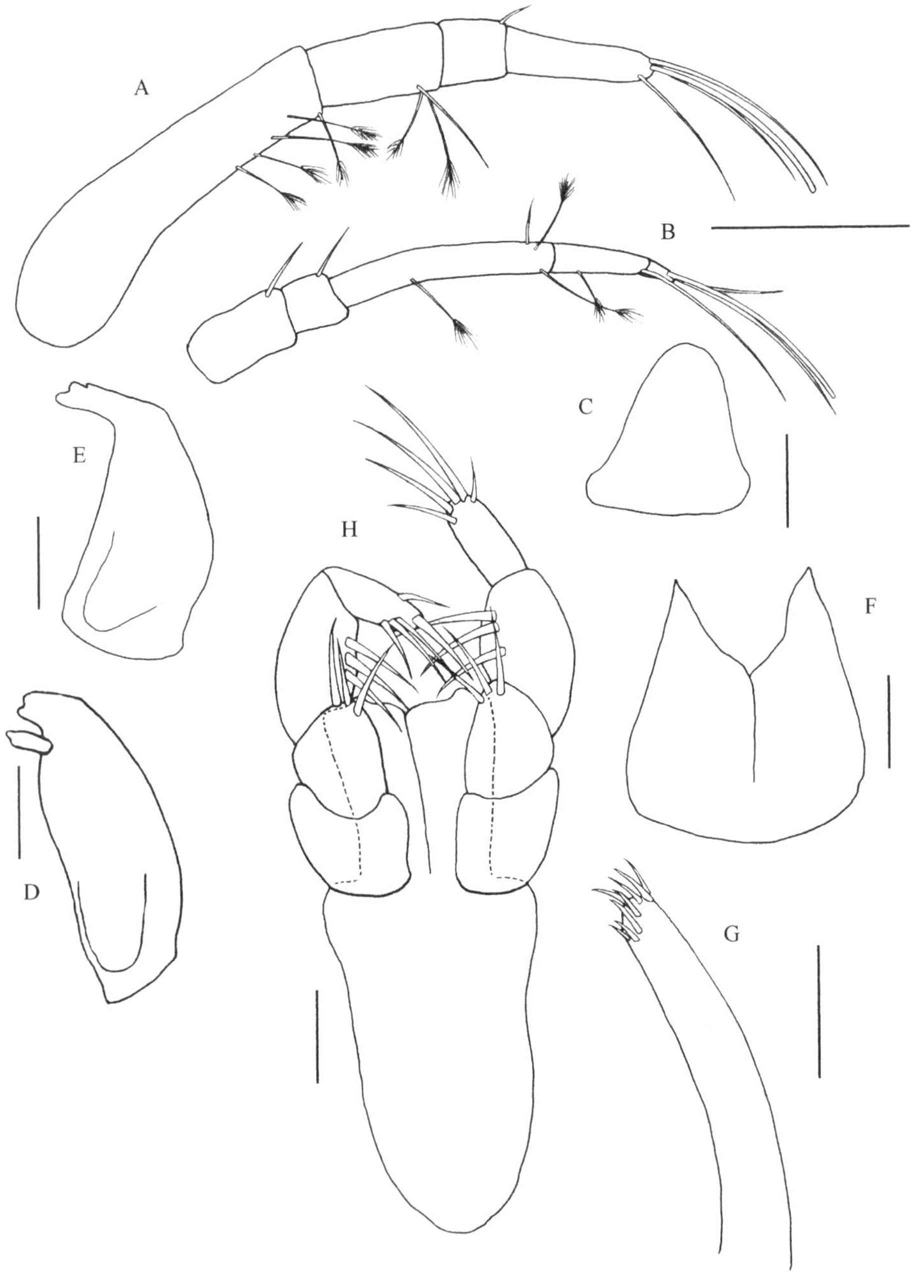

Fig. 29. *Paranarthrura bispinosa* sp. nov., female; A, antennule; B, antenna; C, labrum; D, left mandible; E, right mandible; F, labium; G, maxillule; H, maxilliped. Scale bars equal: A-B, 0.1 mm; C-H, 0.05 mm.

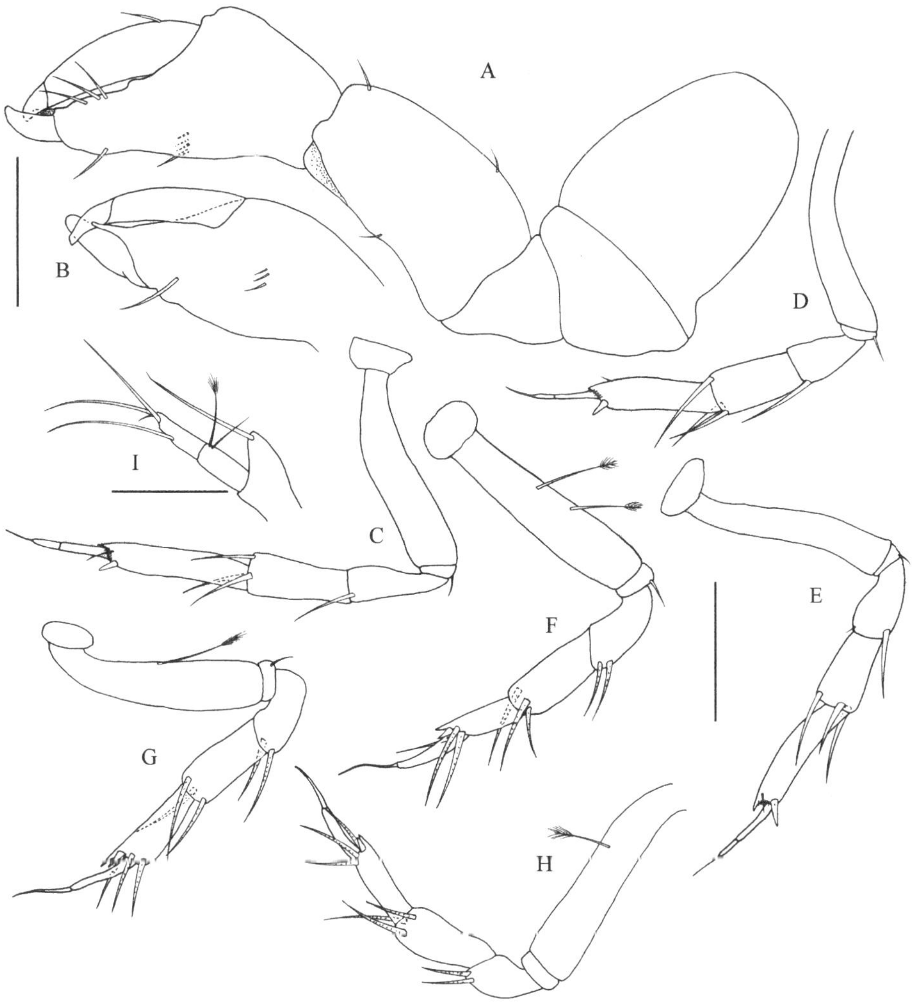

Fig. 30. *Paranarthrura bispinosa* sp. nov., female; A, left cheliped, outer view; B, right cheliped, inner view; C, pereopod 1; D, pereopod 2; E, pereopod 3; F, pereopod 4; G, pereopod 5; H, pereopod 6; I, uropod. Scale bars equal: 0.1 mm.

inner row of robust setae. Fixed finger with one seta ventrally and three setae on inner margin. Dactylus as long as fixed finger and apex not thickened.

Pereopods 1–3 (fig. 30C-E) with coxa rounded. Basis as long as the three succeeding articles combined, smooth. Ischium with one simple seta. Merus more than 0.5 times as long as carpus, widening distally, with one simple seta ventro-distally. Carpus 0.5 times as long as basis, with three setae distally. Propodus with one stout spiniform seta ventro-distally, one simple seta, a spine dorsally, and row of setules. Dactylus and unguis combined shorter than propodus and not fused, unguis with abrupt decline in width halfway its length.

Pereopod 4 (fig. 30F) with coxa rounded. Basis longer than the three succeeding articles combined, with two setulated setae medially. Ischium with one seta. Merus more than 0.5 times carpus, widening distally, and with two denticulate setae distally. Carpus shorter than 0.5 times basis, with three denticulate setae distally. Propodus longer than carpus, with three denticulate setae distally, one short, stout spiniform seta, and one spine dorsally. Dactylus and unguis combined longer than propodus, not fused, and unguis without decline in width at midlength.

Pereopod 5 (fig. 30G) as pereopod 4 except: basis with one setulated seta medially.

Pereopod 6 (fig. 30H) as pereopod 5 except: ischium naked.

Pleopods not present in females.

Uropod (fig. 30I) longer than pleotelson. Basal article smooth, as long as endopod article 1 (without exopod process). Endopod article 1 with one setulated and one simple setae distally; article 2 with three long and one short simple setae distally. Exopod about 0.25 times as long as endopod, with one seta distally.

Adult male (where different from female). Body (fig. 28B) pleon as wide as pereon and pleotelson.

Antennule (fig. 28C) thicker than that of female. Article 1 shorter than rest of antennule. Article 4 longer than in female.

Antenna (fig. 28D) as in female.

Mouthparts (fig. 31A-G) identical to those of female except: maxillule (fig. 31D) at least one of the eight terminal setae is bifurcate and one other serrated. This may also be the case in the female, but could not be observed owing to the closeness of the setae; palp only 0.5 times as long as endite, with two setae of unequal length. Maxilliped (fig. 31G), with one endite that was observed with one seta distally, but without setae on the other endite. Observations could not determine if this had been broken off.

Cheliped (fig. 31H) with pseudocoxa and merus even more prominent than in female.

Pereopods (fig. 32A-F) except for a few odd setae, the pereopods are identical to those of the female.

Pleopods (fig. 28E), all pairs subequal, weakly coalesced and with simple setae only.

Uropod (fig. 28F) as in female.

Fig. 31. *Paranarthrura bispinosa* sp. nov., male; A, labrum; B, left mandible; C, right mandible; D, maxillule; E, maxilla; F, labium; G, maxilliped; H, cheliped. Scale bars equal: A-G, 0.05 mm; H, 0.1 mm.

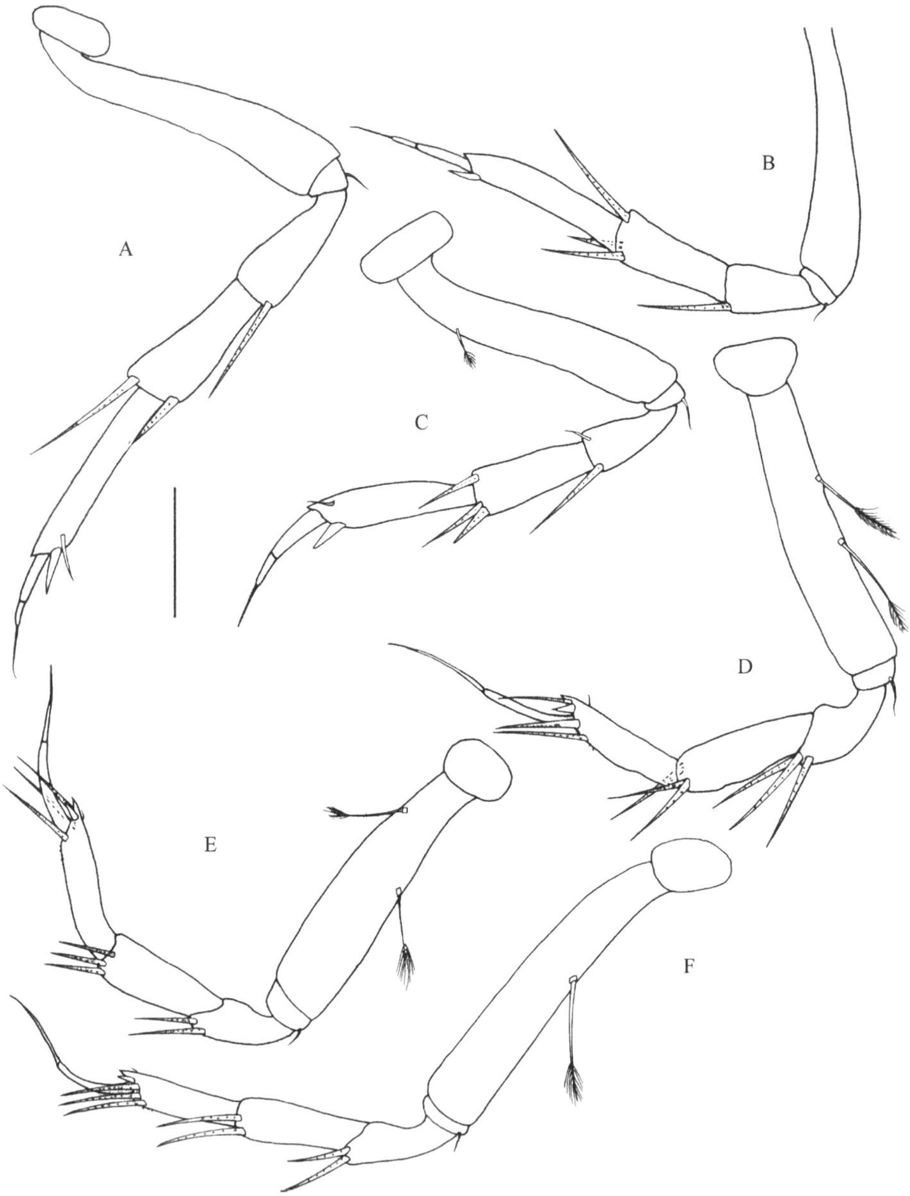

Fig. 32. *Paranarthrura bispinosa* sp. nov., male; A, pereopod 1; B, pereopod 2; C, pereopod 3; D, pereopod 4; E, pereopod 5; F, pereopod 6. Scale bar equals 0.1 mm.

Distribution. — Only found in the deeper part of the Gulf. Bathymetric range: 2070–2750 m. See map 22.

Remarks. — This species can be distinguished from other species in the Gulf of Mexico by the uropodal exopod being shorter than the first endopod article. Otherwise this species is rather similar to *P. tenuimanus*. *Paranarthrura bispinosa* can be separated from *P. tenuimanus* by the additional characters: (1) cheliped propodus with inferior margin almost twice as long as carpus; (2) pereopod 1–3 ungues without dorsal process. There appears to be a clear bathymetric boundary between these two species. The stout spiniform seta next to the dorsal spine on pereopods 4–6 is a character also found in *P. tenuimanus*, *P. crassa*, and *P. intermedia* from the North Atlantic. However, *P. bispinosa* can easily be separated from *P. crassa* by the longer uropod, the thinner cheliped apex, as well as the longer carapace. The short uropodal exopod separates *P. bispinosa* from both *P. crassa* and *P. tenuimanus*. *Paranarthrura bispinosa* can further be separated from *P. intermedia* by the bi-articulate uropod and the wider pleon.

A weak difference was found in the left and right cheliped of the female. This is attributed to a re-growth, as seen in other tanaids (Larsen, 1999a).

Paranarthrura cf. **insignis** Hansen, 1913

Paranarthrura insignis Hansen, 1913, Danish Ingolf Expedition, **3**(3): 121.

Material examined. — 1 non-ovigerous ♀, NGoMCSS Sta. W4-1. 1 non-ovigerous ♀, NGoMCSS Sta. W4-2. 1 non-ovigerous ♀, NGoMCSS Sta. WC11-2. 1 non-ovigerous ♀, NGoMCSS Sta. E3A-1. 1 non-ovigerous ♀, NGoMCSS Sta. E3-3. 1 non-ovigerous ♀, NGoMCSS Sta. E3-6. 1 non-ovigerous ♀, NGoMCSS Sta. E3-1. 1 non-ovigerous ♀, NGoMCSS Sta. E1B-1. 1 non-ovigerous ♀, NGoMCSS Sta. E3D-1. 1 non-ovigerous ♀, NGoMCSS Sta. E3D-6. 1 non-ovigerous ♀, NGoMCSS Sta. C3-4. 1 non-ovigerous ♀, NGoMCSS Sta. C9-1. 1 non-ovigerous ♀, NGoMCSS Sta. C9-4. 1 non-ovigerous ♀, NGoMCSS Sta. C2-1. 1 non-ovigerous ♀, NGoMCSS Sta. C3-6. 1 non-ovigerous ♀, NGoMCSS Sta. C2-1. 2 non-ovigerous ♀♀, NGoMCSS Sta. C2-6. 1 non-ovigerous ♀, NGoMCSS Sta. E3E-1. 1 non-ovigerous ♀, NGoMCSS Sta. E4-1. 2 non-ovigerous ♀♀, NGoMCSS Sta. E4-2. 1 non-ovigerous ♀, NGoMCSS Sta. E4-3. 1 non-ovigerous ♀, NGoMCSS Sta. W2-1. 1 non-ovigerous ♀, NGoMCSS Sta. E2-3. 2 non-ovigerous ♀♀, NGoMCSS Sta. C2-6. 2 non-ovigerous ♀♀, NGoMCSS Sta. E2E-1. 2 non-ovigerous ♀♀, NGoMCSS Sta. IC2-1. 1 non-ovigerous ♀, NGoMCSS Sta. IC3-1. 1 non-ovigerous ♀, NGoMCSS Sta. IC1-2.

Diagnosis. — (Modified after Lang, 1971b.) Female. Carapace as long as pereonites 1 and 2 combined. Pereonite 1 wider than long. Antennular article 1 longer than rest of antennule. Cheliped inferior margin hardly longer than carpus; fixed finger and dactylus apex thickened. Pereopod 1–3 ungues without dorsal process. Uropodal exopod as long as first endopod article. Uropods longer than pleotelson, endopod bi-articulate. Pereopod 4 propodus with one short, very stout distal spiniform seta dorsally next to dorsal spine. Pereopod 4–6 dactyli with setae. Pleon narrower than pereon and pleotelson. Pleotelson rectangular with pointed apex.

Male. Essentially as female except for the presence of pleopods with simple setae.

Distribution. — Davis Strait, Antarctica. Bathymetric range: 582–5000 m. In the Gulf of Mexico this species (if it is *P. insignis*) was found widespread in the northern part of the Gulf in comparatively shallow water, 320–2541 m. See map 23.

Remarks. — *Paranarthrura insignis* was reported from the NGoMCSS program conducted by Texas A&M University in the 1980's. The tanaidacean material from the NGoMCSS study is currently located at the Gulf Coast Research Laboratory, Mississippi, and was examinated by the present author. However, while this material does correspond to *P. insignis*, it seems to have crystallized and does not allow for reliable identification. Species assignment to *P. insignis* is possibly a misidentification owing to the limited number of characters found in this species. It is unlikely that these specimens should really be *P. insignis*, because the type locality is Polar: reported from the Davis Strait by Hansen (1913) and from Antarctica by Kudinova-Pasternak (1975). Such global distribution is highly improbable, but even if *P. insignis* were a bipolar species, its presence in relatively shallow (320 m) warm waters in the Gulf is suspicious. Nevertheless, *Paranarthrura insignis* was also reported from south of Cuba by Kudinova-Pasternak & Pasternak (1978).

Paranarthrura tenuimanus sp. nov.
(figs. 33–37)

Material examined. — Holotype, non-ovigerous ♀ (USNM 1068869), body length 3.1 mm, DGoMB Sta. MT3-1. Paratypes: 1 ♂ (USNM 1068870), body length 2.35 mm, DGoMB Sta. W2-2. 1 non-ovigerous ♀ (USNM 1068871), body length 2.75 mm, DGoMB Sta. C4-3. 1 non-ovigerous ♀ (USNM 1068872), DGoMB Sta. MT3-1 (dissected). 1 ♂ (USNM 1068873), DGoMB Sta. W2-2 (dissected).

Additional material. — 4 mancas, DGoMB Sta. MT3-2. 3 mancas, DGoMB Sta. MT3-2. 1 non-ovigerous ♀, DGoMB Sta. W2-2. 1 manca, DGoMB Sta. S36-1. 1 ♂ (immature), 1 manca, DGoMB Sta. W3-1. 1 ♂, DGoMB Sta. W3-2. 1 manca, DGoMB Sta. C4-3. 1 non-ovigerous ♀, DGoMB Sta. S42–3. 2 non-ovigerous ♀♀, DGoMB Sta. S42-2. 1 ♂, DGoMB Sta. W2-3. 1 non-ovigerous ♀, 6 mancas, DGoMB Sta. S35-2. 1 non-ovigerous ♀, 1 manca, DGoMB Sta. S43-2. 1 non-ovigerous ♀, DGoMB Sta. W3-2. 2 non-ovigerous ♀♀, DGoMB Sta. WC5-1.

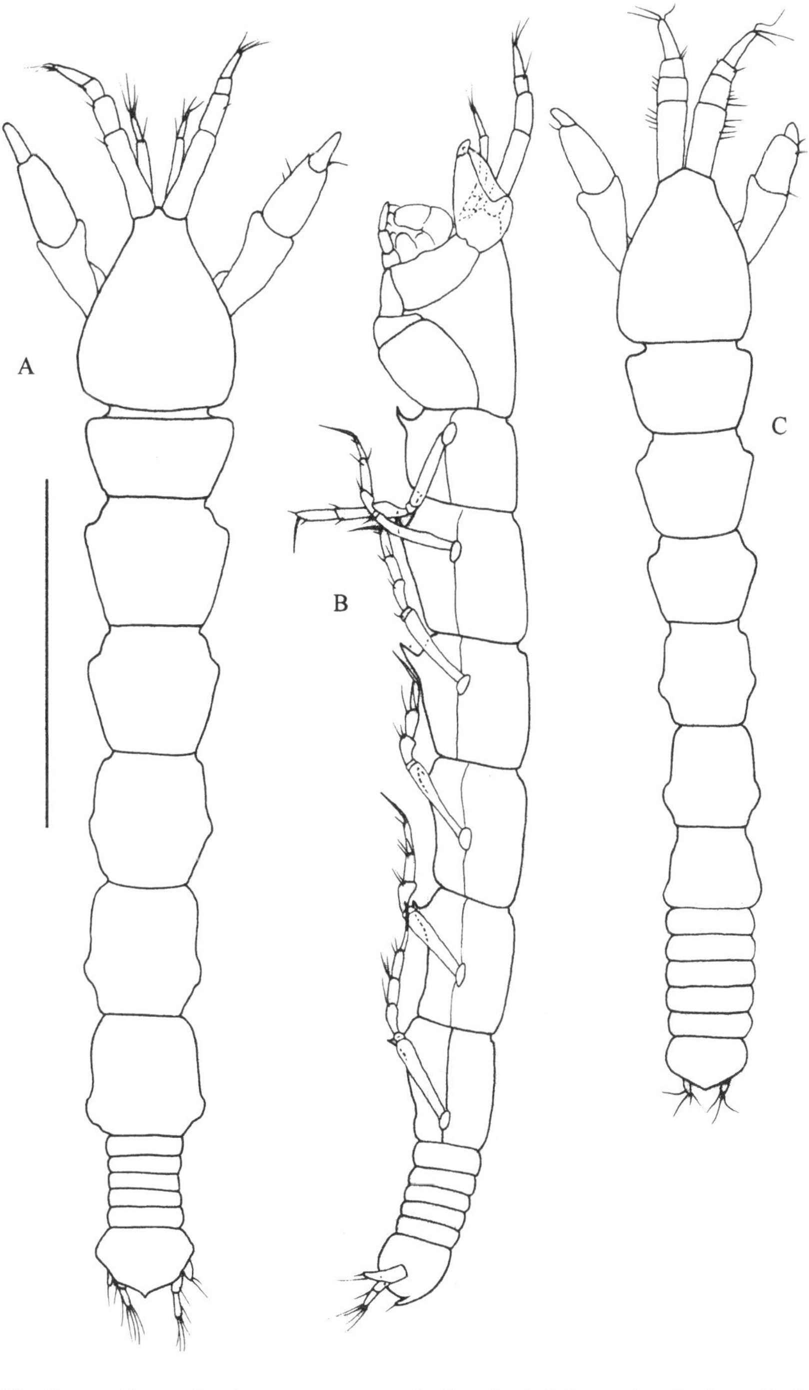

Fig. 33. *Paranarthrura tenuimanus* sp. nov.; A, female, holotype, dorsal view; B, female, holotype, lateral view; C, male, paratype, dorsal view. Scale bar equals 1.0 mm.

Diagnosis. — Female. Carapace in dorsal view with smooth lateral edges; longer than wide; as long as pereonites 1 and 2 combined. Pereonites 1 shorter than wide. Cheliped with inferior margin of propodus less than twice as long as carpus; fixed finger and dactylus apices not thickened. Pereopods 1–3 with unguis with dorsal process. Pereopod 4–6 propodi with one short, very stout spiniform seta dorso-distally next to dorsal spine: dactylus without setae. Pleon narrower than pereon. Pleotelson acorn-shaped with pointed apex. Uropodal endopod bi-articulate; first article widening distally; longer than basal article. Uropodal exopod process about 0.5 times as long as endopod.

Male. Pleonites as wide as pereon and pleotelson. Pleopods with simple setae present.

Etymology. — Named after the slender nature of the cheliped, compared to the otherwise very similar *P. fortispina* (Latin, tenuis = slender, manus = hand). The species name consequently is a noun in apposition to the generic name.

Type locality. — DGoMB Sta. MT3-1, 28°13'2246"N 89°29'7679"W, depth 983 m.

Description. — Adult female. Body (fig. 33A, B) 6.5 times as long as wide. Lateral shoulders weakly defined.

Carapace with smooth lateral edges in dorsal view. Longer than wide. As long as lengths of pereonites 1 and 2 combined.

Pereon with pereonites 1 and 2 wider than long. Pereonites 3, 4, and 5 longer than wide. Pereonite 6 as long as wide. All pereonites with one anteriorly pointing ventral hyposphenian.

Pleon short (including pleotelson only 0.15 times total body length). All pleonites subequal. Pleotelson longer than the lengths of three free pleonites combined, acorn-shaped and apex pointed and covered by a dorsal plate.

Antennule (fig. 34A) shorter than carapace, with four articles. Article 1 longer than rest of antennule, with one simple seta distally. Article 2 longer than article 3, with two setulose setae distally. Article 3 shorter than other articles, with two simple setae distally. Article 4 shorter than article 2, with five simple setae, one setulose seta, and one aesthetasc distally.

Antenna (fig. 34A) with five articles, 0.67 times as long as antennule. Article 1 longer than article 2, with one simple seta dorsally. Article 2 shorter than article 4, with one seta dorso-distally. Article 3 longer than other articles, with one seta medially and two setulose setae distally. Article 4 as long as article 1, with two simple setae distally. Article 5 minute, with three long setae distally.

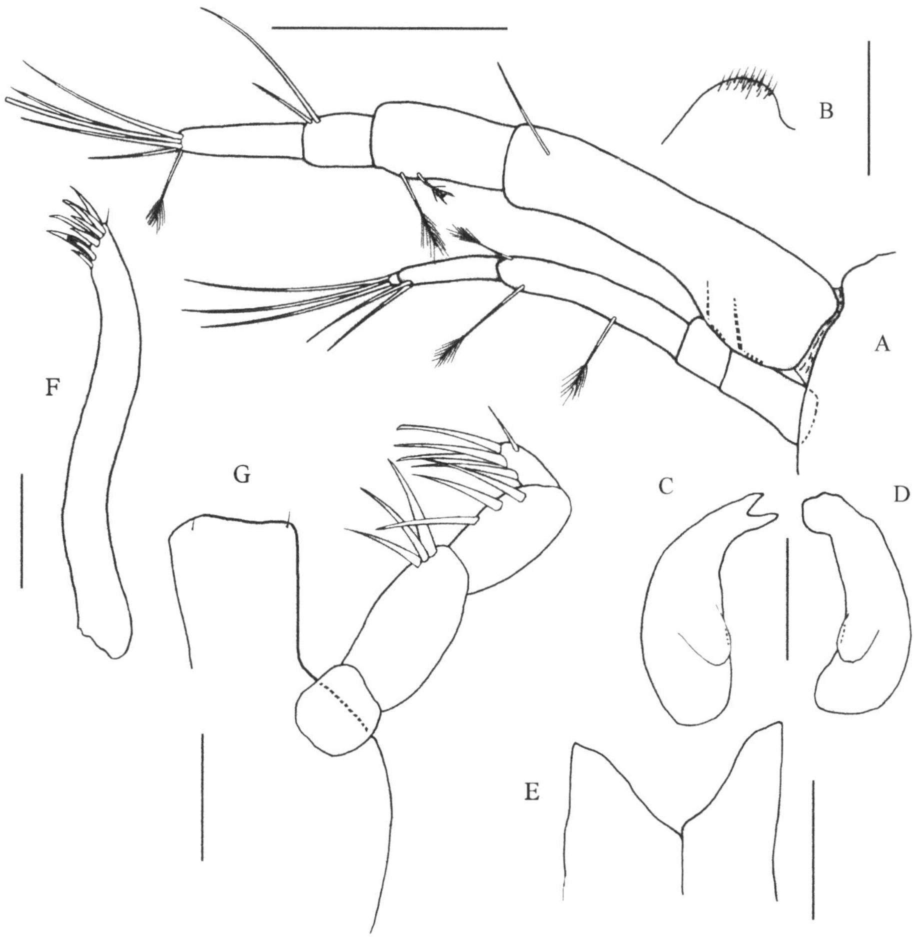

Fig. 34. *Paranarthrura tenuimanus* sp. nov., female; A, antennae; B, labrum; C, left mandible; D, right mandible; E, labium; F, maxillule; G, maxilliped. Scale bars equal: A, 0.1 mm; B-G, 0.05 mm.

Mouthparts. Labrum (fig. 34B) finely setose and hood-shaped. Mandibular molar indistinct. Left mandible (fig. 34C), lacinia mobilis longer than incisor; incisor narrow, consisting of a single spine. Right mandible (fig. 34D) incisor broad and blunt. Labium (fig. 34E) consisting of one pair of triangular lobes. Maxillule (fig. 34F) endite with nine spiniform setae distally; palp not recovered. Maxilla not recovered. Maxilliped (fig. 34G) endites with two small distal setae; palp article 1 naked, articles 2 and 3 with three setae on inner margin, article 4 only 0.5 times as wide as article 3 and with five setae. Epignath not recovered.

Fig. 35. *Paranarthrura tenuimanus* sp. nov., female; A, pereopod 1; B, pereopod 2; C, pereopod 3; D, pereopod 4; E, pereopod 5; F, pereopod 6; G, cheliped; H, uropod. Scale bar equals 0.1 mm.

Cheliped (fig. 35G) pseudocoxa more than 3 times as long as basis. Merus prominent and naked. Carpus almost as long as propodus including fixed

Fig. 36. *Paranarthrura tenuimanus* sp. nov., male; A, antennnule; B, antenna; C, labrum; D, right mandible; E, labium; F, maxillule; f, same, apex, enlarged; G, maxilliped; H, cheliped. Scale bars equal: A, B, H, 0.1 mm; C-G, 0.05 mm.

finger, with two setae ventrally and two dorsally. Propodus with inner row of robust setae. Fixed finger with one seta ventrally and two on inner margin, apex not thickened. Dactylus as long as fixed finger, apex not thickened.

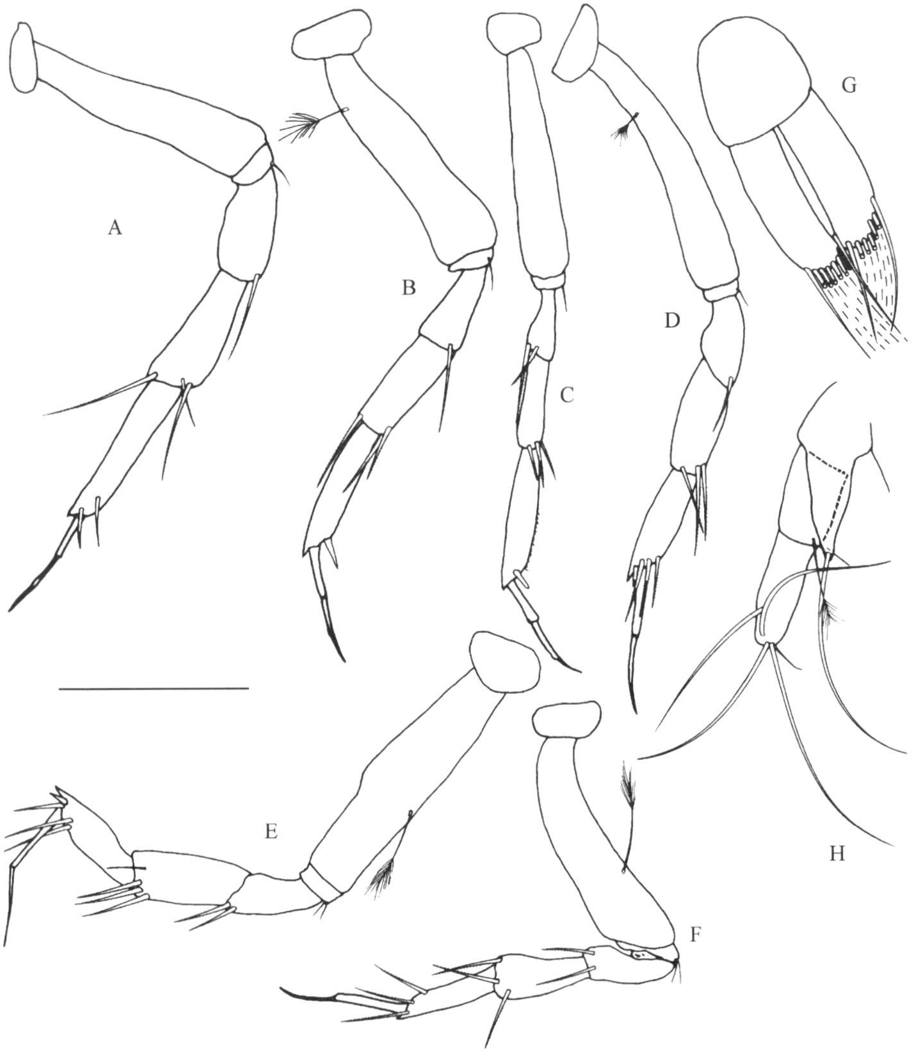

Fig. 37. *Paranarthrura tenuimanus* sp. nov., male; A, pereopod 1; B, pereopod 2; C, pereopod 3; D, pereopod 4; E, pereopod 5; F, pereopod 6; G, pleopod; H, uropod. Scale bar equals 0.1 mm.

Pereopod 1 (fig. 35A) coxa rounded. Basis as long as the three succeeding articles combined, smooth. Ischium with one simple seta. Merus more than 0.5 times as long as carpus, widening distally and with two setae ventro-distally. Carpus 0.5 times as long as basis, with three setae distally. Propodus with two setae distally but no dorsal spine. Dactylus and unguis combined as long as propodus and not fused, unguis with process dorso-medially.

Pereopod 2 (fig. 35B) as pereopod 1 except: basis with one setulose seta. Merus with one seta distally. Propodus with one seta distally and no dorsal spine.

Pereopod 3 (fig. 35C) as pereopod 1 except: propodus with one seta distally and no dorsal spine.

Pereopod 4 (fig. 35D) coxa rounded. Basis longer than the three succeeding articles combined, with one setulose seta medially. Ischium with two setae. Merus more than 0.5 times as long as carpus, widening distally and with two setae distally. Carpus shorter than 0.5 times basis, with three setae distally. Propodus as long as carpus, with two setae distally and one short, stout spiniform seta next to the dorsal spine. Dactylus and unguis combined longer than propodus, not fused, and unguis without dorsal process.

Pereopods 5 and 6 (fig. 35E, F) as pereopod 4 except: basis naked. Carpus with two setae distally. Propodus with three setae distally, one short, stout spiniform seta and one spine dorsally.

Pleopods not present in females.

Uropod (fig. 35H) shorter than pleotelson. Basal article smooth, as long as endopod article 1 (excluding exopod process). Endopod article 1 widening distally, with two setulose setae distally; article 2 with four long, two short simple, and one setulose setae distally. Exopod about 0.5 times as long as endopod.

Male (figs. 33–34), where differing from female.

Antennule (fig. 36A) thicker than in female.

Mouthparts. Maxilliped (fig. 36G) palp article 4 with three setae and two apical heavy setae (twice as wide as normal palp setae).

Cheliped (fig. 36H) propodus without dorsal crenulations.

Pereopods 1–3 (fig. 37A-C) with dorsal projection of unguis less conspicuous than in female.

Pleopods (fig. 37F), endopod and exopod subequal in size. Endopod with eight simple setae, exopod with seven.

Distribution. — Only found in the northern part of the Gulf. Bathymetric range: 625–1825 m. See map 24.

Remarks. — This species can only be identified by the character combination of: pereopod 4–6 propodi with one short very stout spiniform seta dorsodistally next to dorsal spine, uropodal exopod as long as or longer than first endopod article, and antennular article 1 longer than rest of antennule. The dorsal projections on the pereopod 1–3 ungues are unique to this species but appear to be less stable. Otherwise this species is rather similar to *P. insignis* and *P. angolensis* Guerrero-Kommritz et al., 2002. *Paranarthrura tenuimanus* can be separated from *P. insignis* by the following additional characters: uropod shorter than pleotelson, cheliped fixed finger and dactylus blunt; and from *P. angolensis* by the length of the exopod process of the uropod.

Paranarthrura cf. **undulata** Lang, 1968
(fig. 38)

Paranarthrura undulata Lang, 1968, Galathea Report, **9**: 346.

Material examined. — 1 manca III, DGoMB Sta. MT3-1.

Distribution. — Gulf of Panama. One station in the northern Gulf of Mexico. Bathymetric range: 983–3230 m. See map 25.

Remarks. — This species was described by Lang (1968) from the Gulf of Panama (Pacific). The material from the Gulf of Mexico may represent an undescribed sibling species. This specimen (fig. 38) appears to have conspicuously long legs compared to other species in the genus. Whether that is because the specimen is a manca is not known, since we have no information on ontogenetic variation in this group. Lang (1971b) does not illustrate the relative length of the legs nor does he mention this in the descriptive text.

KEY TO THE SPECIES OF *PARANARTHRURA*, FEMALES ONLY (GLOBAL)

1. Uropodal endopod bi-articulate 2
– Uropodal endopod uni-articulate 8
2. Pereonite 1 as long as wide .... *P. lusitanus* Bird & Holdich, 1989 (northeast Atlantic)
– Pereonite 1 shorter than wide 3
3. Uropodal exopod 0.5 times as long as first endopod article 4
– Uropodal exopod as long as first endopod article or longer 5
4. Pereopods 4–6 with one short stout spiniform seta next to, and of similar size as, dorsal spine. Pereopod 1 and 2 ungues with abrupt decrease in width at midlength .... *P. bispinosa* sp. nov. (Gulf of Mexico)
– Pereopods 4–6 without short stout spiniform setae next to dorsal spine. Pereopod 1 and 2 ungues without abrupt decrease in width at midlength .... *P. angolensis* Guerrero-Kommritz, Schmidt & Brandt, 2002 (equatorial east-Atlantic)
5. Carapace shorter than pereonites 1 and 2 combined .... *P. crassa* Bird & Holdich, 1989 (northeast Atlantic)
– Carapace as long as or longer than pereonites 1 and 2 combined 6

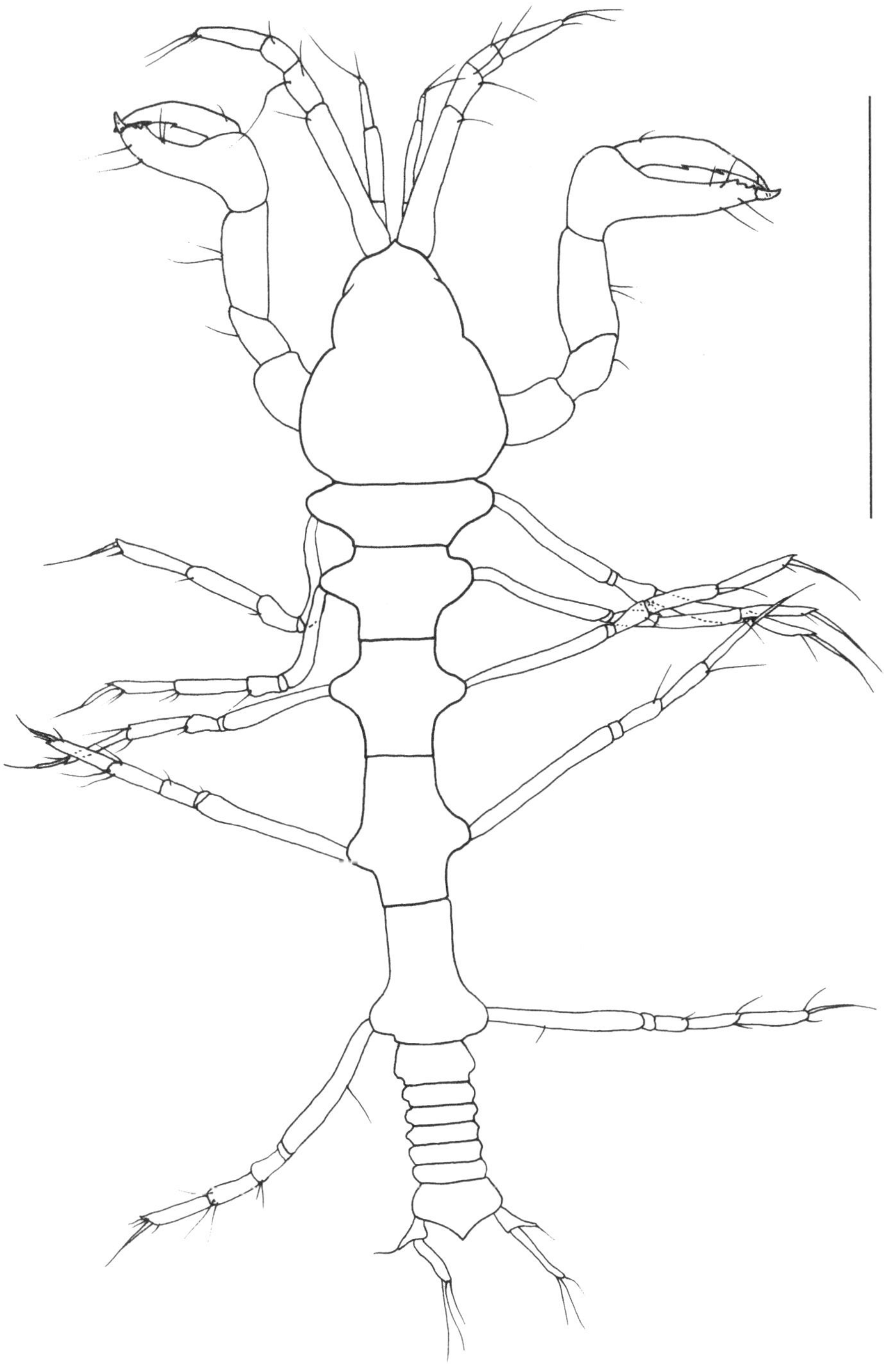

Fig. 38. *Paranarthrura* cf. *undulata* Lang, 1968, manca III, body, dorsal view. Scale bar equals 1.0 mm.

6. Pereopod 4–6 propodi with single dorsal spine .... *P. insignis* Hansen, 1913 (North Atlantic)
– Pereopod 4–6 propodi with dorsal spine and similar-sized spiniform seta 7
7. Antennular article 1 longer than rest of antennule. Cheliped carpus slender (length/width ratio of carpus > 2.0) .... *P. tenuimanus* sp. nov. (Gulf of Mexico)
– Antennular article 1 as long as or shorter than rest of antennule. Cheliped robust (length/width ratio of carpus < 1.5) .... *P. fortispina* Sieg, 1986 (Antarctica)
8. Uropod almost twice as long as pleotelson .... *P. undulata* Lang, 1968 (equatorial east Pacific)
– Uropod about the length of pleotelson or shorter 9
9. Pereopod 4–6 dactyli with two setae .... *P. tridens* Bird & Holdich, 1989 (northeast Atlantic)
– Pereopod 4–6 dactyli without setae 10
10. Uropod only 0.5 times as long as pleotelson 11
– Uropod more than 0.5 times as long as pleotelson 12
11. Pereonite 6 as long as wide .... *P. kurchatovi* Kudinova-Pasternak, 1975 (equatorial east Atlantic)
– Pereonite 6 shorter than wide .... *P. zevinae* Kudinova-Pasternak, 1970 (northwest Pacific)
12. Uropodal endopod as long as basal article .... *P. subtilis* Hansen, 1913 (northeast Atlantic)
– Uropodal endopod longer than basal article 13
13. Uropodal exopod tiny and barely visible 14
– Uropodal exopod 0.25–0.4 times the length of the endopod 15
14. Uropod almost as long as lengths of all pleonites combined .... *P. vitjazi* Kudinova-Pasternak, 1970 (northwest Pacific)
– Uropod shorter than 0.5 times as long as all pleonites combined .... *P. borealis* Bird & Holdich, 1989 (northeast Atlantic)
15. Distal edge of pleotelson almost flat in dorsal view. Uropodal exopod process pointing straightly out from the basal article in a right angle .... *P. similis* Lang, 1971 (northeast Atlantic)
– Distal edge of pleotelson rounded or pointed. Uropodal exopod process pointing posteriorly from the basal article in a 45° angle 16
16. Pereonites 4 and 5 longer than wide .... *P. intermedia* Kudinova-Pasternak, 1982 (northeast Atlantic)
– Pereonites 4 and 5 wider than long or as wide as long .... *P. meridionalis* (Antarctica)

Remarks on the key. — The single-character couplets in this key indicate that the members of this genus are frequently very similar. The key should therefore be applied in concert with geographical data. *Paranarthrura abbreviata* Vanhöffen, 1914 (Antarctica) is considered incertae sedis, since the description is inadequate for comparison and the material appears to be lost. Also Sieg (1983a: 297) considered this species as incertae sedis. *Paranarthrura bacescui* Kudinova-Pasternak, 1986 (Indian Ocean) is not included in the key since this species was only superficially described and the type material could not be obtained.

Family ANARTHRURIDAE Lang, 1971

Anarthruridae Lang, 1971, Arkiv för Zoologi, **2**(23): 366.

Diagnosis. — (Modified after Larsen & Wilson, 2002; Bird, 2004b.) Female and preparatory male. Medium body calcification. No plates in carapace; eyes absent. Pereonite 1 not reduced. Pereon with six free pereonites. Pleon with five free pleonites, or all pleonites and telson fused; as wide as pereon. No articulated setae on pleonites. Antennule with four articles, antenna with six articles, article 3 without dorsal spiniform setae. Mandibles reduced and molar processes reduced or absent. Labium with 1 pair of lobes, medial spiniform seta absent. Maxilliped bases partly fused; endites unfused, conical or subrectangular, without flat setae or denticles. Cheliped attachment through ventral pseudocoxa. Marsupium (where known) of four pairs of oostegites. Pereopods 1–3 with coxa; only simple setae on merus, carpus, and propodus. Pereopods 4–6 without coxa; spiniform setae present, dactylus and unguis not fused. Pleopods present or absent. Uropodal exopod fused to basal article; endopod with 1 or 2 articles.

Male. With extensive sexual dimorphism, mouthparts reduced, pleopods present, longer than in female. Shares no species level characters with female.

Type genus. — *Anarthrura* G. O. Sars, 1882.

Genera included. — *Anarthrura* G. O. Sars, 1882; *Anarthrurella* Bird, 2004; *Anarthruropsis* Lang, 1968; *Anisopechys* Bird, 2004; *Ithyomus* Bird, 2004; *Siphonolabrum* Lang, 1972, *Synanarthrura* Bird, 2004, *Thorkelius* Bird, 2004.

Remarks. — The Anarthruridae were erected by Lang (1971b), redefined by Sieg (1983c), and expanded in 1986 (Sieg, 1986b). Later, Larsen & Wilson (2002) restricted the family in the only computer-assisted phylogenetic analysis to date, while Bird (2004b) expanded the family again with the description of five new genera.

Genus **Anarthruropsis** Lang, 1968

Anarthruropsis Lang, 1968, Galathea Report, **9**: 183.

Diagnosis. — (Modified from Lang, 1968.) Female. Eyes and eye-lobes absent. Pleon with five free pleonites, at least as wide as pereon. Antennule with four articles. Antenna with 5–6 articles. Labrum protruding from under the carapace to reach third antennal article, without obvious siphon-shape. Mandibles reduced (or absent); if present, molar process non-functional, without chitinization. Labium inconspicuous, with 1 pair of lobes. Maxillule with five short, stout, and blunt terminal setae. Maxillipedal endites rounded and with one process or smooth. Epignath elongate. Cheliped attached via large (3 times larger than basis) pseudocoxa. Marsupium consisting of four oostegites. Pereopods 1–3 with coxa. Pereopods 4–6 without coxa, and dactylus and unguis not fused to a claw. Pleopods reduced or absent. Uropod uniramous, endopod with one article, exopod reduced to a 'bump' on outer side of basal article.

Male. Antennule wider than that of female. Pleopods present, but mainly with simple setae only.

Type species. — *Anarthruropsis galatheae* Lang, 1968.

Gender of generic name. — Feminine. Lang (1968) did not mention the gender of the generic name but Sieg (1983a: 292) listed the gender as feminine and the Code (ICZN, 1999: 35, Art. 30.1.2) also indicates that names ending in -opsis are feminine.

Remarks. — This genus is recognizable by the enlarged labrum, the stout blunt spiniform terminal setae of the maxillule, the prominent cheliped pseudocoxa, and the uniramous uropod. However, only the number and special shape of the maxillule setae (and a minor labrum shape difference) distinguish this genus from *Siphonolabrum.* The following characters suggest a close relationship between *Anarthruropsis* and *Siphonolabrum*: (1) The labrum has an elongate shape in both genera, although not quite identical. (2) The pseudocoxa of the cheliped is prominent in both genera. (3) The fixed finger of the cheliped is equipped with a large, prominent process at midlength. (4) The structure of the uropods is also the same.

Species currently assigned to this genus. — *A. edentula* sp. nov.; *A. galatheae* Lang, 1968; *A. langi* Kudinova-Pasternak, 1976; *A. longa* Kudinova-Pasternak, 1984; *A. similis* Kudinova-Pasternak, 1990.

Anarthruropsis edentula sp. nov.
(figs. 39, 40)

Material examined. — Holotype, manca II (USNM 1068874), body length 1.75 mm, DGoMB Sta. RW4-4. Paratypes: 9 mancas II (USNM 1068874), DGoMB Sta. RW4-4. 1 manca II (USNM 1068874), same locality (dissected).

Additional material. — 2 mancas II, DGoMB Sta. NB4-4. 1 manca II, DGoMB Sta. MT4-4. 1 manca II, DGoMB Sta. S36-4.

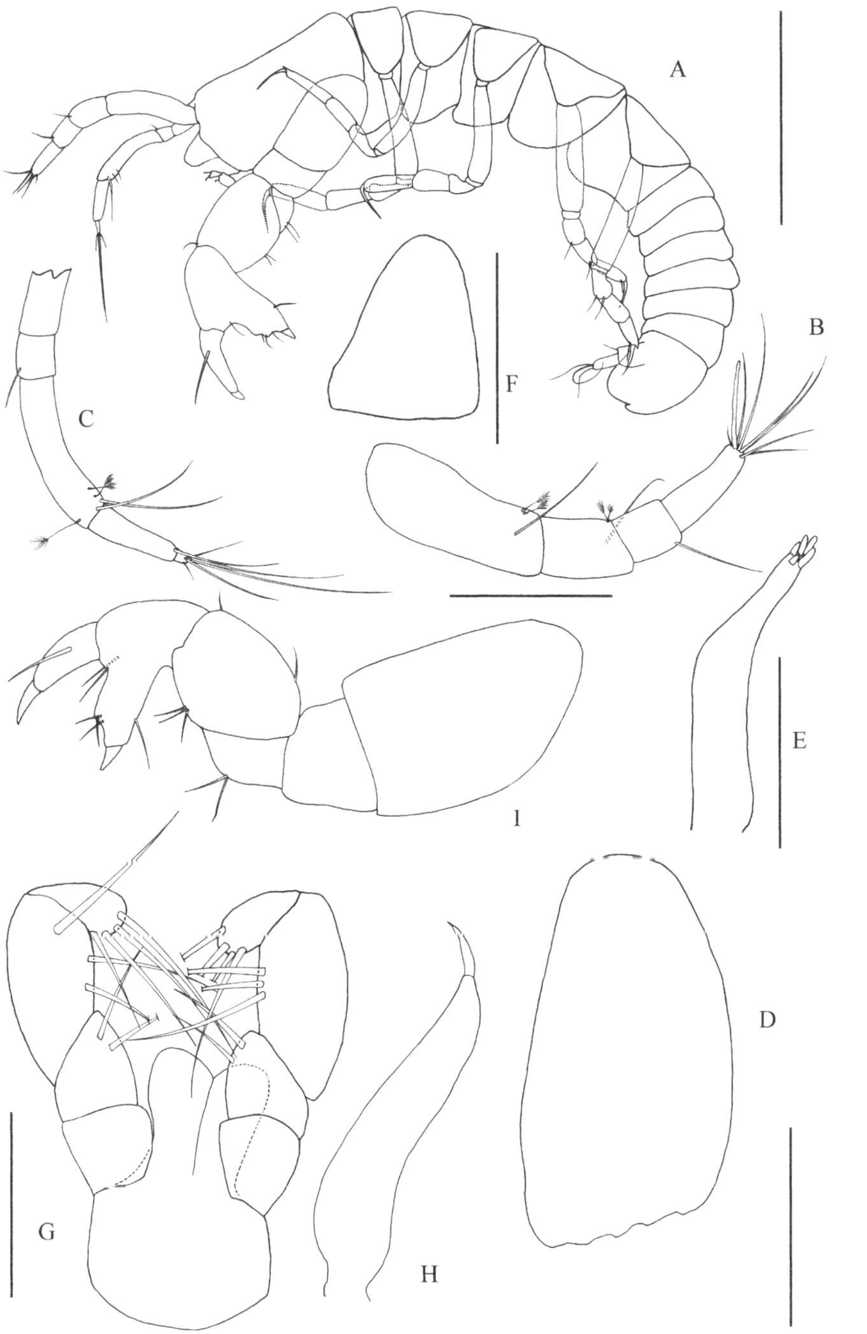

Fig. 39. *Anarthruropsis edentula* sp. nov., manca II; A, holotype, lateral view; B, antennule; C, antenna; D, labrum; E, maxillule; F, maxilla; G, maxilliped; I, cheliped; H, epignath. Scale bars equal: A, 0.5 mm; B-C, 0.1 mm; D-H, 0.05 mm.

Diagnosis. — Manca II. Pereon lateral edge uneven (large gaps between pereonite plates of dorsal shield). Pereonite corners without setae. Antenna with five articles. Maxillipedal endites naked. Pereopod 1–3 dactyli without long seta.

Etymology. — Named after the unidentifiable mandibles of this species (Latin, edentula = feminine form of 'toothless'). The name thus is an adjective agreeing in gender with the generic name.

Type locality. — DGoMB Sta. RW4-4, 26°45'1621"N 95°15'0741"W, depth 1570 m.

Description. — Manca II. Body (fig. 39A) cylindrical, 5.5 times as long as wide.

Carapace hardly shorter than the lengths of pereonites 1–3 combined. Eyes and eye-lobes absent.

Pereon lateral edge uneven (large gaps between pereonite plates of dorsal shield). Pereonite corners without setae.

Pleon rather long (0.25 times total body length, including pleotelson but excluding the reduced 6th pereonite). All pleonites subequal. Pleotelson as long as lengths of three free pleonites combined.

Antennule (fig. 39B) as long as carapace. Article 1 shorter than rest of antennule, with two setulose setae and one simple seta distally. Article 2 less than 0.5 times as long as article 1, with two setulose setae and one simple seta distally. Article 3 shorter than 0.5 times article 4, with one simple seta distally. Article 4 longer than article 2, with seven setae and one aesthetasc distally.

Antenna (fig. 39C) as long as antennule. Article 1 not wider than following articles, naked. Article 2 shorter than article 1, with one simple seta dorso-distally. Article 3 longer than other articles, with two simple and three setulose setae distally. Article 4 longer than article 1, with one simple seta distally. Article 5 decreasing in width at midlength, with three setae at midlength and two distally.

Mouthparts. Labrum (fig. 39D) elongate, protruding from under the carapace and reaching third antenna article. Mandibles not recovered. Labium not recovered. Maxillule (fig. 39E) endite with five spiniform but short, blunt terminal setae; palp lost during dissection. Maxilla (fig. 39F) stout. Maxilliped (fig. 39G) basis naked; endites naked, narrower than basis; palp article 1 naked, article 2 with two simple inner setae, article 3 with three long (longer than article) inner setae, article 4 with four long (longer than two articles) setae. Epignath (fig. 39H) elongate, terminal seta stout, short, and naked.

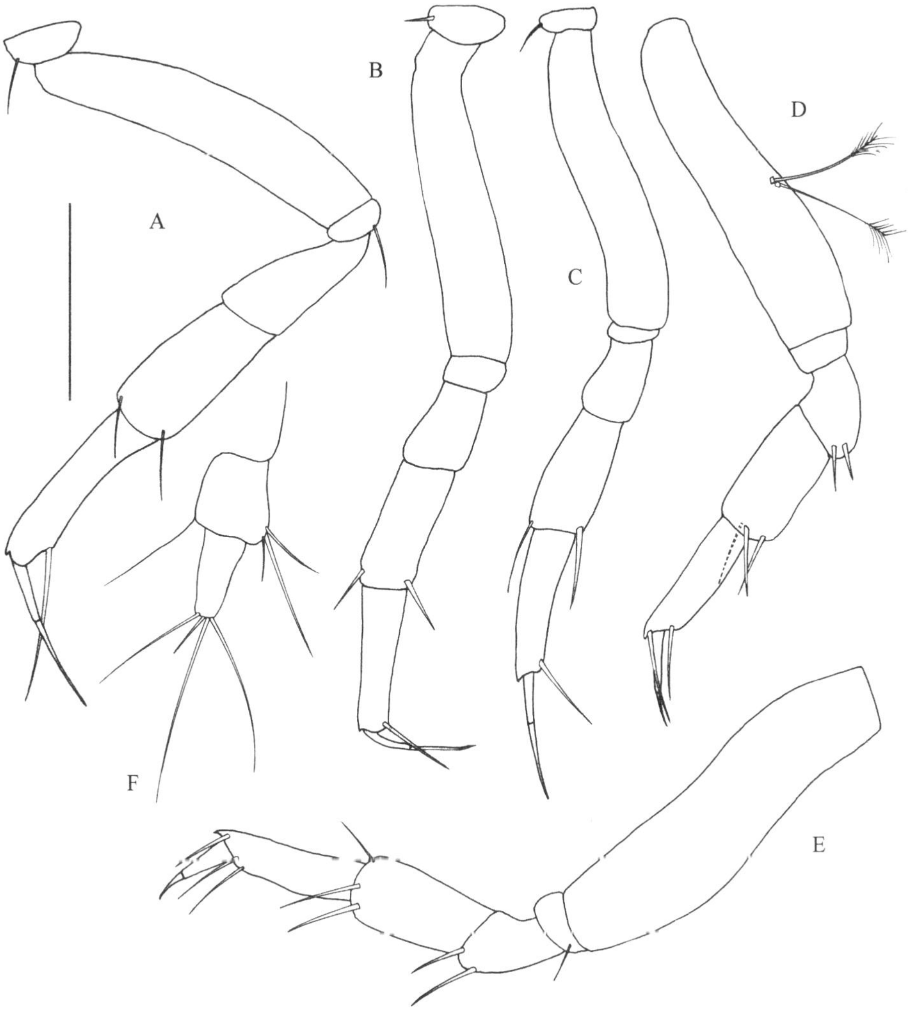

Fig. 40. *Anarthruropsis edentula* sp. nov., manca II; A, pereopod 1; B, pereopod 2; C, pereopod 3; D, pereopod 4; E, pereopod 5; F, uropod. Scale bar equals 0.1 mm.

Cheliped (fig. 39I) pseudocoxa prominent (3 times as long as basis). Basis as long as merus. Merus with two setae ventrally. Carpus shorter than propodus including fixed finger, with three setae ventrally and two dorsally. Propodus with two setae at dactylus insertion. Fixed finger with one seta ventrally and three on inner margin, with prominent process on inner margin. Dactylus as long as fixed finger, with stout seta at midlength.

Pereopod 1 (fig. 40A) coxa with one simple seta. Basis longer than the three succeeding articles combined, naked. Ischium with one seta ventrally. Merus as long as carpus, widening distally, naked. Carpus 0.75 times as long as propodus, with two setae distally. Propodus more than 0.5 times as long as basis, with one seta ventrally, which is as long as dactylus/unguis, and small spine dorsally. Dactylus and unguis combined shorter than propodus and not fused to a claw.

Pereopods 2 and 3 (fig. 40B, C) as pereopod 1 except: ischium naked.

Pereopod 4 (fig. 40D) no visible coxa. Basis with two setulose setae ventrally. Ischium naked. Merus with two robust setae. Carpus longer than merus, with two robust and one simple setae distally. Propodus with two long simple setae distally and small dorsal spine. Dactylus and unguis combined shorter than propodus and not fused to a claw.

Pereopod 5 (fig. 40E) as pereopod 4 except: basis naked. Ischium with one seta ventrally. Propodus with three simple setae and a dorsal spine.

No pleopods are present.

Uropod (fig. 40F) basal article as long as endopod. Endopod with one article, with four setae distally. Exopod reduced to a tiny bump on basal article, with three setae.

Distribution. — Only occurring in the northwestern part of the Gulf. Bathymetric range narrow: 1402–2050 m. See map 26.

Remarks. — This species can be separated from *A. galatheae*, *A. longa* Kudinova-Pasternak, 1984, and *A. langi* Kudinova-Pasternak, 1976 by not having a seta on the dactylus of pereopods 1–3; from *A. galatheae* by the lack of maxillipedal endite structures; and from *A. similis* Kudinova-Pasternak, 1990 by the uropod being shorter than the pleotelson (while this is longer in *A. similis*) and the much more slender antennule (article 1 more than twice as long as wide, while this article is almost square in *A. similis*).

However, given the proximity in sampling locality of *A. edentula* and *A. galatheae*, the morphological similarity, and the developmental stage of the collected specimens of *A. edentula* (all mancas II), the possibility remains that the recorded morphological differences between *A. edentula* and *A. galatheae* could be ontogenetic. Thus, it can not be excluded that *A. edentula* might be the manca of *A. galatheae*.

The specimen described has all the characters of a manca. However, 13 specimens were found, all of them stage II mancas. With the relatively high number of specimens found, it seems unusual that only mancas, and then only in the developmental stage of mancas II, were found. Combined with the relatively large size (for a manca II), a situation involving neoteny is supposed, although no conclusive evidence (such as ovigerous specimens) was found.

Also surmised, but not determined with certainty, is the complete lack of functional mandibles. The mandibles of *Anarthruropsis* are supposed to be very small, reduced, without chitinization, and easily lost during dissection (Lang, 1968). Six specimens were dissected. All twelve mandibles appeared lost, suggesting that these are very rudimentary indeed. Also, in *A. langi*, no mandibles are found either (Kudinova-Pasternak, 1976). In *A. similis*, the mandibles are illustrated as an elongate structure not resembling mandibles at all (Kudinova-Pasternak, 1984: 835, fig. 3) and only in the two remaining species, *A. galatheae* and *A. longa*, do the mandibles, albeit reduced, resemble genuine mandibles.

Anarthruropsis longa is very superficially described but the maxillule is illustrated as having the 'normal' elongate, spiniform terminal setae (Kudinova-Pasternak, 1984), while this is not the case for *A. galatheae* or *A. edentula*. Therefore, it is possible that *A. longa* does not belong in this genus. For *A. langi*, the maxillule is not described and the species cannot be generically assigned with certainty (Kudinova-Pasternak, 1976). Neither *A. longa* nor *A. langi* could be obtained for study.

KEY TO THE SPECIES OF *ANARTHRUROPSIS*, FEMALES ONLY (GLOBAL)

1. Uropodal endopod at most 0.33 times as long as pleotelson 2
– Uropodal endopod at least 0.67 times as long as pleotelson 4
2. Antennule first article slender (more than twice as long as wide) 3
– Antennule first article stout (almost square) .... *A. similis* Kudinova-Pasternak, 1990
3. Pereopod 4–6 propodi with specialized "asparagus-shaped" setae .... *A. langi* Kudinova-Pasternak, 1976
– Pereopod 4–6 propodi without specialized "asparagus-shaped" setae .... *A. longa* Kudinova-Pasternak, 1984
4. Pereopod 1–3 dactyli with setae. Pereopod 1 propodus with two bifurcate setae ventrally .... *A. galatheae* Lang, 1968
– Pereopod 1–3 dactyli without setae. Pereopod 1 propodus without bifurcate setae ventrally .... *A. edentula* sp. nov.

Family COLLETTEIDAE Larsen & Wilson, 2002

Colletteidae Larsen & Wilson, 2002, Journal of Zoological Systematics and Evolutionary Research, **40**(4): 205–222.

Diagnosis. — (Modified from Larsen & Wilson, 2002.) Female. Medium body calcification. No plates in carapace. Eyes absent but eye-lobes may be present. Pereon with six free pereonites. Pereonite 1 not reduced. Pleon with five free pleonites, or all fused with pleotelson (*Isopodidus*), as wide as, or narrower (*Macrinella*) than pereon, no articulated setae on pleonites. Antennule with 4–5 articles. Antenna with six articles (or seven including fusion line), article 3 without spiniform setae dorsally. Molar process of mandible broad or pointed. Labium with one pair of lobes, medial spiniform setae absent. Maxillipedal bases partly fused, endites not fused, narrower than basis and with one (*Leptognathiella*) or no flat setae or processes. Cheliped attachment via sclerite. Marsupium (where known) with four pairs of oostegites. Pereopod coxae present or absent, dactylus and unguis not fused. Pleopods well developed, rudimentary, or absent. Uropodal endopod with two articles, exopod with one article or reduced (*Libanius*).

Male (where known). Body similar to that of female. Antennule with 4–5 articles and only a single aesthetasc. Functional mouthparts retained. Pleopods present but reduced or with simple setae only. Pleon as wide as pereon.

Type genus. — *Collettea* Lang, 1973.

Genera included. — *Cetiopyge* Larsen & Heard, 2002; *Collettea* Lang, 1973; *Cristotanais* Kudinova-Pasternak, 1990; *Caudalonga* n. gen.; *Filitanais* Kudinova-Pasternak, 1973; *Haplocope* G. O. Sars, 1882; *Isopodidus* Larsen & Heard, 2002; *Leptognathiella* Bird & Holdich, 1984; *Leptognathiopsis* Holdich & Bird, 1986; *Libanius* Lang, 1971b; *Macrinella* Lang, 1971; *Mirandotanais* Kussakin & Tzareva, 1974; *Nematotanais* Bird & Holdich, 1985; *Pseudoleptognathia* Sieg, 1986; *Subulella* Holdich & Bird, 1986; *Tumidochelia* Knight, Larsen & Heard, 2003.

Remarks. — This family diagnosis closely resembles those of the Tanaellidae and Leptognathiidae. The Colletteidae can, however, be separated from Tanaellidae by never having the character combination of a reduced uropodal exopod together with lacking pereopod 4–6 coxae. The Colletteidae can be separated from the Leptognathiidae by not having a maxillipedal basis completely fused and by having (where known) only limited sexual dimorphism. Considering the many ‘missing males’ and problems in interpreting the level of fusion of the maxillipedal basis, this separation is weak. In the Leptognathiidae, the mandibles are always pointed and the pleon is comparative short, while these characters vary in the Colletteidae.

KEY TO THE GENERA OF COLLETTEIDAE IN THE GULF OF MEXICO

1. Pleon completely fused with pleotelson. Pleopods divided by inner septum .... *Isopodidus* Larsen & Heard, 2002
– Pleon not fused with pleotelson. Pleopods (if present) not divided by septum 2
2. Uropod with endopod, exopod, and inner spiniform process .... *Tumidochelia* Knight, Larsen & Heard, 2003
– Uropod without inner spiniform process 3
3. Maxilliped palp article 2 with robust multifurcate seta .... *Caudalonga* n. gen.
– Maxilliped palp article 2 without robust multifurcate seta 4
4. Pleon enlarged and laterally compressed .... *Cetiopyge* Larsen & Heard, 2002
– Pleon not laterally compressed 5
5. Mandible molar broad .... *Collettea* Lang, 1973
– Mandible molar pointed 6
6. Pleonites almost as long as individual pereonites and pleotelson .... *Filitanais* Kudinova-Pasternak, 1973
– Pleonites less than half as long as individual pereonites .... *Leptognathiella* Hansen, 1913

Genus **Caudalonga** n. gen.

Diagnosis. — Female. Body almost completely cylindrical. Carapace shorter than first two pereonites combined. Pleon last somite and pleotelson partly fused. Antennule with four articles; 0.5 times as long as carapace. Antenna with six articles. Mandibular molar process tapering, with few terminal denticles. Maxillule with seven spiniform terminal setae. Maxillipedal bases fused, endites not fused, distal edge with medial process. Maxilliped palp article 2 with one conspicuously spiniform, multifurcate seta. Chelipeds slender, ventrally attached via sclerite. Pereopods long (longer than pleon). All pereopods with coxae. Pereopod 1–3 ungues longer than dactyli. Pereopod 4–6 dactyli longer than ungues and not fused. Pleopods absent. Uropods biramous; considerably longer than pleotelson; endopod with two articles; exopod with one minute, partly fused article.

Male. Unknown.

Etymology. — Named after the long uropodal endopod (Latin, cauda + longus = tail + long).

Gender of generic name. — Feminine.

Remarks. — This genus has an akanthophoreinid-like cheliped/cephalothorax articulation, coupled with an agathotanaid-like pleon and a uropodal exopod fused to the basal article. This genus appears close to *Libanius* in the structure of the pereopods and uropods, but not in body shape. The character of the unguis of the pereopods, being longer than the dactylus in pereopods

1–3 but shorter than the dactylus in pereopods 4–6, is seen in several taxa within the Tanaellidae, which are also diagnosed by a reduced uropodal exopod (Larsen & Heard, 2004). However, a coxa on pereopods 4–6 is not found in the Tanaellidae. Fusion in the pleon is seen both in the Agathotanaidae (*Metagathotanais*), Anarthruridae (*Anarthrura*), and Colletteidae (*Isopodidus*) and is thus not informative for family assignment. This genus may also be a sister-group to the agathotanaids, which it resembles in body shape and in the heavy cuticle, but not in the structure of the uropod.

From this list of characters it can be concluded that family designation of this genus is far from obvious, and thus it should only be considered assigned tentatively to the Colletteidae.

Species currently assigned to this genus. — *C. quatropleon* sp. nov.

Caudalonga quatropleon sp. nov.
(figs. 41–43)

Material examined. — Holotype, non-ovigerous ♀ (USNM 1068877), body length 2.0 mm, DGoMB Sta. W2-1. Paratypes: 1 non-ovigerous ♀ (USNM 1068878), body length 2.3 mm, DGoMB Sta. W2-2. 1 non-ovigerous ♀ (USNM 1068879), DGoMB Sta. MT4-4 (dissected).

Additional material. — 1 ovigerous ♀, DGoMB Sta. S35-4. 1 manca, DGoMB Sta. RW3-1. 1 non-ovigerous ♀, DGoMB Sta. S41-2. 1 manca, 1 ♂ (badly damaged), DGoMB Sta. MT6-1.

Diagnosis. — Monotypic genus, no species diagnosis given.

Etymology. — Named after the partly fused pleon (Latin, quattuor + Greek, pleon = four + pleon). Quatropleon is chosen as it is easier to pronounce and it is an arbitrary combination of letters that can be treated as a Latin word. It should be considered a noun in apposition to the generic name.

Type locality. — DGoMB Sta. W2-1, 27°24'8008"N 93°20'2579"W, depth 625 m.

Description. — Adult female. Body (fig. 41A, B) about 8.5 times as long as wide.

Carapace lateral margins gently curved. Almost 1.5 times as long as wide. Shorter than the lengths of pereonites 1 and 2 combined. Eye-lobes absent. Carapace covers basal part of antennules.

Pereon lateral margins gently curved. Pereonite 1 wider than long. Pereonites 2–5 longer than wide. Pereonite 6 as wide as long.

Pleon short, including pleotelson about 0.15 times total body length. Pleonite 5 partly fused to pleotelson. Pleotelson longer than the lengths of four pleonites combined, apex pointed and covered by dorsal plate.

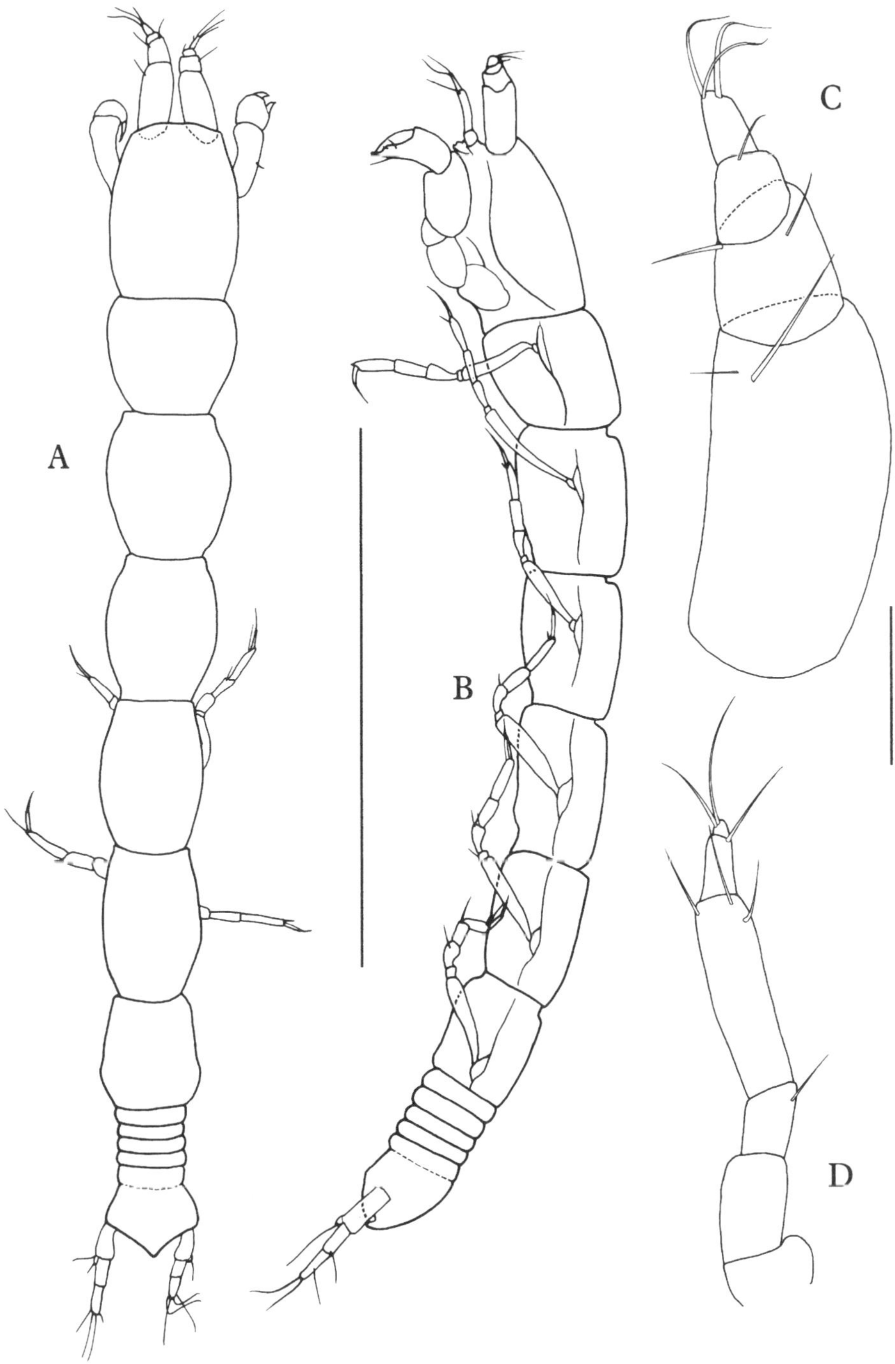

Fig. 41. *Caudalonga quatropleon* gen. et sp. nov., female; A, holotype, dorsal view; B, holotype, lateral view; C, antennule; D, antenna. Scale bars equal: A, B, 1.0 mm; C-D, 0.1 mm.

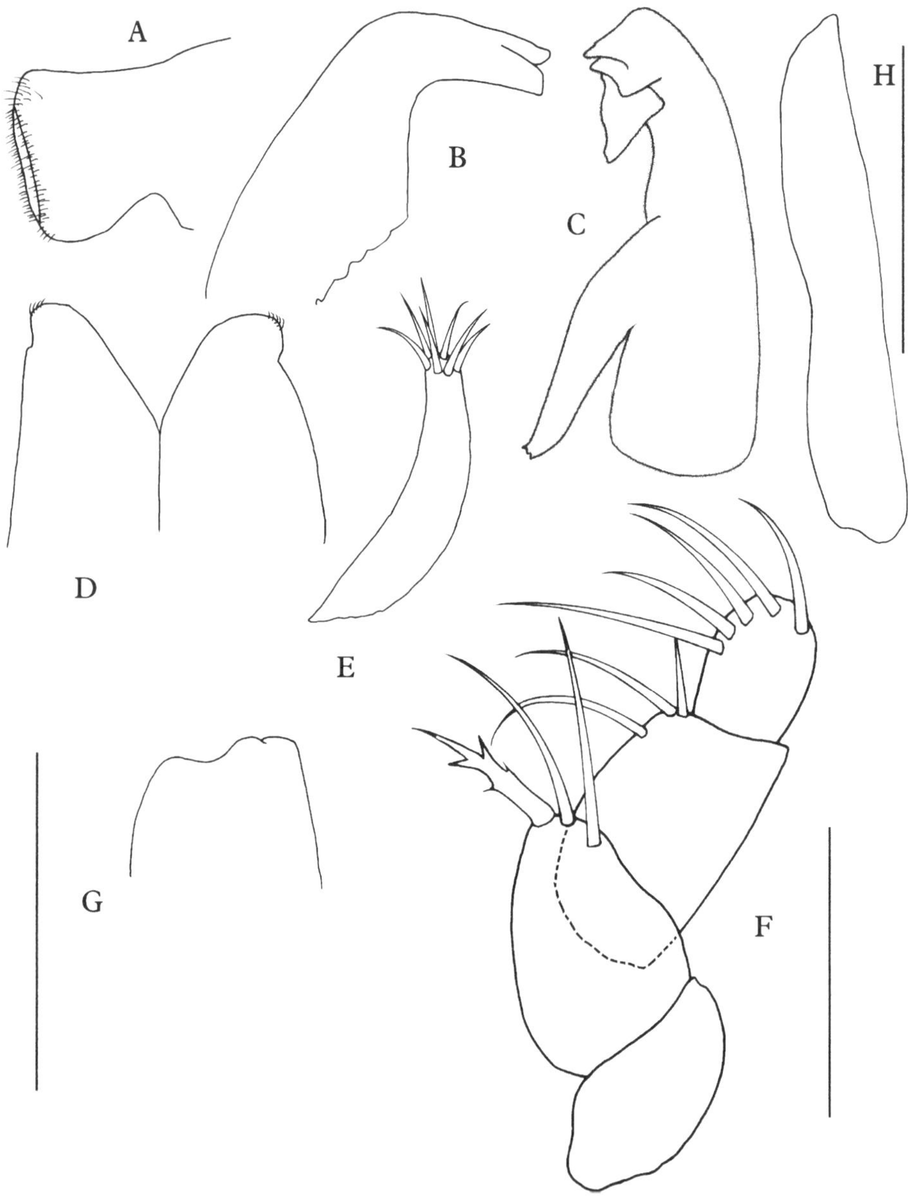

Fig. 42. *Caudalonga quatropleon* gen. et sp. nov., female; A, labrum; B, right mandible; C, left mandible; D, labium; E, maxillule; F, maxilliped palp; G, maxilliped endite; H, epignath. Scale bars equal: 0.05 mm.

Antennule (fig. 41C) only hardly more than 0.5 times as long as carapace, with four articles. Article 1 clearly longer than rest of antennule, with two simple setae distally. Article 2 longer than article 3, with two simple setae distally. Article 3 as long as article 4, with one simple seta distally. Article 4 with three thick simple setae distally and apparently no aesthetascs.

Antenna (fig. 41D) with six articles. About 0.1 times as long as antennule. Article 1 wider than other articles, naked. Article 2 longer than article 3, naked. Article 3 as long as article 5, with one simple seta distally. Article 4 longer than other articles, with three simple setae distally. Article 5 shorter than article 2, with one simple seta distally. Article 6 minute, with two setae distally.

Mouthparts. Labrum (fig. 42A) finely setose, with flat apex. Mandibular molar tapering distally, with 2–3 denticles distally. Left mandible (fig. 42C) lacinia mobilis longer and wider than incisor, with flat apex; incisor broad, with three widely separated denticles. Right mandible (fig. 42B) incisor with two widely separated denticles distally. Labium (fig. 42D) as wide as maxillipedal endites, outer corners setose. Maxillule (fig. 42E) endite with seven spiniform setae distally; palp lost during dissection. Maxilla not recovered. Maxillipedal (fig. 42G) endites with one inner denticle; palp (fig. 42F) article 1 naked, article 2 with two inner simple setae and one large spiniform multifurcate seta, article 3 with three setae on inner margin, article 4 with five setae. Epignath (fig. 42H) naked and apparently without terminal seta.

Cheliped (fig. 43G) basis short (less than 0.5 times as long as carpus), attached to elongate sclerite, naked. Merus with two simple setae ventrally. Carpus longer than propodus including fixed finger, naked. Propodus slender, with one seta at dactylus insertion. Fixed finger with two setae ventrally and three on inner margin as well as four denticles distally. Dactylus as long as fixed finger.

Pereopods 1–3 (fig. 43A-C) with coxa small and naked. Basis longer than the three succeeding articles combined, naked. Ischium naked. Merus shorter than carpus, widening distally and with one stiff seta ventro-distally. Carpus shorter than 0.5 times basis, with one stiff seta distally on each margin. Propodus with one stiff seta distally and a spine dorsally. Dactylus and unguis combined shorter than propodus and not fused, dactylus shorter than unguis.

Pereopod 4 (fig. 43D) as pereopod 1 except: merus with two stiff setae distally. Propodus longer than carpus, with three stiff setae distally and a dorsal spine. Dactylus and unguis combined almost as long as propodus, not fused, dactylus longer than unguis and with setulated ventral margin.

Pereopods 5 and 6 (fig. 43E, F) as pereopod 4 except: carpus with three stiff setae distally. Propodus longer than carpus, with one setulated seta dorsomedially, three stiff setae distally, and dorsal spines.

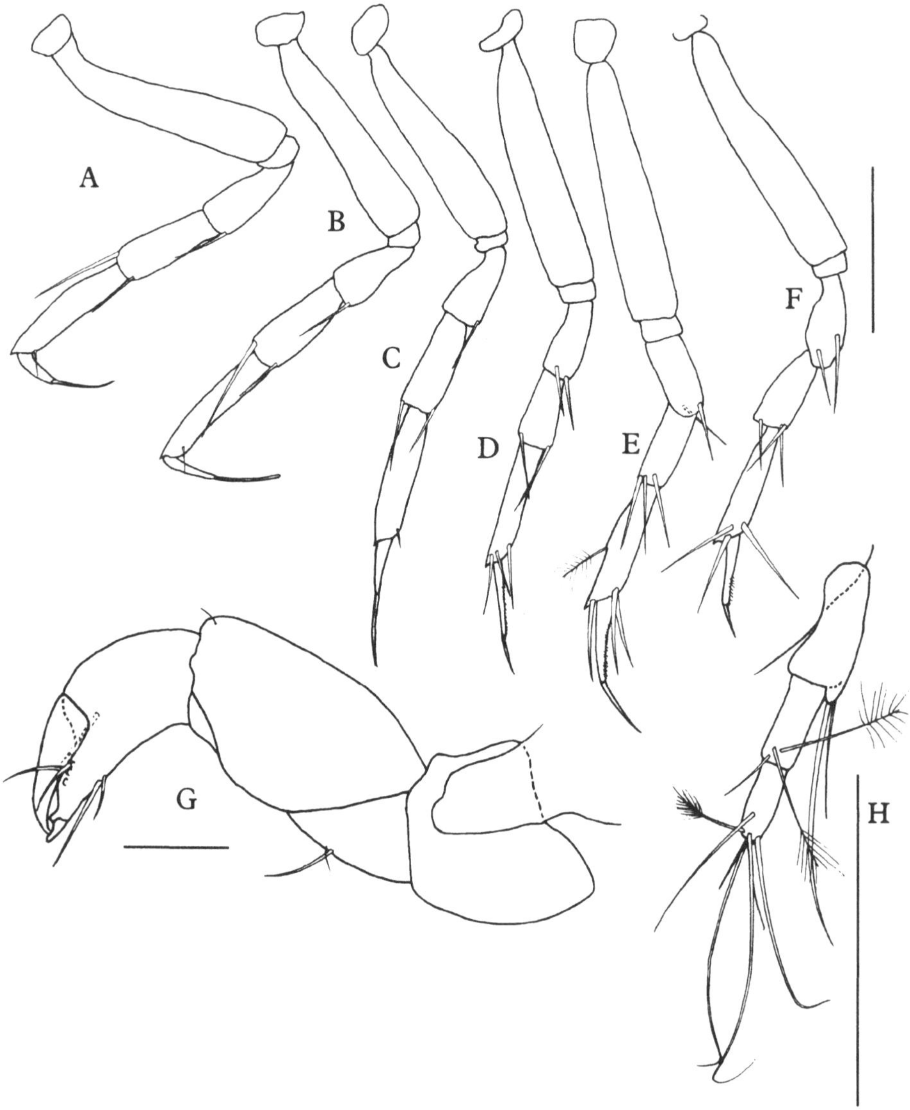

Fig. 43. *Caudalonga quatropleon* gen. et sp. nov., male; A, pereopod 1; B, pereopod 2; C, pereopod 3; D, pereopod 4; E, pereopod 5; F, pereopod 6; G, cheliped; H, uropod. Scale bars equal: 0.1 mm.

Pleopods not present in females.

Uropod (fig. 43H) longer than pleotelson and last two pleonites combined. Basal article naked and longer than first endopod article. Exopod minute, partly fused to basal article, with two simple setae distally. Endopod article 1 longer than article 2, with one simple and two setulose setae distally; article 2 with four long setae distally.

Distribution. — Only found in the northern part of the Gulf. Bathymetric range: 625–2974 m. See map 27.

Remarks. — The slender, stick-like pereopods are interpretated as a plesiomorphic condition. A conspicuously strong, complex seta on the maxilliped palp article 2 is also seen in *Paratanais spinotandus* Sieg, 1982, but this is serrated on the margin rather than multifurcated as in *C. quatropleon*. Aside from this, *C. quatropleon* obviously does not belong in the Paratanaidae. The long pereopods are seen in *Collettea longipedia* Kudinova-Pasternak, 1986, but *Caudalonga* lacks the typical '*Collettea*' pleotelson.

The fused fifth pleonite of the type species is a character that should not be given too much weight, since considerable variation was observed, from the point of almost no fusion to almost complete fusion.

Genus **Cetiopyge** Larsen & Heard, 2002

Cetiopyge Larsen & Heard, 2002, Proceedings of the Biological Society of Washington, **115**(2): 407.

Diagnosis. — (Modified after Larsen & Heard, 2002.) Female. Carapace as long as two pereonites combined. Pleon and pleotelson not fused. Pleon enlarged and laterally compressed. Antennule with five articles, distal article being minute. Antenna with six articles. Mandible molar broad. Maxillule with some bifurcate terminal setae. Maxillipedal bases not fused distally, endites not fused, without denticles, and narrower than basis; palp article 2 without multifurcate setae. Chelipeds attached laterally by sidepiece. Pereopods with coxae; propodus never longer than dactylus and unguis combined. Pereopod 4–6 dactyli and ungues not fused to hook-like claws. Distal pereon and pleotelson greatly enlarged and laterally compressed. Pleopods absent in female. Uropods small, exopod minute, endopod consisting of two partially fused articles.

Male. Unknown.

Type species. — *Cetiopyge mira* Larsen & Heard, 2002.

Gender of generic name. — Neuter.

Remarks. — An enlarged pleon is also seen in *Expina* and *Mirandotanais*, but only in *Cetiopyge* is this laterally compressed.

Species currently assigned to this genus. — *C. mira* Larsen & Heard, 2002.

Cetiopyge mira Larsen & Heard, 2002

Cetiopyge mira Larsen & Heard, 2002, Proceedings of the Biological Society of Washington, **115**(2): 407.

Diagnosis. — Monotypic genus, no species diagnosis given.

Distribution. — Probably one of the most common species in the Gulf. Bathymetric range: 213–2060 m. See map 28.

Genus **Collettea** Lang, 1973

Strongylura G. O. Sars, 1882, Archiv for Matematik og Naturvidenskab, **7**: 52.
Collettea Lang, 1973, Zoologica Scripta, **2**(5–6): 227.

Diagnosis. — (Modified after Lang, 1971c; Larsen, 2000b.) Female. Body almost completely cylindrical. Carapace longer than wide. Eye-lobes present but without visual elements. Pleonites without lateral edges of dorsal shield. Pleon and pleotelson not fused. Pleotelson mostly longer than last three free pleonites combined, terminating in dorsal plate covering uropods. Antennule with 4–5 articles. Antenna with 6–7 (fusion line) articles. Mandibles with broad molar. Maxillule setal number varying, some setae may be bifurcate or setulose. Maxillipedal endites without distal setae. Maxilliped palp article 2 without multifurcate seta. Chelipeds attached via lateral sclerite. Pereopods slender, with coxae, and with dactylus and unguis not fused to hook. Pleopods absent in females. Uropods biramous; endopod with 2 articles; exopod uniarticulate.

Male. Similar to female. Functional mouthparts retained in adult. Antennule thicker than in female. Pleopods present, most often with simple setae.

Type species. — *Strongylura cylindrata* G. O. Sars, 1882.

Gender of generic name. — Feminine.

Remarks. — Most of the diagnostic characters are stable in this genus and it is usually not hard to identify it from other genera. Characters useful in species identification include: (1) Length of uropods. (2) Length/width ratios of pereonites. (3) Length of the dactylus of the cheliped relative to fixed finger. (4) Number of denticles of the fixed finger of the cheliped. (5) Presence and position of fusion line on antenna article 4.

The mouthparts of the species in this genus rarely display any useful information for species separation.

The genus was described under the name *Strongylura* G .O. Sars, but Lang (1973) changed the name to *Collettea* since the name *Strongylura* was preoccupied. This is a cosmopolitan genus found in all major oceans. A key to the genus is given in Larsen (2000b).

Species currently assigned to this genus. — *C. antarctica* (Vanhöffen, 1914); *C. arnaudi* (Shiino, 1978); *C. cylindrata* (G. O. Sars, 1882a); *C. cylindratoides* Larsen, 2000; *C. elongata* Larsen, 2002; *C. humbolti* Larsen, 2000; *C. larviformis* (Kudinova-Pasternak, 1973); *C. longipedia* Kudinova-Pasternak, 1986; *C. minima* (Hansen, 1913); *C. pegmata* Bamber, 2000; *C. rotundicauda* Kudinova-Pasternak, 1983; *C. subtilis* Kudinova-Pasternak, 1981; *C. vermiformis* (Lang, 1971); *C. wilsoni* Larsen, 1999.

Collettea elongata Larsen, 2002

Collettea elongata Larsen, 2002, Gulf and Caribbean Research, **14**: 57.

Diagnosis. — (Modified after Larsen, 2002.) Female. Body 12 times as long as wide. Antenna article 4 without fusion line. Maxillipedal endite without process. Cheliped fixed finger without processes. Cheliped dactylus not longer than fixed finger. Heavy armament with spiniform setae only on pereopods 4–6. Uropods long (> 0.3 times as long as pleotelson).

Male. Pleonites longer than those of female, all with pleopods. Pleotelson shorter than the lengths of three pleonites combined.

Distribution. — Species only recorded from the northern part of the Gulf. Bathymetric range: 291–3015 m. See map 29.

Remarks. — This species was one of a large number of filiform species found in the deep-sea Gulf samples (see under Filiform species) and can be distinguished from other *Collettea* species by the elongate body shape. The difference in length of the pleonites in the two sexes is puzzling. It cannot be excluded that a 'colletteid' species complex of small, elongate forms exists.

Collettea sp. A
(fig. 44)

Material examined. — 1 non-ovigerous ♀, DGoMB Sta. MT3-4.

Distribution. — Species only recorded from one station south of New Orleans. Bathymetric range: 988 m. See map 30.

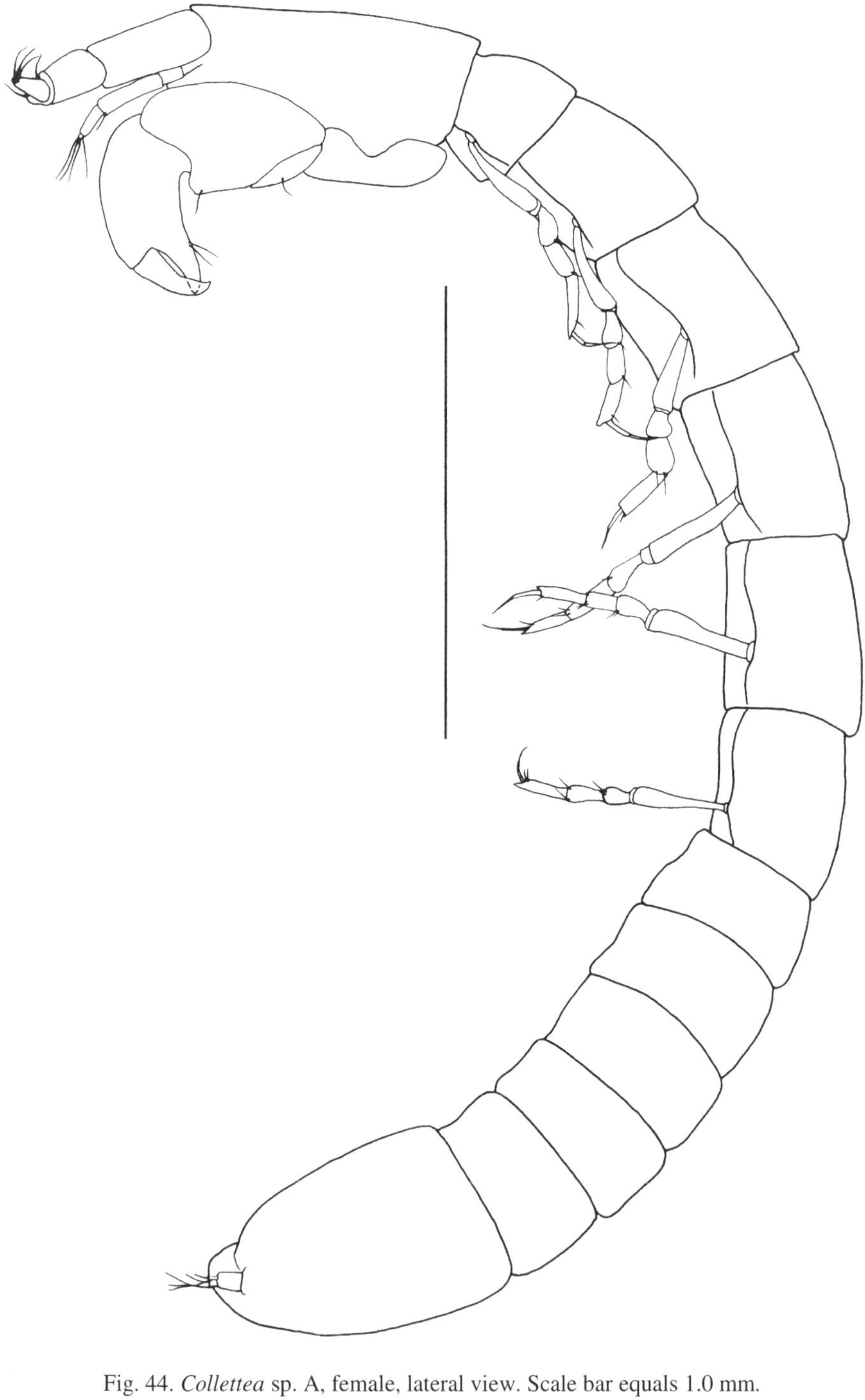

Fig. 44. *Collettea* sp. A, female, lateral view. Scale bar equals 1.0 mm.

Remarks. — This species belongs without doubt to the genus *Collettea*. Unfortunately, only one specimen was found and the species is thus not described. *Collettea* sp. A can be distinguished from the other colletteid species in the Gulf, *Collettea elongata*, by having short uropods (about 0.15 times as long as pleotelson). The character combination of short uropods, short pleon (relative to most other *Collettea*) and pereonites 4 and 5 longer than wide, separates *Collettea* sp. A from most other *Collettea* spp. (including *C. cylindrata* and *C. wilsoni* from the North Atlantic). *Collettea* sp. A does show similarities with *Collettea* sp. reported from the deep Pacific by Larsen (2000), but can be separated from this by having a pleotelson not longer than three pleonites combined, while the pleotelson of *Collettea* sp. (Pacific), is longer than four pleonites combined.

Genus **Filitanais** Kudinova-Pasternak, 1973

Filitanais Kudinova-Pasternak, 1973, Trudy Instituta Okeanologii, Akademiya Nauk SSSR, **91**: 158.

Diagnosis. — (After Kudinova-Pasternak, 1973.) Female. Body almost completely cylindrical and filiform. Carapace longer than wide. Eye-lobes and visual elements absent. Pleonites long (almost as long as pereonites and pleotelson), without lateral edges of dorsal shield. Pleon and pleotelson not fused. Pleotelson terminating in dorsal plate covering uropods. Antennule with four articles. Antenna with 5–6 articles. Mandibles with pointed molar. Maxillule setal number varying. Maxillipedal endites without setae distally; palp article 2 without multifurcate seta. Chelipeds attached via lateral sclerite. Pereopods slender, with coxae, and with dactylus and unguis not fused to a hook. Pleopods absent in female. Uropods usually long (about 0.5 times as long as pleotelson), biramous; endopod with two articles; exopod with 1–2 articles.

Male. Similar to female. Functional mouthparts retained in adults. Antennule thicker than in female. Pleopods present, most often with simple setae.

Type species. — *Filitanais moskalevi* Kudinova-Pasternak, 1973.

Gender of generic name. — Masculine.

Remarks. — The mouthparts of the species in this genus do not display any reliable characters for species separation. The shape of the pleotelson and pleonites and the length of the uropodal exopod appear to be the easiest way to identify the species. The uropodal exopod articulation is a less reliable character.

This genus is poorly defined and the type species, *Filitanais moskalevi*, incompletely described, as description of the mandibles is lacking (Kudinova-Pasternak, 1973). The type specimen is apparently lost (M. Błażewicz-Paszkowycz, pers. comm.) thus making revision impossible. However, it is the opinion of the present author that the species *Filitanais vulgaris* Kudinova-Pasternak, 1981, despite the long pleon, does not belong to *Filitanais* but rather to *Collettea* by virtue of the presence of a broad mandibular molar (Kudinova-Pasternak, 1981: 98, fig. 3). Also, *Filitanais curticaudus* [sic] Kudinova-Pasternak, 1993 probably does not belong to *Filitanais*, owing to its short pleon (Kudinova-Pasternak, 1993: 140, fig. 2) and is more likely to be a leptognathiid.

It is here considered that Kudinova-Pasternak (1984) incorrectly constructed the name *curticaudus*, i.e., possibly intended but not formally explained as a derivation from Latin 'cauda', tail. As cauda is a noun, the name should be considered a noun in apposition to the generic name, the more so since the corresponding adjective would be 'caudatus'. Thus, the name cannot be declined to agree in gender with the (masculine) generic name. The name *curticaudus* is hence considered an "incorrect latinization" and, since such does not constitute a ground for justified emendation, the incorrect name cannot be changed and must de retained. As a result, *curticaudus* thus has to be considered "an arbitrary combination of letters" and remains a noun in apposition that, consequently, will not change its ending when combined with a feminine or neuter generic name.

Species currently assigned to this genus. — *F. curticaudus* Kudinova-Pasternak, 1993 [sic]; *F. filiformis* sp. nov.; *F. moskalevi* Kudinova-Pasternak, 1973; *F. rebainsi* Kudinova-Pasternak, 1975; *F. vulgaris* Kudinova-Pasternak, 1981.

Filitanais filiformis sp. nov.
(figs. 45, 46)

Material examined. — Holotype, non-ovigerous ♀ (USNM 1068880), body length 2.15 mm, DGoMB Sta. C7-4. 1 non-ovigerous ♀ (USNM 1068881), same locality. 1 non-ovigerous ♀ (USNM 1068882), same locality (dissected).

Additional material. — 1 manca III, DGoMB Sta. NB4-4. 1 non-ovigerous ♀, DGoMB Sta. MT4-4. 1 non-ovigerous ♀, DGoMB Sta. S36-2.

Diagnosis. — Female. Body more than 20 times as long as wide. Antenna article 4 without fusion line. Maxillipedal endite without setae. Cheliped fixed finger with conspicuous M-shaped process. Cheliped dactylus not longer than fixed finger; with two tiny inner spines. Pereopod 1–3 propodi with one ventral and one dorsal spiniform seta. Uropodal exopod uni-articulate and as long as first endopod article.

Male. Unknown.

Etymology. — Named to reflect the extremely elongate body-shape (Latin, filiformis = thread-like). The name is thus an adjective agreeing with the gender of the generic name.

Type locality. — DGoMB Sta. C7-4, 27°43'8089"N 89°59'1213"W, depth 1070 m.

Description. — Adult female. Body (fig. 45A) 20.5 times as long as wide.

Carapace shorter than the lengths of pereonites 1 and 2 combined.

Pereon with pereonite 1 as wide as long. Pereonites 2–6 longer than wide.

Pleon more than 0.5 times total body length. All pleonites subequal, as long as, or longer than, pereonites. Pleotelson hardly longer than last pleonite.

Antennule (fig. 45B) with four articles. As long as carapace. Article 1 shorter than rest of antennule, with two simple setae distally. Article 2 as long as article 4, with two simple setae distally. Article 3 hardly shorter than article 2, with one simple seta distally. Article 4 less than 0.5 times as long as article 1, with five simple setae and one aesthetasc distally.

Antenna (fig. 45C) 0.7 times as long as antennule. Article 1 longer than wide, with spiniform seta dorsally. Article 2 shorter than article 1, with one spiniform seta dorso-distally. Article 3 longer than other articles, with one setulose seta medially and one simple seta distally. Article 4 longer than article 1, with one simple seta distally. Article 5 minute, with four simple setae distally.

Mouthparts. Labrum (fig. 45D) hood-shaped, naked. Mandibular molar process longer than incisor and tapering, but with blunt apex. Left mandible (fig. 45E) lacinia mobilis narrow and shorter than incisor; incisor with blunt apex. Right mandible (fig. 45F) incisor broad, with three denticles. Maxillule (fig. 45H) endite with ten spiniform setae distally; palp not recovered. Maxilla (fig. 45H) ovoid. Maxilliped (fig. 45I) basis with one small seta at palp insertion; endites naked, with small medial process; palp article 1 naked, article 2 with two bipinnate setae, article 3 with three bipinnate setae, article 4 with four bipinnate setae. Epignath not recovered.

Cheliped (fig. 46G) basis widening distally, shorter than carpus. Merus with one seta ventrally. Carpus shorter than propodus including fixed finger, with

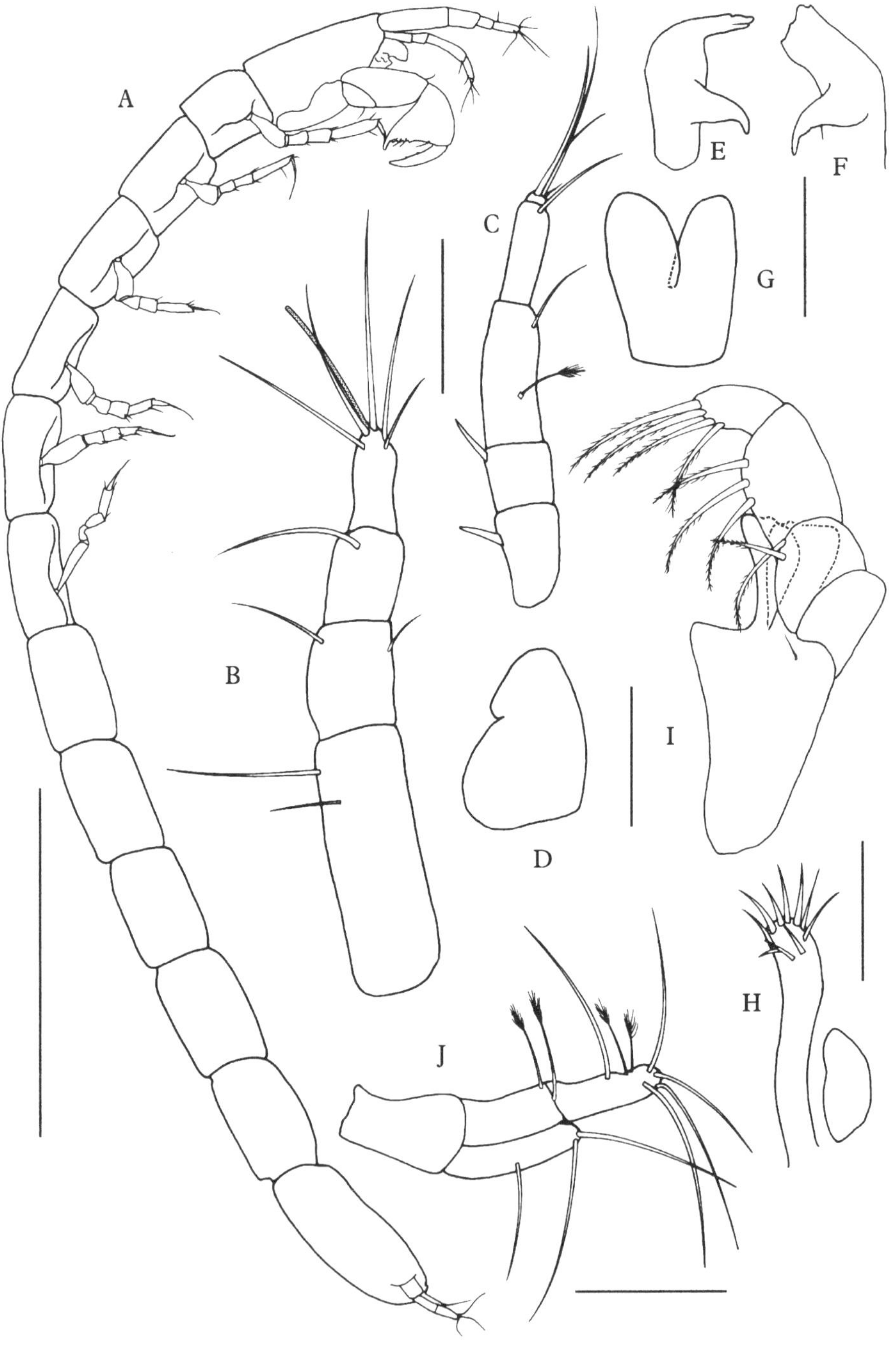

Fig. 45. *Filitanais filiformis* sp. nov., female; A, holotype, lateral view; B, antennule; C, antenna; D, labrum; E, left mandible; F, right mandible; G, labium; H, maxillula and maxilla; I, maxilliped; J, uropod. Scale bars equal: A, 0.5 mm; B-C, J, 0.1 mm; D-I, 0.05 mm.

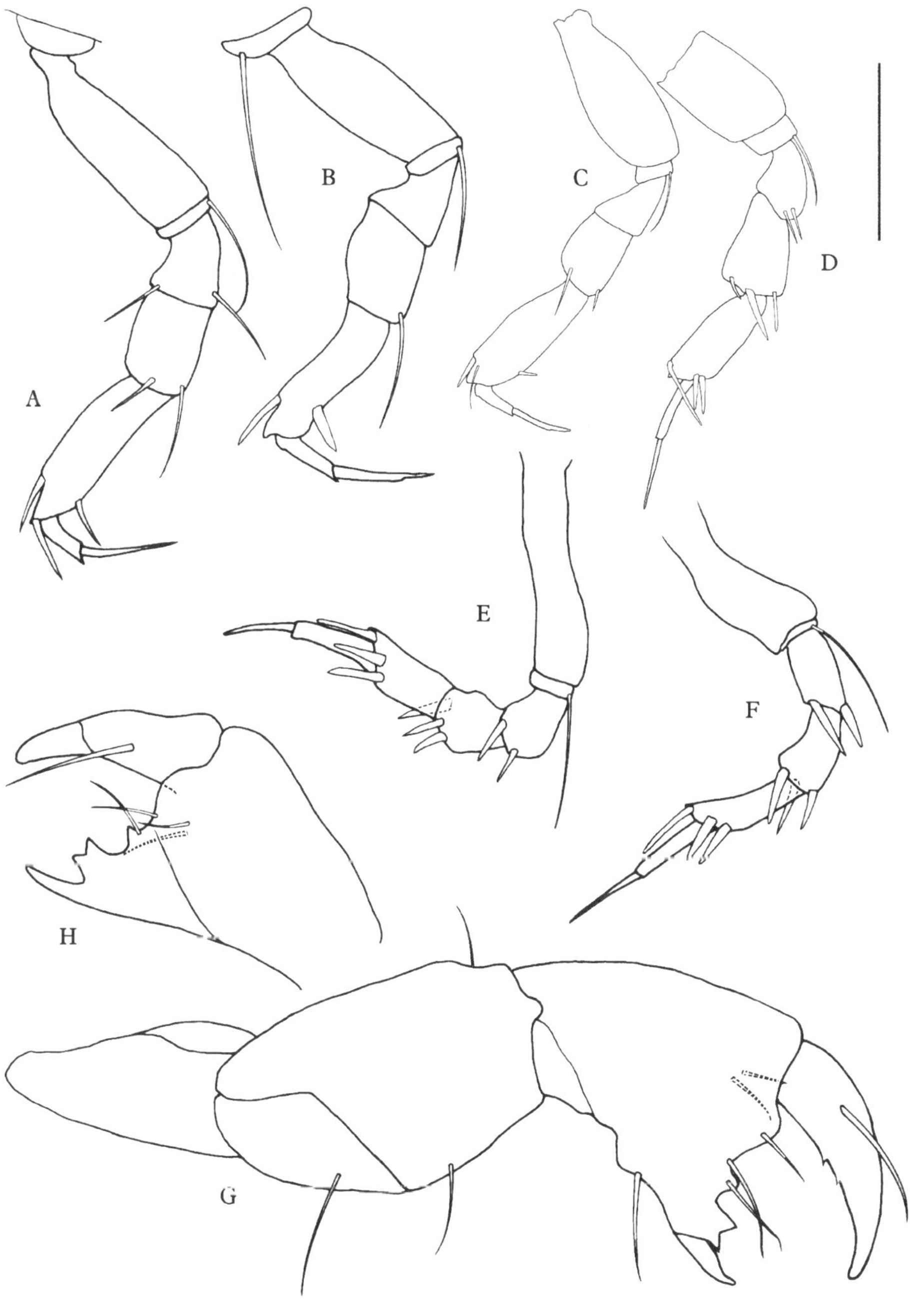

Fig. 46. *Filitanais filiformis* sp. nov., female; A, pereopod 1; B, pereopod 2; C, pereopod 3; D, pereopod 4; E, pereopod 5; F, pereopod 6; G, right cheliped; H, propodus and dactylus, left cheliped. Scale bar equals 0.1 mm.

one simple seta ventro-medially and one simple seta dorso-distally. Propodus with three simple setae at dactylus insertion. Fixed finger with a medial and a conspicuous M-shaped distal process, with one seta ventrally and two setae on inner margin. Dactylus with one prominent seta medially and two small ventral teeth.

Pereopod 1 (fig. 46A) coxa naked. Basis longer than the three succeeding articles combined, naked. Ischium with one long (as long as merus) simple seta. Merus shorter than carpus, widening distally, with two simple setae distally. Carpus 0.5 times as long as propodus, with two setae distally. Propodus 0.8 times as long as basis, with one simple seta dorso-subdistally, one seta ventro-subdistally, and one spiniform seta dorso-distally, with dorsal spine. Dactylus/unguis combined shorter than propodus, dactylus with dorso-distal spine. Unguis longer than dactylus.

Pereopod 2 (fig. 46B) as pereopod 1 except: coxa with long (basis length) seta. Merus apparently naked. Propodus 0.8 times as long as basis, with one simple seta dorsally and one spiniform seta ventro-subdistally, with dorsal spine. Dactylus/unguis combined as long as propodus, dactylus with dorso-distal spine. Unguis longer than dactylus and with dorsal decrease in width on the distal third of the unguis.

Pereopod 3 (fig. 46C) as pereopod 2 except: carpus with two simple setae distally. Propodus with one seta dorsally, one spiniform seta ventro-subdistally, and simple seta distally, with dorsal spine.

Pereopod 4 (fig. 46D) as pereopod 3 except: merus with two spiniform setae. Carpus with three spiniform setae distally. Propodus with two ventral and one dorsal spiniform setae distally. Dactylus/unguis combined longer than propodus. Dactylus as long as unguis.

Pereopods 5 and 6 (fig. 46E, F) as pereopod 4 except: dactylus longer than unguis.

No pleopods present in female.

Uropod (fig. 45J) basal article naked, as long as exopod. Endopod article 1 as long as article 2, with two setulated setae; article 2 with one seta subdistally, and four long simple setae and two short setulated setae distally. Exopod as long as endopod article 1, with one seta medially and two simple setae distally.

Distribution. — Widespread in the northern part of the Gulf. Bathymetric range wide: 1079–2050 m. See map 31.

Remarks. — The shape of the articles of pereopods 5–6 should not be given too much weight as some drying had occurred.

This species was one of a large number of filiform species found in the deep-sea Gulf samples (see under Filiform species, below).

KEY TO THE SPECIES OF *FILITANAIS*, FEMALES ONLY (GLOBAL)

1. Pleonites as long as, or longer than, pereonites .... *F. filiformis* sp. nov.
– Pleonites shorter than pereonites 3
2. Pleonites and pleotelson rounded. Uropodal exopod shorter than 0.5 times first endopod article.. *F. moskalevi* Kudinova-Pasternak, 1973
– Pleonites and pleotelson rectangular or square. Uropodal exopod as long as 0.5 times first endopod article .... *F. rebainsi* Kudinova-Pasternak, 1975

Remarks on the key. — As stated under generic remarks, it is the present author's opinion that neither *Filitanais vulgaris* Kudinova-Pasternak, 1981 nor *F. curticaudus* Kudinova-Pasternak, 1993 belong to *Filitanais*, and those species are thus not included in the key.

Genus **Isopodidus** Larsen & Heard, 2002

Isopodidus Larsen & Heard, 2002, Proceedings of the Biological Society of Washington, **115**(2): 403.

Diagnosis. — (Modified after Larsen & Heard, 2002.) Female. Body dorsoventrally flattened and semi-cylindrical. Carapace as long as lengths of first five pereonites combined. Eye lobes and visual elements absent. Pleon and pleotelson fused. Antennule with four articles. Antenna with six articles. Mandibular molar well developed. Maxillule with some bifurcate terminal setae. Maxillipedal bases not fused distally, endites not fused; with one minute denticle distally; palp article 2 without multifurcate seta. Chelipeds attached ventrally by sclerite. Pereopods without coxae, with dactylus and unguis combined never longer than propodus. Pereopod 4–6 dactyli and ungues not fused to a claw. Pleopods present, sitting in niches created by inner septum of pleonite wall; with simple setae only. Uropods small, exopod partially fused with basal article, endopod consisting of two partially fused articles.

Male. Unknown.

Type species. — *Isopodidus janum* Larsen & Heard, 2002.

Gender of generic name. — Neuter [as specifically implied and explicitly stated by the authors].

Remarks. — Several tanaid genera show fusion in the pereon. However, except for *Isopodidus*, fusion of all pleonites is always accompanied by the loss of pleopods.

Species currently assigned to this genus. — *I. janum* Larsen & Heard, 2002.

Isopodidus janum Larsen & Heard, 2002

Isopodidus janum Larsen & Heard, 2002, Proceedings of the Biological Society of Washington, **115**(2): 403.

Diagnosis. — Monotypic genus, no species diagnosis given.

Distribution. — A very rare species, only known from a few specimens from the northern part of the Gulf of Mexico and one location in the Caribbean Sea. Bathymetric range: 858–988 m. See map 32.

Genus **Leptognathiella** Hansen, 1913

Leptognathiella Hansen, 1913, Danish Ingolf Expedition, **3**(3): 104.
Leptognathia – Lang, 1968, Galathea Report, **9**: 166.

Diagnosis. — (Modified from Bird & Holdich, 1984.) Female. Body elongate. Carapace length varies. Eye-lobes and visual elements absent. Pleon and pleotelson not fused, short (rarely as long as lengths of last three pereonites combined). Antennule with 4–5 articles; about as long as carapace; article 2 with dorsal projection overlapping basal part of article 3. Antenna with 5–6 articles. Molar process of mandible thin, with few terminal spines. Maxillule with 7–8 spiniform terminal setae (see Remarks). Maxillipedal endite not fused, distal edge with medial, frequently triangular, process; palp article 2 without multifurcate seta. Chelipeds attached via sclerite. Pereopods 1–3 often stout, with or without coxa. Pereopods 4–6 without coxa and not stouter than pereopods 1–3; dactylus and unguis not fused. Pleopods absent or present with simple setae only. Uropods longer than pleotelson, biramous; rami with 1–2 articles.

Male. Pleon and pleotelson longer than in female, often longer than the lengths of last two pereonites combined (but barely so). Pleopods present, with simple setae only.

Type species. — *Leptognathiella abyssi* Hansen, 1913.

Gender of generic name. — Neither Hansen (1913) who erected this genus, nor Bird & Holdich (1984), who re-erected the genus, mentioned the gender of the generic name. However, as the name is ending in -ella, according to the Code (1999: 34, Art. 30.1) the gender is feminine.

Remarks. — The dorsal projection of antennular article 2 (overlapping basal part of article 3) is a stable character within this genus. It is, however, not unique to *Leptognathiella* and should be considered a valuable, but not decisive, diagnostic character.

All species of *Leptognathiella* described so far have consistently displayed 7–8 spiniform terminal setae in the adult. However, the number of spiniform terminal setae on the maxillule has been shown to be ontogenetically dependent in a number of other tanaidacean taxa (Bird & Holdich, 1984; Larsen & Wilson, 1998; Larsen, 2001, 2002a) and should thus not be given too much value.

The medial, frequently triangular process on the distal edge of the endite of the maxilliped seems to be a very useful character for this genus.

The ischial setulation described as diagnostic by Bird & Holdich (1984) seems to be a good character for the Atlantic species. However, these setae are fragile and it is often impossible to determine whether they have been broken off or have never been present (inconsistencies were noted when appendages from one side of the body were compared with those from the other side).

Leptognathiella was erected by Hansen (1913) and later synonymized with *Leptognathia* by Lang (1968), but then recognized as valid by Bird & Holdich (1984). Most recent workers agree that the genus *Leptognathia* is enormously over-burdened and such without uniting, diagnostic characters (Sieg, 1983a, 1986a; Kudinova-Pasternak, 1983; Bird & Holdich, 1984; Larsen & Wilson, 2002). *Leptognathiella* is remarkably well defined for a leptognathiid genus. The genus has so far only been recorded from the North Atlantic. The species described herein are the first to be described outside the northeast Atlantic.

Species currently assigned to this genus. — *L. abyssi* Hansen, 1913; *L. clivicola* Bird & Holdich, 1984; *L. occidentis* sp. nov.; *L. spinicauda* Bird & Holdich, 1984; *L. subaequalis* (Hansen, 1913); *L. vermiformis* sp. nov.

Leptognathiella occidentis sp. nov.
(figs. 47–50)

Material examined. — Holotype, non-ovigerous ♀ (USNM 1068883), body length 1.45 mm, DGoMB Sta. S37-2. Paratypes: 1 non-ovigerous ♀ (USNM 1068884), body length 1.45 mm,

DGoMB Sta. NB5-1. 1 ♂ (USNM 1068885), body length 1.3 mm, DGoMB Sta. WC12-2. 1 ♂ (USNM 1068886), body length 1.25 mm, DGoMB Sta. S37-3. 1 non-ovigerous ♀ (USNM 1068887), DGoMB Sta. S37-2 (dissected). 1 ♂ (USNM 1068888), same locality (dissected).

Additional material. — 2 non-ovigerous ♀♀, DGoMB Sta. S37-1. 2 ♂♂, 3 non-ovigerous ♀♀, 3 mancas, DGoMB Sta. S37-3. 1 non-ovigerous ♀, DGoMB Sta. W5-3. 1 ♂, 1 non-ovigerous ♀, DGoMB Sta. NB2-1. 1 ♂, 1 manca, DGoMB Sta. RW1-1. 1 ♂, 1 non-ovigerous ♀, DGoMB Sta. C4-1. 1 ovigerous ♀, DGoMB Sta. RW3-1. 3 non-ovigerous ♀♀, DGoMB Sta. S37-1. 1 ♂, DGoMB Sta. NB4-1. 2 non-ovigerous ♀♀, 2 mancas, DGoMB Sta. NB5-2. 1 manca, DGoMB Sta. S44-2. 2 non-ovigerous ♀♀, DGoMB Sta. NB3-3. 1 non-ovigerous ♀, DGoMB Sta. RW2–3. 1 ♂, 2 mancas, DGoMB Sta. WC12–3. 1 ♂, 2 non-ovigerous ♀♀, 1 manca, DGoMB Sta. RW5-2. 1 manca, DGoMB Sta. W4-2. 2 ♂♂, 2 non-ovigerous ♀♀, 1 manca, DGoMB Sta. C4-2. NGoMCSS Sta. E2B-4496-6. 2 non-ovigerous ♀♀, DGoMB Sta. W6-4. 2 non-ovigerous ♀♀, DGoMB Sta. S39-4. 4 non-ovigerous ♀♀, DGoMB Sta. C4-4. 1 non-ovigerous ♀, DGoMB Sta. NB3-4. 4 non-ovigerous ♀♀, DGoMB Sta. S36-1. 1 non-ovigerous ♀, DGoMB Sta. S41-4. 4 non-ovigerous ♀♀, DGoMB Sta. W4-4. 3 non-ovigerous ♀♀, DGoMB Sta. WC12-4. 1 non-ovigerous ♀, DGoMB Sta. NB4-4.

Diagnosis. — Female. Body stout relative to other species of the genus, 7 times as long as wide. Pereopods 1–3 with carpus and propodus without small ventral spines. Pereopod 2 apparently without ischial setae. Uropods longer than pleon; basal article longer than pleotelson and exopod minute (0.3 times as long as first endopod article).

Male. Antennule more robust than in female. Pleopods better developed than in female but still with only simple setae.

Etymology. — Named after the 'western' distribution of this species compared to all other species of the genus (Latin occidens = the west). The specific name thus is a noun in the genitive singular.

Type locality. — DGoMB Sta. S37-2, 28°33'4292"N 87°45'6441"W, depth 2382 m.

Description. — Adult female. Body (fig. 47A) slightly dorso-ventrally flattened, 7 times as long as wide.

Carapace shorter than the lengths of pereonites 1 and 2 combined. Eye-lobes absent.

Pereonites 1, 2, 3, 4, and 6 wider than long. Pereonite 5 as wide as long.

Pleon short (including pleotelson only 0.2 times total body length). All pleonites subequal. Pleotelson as long as three free pleonites combined, apex almost flat, and covered by a dorsal plate.

Antennule (fig. 47C) hardly longer than carapace. Article 1 shorter than rest of antennule, naked. Article 2 is 0.25 times as long as dorsal part of article 3, more than 0.5 times as long as article 1, with one simple and one setulated seta distally. Article 3 shorter than article 4, with one simple seta distally. Article 4

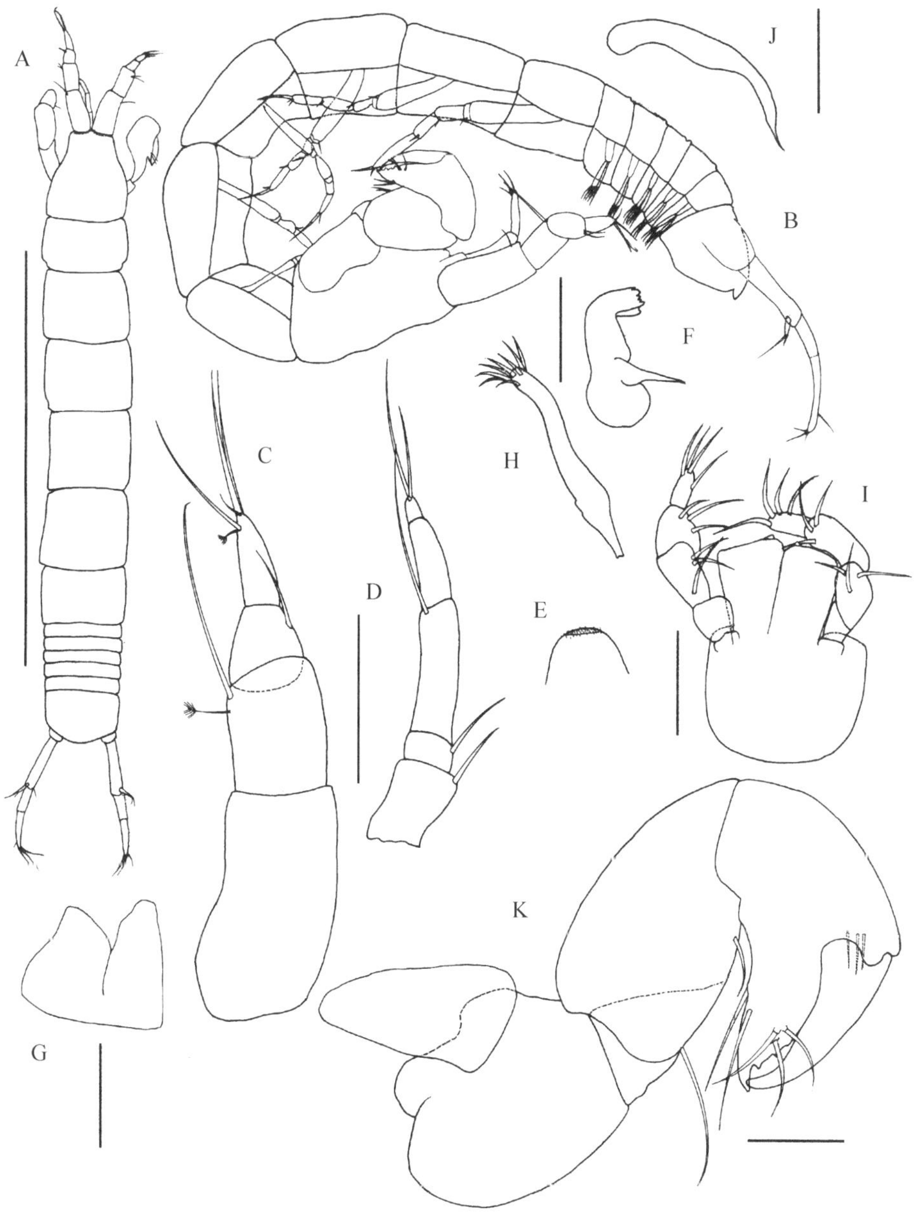

Fig. 47. *Leptognathiella occidentis* sp. nov.; A, holotype, female, dorsal view; B, paratype, male, lateral view; C-K, female; C, antennule; D, antenna; E, labrum; F, left mandible; G, labium; H, maxillule; I, maxilliped; J, epignath; K, cheliped. Scale bars equal: A, B, 1.0 mm; C-D, K, 0.1 mm; E-J, 0.05 mm.

shorter than article 2, with three thick long setae, one tiny seta, and one tiny setulated seta distally.

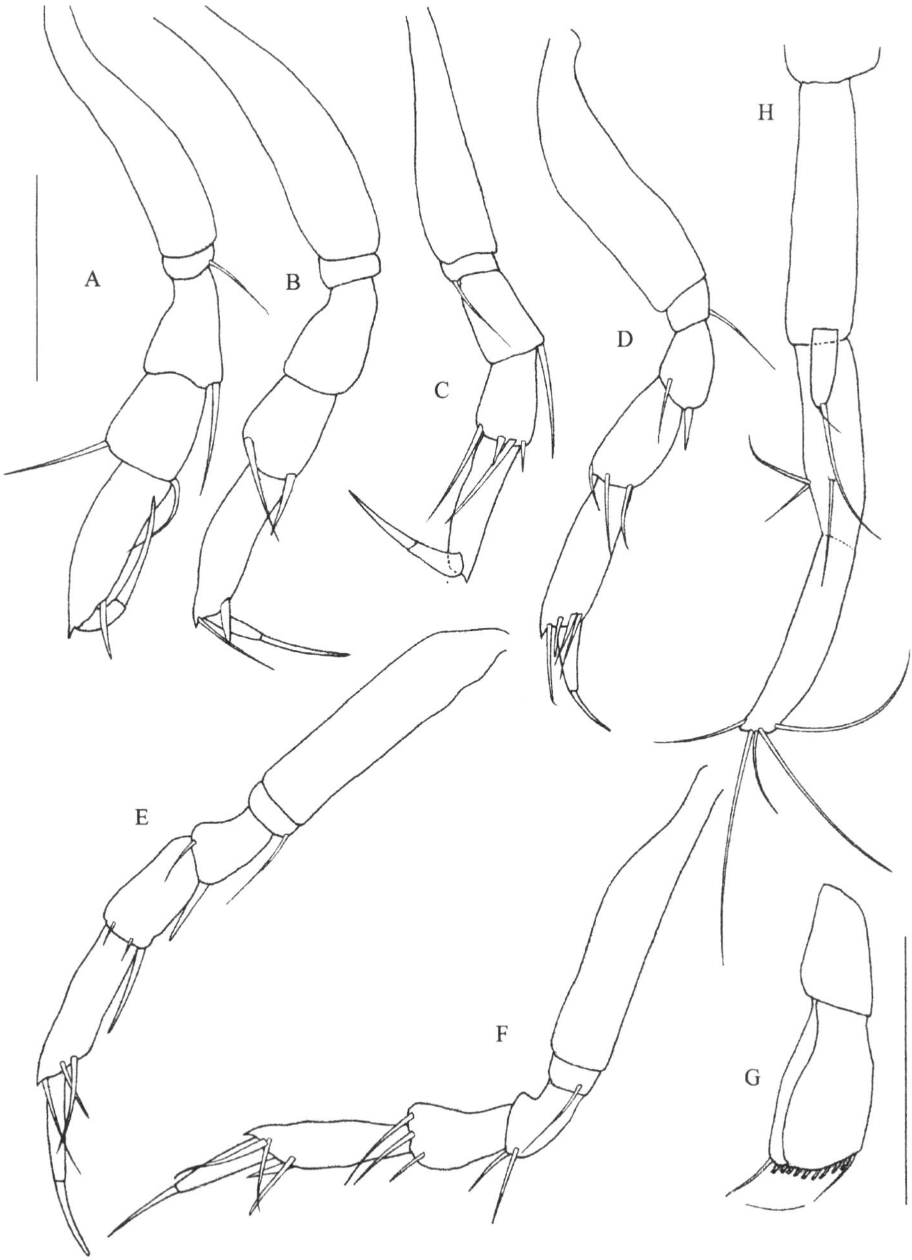

Fig. 48. *Leptognathiella occidentis* sp. nov., female; A, pereopod 1; B, pereopod 2; C, pereopod 3; D, pereopod 4; E, pereopod 5; F, pereopod 6; G, pleopod; H, uropod. Scale bars equal 0.1 mm.

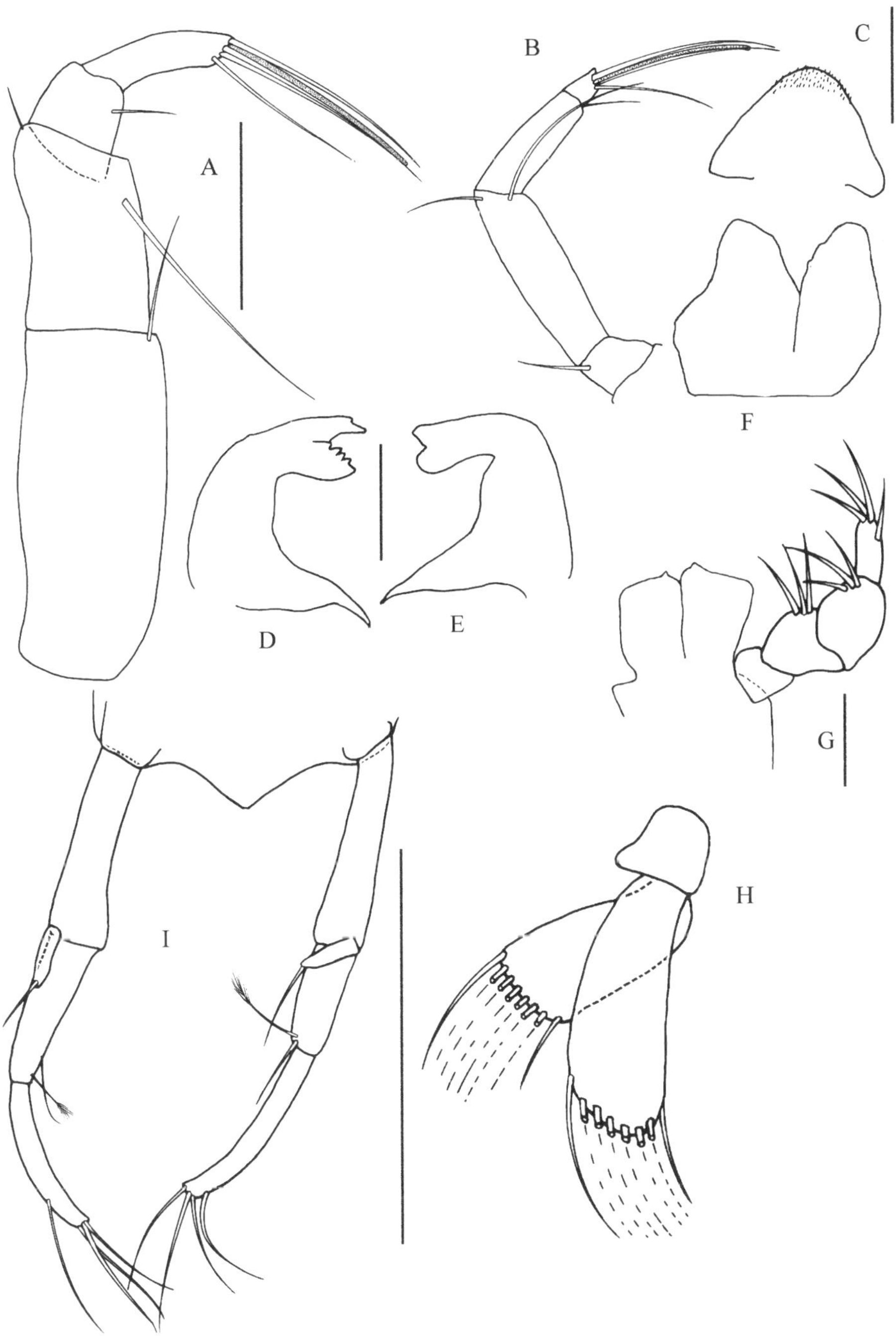

Fig. 49. *Leptognathiella occidentis* sp. nov., male; A, antennule; B, antenna; C, labrum; D, left mandible; E, right mandible; F, labium; G, maxilliped; H, pleopod; I, pleotelson and uropods. Scale bars equal: A, B, H, I, 0.1 mm; C-G, 0.05 mm.

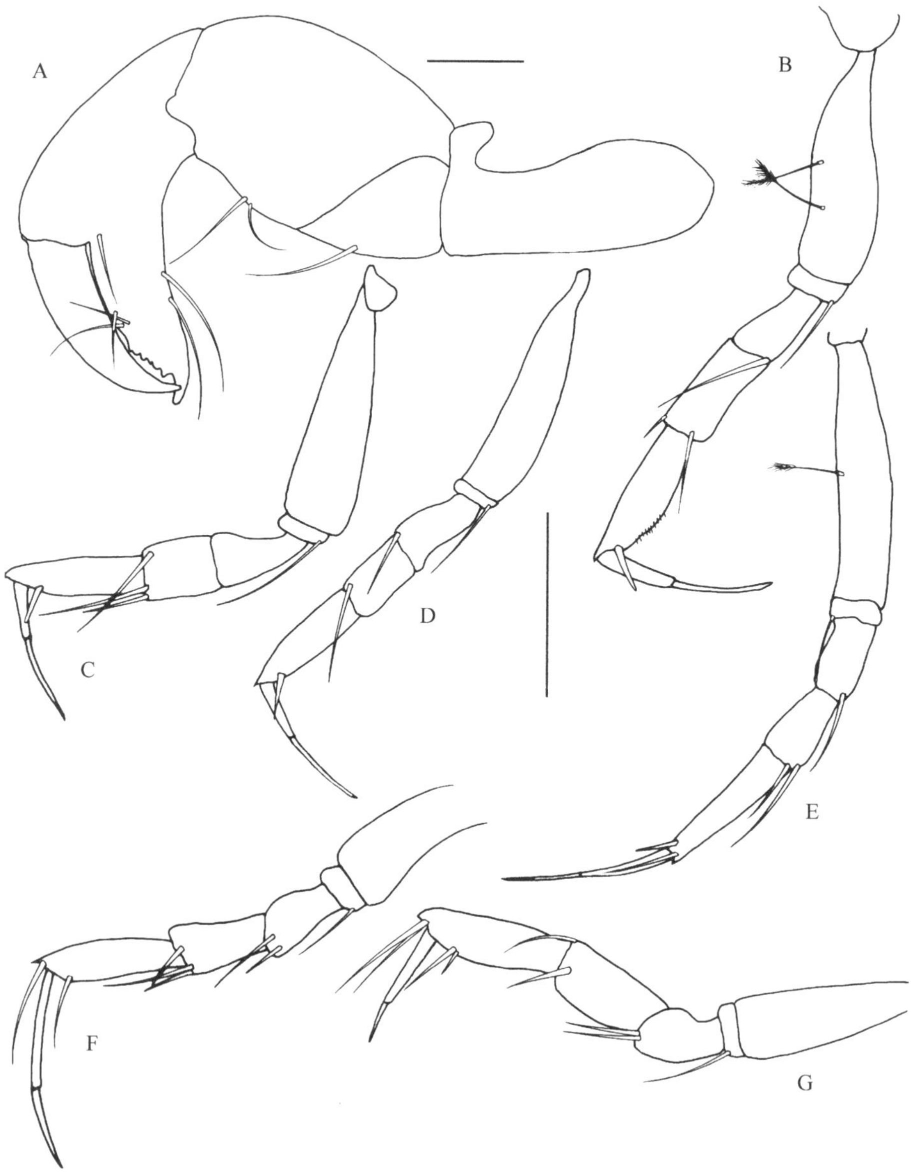

Fig. 50. *Leptognathiella occidentis* sp. nov., male; A, cheliped; B, pereopod 1; C, pereopod 2; D, pereopod 3; E, pereopod 4; F, pereopod 5; G, pereopod 6. Scale bars equal 0.1 mm.

Antenna (fig. 47D) 0.67 times as long as antennule. Article 1 wider than following articles, with one seta dorso-distally. Article 2 longer than article 5, with one seta dorso-distally. Article 3 longer than other articles, with one simple long seta distally. Article 4 longer than article 1, naked. Article 5 minute, with two long setae distally.

Mouthparts. Labrum (fig. 47E) finely setose and hood-shaped. Mandibular molar process pointed and longer than incisor, with small distal spines. Left mandible (fig. 47F) lacinia mobilis shorter than incisor; incisor broad, with five denticles. Labium (fig. 47G) lobes naked. Maxillule (fig. 47H) endite with eight spiniform setae distally; palp not recovered. Maxilla not recovered. Maxilliped (fig. 47I) endites with one small triangular denticle medially on distal margin, almost as wide as basis; palp article 1 naked, articles 2 and 3 with distal expansion and three setae on inner margin, article 4 only 0.5 times as wide as article 3, with four setae. Epignath (fig. 47J) feeble and with naked apex.

Cheliped (fig. 47K) basis divided by prominent sclerite, shorter than carpus. Merus triangular, with one seta ventrally. Carpus shorter than propodus including fixed finger, with one seta ventrally. Propodus with three setae near dactylus insertion on inner margin. Fixed finger with two setae ventrally and three on inner margin, with three denticles on inner margin. Dactylus as long as fixed finger.

Pereopod 1 (fig. 48A) without visible coxa. Basis longer than the three succeeding articles combined, naked. Ischium with one long (as long as merus) seta. Merus as long as carpus, widening distally and with one seta distally. Carpus 0.5 times as long as propodus, with one seta distally. Propodus more than 0.5 times as long as basis, with one spiniform seta distally and a dorsal spine. Dactylus and unguis together only hardly shorter than propodus and not fused to a claw.

Pereopod 2 (fig. 48B) as pereopod 1 except: merus naked. Propodus with one simple and one spiniform seta distally.

Pereopod 3 (fig. 48C) as pereopod 1 except: carpus with four setae distally. Propodus with only a dorsal spine.

Pereopod 4 (fig. 48D) as pereopod 3 except: merus with one simple and one spiniform seta distally. Carpus with three setae distally. Propodus with four simple setae distally and a dorsal spine.

Pereopod 5 (fig. 48E) as pereopod 4 except: carpus with three setae distally. Propodus with three setae distally and a dorsal spine. Dactylus and unguis together longer than propodus.

Pereopod 6 (fig. 48F) similar to pereopod 5 except: merus with two simple setae distally. Carpus with three setae distally. Propodus with three simple and one spiniform setae distally, and a dorsal spine.

Pleopods (fig. 48G) small, rather stout. All pairs similar. Armed with simple setae only.

Uropod (fig. 48H) basal article longer than pleotelson, naked. Endopod divided into two pseudoarticles, longer than basal article, with three simple setae medially near pseudoarticulation and five long simple setae distally. Exopod shorter than 0.5 times first endopod 'pseudoarticle', with two unequal simple setae distally.

Male (figs. 47B, 49, 50), where differing from female.

Mouthparts. Left mandible (fig. 49D) incisor narrower than in female and with only two denticles, gap between incisor and lacinia mobilis larger than in female; lacinia mobilis with three or four denticles.

Cheliped (fig. 50A) basis narrower than in female (possibly an artifact). Fixed finger with multiple denticles.

Pleopods (fig. 49H) larger than in female, but still with simple setae only.

Distribution. — Widespread in the northern part of the Gulf. Bathymetric range wide: 213–3145 m. See map 33.

Remarks. — This species is very similar to the type-species of the genus, *L. abyssi*, but can be separated from this by the body being only 7 times as long as wide; by the lack of small ventral spines on pereopods 1–3; and by the short uropodal exopod (less than 0.5 times as long as first endopod article). It also differs in the lack of ischial setae on pereopod 2, but this is likely to be an artifact. This species was found in tubes consisting of very fine detritus, without any solid objects incorporated.

Leptognathiella vermiformis sp. nov.
(figs. 51–54)

Material examined. — Holotype, non-ovigerous ♀ (USNM 1068889), body length 1.45 mm, DGoMB Sta. MT3-1. Paratypes: 2 non-ovigerous ♀♀, body length 1.45 mm and 1.4 mm, 1 ♂, body length 1.25 mm (USNM 1068890), same locality. 1 non-ovigerous ♀ (USNM 1068892), same locality (dissected). 1 ♂ (USNM 1068891), same locality (dissected).

Additional material. — 9 non-ovigerous ♀♀, DGoMB Sta. MT3-2. 2 non-ovigerous ♀♀, DGoMB Sta. AC1-2. 1 non-ovigerous ♀, DGoMB Sta. S35-1. 1 ♂, 1 non-ovigerous ♀, DGoMB Sta. NB2-1. 2 non-ovigerous ♀♀, DGoMB Sta. RW4-2. 1 ♂, DGoMB Sta. MT4-4. 1 ♂, DGoMB Sta. C1-4. 1 non-ovigerous ♀, DGoMB Sta. C1-2. 1 non-ovigerous ♀, DGoMB Sta. MT2-2. 2 non-ovigerous ♀♀, DGoMB Sta. RW5-2. 1 non-ovigerous ♀, DGoMB Sta. W3-2. 3 non-ovigerous ♀♀, DGoMB Sta. B3-2. (Most of this material is in very poor condition.)

Diagnosis. — Female. Body greatly elongate relative to other species of the genus (17 times as long as wide). Pereopods 1–3 with carpus and propodus without small ventral spines. Uropods longer than pleotelson, basal article shorter than pleotelson, and exopod longer than first endopod article.

Male. Antennule more robust than in female. With reduced pleopods.

Etymology. — Named after the elongate, worm-like body shape. (Latin, vermiformis = worm-like). The name is an adjective agreeing in gender with the generic name.

Type locality. — DGoMB Sta. MT3-1, 28°13'2246"N 89°30'7679"W, depth 983 m.

Description. — Adult female. Body (fig. 51A) extremely elongate, 17 times as long as wide.

Carapace as long as pereonite 2. Eye-lobes absent.

Pereonites all longer than wide.

Pleon short (including pleotelson only 0.15 times total body length). All pleonites subequal. Pleotelson only hardly longer than the individual, free pleonites, apex rounded.

Antennule (fig. 51B) hardly shorter than carapace. Article 1 shorter than rest of antennule, with one simple and two setulose setae distally. Article 2 more than 0.5 times as long as article 1, with one simple seta. Article 3 longer than article 4, naked. Article 4 is 0.25 times as long as article 2, with one thick seta distally. Article five minute, with two long setae and one aesthetasc.

Antenna (fig. 51C) short (hardly longer than antennular article 1). Article 1 wider than following articles, with one seta dorso-distally. Article 2 longer than article 5, with one seta dorso-distally. Article 3 longer than other articles, with two simple and one setulate setae distally. Article 4 longer than article 1, with one simple seta distally. Article 5 minute, with three setae distally.

Mouthparts. Labrum (fig. 51D) finely setose and hood-shaped. Mandibular molar process pointed, as long as incisor. Left mandible (fig. 51E) lacinia mobilis shorter than incisor; incisor narrow, without denticles. Right mandible (fig. 51F) incisor wider than that of left mandible, with three denticles. Labium (fig. 51G) lobes naked. Maxillule (fig. 51H) endite with seven spiniform setae, of which a distal one is bifurcated; palp not recovered. Maxilla (fig. 51I) tapering in anterior direction. Maxillipedal endites (fig. 51J) with one small triangular denticle medially on distal margin (but apparently only on one endite), almost as wide as basis; palp (fig. 51K) article 1 naked, articles 2 and 3 with three setae on inner margin, article 4 only 0.5 times as wide as article 3, with five setae. Epignath not recovered.

Fig. 51. *Leptognathiella vermiformis* sp. nov., female; A, holotype, lateral view; B, antennule; C, antenna; D, labrum; E, left mandible; F, right mandible; G, labium; H, maxillule; I, maxilla; J, maxilliped, endites; K, maxilliped, palp. Scale bars equal: A, 0.5 mm; B-C, 0.1 mm; D-K, 0.05 mm.

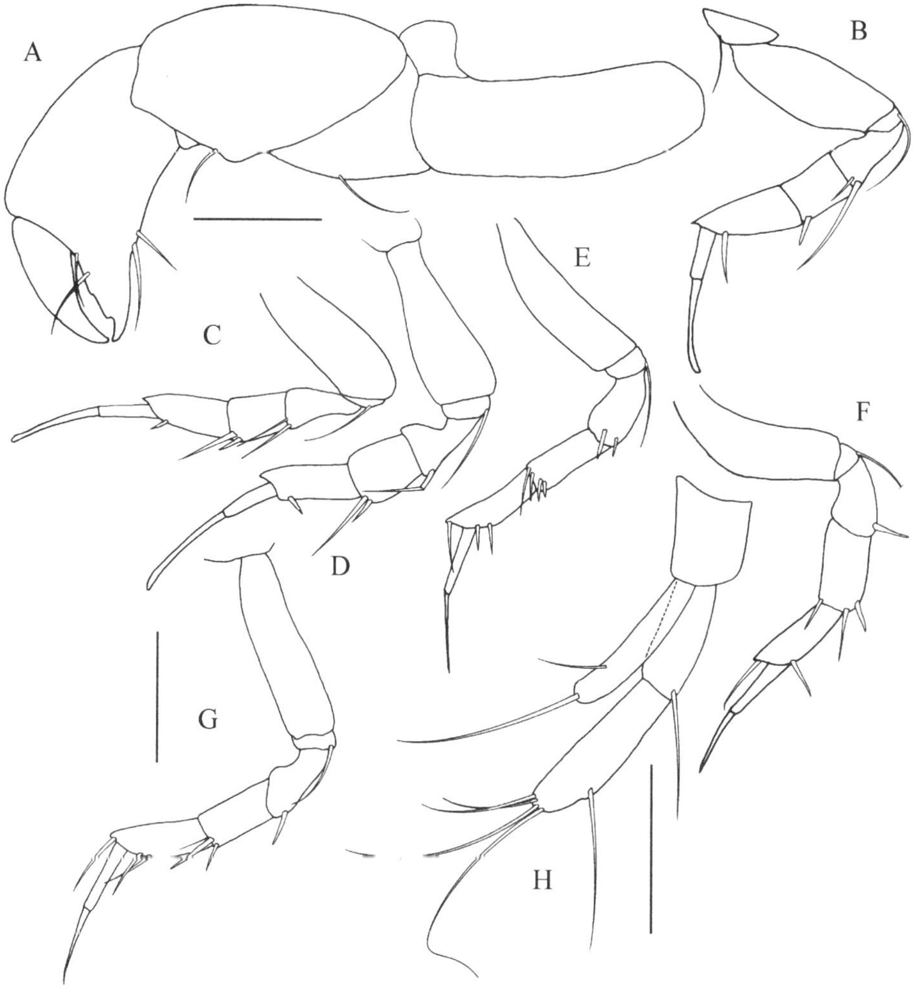

Fig. 52. *Leptognathiella vermiformis* sp. nov., female A, cheliped; B, pereopod 1; C, pereopod 2; D, pereopod 3; E, pereopod 4; F, pereopod 5; G, pereopod 6; H, uropod. Scale bars equal 0.1 mm.

Cheliped (fig. 52A) attachment via elongated sclerite. Basis narrower than other articles, as long as carpus. Merus triangular, with one seta ventrally. Carpus longer than propodus including fixed finger, with one seta ventrally. Propodus naked. Fixed finger with two setae ventrally and two on inner margin, with two denticles on inner margin. Dactylus as long as fixed finger. Both dactylus and fixed finger without scleretorized unguis.

Pereopod 1 (fig. 52B) coxa partly fused to pereonite, with one simple seta. Basis longer than the three succeeding articles combined, naked. Ischium with long (as long as merus) seta. Merus as long as carpus, widening distally, and with two setae of unequal length distally. Carpus more than 0.5 times as long as propodus, with one short spiniform seta distally. Propodus shorter than 0.5 times basis, with one spiniform seta distally and a dorsal spine. Dactylus and unguis together longer than propodus, and not fused to a claw. Unguis longer than dactylus, with spatulate tip.

Pereopod 2 (fig. 52C) as pereopod 1 except: merus with two equally long setae distally. Carpus with two spiniform setae of unequal length distally.

Pereopod 3 (fig. 52D) as pereopod 2 except: merus with one seta distally.

Pereopod 4 (fig. 52E) not wider than pereopods 1–3. No visible coxa. Basis longer than the three succeeding articles combined, naked. Ischium with long (as long as merus) seta. Merus as long as carpus, widening distally, and with two equally long small setae distally. Carpus more than 0.5 times as long as propodus, with four short spiniform setae distally. Propodus shorter than 0.5 times basis, with three spiniform setae distally and a dorsal spine. Dactylus and unguis together longer than propodus, and not fused to a claw. Dactylus longer than unguis. Unguis without spatulate tip.

Pereopod 5 (fig. 52F) as pereopod 4 except: merus with one spiniform seta. Carpus with three spiniform setae distally. Propodus with two setae distally and a dorsal spine.

Pereopod 6 (fig. 52G) as pereopod 5 except: propodus with three spiniform setae distally and a dorsal spine.

Pleopods not present in female.

Uropod (fig. 52H) basal article shorter than 0.5 times pleotelson, naked. Endopod article 1 longer than basal article, with one simple seta distally; article 2 longer than first article, with four setae distally and one subdistally. Exopod longer than first endopod article, with two unequally long simple setae distally.

Male (figs. 53, 54), where differing from female. Body (fig. 53A) less extremely elongate than in female, about 15 times as long as wide.

Pleon longer than in female (including pleotelson about 0.23 times total body length). All pleonites subequal. Pleotelson (fig. 54I) as long as two free pleonites combined, almost square.

Antennule (fig. 53B) conspicuously stouter than in female but otherwise identical in structure.

Antenna (fig. 53C) identical to that of female.

Mouthparts (fig. 53D-H) all identical to those of female.

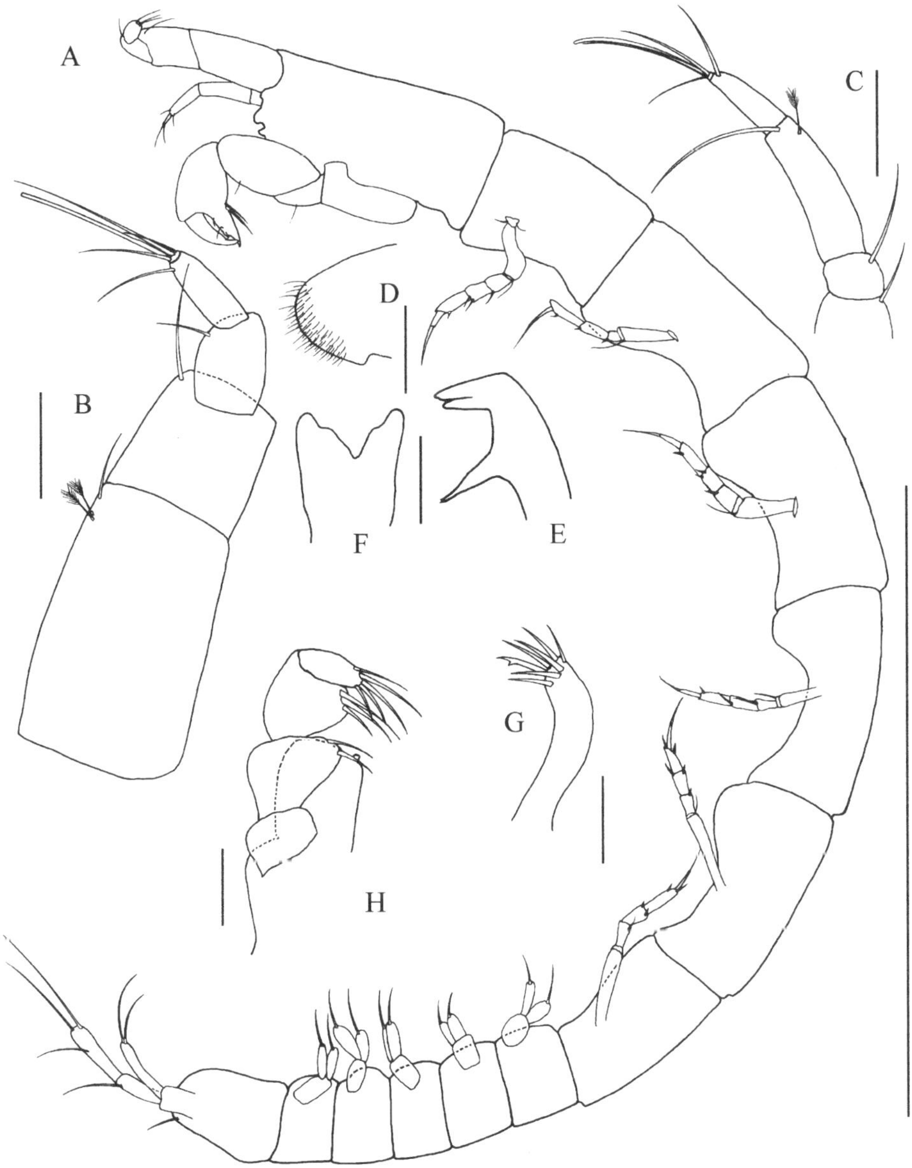

Fig. 53. *Leptognathiella vermiformis* sp. nov., male; A, paratype, lateral view; B, antennule; C, antenna; D, labrum; E, left mandible; F, labium; G, maxillule; H, maxilliped. Scale bars equal: A, 0.5 mm; B-C, 0.1 mm; D-H, 0.05 mm.

Cheliped (fig. 54A) as female, except basis shorter than in female. Carpus with two setae ventrally.

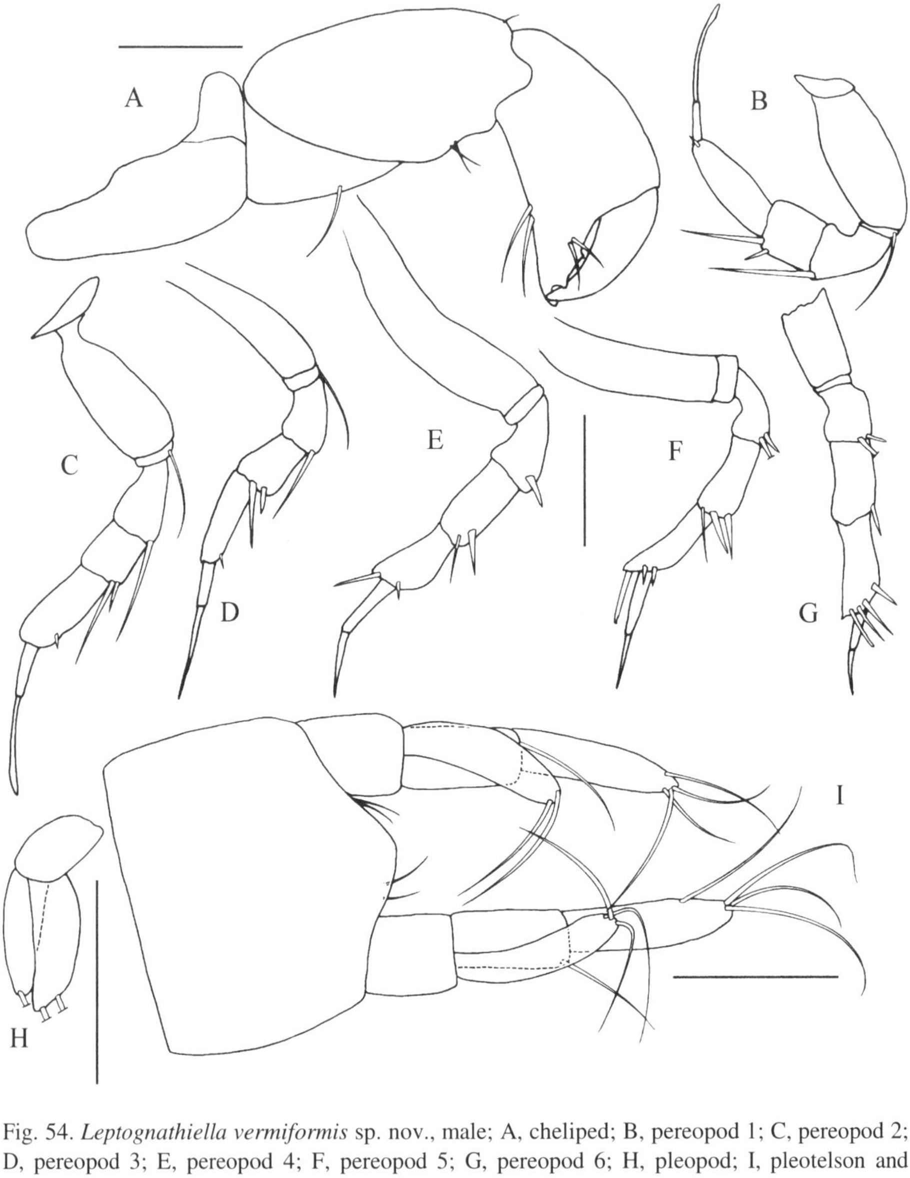

Fig. 54. *Leptognathiella vermiformis* sp. nov., male; A, cheliped; B, pereopod 1; C, pereopod 2; D, pereopod 3; E, pereopod 4; F, pereopod 5; G, pereopod 6; H, pleopod; I, pleotelson and uropods. Scale bars equal 0.1 mm.

Pereopods (fig. 54B-G) essentially as the female pereopods, except for some differences in setal numbers. These setal numbers should not be considered fixed, as the male pereopods were in very poor condition.

Pleopods (fig. 54H) small and reduced. Endo- and exopod with only one or two simple setae.

Uropod (fig. 54I) as in female.

Distribution. — Widespread in the northern part of the Gulf. Bathymetric range wide: 336–2580 m. See map 34.

Remarks. — This species is one of the 'filiform' species group (see under Filiform species, below) and can be separated from all other species of *Leptognathiella* by the filiform, extremely elongate body shape, 17 times as long as wide.

The maxilliped apparently has a denticle distally on only one endite. Normally this would be considered an artifact, but the same condition is found in all four dissected specimens (albeit the material is in very poor condition). Also, owing to the bad condition of the material, setal characters should be treated with caution, as the risk for artifacts seems rather high, particularly in the male.

KEY TO THE SPECIES OF *LEPTOGNATHIELLA*, FEMALES ONLY (GLOBAL)

1. Body extremely elongate (length/width ratio > 15) .... *L. vermiformis* sp. nov.
– Body less elongate (length/width ratio < 11) 2
2. Uropod basal article longer than pleotelson. Total length of uropods greater than length of pleon 3
– Uropod basal article shorter than pleotelson. Total length of uropods less than length of pleon 4
3. Pereopod 1–3 carpus/propodus with row of small ventral spines. Uropodal exopod more than 0.5 times as long as first endopod article .... *L. abyssi* Hansen, 1913
– Pereopods 1–3 carpus/propodus without row of small ventral spines. Uropodal exopod less than 0.5 times as long as first endopod article .... *L. occidentis* sp. nov.
4. Uropod basal article with projection adjacent to exopod. Pleotelson with demarcated apical projection .... *L. spinicauda* Bird & Holdich, 1984
– Uropod basal article without projection. Pleotelson tapering steadily towards apex, without projection 5
5. Cheliped fixed finger with four denticles. Uropodal exopod uni-articulate, reaching to middle of endopod article 2 .... *L. subaequalis* (Hansen, 1913)
– Cheliped fixed finger with two or less denticles. Uropodal exopod bi-articulate, reaching to beginning of endopod article 2 only .... *L. clivicola* Bird & Holdich, 1984

Genus **Tumidochelia** Knight, Larsen & Heard, 2003

Tumidochelia Knight, Larsen & Heard, 2003, Proceedings of the Biological Society of Washington, **116**(2): 497.

Diagnosis. — (Modified after Knight et al., 2003.) Female. Body elongate. Carapace longer than wide. Eye-lobes and visual elements absent. Pleon and pleotelson not fused. Antennule with four articles. Antenna with six articles.

Molar process of mandible fairly thin, with few terminal spines. Maxillule with eight spiniform terminal setae, some of which setulose Maxillipedal endite not fused, naked; palp article 2 without multifurcate seta. Cheliped with inflated, "shield-like" expansion of disto-lateral face of carpus. Pereopods 1–3 with dactylus and unguis together shorter than propodus. Pereopods 4–6 with dactylus and unguis together longer than propodus. Pleopods present, with simple setae. Uropod biramous, bearing distal process on basal article; endopod and exopod each with 1 article.

Male. Unknown.

Type species. — *Tumidochelia randyi* Knight, Larsen & Heard, 2003.

Gender of generic name. — Feminine.

Remarks. — The combination of a cheliped carpus shield and the uropod bearing a distal process on the basal article, as well as both endo- and exopod well developed, make this genus unique.

Species currently assigned to this genus. — *T. randyi* Knight, Larsen & Heard, 2003.

Tumidochelia randyi Knight, Larsen & Heard, 2003

Tumidochelia randyi Knight, Larsen & Heard, 2003, Proceedings of the Biological Society of Washington, **116**(2): 498.

Diagnosis. — Monotypic genus, no species diagnosis given.

Distribution. — Relatively common in the northern part of the Gulf. Bathymetric range: 482–2974 m. See map 35.

Family LEPTOCHELIIDAE Lang, 1973

Leptocheliidae Lang, 1973, Zoologica Scripta, **2**(5–6): 197.
Paratanaidae – Sieg, 1983, Crustaceorum Catalogus, **6**: 419.

Diagnosis. — (Modified after Larsen & Wilson, 2002.) Female. Medium body calcification. No plates in carapace. Eyes present (except *Mesotanais*). Pereon with 6 free pleonites. Pereonite 1 not reduced. Pleon with 5 free pleonites, as wide as pereon, no articulated setae on pleonites. Antennule with 3–5 articles. Antenna with six articles, articles 2 and 3 with or without (*Heterotanaoides*) spiniform setae dorsally. Molar process of mandible broad and heavily calcified. Labium with two pairs of lobes, medial spiniform setae absent. Maxillipedal bases not fused; endites not fused, narrower than basis and with two or (frequently) three, flat setae. Cheliped attachment via sclerite. Marsupium (where known) with four pairs of oostegites. Pereopod coxa present on pereopods 1–3 but absent on pereopods 4–6. Pereopods 4–6 with dactylus and unguis fused to a hook. Pleopods well developed. Uropodal endopod with 3–6 articles, exopod with 1–2 articles.

Male (where known) sexual polymorphism high, with several morphologically different male morphs. Eyes larger than in female. Pleon slightly wider than pereon, no articulated setae on pleonites. Antennule with more than four articles and several aesthetascs. Mouthparts reduced and not functional. Cheliped always significantly enlarged and elongate. Pleopods well developed.

KEY TO THE DEEP-SEA GENERA OF LEPTOCHELIIDAE IN THE GULF OF MEXICO

1. Eyes present .... *Bathyleptochelia* Larsen, 2003
– Eyes absent .... *Mesotanais* Dollfus, 1897

Subfamily HETEROTANAIDINAE Larsen & Wilson, 2002

Heterotanaidinae Larsen & Wilson, 2002, Journal of Zoological Systematics and Evolutionary Research, **40**(4): 213.

Diagnosis. — (Modified after Larsen & Wilson, 2002.) Female. Antennule with three articles. Antenna articles 2–3 with or without spiniform setae dorsally. Eyes and eye-lobes present or absent (*Mesotanais*).

Male. Cheliped fixed finger and dactylus often of strikingly dissimilar size and shape.

Type genus. — *Heterotanais* G. O. Sars, 1882.

Genera included. — *Bathyleptochelia* Larsen, 2003; *Heterotanais* G. O. Sars, 1882; *Heterotanaoides* Sieg, 1976; *Mesotanais* Dollfus, 1897.

Genus **Bathyleptochelia** Larsen, 2003

Bathyleptochelia Larsen, 2003, Journal of Crustacean Biology, **3**(4): 785.

Diagnosis. — (After Larsen, 2003b.) Female. Body somites not fused. Eyes well developed. Antennule with three articles. Antenna with six articles; articles 2–3 without spiniform setae dorsally. Maxillule with six short spiniform terminal setae; endite shaft with several setae on inner margin. Maxillipedal endites with three short, flat, unequal-sized setae and one simple seta; basis with two long simple setae. Chelipeds attached via sclerite. Pereopod 1 almost 2 times as long as following pereopods; dactylus/unguis combined longer than propodus. Other pereopods with dactylus/unguis combined shorter than propodus. Pereopods 4–6 with basis thicker than in pereopods 1–3, dactylus/unguis fused to an elongate claw. Uropods biramous; exopod uni-articulate; endopod with three or more articles.

Male. Unknown.

Type species. — *Bathyleptochelia oculata* Larsen, 2003.

Gender of generic name. — Feminine.

Remarks. — This genus is remarkably similar to other genera in the family Leptocheliidae, but can be separated by having no spiniform setae on the antennal articles 2 and 3, and by the conspicuously shaped maxillule. *Bathyleptochelia* can easily be separated from the only other genus of Leptocheliidae found in deep water, *Mesotanais*, by the presence of large eyes.

Species currently assigned to this genus. — *B. oculata* Larsen, 2003.

Bathyleptochelia oculata Larsen, 2003

Bathyleptochelia oculata Larsen, 2003, Journal of Crustacean Biology, **3**(4): 785.

Diagnosis. — Monotypic genus, no species diagnosis given.

Distribution. — This species was found in only one sample, from a cold seep: GB 425, Sta. BL7, 27°33'60"N 92°32'33"W. Bathymetric range: 570 m. See map 36.

Genus **Mesotanais** Dollfus, 1897

Mesotanais Dollfus, 1897, Bulletin de la Société Zoologique de France, **21**: 213.

Diagnosis. — (Modified from Dollfus, 1897; Sieg & Heard, 1989.) Female. Carapace without eyes and eye-lobes. Antennule with three articles. Antenna with six articles; at least article 2 with spiniform setae dorsally. Maxillule with nine spiniform setae. Maxillipedal basis with two long simple setae at palp insertion; endites with 2–3 modified distal setae and sometimes tiny coupling hooks. Chelipeds attached via sclerite. Pereopod 1 longer than following pereopods; dactylus/unguis combined longer than propodus. Pereopods 4–6 with basis thicker than in pereopods 1–3, dactylus/unguis fused to an elongate claw. Uropodal exopod with two articles.

Male. Antennule with seven articles densely covered with aesthetascs. Mouthparts reduced. Basis of cheliped elongate (no completely intact male chelipeds have been described).

Type species. — *Mesotanais dubius* Dollfus, 1897.

Gender of generic name. — Masculine.

Remarks. — This genus is easily recognizable as a leptocheliid by the multi-articulate, biramous uropod, two paired labium lobes, and unfused maxillipedal basis; and it is differentiated from other Leptocheliidae by the lack of eyes, combined with the two long simple setae on the maxillipedal basis.

Sieg & Heard (1989) noted the almost complete lack of males (only one damaged male ever found) in their huge collection of *Mesotanais*. Since males of Leptocheliidae are usually fairly common compared to those of other taxa, Sieg & Heard (1989) speculated that *Mesotanais* had seasonal breeding (and seasonal males) and that the lack of collected males was due to discontinuous sampling (Sieg & Heard, 1989: 75). During the DGoMB project, numerous *Mesotanais* specimens were collected, but again no males were found. Since the DGoMB sampling was exclusively executed during the summer months (May–August) and since the collection by Sieg & Heard (1989) continued from April to November, breeding of *Mesotanais* is suggested to happen during winter.

Species currently assigned to this genus. — *M. dubius* Dollfus, 1897; *M. elongatus* Sieg & Bird, 1989; *M. longisetosus* Sieg & Heard, 1989; *M. vadicola* Sieg & Heard, 1989.

Mesotanais longisetosus Sieg & Heard, 1989

Mesotanais longisetosus Sieg & Heard, 1989, Gulf Research Report, **8**(2): 86.

Diagnosis. — (Modified after Sieg & Heard, 1989.) Female. Antennule with long (almost as long as antennule) setae. Mandibular molar not tapering distally. Maxillipedal endites with two large and one small modified setae distally, coupling hooks absent.

Male. Cheliped basis elongate.

Distribution. — This species is only recorded from the Gulf of Mexico. Bathymetric range: 605–1386 m. See map 37.

Mesotanais vadicola Sieg & Heard, 1989

Mesotanais vadicola Sieg & Heard, 1989, Gulf Research Report, **8**(2): 75.

Diagnosis. — (Modified after Sieg & Heard, 1989.) Female. Antennule without long (almost as long as antennule) setae. Mandibular molar slightly tapering distally. Maxillipedal endites with two small, modified setae distally and tiny coupling hooks.

Male. Unknown.

Distribution. — This species was recorded in the northern part of the Gulf of Mexico as well as in the Atlantic off Florida and South Carolina. Bathymetric range: 22–865 m. See map 38.

KEY TO THE SPECIES OF *MESOTANAIS*, FEMALES ONLY (GLOBAL)

1. Antennae with long terminal setae (about 0.67 times as long as antennae proper) 2
– Antennae with short terminal setae (about 0.33 times as long as antennae proper) 3
2. Pereopod 1 carpus with several short setae distally and one longer seta almost reaching halfway propodus. Pereopod 4–6 meri with two small spiniform setae .... *Mesotanais longisetosus* Sieg & Heard, 1989
– Pereopod 1 carpus with only several short setae distally, but no longer seta. Pereopod 4–6 meri with one small spiniform and one simple seta .... *Mesotanais dubius* Dollfus, 1897
3. Chelipeds slender (carpus longer than propodus including fixed finger). Labium lateral corners without spiniform setae .... *Mesotanais elongatus* Sieg & Bird, 1989
– Chelipeds robust (carpus as long as propodus including fixed finger). Labium lateral corners with spiniform setae .... *Mesotanais vadicola* Sieg & Heard, 1989

Subfamily LEPTOCHELIINAE Lang, 1973

Leptocheliinae – Larsen & Wilson, 2002, Journal of Zoological Systematics and Evolutionary Research, **40**(4): 213.

Diagnosis. — (Modified after Larsen & Wilson, 2002.) Female. Antennule with four articles, articles 2 and 3 with spiniform setae dorsally. Eyes present. Maxillipedal endites with two (*Intermedichelia*) or three flat, short setae.

Male. Cheliped always significantly enlarged, basis and carpus elongated in certain species.

Type genus. — *Leptochelia* Dana, 1852.

Genera included. — *Hargeria* Lang, 1973; *Intermedichelia* Guţu, 1996; *Leptochelia* Dana, 1852; *Pseudoleptochelia* Lang, 1973; *Pseudonototanais* Lang, 1973.

Remarks. — This subfamily is represented in the Gulf by *Hargeria*, *Leptochelia*, and *Pseudoleptochelia*, but none of these genera is found below 50 m. However, *Intermedichelia* has been recorded from 70 m off Brazil, and may be present in the Gulf.

Family LEPTOGNATHIIDAE Sieg, 1973

Leptognathidae Sieg, 1973, Ein Beitrag zum natürlichen System der Dikonophora Lang: 46.
Leptognathiidae – Sieg, 1976, Zeitschrift für Zoologische Systematik und Evolutionsforschung, **14**: 185.
Anarthruridae – Sieg, 1986, Mitteilungen aus dem Zoologischen Museum der Universität Kiel, **2**(4): 18, 44.

Diagnosis. — (Modified from Larsen & Wilson, 2002.) Female. Medium body calcification. No plates in carapace. Eyes absent. Pereon with six free pereonites. Pereonite 1 not reduced. Pleon with five free pleonites, as wide as pereon, no articulated setae on pleonites. Antennule with four articles. Antenna with 5–6 articles, article 3 without spiniform setae dorsally. Molar process of mandible pointed. Labium with one pair of lobes, medial spiniform setae absent. Maxillipedal bases fused, endites not fused, narrower than basis, without setose or serrated antero-lateral corners, flat setae, or denticles. Cheliped slender and attached via sclerite. Marsupium consisting of four pairs of oostegites. Pereopod coxae present or absent on pereopods 1–3, but always absent on pereopods 4–6, dactylus and unguis not fused. Pleopods well developed or absent. Uropodal endopod with two articles, exopod with one article.

Male. Body significantly shorter than in female. Pereonites less, pleonites more developed than in female. Antennule with more than seven articles, densely packed with aesthetascs. Mouthparts reduced. Cheliped not enlarged. Pleopods well developed.

Type genus. — *Leptognathia* G. O. Sars, 1882.

Genera included. — *Leptognathia* G. O. Sars, 1882.

Remarks. — This family has been a focal point of systematic confusion and controversy (Lang, 1968; Bird & Holdich, 1984; Sieg, 1986a, b; Larsen & Wilson, 2002). The Leptognathiidae were re-erected by Larsen & Wilson (2002), because the type genus on which the family name is based, *Leptognathia*, can not be assigned to incertae sedis.

Genus **Leptognathia** G. O. Sars, 1882

Tanais Lilljeborg, 1864, (in part), Inbjudingsskrift, University of Uppsala: 12.
Haplocope G. O. Sars, 1882, Archiv for Matematik og Naturvidenskab, **7**: 51.
Leptognathia G. O. Sars, 1882, Archiv for Matematik og Naturvidenskab, **7**: 40.
Strongylurella Hansen, 1913, Danish Ingolf Expedition, **3**(3): 120.

Diagnosis. — Monotypic family, no generic diagnosis given.

Type species. — *Tanais breviremis* Lilljeborg, 1864.

Gender of generic name. — Feminine.

Remarks. — This genus is badly in need of revision. Although most of the diagnostic characters in this genus (family) are actually stable, they should nonetheless be regarded with the greatest mistrust, as the status of this genus or its components most certainly will change in the future. Useful characters for species identification are: (1) length of uropods, (2) relative length of the endopod and exopod of the uropod, (3) length/width ratios of pereonites, and (4) number of denticles on the fixed finger of the cheliped. However, the large number of species assigned to this genus, the inadequate nature of the older descriptions, and the systematic quagmire of this taxon make species identifications inherently dubious. For the same reasons, the key to this genus here given applies only to the species found in the Gulf of Mexico.

The males of this family are reported to be of the non-feeding type, that shares no species-specific characters with the female (Sieg, 1984). While nobody contests that these males exist, the question remains how one can ascertain that they really are referable to the genus *Leptognathia*. That these males are found in the same samples as female *Leptognathia* may indicate that they are conspecific, but since the tanaid diversity in most deep-sea samples is very high, this coexistence is no guarantee. Only genetic examination can justify such coupling when no species-specific characters shared by both sexes exist (Larsen, 2001). However, such studies are yet to be conducted on *Leptognathia*.

Species currently assigned to this genus. — *L. abyssorum* (Dollfus, 1897); *L. acanthifera* Hansen, 1913; *L. angustocephala* Kudinova-Pasternak, 1975; *L. arctophylax* (Norman & Stebbing, 1886); *L. birsteini* Kudinova-Pasternak, 1965; *L. brachiata* Hansen, 1913; *L. breviremis* (Lilljeborg, 1864); *L. breviremoides* Sieg, 1986; *L. crassa* Hansen, 1913; *L. dentifera* G. O. Sars, 1896; *L. distincta* Kudinova-Pasternak, 1981; *L. elegans* Kudinova-Pasternak, 1965; *L. glacialis* Hansen, 1913; *L. glandiceps* Shiino, 1978; *L. greveae* Kudinova-Pasternak, 1976; *L. gyreae* sp. nov.; *L. indivisa* (Hansen, 1913); *L. langi* Kudinova-Pasternak, 1970; *L. latiremis* Hansen, 1913; *L. linearis*

(Hansen, 1913); *L. lineata* Shiino, 1978; *L. longa* (Kudinova-Pasternak, 1982); *L. longimanus* sp. nov.; *L. luykeni* Vanhöffen, 1914; *L. manca* G. O. Sars, 1882; *L. microcephala* Kudinova-Pasternak, 1978; *L. mironovi* Kudinova-Pasternak, 1981; *L. parabranchiata* Kudinova-Pasternak, 1978; *L. paramanca* Lang, 1958; *L. parelegans* Kudinova-Pasternak, 1970; *L. profunda* Hansen, 1913; *L. rotundicauda* Kudinova-Pasternak, 1970; *L. tenella* Hansen, 1913; *L. tuberculata* Hansen, 1913; *L. uncinata* Hansen, 1913; *L. vanhoeffeni* Guţu, 1972; *L. ventralis* Hansen, 1913; *L. vinogradovae* Kudinova-Pasternak, 1970; *L. vitjazi* Kudinova-Pasternak, 1982; *L. voeringi* (G. O. Sars, 1877); *L. zenkevitchi* Kudinova-Pasternak, 1970; *L. zezinae* Kudinova-Pasternak, 1973.

Leptognathia gyreae sp. nov.
(figs. 55–57)

Material examined. — Holotype, non-ovigerous ♀ (USNM 1068893), body length 1.6 mm, DGoMB Sta. MT2-1. Paratypes: 1 non-ovigerous ♀ (USNM 1068894), body length 1.6 mm, same locality. 1 non-ovigerous ♀ (USNM 1068895), same locality (dissected). 1 juvenile ♂ (USNM 1068896), body length 1.1 mm, DGoMB Sta. MT2-2.

Additional material. — 1 non-ovigerous ♀, DGoMB Sta. MT2-2.

Diagnosis. — Female. Two of the terminal maxillule setae with conspicuous serration. Cheliped dactylus as long as fixed finger. Uropods long (longer than pleotelson and last pleonite combined). Uropodal exopod uni-articulate and almost as long as first endopod article.

Male. Not dissected, only juvenile males known, so no diagnosis is given for the male.

Etymology. — Named after the R/V "Gyre", which collected the material for the DGoMB program. The species name '*gyreae*' is derived as the feminine form of the name Gyre in the genitive singular.

Type locality. — DGoMB Sta. MT2-1, 28°27'0646"N 89°40'3563"W, depth 676 m.

Description. — Adult female. Body (fig. 55A, B) about 10 times as long as wide. (Owing to the curved nature of the holotype, morphometric calculations of the body are based on the lateral view in fig. 55B.)

Carapace lateral margins gently curved. Less than 1.5 times as long as wide. Shorter than the lengths of pereonites 1 and 2 combined. Eye-lobes absent. Carapace covers basal part of antennules.

Pereon lateral margins almost straight. Pereonites 1 and 6 wider than long. Pereonites 2–5 longer than wide.

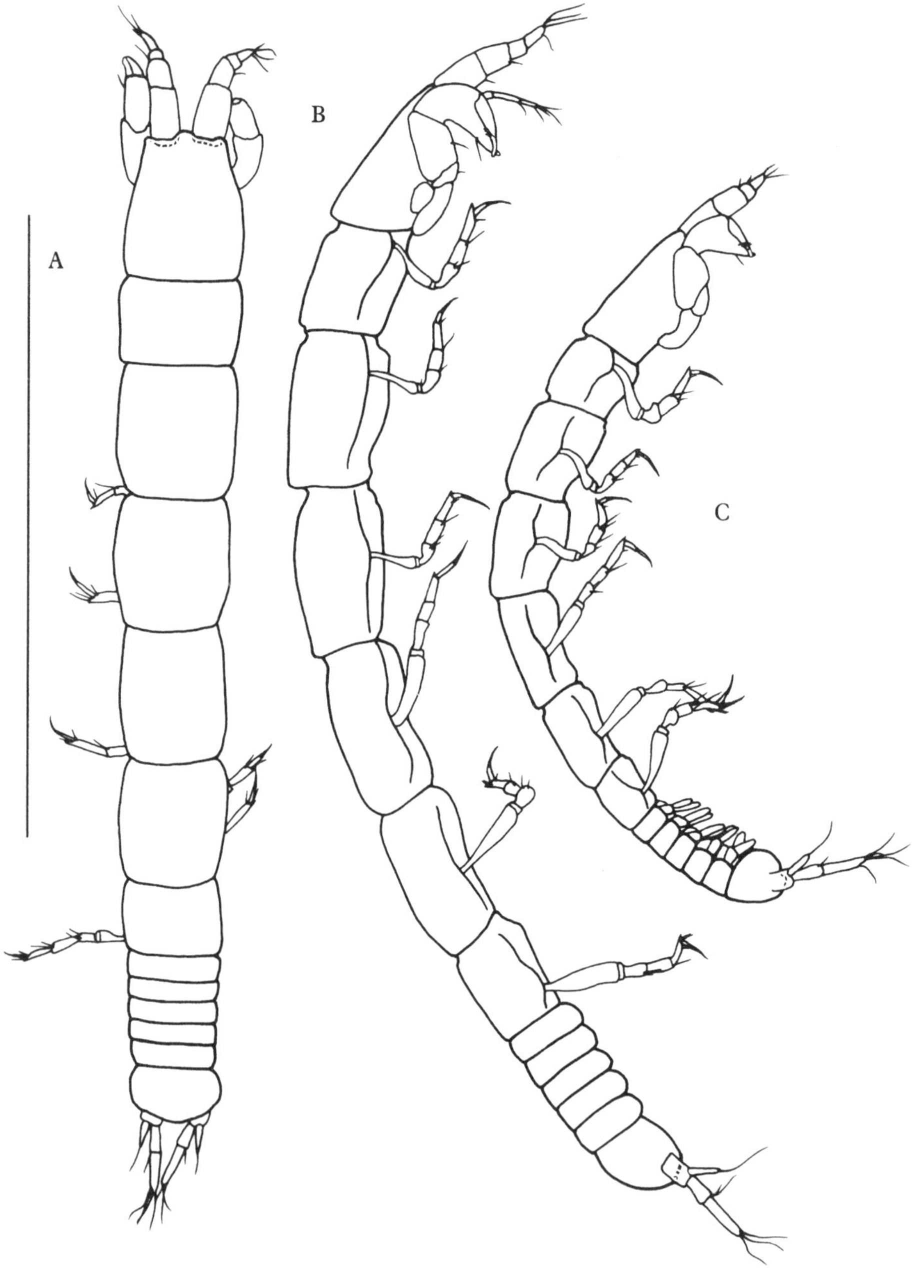

Fig. 55. *Leptognathia gyreae* sp. nov.; A, female holotype, dorsal view; B, holotype, female, lateral view; C, paratype, juvenile male, lateral view. Scale bar equals 1.0 mm.

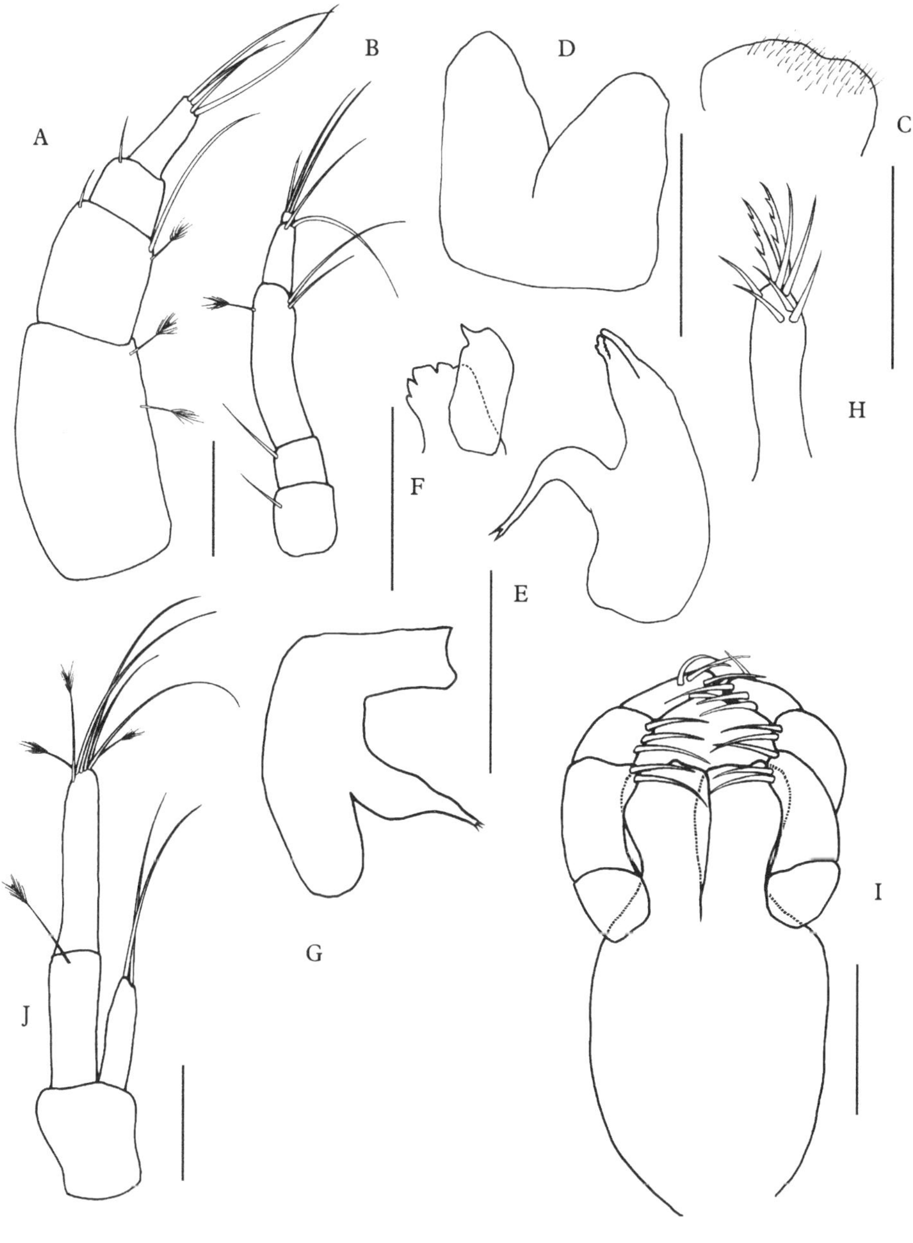

Fig. 56. *Leptognathia gyreae* sp. nov., female; A, antennule; B, antenna; C, labrum; D, labium; E, left mandible; F, left mandible, incisor, ventral view; G, right mandible; H, maxillule; I, maxilliped; J, uropod. Scale bars equal: A-B, J, 0.1 mm; C-I, 0.05 mm.

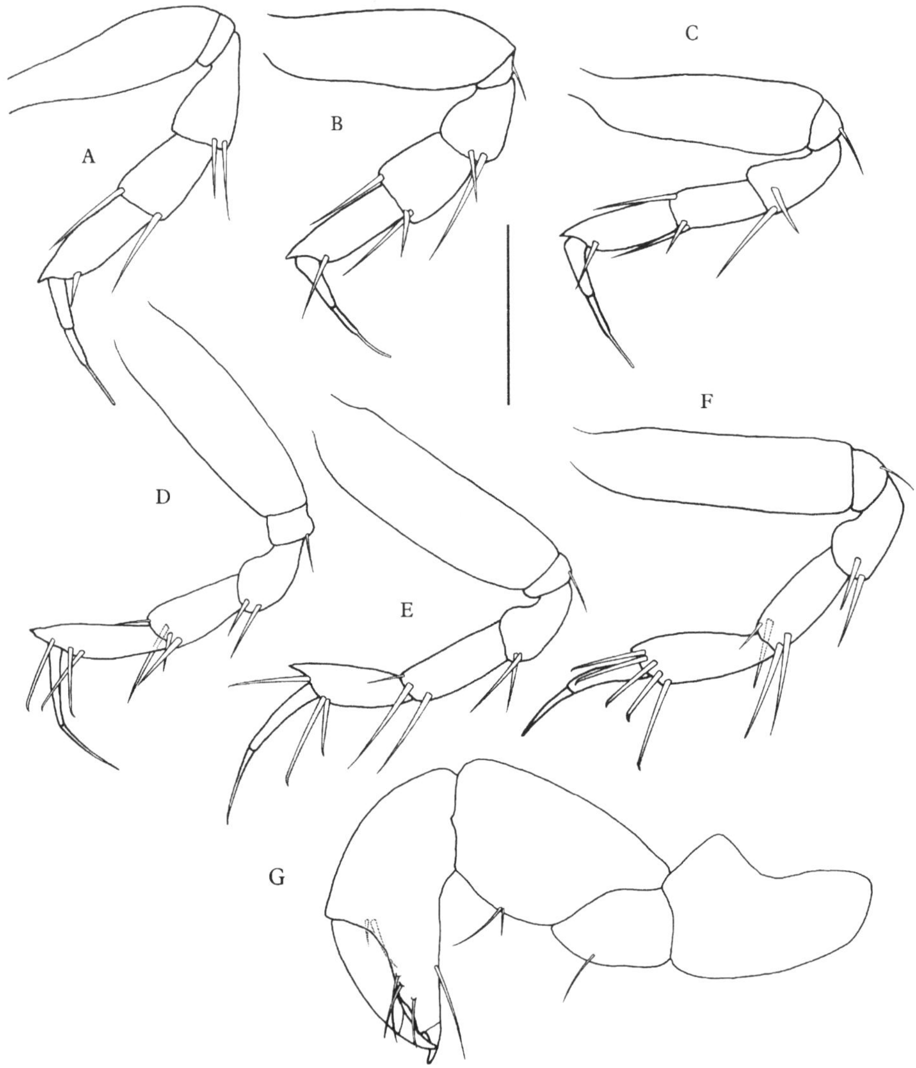

Fig. 57. *Leptognathia gyreae* sp. nov., female; A, pereopod 1; B, pereopod 2; C, pereopod 3; D, pereopod 4; E, pereopod 5; F, pereopod 6; G, cheliped. Scale bar equals 0.1 mm.

Pleon short (including pleotelson about 0.17 times total body length). Pleotelson shorter than the lengths of three free pleonites combined, apex pointed and covered by a dorsal plate.

Antennule (fig. 56A) 0.67 times as long as carapace, with four articles. Article 1 shorter than rest of antennule, with two setulated setae. Article 2 longer than article 3, with two dissimilar simple and one setulose setae distally. Article 3 shorter than article 4, with one simple seta distally. Article 4 with four thick simple setae distally and apparently no aesthetascs.

Antenna (fig. 56B) with five articles. About 0.75 times as long as antennule. Article 1 only hardly wider than other articles, with one simple seta dorsally. Article 2 shorter than article 1, with one simple seta dorsally. Article 3 longer than other articles, with two simple and one setulose setae distally. Article 4 longer than article 2, with two simple setae distally. Article 5 minute, with three setae distally.

Mouthparts. Labrum (fig. 56C) finely setose, with flat apex. Mandibular molar pointed, with two to three denticles distally. Left mandible (fig. 56E) lacinia mobilis longer than, but as wide as incisor, with one apical spine; incisor (fig. 56F) broad, with four denticles distally. Right mandible (fig. 56G) incisor with medial depression. Labium (fig. 56D) as wide as maxillipedal endites, outer corners naked. Maxillule (fig. 56H) endite with eight spiniform setae distally, two of which are serrated; palp lost during dissection. Maxilla not recovered. Maxilliped (fig. 56I) endites with one distal denticle; palp article 1 naked, article 2 with two or three simple inner setae, article 3 with 2–3 setae on inner margin, article 4 with 4–5 setae. Epignath not recovered.

Cheliped (fig. 57G) basis shorter than carpus, attached to square sclerite, naked. Merus with one simple seta ventrally. Carpus as long as propodus including fixed finger, with two simple setae ventrally. Propodus slender, with two setae at dactylus insertion. Fixed finger with only one seta ventrally and three on inner margin, and one large distal denticle. Dactylus as long as fixed finger.

Pereopod 1 (fig. 57A) without visible coxa. Basis longer than the three succeeding articles combined, naked. Ischium naked. Merus as long as carpus, widening distally and with two stiff setae ventro-distally. Carpus shorter than 0.5 times basis, with one stiff seta distally on each margin. Propodus with one spiniform seta ventro-distally and a dorsal spine. Dactylus and unguis together longer than propodus and not fused, dactylus shorter than unguis, unguis with abrupt decrease in width medially.

Pereopods 2 and 3 (fig. 57B, C) as pereopod 1 except: ischium with one simple seta. Carpus with three stiff setae distally.

Pereopod 4 (fig. 57D) without visible coxa. Basis longer than the three succeeding articles combined, naked. Ischium with one simple seta. Merus with two stiff setae distally. Carpus with three stiff and one simple setae distally. Propodus longer than carpus, with three stiff setae distally and a dorsal spine.

Dactylus and unguis together longer than propodus, not fused, dactylus longer than unguis and unguis without decrease in width medially.

Pereopod 5 (fig. 57E) as pereopod 4 except: carpus with two stiff setae distally.

Pereopod 6 (fig. 57F) as pereopod 4 except: propodus with five stiff setae distally.

Pleopods absent in female.

Uropod (fig. 56J) about twice as long as pleotelson. Basal article naked and shorter than first endopod article. Exopod hardly shorter than first endopod article, with two setae distally. Endopod article 1 shorter than article 2, with one setulose seta distally; article 2 with four long simple and three setulose setae distally.

Juvenile male. — (Not dissected.)

Body (fig. 55C) except for the presence of budding pleopods (without setae) the male appears identical to the female. Functional mouthparts retained.

Distribution. — A very rare species, only found in the northern part of the Gulf. Bathymetric range: 675–677 m. See map 39.

Remarks. — The retained mouthparts and the budding pleopods suggest that the male is a juvenile. However, as seen in the male of *L. longimanus* sp. nov. (described below) the development of pleopods may not be dependent on (or occur at the same time as) mouthpart reduction, if such reductions occur in all species.

The specific serration of the setae on the maxillule endite can separate this species from all other *Leptognathia*, although it is stressed that this is a hazardously detailed character when applied to the older literature, in which the descriptions are often of unsatisfactory quality. Other characters, as the combination of the elongate body shape and the uni-articulate uropodal exopod, separate this species from other Atlantic *Leptognathia*.

Leptognathia longimanus sp. nov.
(figs. 58–60)

Material examined. — Holotype, non-ovigerous ♀ (USNM 1068897), body length 0.8 mm, DGoMB Sta. C1-4. Paratypes: 1 non-ovigerous ♀, body length 0.75 mm, 1 ♂, body length 1.6 mm (USNM 1068898), DGoMB Sta. W4-4. 1 ♀ (USNM 1068899), DGoMB Sta. C1-4 (dissected).

Additional material. — 1 ♀, DGoMB Sta. MT4-4. 4 ♀♀, 2 juvenile ♂♂, DGoMB Sta. NB3-4. 1 ♀, DGoMB Sta. S36-1.

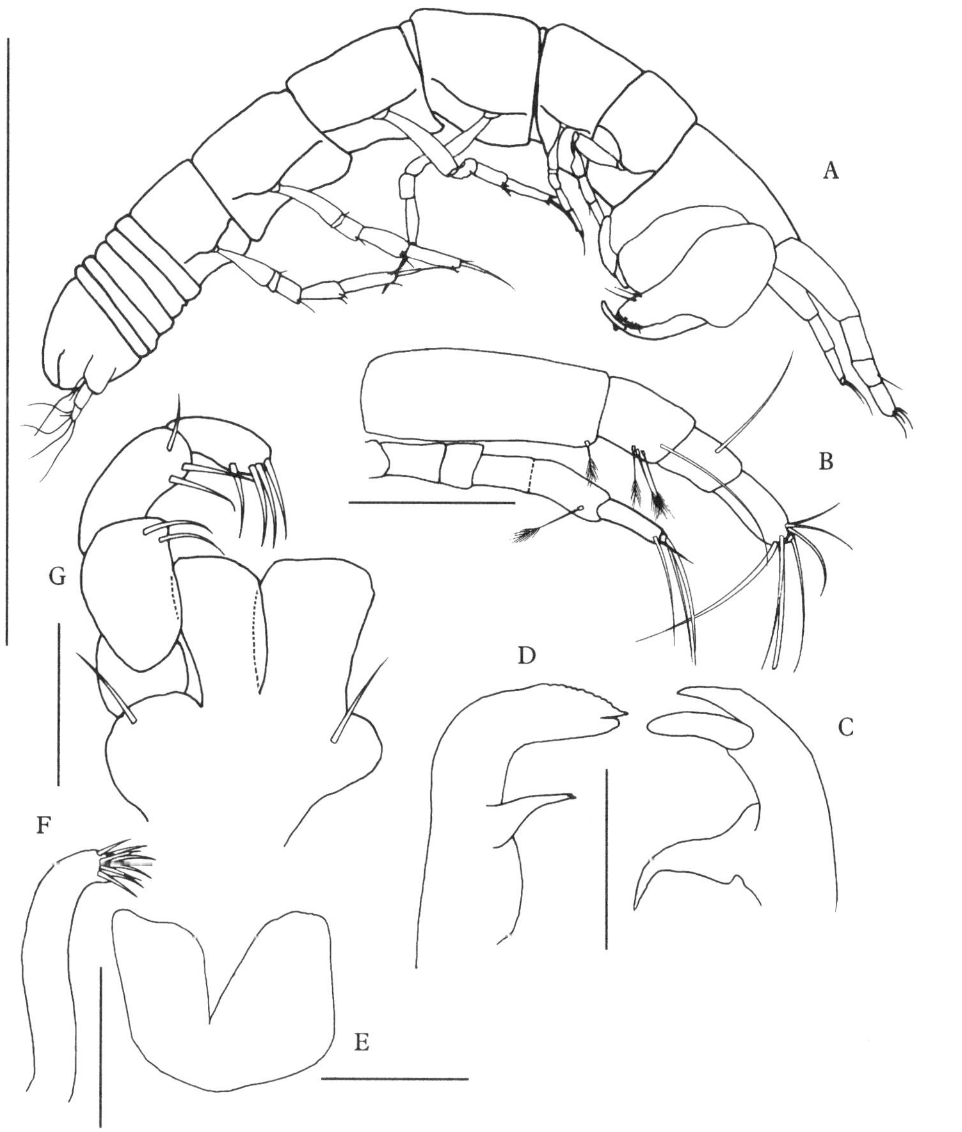

Fig. 58. *Leptognathia longimanus* sp. nov.; A, female, holotype, lateral view; B, antennae; C, left mandible; D, right mandible; E, labium; F, maxillule; G, maxilliped. Scale bars equal: A, 0.5 mm; B, 0.1 mm; C-G, 0.05 mm.

Diagnosis [juvenile female (?); see Remarks]. — Maxillule setae not setulose. Cheliped propodus enlarged; dactylus longer than fixed finger. Uropods short (shorter than pleotelson). Uropodal exopod uni-articulate and no more than 0.5 times as long as first endopod article.

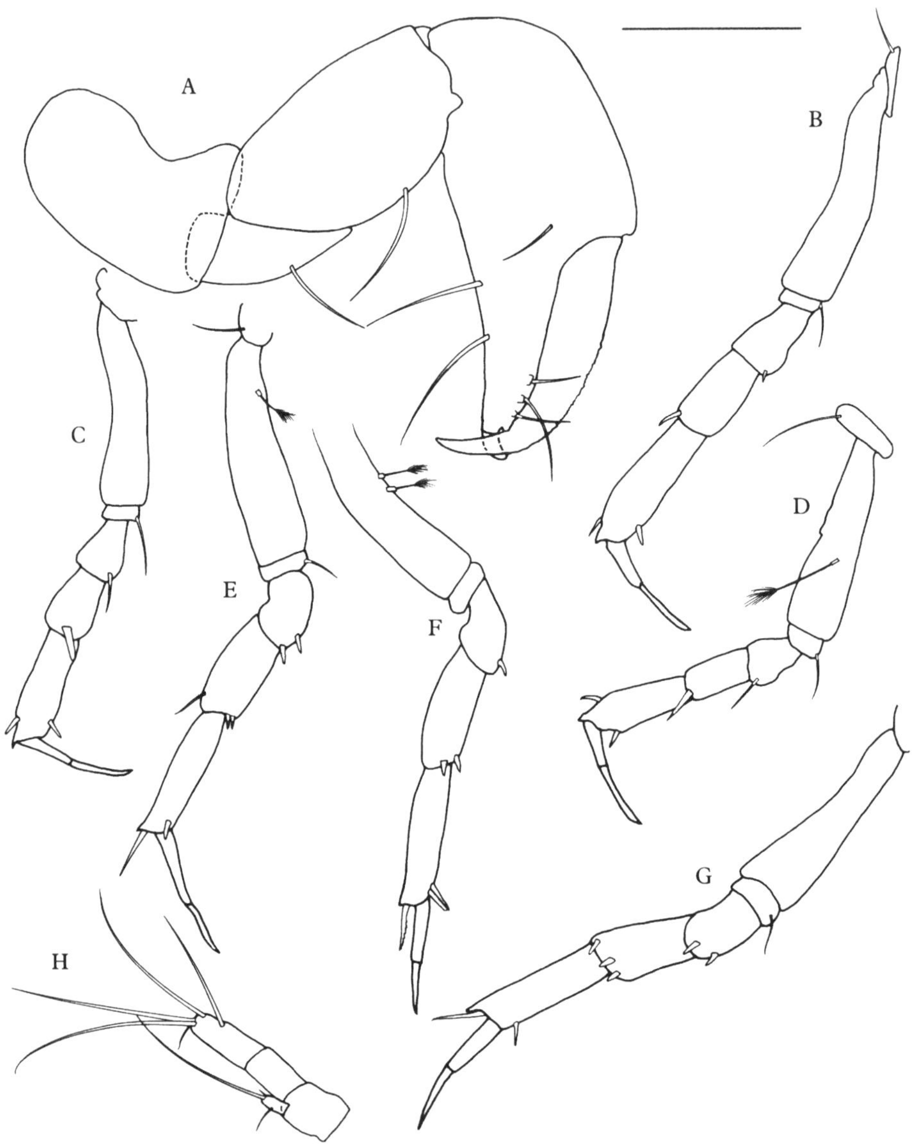

Fig. 59. *Leptognathia longimanus* sp. nov., female; A, cheliped; B, pereopod 1; C, pereopod 2; D, pereopod 3; E, pereopod 4; F, pereopod 5; G, pereopod 6; H, uropod. Scale bar equals 0.1 mm.

Male. Not dissected, no diagnosis given for the male.

Etymology. — Named after the elongate cheliped propodus (from Latin, longus = long + manus = hand) as a noun in apposition to the generic name.

Type locality. — DGoMB Sta. C1-4, 28°03'5830"N 90°14'9610"W, depth 336 m.

Description. — Juvenile female. (Owing to the curved nature of the holotype body, morphometric calculations are based on the lateral view in fig. 58A.)

Body (fig. 58A) about 7 times as long as wide.

Carapace lateral margins gently curved; less than 1.5 times as long as wide; as long as lengths of pereonites 1 and 2 combined. Eye-lobes absent.

Pereon lateral margins almost straight. Pereonites 1, 2, and 6 wider than long. Pereonite 3 as long as wide. Pereonites 4 and 5 longer than wide.

Pleon short (including pleotelson about 0.2 times total body length). Pleotelson longer than the lengths of four free pleonites combined, apex blunt and covered by a dorsal plate.

Antennule (fig. 58B) longer than carapace, with four articles. Article 1 longer than rest of antennule, with one setulated seta distally. Article 2 longer than article 3, with three setulose and one simple setae distally. Article 3 shorter than article 4, with one simple seta medially. Article 4 with six simple setae and one aesthetasc distally.

Antenna (fig. 58B) with five articles. About 0.75 times as long as antennule. Article 1 longer than article 2, naked. Article 2 shorter than article 4, naked. Article 3 with medial fusion line, longer than other articles, with one setulose seta distally. Article 4 longer than article 2, with two simple setae distally. Article 5 minute, with three setae distally.

Mouthparts. Labrum (not illustrated) naked, with flat apex. Mandibular molar pointed, with 2–3 denticles distally. Left mandible (fig. 58C) lacinia mobilis longer and wider than incisor; incisor narrow, without denticles. Right mandible (fig. 58D) incisor with medial depression and dorsal margin serrated. Labium (fig. 58E) as wide as maxillipedal endites, outer corners naked and blunt. Maxillule (fig. 58F) endite with nine spiniform setae distally; palp lost during dissection. Maxilla not recovered. Maxilliped (fig. 58G) endites smooth; basis with one seta at palp insertion; palp article 1 naked, article 2 with two simple inner setae, article 3 with three setae on inner margin, article 4 with four setae. Epignath not recovered.

Cheliped (fig. 59A) basis shorter than carpus, attached laterally via sclerite, naked. Merus with one simple seta ventrally. Carpus only 0.7 times as long as propodus including fixed finger, with one simple seta ventrally. Propodus slender, with one seta at dactylus insertion. Fixed finger with two setae

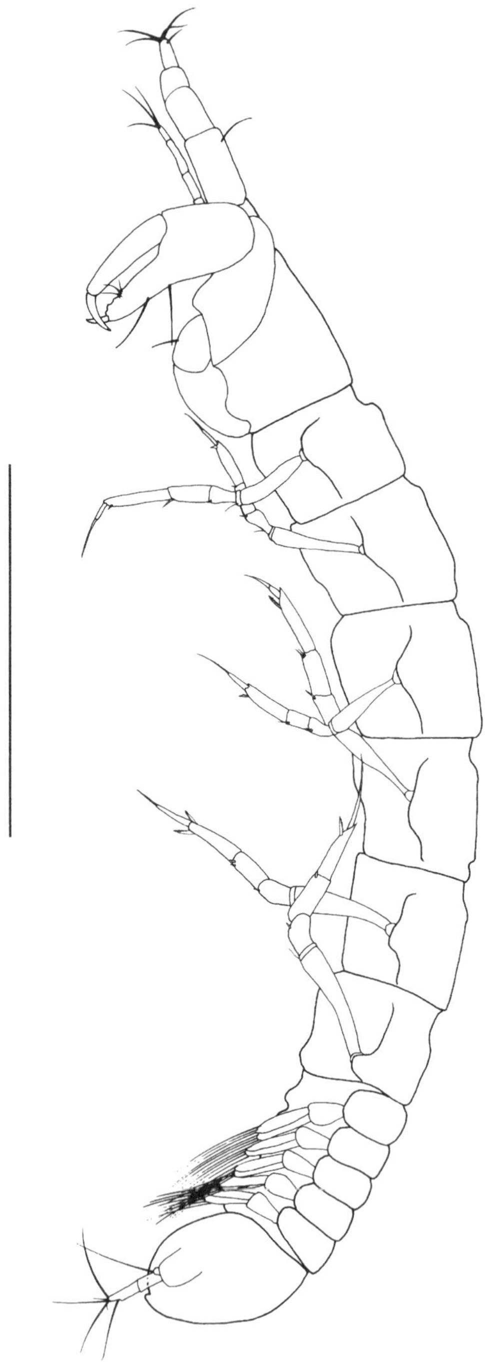

Fig. 60. *Leptognathia longimanus* sp. nov., male, paratype, lateral view. Scale bar equals 0.5 mm.

ventrally and three on inner margin, and one distal denticle. Dactylus longer than fixed finger, dorso-distal edge uneven.

Pereopod 1 (fig. 59B) coxa with one simple seta. Basis longer than the three succeeding articles combined, naked. Ischium with one simple seta. Merus as long as carpus, widening distally and with one tiny spiniform seta ventro-distally. Carpus shorter than 0.5 times basis, with one spiniform seta dorso-distally. Propodus with one spiniform seta subdistally on each margin and dorsal spine. Dactylus and unguis together shorter than propodus and not fused, dactylus shorter than unguis, unguis with wide, spatulate apex.

Pereopod 2 (fig. 59C) as pereopod 1 except: merus with one simple seta.

Pereopod 3 (fig. 59D) as pereopod 2 except: basis with one setulated seta.

Pereopod 4 (fig. 59E) as pereopod 2 except: merus with two small spiniform distal setae. Carpus with two small spiniform and one simple setae distally. Propodus with one spiniform seta ventro-subdistally and one spiniform seta dorso-distally.

Pereopod 5 (fig. 59F) as pereopod 4 except: basis with two setulated setae medially. Ischium naked. Merus with one small spiniform seta distally. Propodus with two spiniform setae ventro-subdistally and one dorso-distally.

Pereopod 6 (fig. 59G) as pereopod 4 except: basis naked. Carpus with three small spiniform setae distally. Propodus with one spiniform seta subdistally and one dorso-distally.

Pleopods not present in female.

Uropod (fig. 59H) about 0.6 times as long as pleotelson. Basal article naked and as long as first endopod article. Exopod less than 0.5 times as long as first endopod article, with two small and one long setae distally. Endopod article 1 shorter than article 2, naked; article 2 with four long simple and one small setae distally.

Adult male. — Not dissected (see Remarks). Body (fig. 60A) with pleon with lateral shield present on all pleonites.

Mouthparts. Functional mouthparts present.

Cheliped fixed finger with more demarcated denticles than in female.

Pereopods with the spiniform setae ventro-subdistally on the propodi of pereopods 4–6 serrated.

Pleopods well developed, with plumose setae.

Juvenile male essentially as female except for presence of budding pleopods.

Distribution. — A rare species only found in the northern part of the Gulf. Bathymetric range: 336–1875 m. See map 40.

Remarks. — The holotype and dissected specimen are perceived as juveniles, as the juvenile male (male with budding pleopods) is of the same size as the female, whereas the male paratype (adult, with well developed pleopods) is twice as long. There is some uncertainty about the conspecificity of the male (if adult), since it displays functional mouthparts (in conflict with the diagnoses of the genus and family). However, the conspecificity of the male can not be asserted with certainty without molecular study. An alternative possibility is that the male has yet to undergo the final moult to become a non-feeding terminal male. Currently, there is no sufficient knowledge about the development of non-feeding terminal males (Larsen, 2001, 2002a) and this species (including the male) is placed in *Leptognathia* until more material becomes available.

Leptognathia longimanus sp. nov. shows great resemblance to *L. rotundicauda* Kudinova-Pasternak, 1970, in particular in the mandibles and chelipeds. Another interesting similarity is the serrated spiniform seta on the propodus of pereopod 5. Kudinova-Pasternak (1970), however, figured this seta as on pereopod 4 in her illustration (Kudinova-Pasternak, 1970: 364, fig. 15) and the description is only superficial, to say the least. As the type material of *L. rotundicauda* is lost, one cannot completely exclude conspecificity of these two species, but the fact that *L. rotundicauda* was collected from the Kurile-Kamchatka Trench makes this unlikely.

Another, and more likely explanation was brought to my attention by Dr. G. Bird (Penponds), i.e., that this possibly constitutes a new genus, including *L. longimanus*, *L. rotundicauda*, and a third and very similar species found in the North Atlantic, collected by Dr. Bird during the BIOGAS (as *Leptognathia* sp. G) and BIOICE (as leptognathiid genus H) projects.

KEY TO THE SPECIES OF *LEPTOGNATHIA* IN THE GULF OF MEXICO, FEMALES

1. Uropods longer than pleotelson and last pleonite combined. Cheliped dactylus as long as fixed finger. Uropodal exopod almost as long as first endopod article .... *L. gyreae* sp. nov.
– Uropods shorter than pleotelson. Cheliped dactylus longer than fixed finger. Uropodal exopod no more than 0.5 times as long as first endopod article .... *L. longimanus* sp. nov.

Family NOTOTANAIDAE Sieg, 1976

Nototanaidae Sieg, 1976, Zeitschrift für Zoologische Systematik und Evolutionsforschung, **14**: 186.

Nototanaididae Larsen & Wilson, 2002, Journal of Zoological Systematics and Evolutionary Research, **40**(4): 217.

Diagnosis. — (Modified after Larsen & Wilson, 2002.) Female. Medium body calcification. No plates in carapace. Eyes or eye-lobes present or absent. Pereon with six free pereonites. Pereonite 1 not reduced. Pleon with five free pleonites, as wide as pereon, no articulated setae on pleonites. Antennule with three articles. Antenna with 5–6 articles, article 3 with or without spiniform setae dorsally. Molar process of mandible broad or pointed. Labium with one pair of lobes, medial spiniform setae absent. Maxillipedal bases at least partially fused, endites either not fused, or fused (*Tanaissus*), narrower than basis, and with one, two, or no flat setae or denticles. Cheliped attachment via sclerite or ventral coxa (*Nesotanais*). Marsupium (where known) with four pairs of oostegites. Pereopods 1–3 with coxae present or absent. Pereopods 4–6 with coxae absent, dactylus and unguis fused to claw. Pleopods well developed. Uropodal endopod and exopod with 1–2 articles.

Male (where known). Sexual dimorphism extensive, with several different male morphs. Antennule with more than five articles and multiple aesthetascs. Pereonites less, pleonites more developed than in female. Mouthparts reduced. Pleopods well developed. Uropodal endopod with 2–3 articles.

Type genus. — *Nototanais* Richardson, 1906

Genera included. — *Bathytanaissus* Bird & Holdich, 1989; *Metatanais* Shiino, 1952; *Nesotanais* Shiino, 1968; *Nototanais* Richardson, 1906; *Paratyphlotanais* Kudinova-Pasternak & Pasternak, 1978; *Peraeospinosus* Sieg, 1986; *Protanaissus* Sieg, 1983; *Tanaissus* Norman & Scott, 1906; *Teleotanais* Lang, 1956; *Typhlotanais* G. O. Sars, 1882; *Typhlotanaoides* Sieg, 1983.

Remarks. — The families Nototanaidae and Typhlotanaidae were combined by Larsen & Wilson (2002) as the result of a phylogenetic analysis. Intuitively, this decision seemed dubious and also the support values were weak (Larsen & Wilson, 2002: 213). It is obvious that the data (characters) are currently too limited to reveal the true nature of tanaidacean phylogeny. These families are likely to be split up again when a higher systematic resolution is attained, possibly into a 'nototanaid' component and a 'typhlotanaid' componen, but probably with some genera 'shuffled around'.

As Sieg (1983) writes the generic name both as *Typhlotanaoides* (with two "a"-s: two times, in bold face, both in the generic heading and in the species heading) and as *Typhlotanoides* (without the second "a": four times in all, i.e., twice in the abstract and twice in the captions), these two different spellings can be considered different, synonymous names. Hence Articles 24.2.1 and 24.2.2 (Determination by the First Reviser) of the Code (1999) apply. Until now no other author has explicitly chosen between these names as to which is the correct one, though Sieg himself did use the spelling *Typhlotanoides* (without second "a") in two later papers (Sieg, 1986a: 21, 72, 77; 1986c: 102). However, acting as the First Reviser, I here choose the name *Typhlotanaoides* (with the second "a") as the correct name for this genus. The reasons for this choice are as follows: (1) in deriving a generic name from *Tanais* by adding the suffix -oides, an author most probably applies that suffix to -tana-, rather than to -tan-; (2) mention of this name in the two bold face headings in my opinion indicates that Sieg (1983) most probably meant this spelling to represent his intention; and (3) the Nomenclator Zoologicus (Neave) gives only this spelling, as does The Zoological Record reporting on the publications of the year 1983. As a result, the name *Typhlotanoides* Sieg, 1983, without second "a", is from now on to be regarded a junior synonym of *Typhlotanaoides* Sieg, 1983. Both names, of course, are available according to the criteria of the Code, whereas only *Typhlotanaoides* is valid at the moment.

Genus **Meromonakantha** Sieg, 1986

Meromonakantha Sieg, 1986, Mitteilungen aus dem Zoologischen Museum der Universität Kiel, **2**(4): 72.

Diagnosis. — (Modified after Sieg, 1986c.) Female. Body almost completely cylindrical. Carapace wider than pereon. Antennule with three articles. Antenna with 5–6 articles. Molar process of mandible broad, with several terminal denticles. Maxillule with 7–8 spiniform terminal setae. Maxillipedal bases partially fused, endites not fused; distal edge with lateral process. Chelipeds ventrally attached via sclerite (see Remarks). Pereopods 1–3 with coxae; merus and carpus with simple setae only. Pereopods 4–6 with or without coxae; basis stouter or not stouter than in pereopods 1–3; merus and carpus with at least one spiniform seta; dactylus and unguis fused. Pleopods present, with simple or plumose setae. Uropods biramous; endopod with two articles; exopod with one or two articles.

Male. Unknown.

Gender of generic name. — Feminine.

Remarks. — Sieg (1986c) defined the genus as having a bi-articulate uropodal exopod. The species described below is equipped with a uni-articulate uropodal exopod, but otherwise in accordance with the generic diagnosis. However, as mentioned above, the family group is considered unstable and none of the generic characters are necessarily stable. At present the uneven lateral margin (dorsal view), the rounded carapace, and the armament of the pereopod merus and carpus, seem best to define the species group currently residing in *Meromonakantha*.

It is the opinion of the present author that the genus *Typhlotanais* and its most closely allied genera (*Peraeospinosus*, *Paratyphlotanais*, *Typhlotanaoides*, and *Meromonakantha*), while likely to constitute a monophyletic taxon, are so intermingled and poorly defined that only a thorough phylogenetic analysis can resolve the systematics of this group. Such a systematic revision of *Typhlotanais* is currently in preparation (M. Błażewicz-Paszkowycz, pers. comm.) and therefore no key is given to *Meromonakantha*.

Species currently assigned to this genus. — *M. macrocephala* (Hansen, 1913); *M. gilrowei* sp. nov.

Meromonakantha gilrowei sp. nov.
(figs. 61–63)

Material examined. — Holotype, non-ovigerous ♀ (USNM 1068900), body length 1.05 mm, DGoMB Sta. NB2-1. Paratypes. 1 non-ovigerous ♀ (USNM 1068901), body length 1.4 mm, DGoMB Sta. NB3-4. 1 non-ovigerous ♀ (USNM 1068902), DGoMB Sta. RW5-2 (dissected).

Diagnosis. — Female. Pereopods long (longer than pleon). Pereopods 4–6 not stouter than pereopods 1–3; dactylus and unguis fused, but not into a claw. Uropods longer than pleotelson; endopod with two articles; exopod with one article, almost as long as endopod.

Male. Unknown.

Etymology. — Named after Professor Gilbert T. Rowe, a pioneer of deep-sea exploration and the administrator of the Biodiversity Fellowship at Texas A&M University. The name is a noun in the genitive singular.

Type locality. — DGoMB Sta. NB2-1, 27°08'0243"N 91°59'9584"W, 1530 m.

Description. — Female. Body (fig. 61A, B) cylindrical, 5.5 times as long as wide.

Carapace shorter than the lengths of pereonites 1 and 2 combined. Eye-lobes absent.

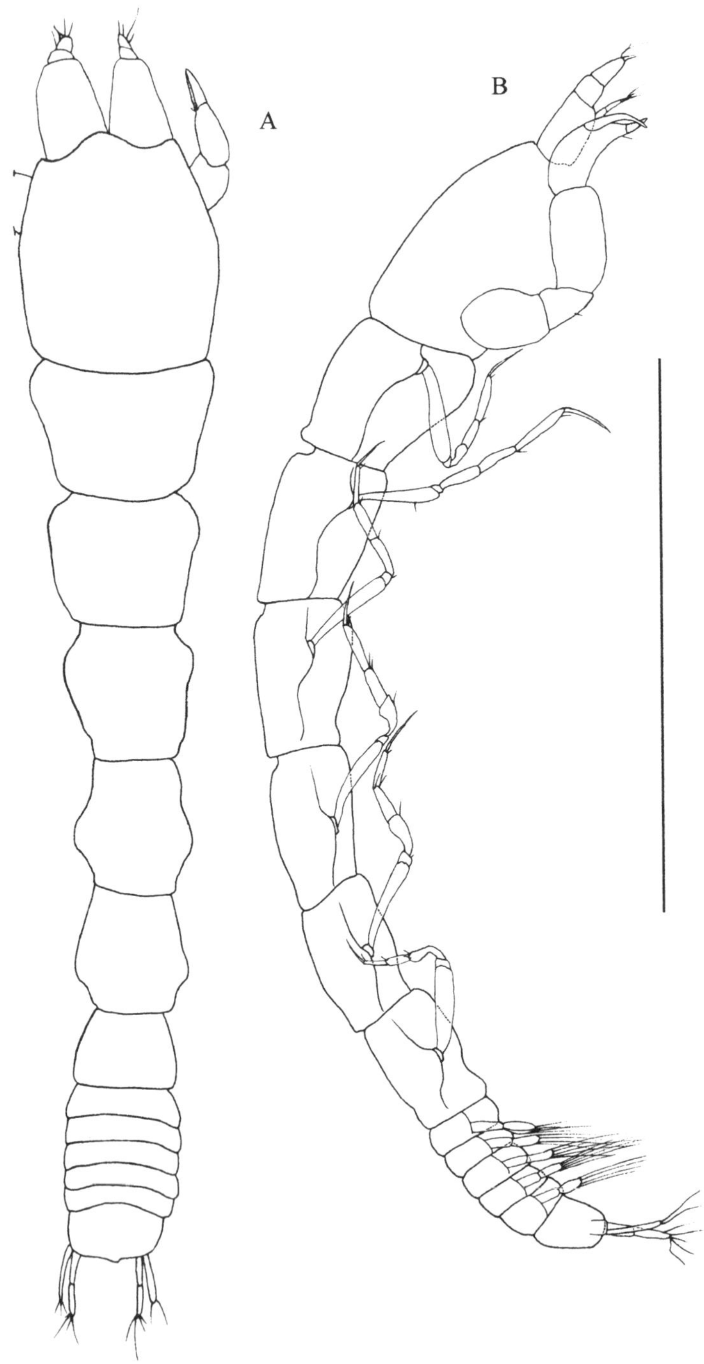

Fig. 61. *Meromonakantha gilrowei* sp. nov., female; A, holotype, dorsal view; B, same, lateral view. Scale bar equals 0.5 mm.

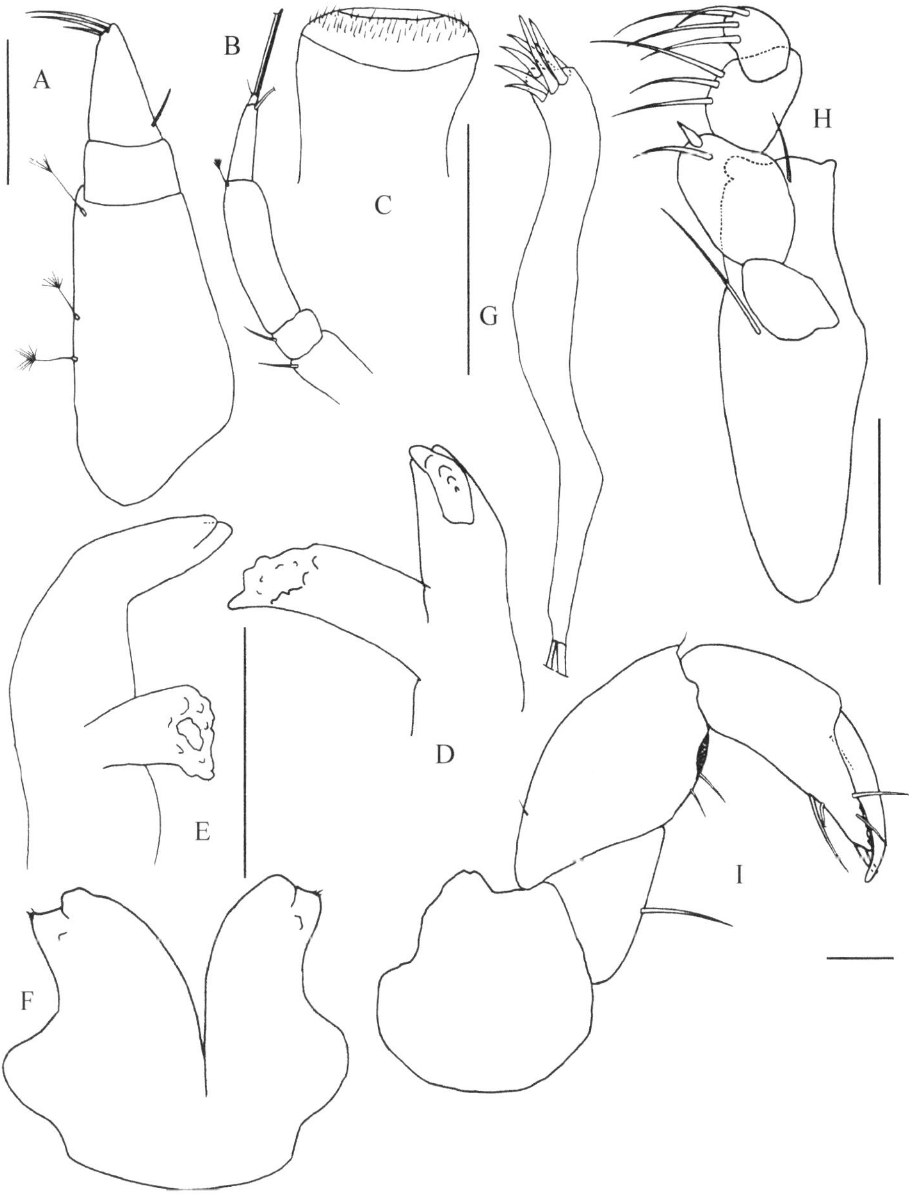

Fig. 62. *Meromonakantha gilrowei* sp. nov., female; A, antennule; B, antenna; C, labrum; D, left mandible; E, right mandible; F, labium; G, maxillule; H, maxilliped; I, cheliped. Scale bars equal: A-B, 0.1 mm; C-I, 0.05 mm.

Pereon tapering distally. Pereonites 1, 2, and 6 wider than long. Pereonite 3 as long as wide. Pereonites 4 and 5 longer than wide.

Pleon short (including pleotelson only 0.15 times total body length). All pleonites subequal. Pleotelson as long as lengths of two free pleonites combined.

Antennule (fig. 62A) about 0.5 times as long as carapace. Article 1 longer than rest of antennule, with three setulose setae. Article 2 less than 0.5 times as long as article 3, naked. Article 3 shorter than 0.5 times article 1, with three simple setae distally.

Antenna (fig. 62B) 0.75 times as long as antennule and 0.33 times as wide. Article 1 not wider than following articles, with one simple seta dorso-distally. Article 2 shorter than article 1, with one seta dorso-distally. Article 3 longer than other articles, with one setulose seta dorso-distally. Article 4 as long as article 1, with one seta ventro-distally. Article 5 minute, with two setae and one aesthetasc distally.

Mouthparts. Labrum (fig. 62C) flat and with several setules. Mandibular molar process broad, and longer than incisor, with terminal ring of denticles. Left mandible (fig. 62D) lacinia mobilis as long as incisor, with five denticles; incisor broad, with two denticles. Right mandible (fig. 62E) incisor broad, with two denticles. Labium (fig. 62F) with inner and outer distal processes, outer processes with a few setules. Maxillule (fig. 62G) endite with seven spiniform setae distally; palp as long as endite. Maxilla not recovered. Maxilliped (fig. 62H) basis with one long simple seta at palp insertion; endites with inner and outer distal processes, almost as wide as basis; palp (twisted by dissection, inner and outer margins appear reversed in illustration) article 1 naked, article 2 with one spiniform and one simple distal inner seta, article 3 with three inner setae, article 4 with four setae. Epignath not recovered.

Cheliped (fig. 62I) basis unequally divided by small sclerite, shorter than carpus. Merus prominent, with one seta ventrally. Carpus as long as, but wider than propodus including fixed finger, with two setae ventrally and two small setae dorsally. Propodus with one seta at dactylus insertion. Fixed finger with two setae ventrally and two on inner margin, with several visible denticles on inner margin. Dactylus as long as fixed finger.

Pereopod 1 (fig. 63A) coxa with one simple seta. Basis longer than the three succeeding articles combined, naked. Ischium with one seta ventrally. Merus as long as carpus, widening distally, naked. Carpus 0.67 times as long as propodus, with one seta distally. Propodus more than 0.5 times as long as basis, naked. Dactylus and unguis combined shorter than propodus and not fused to a claw.

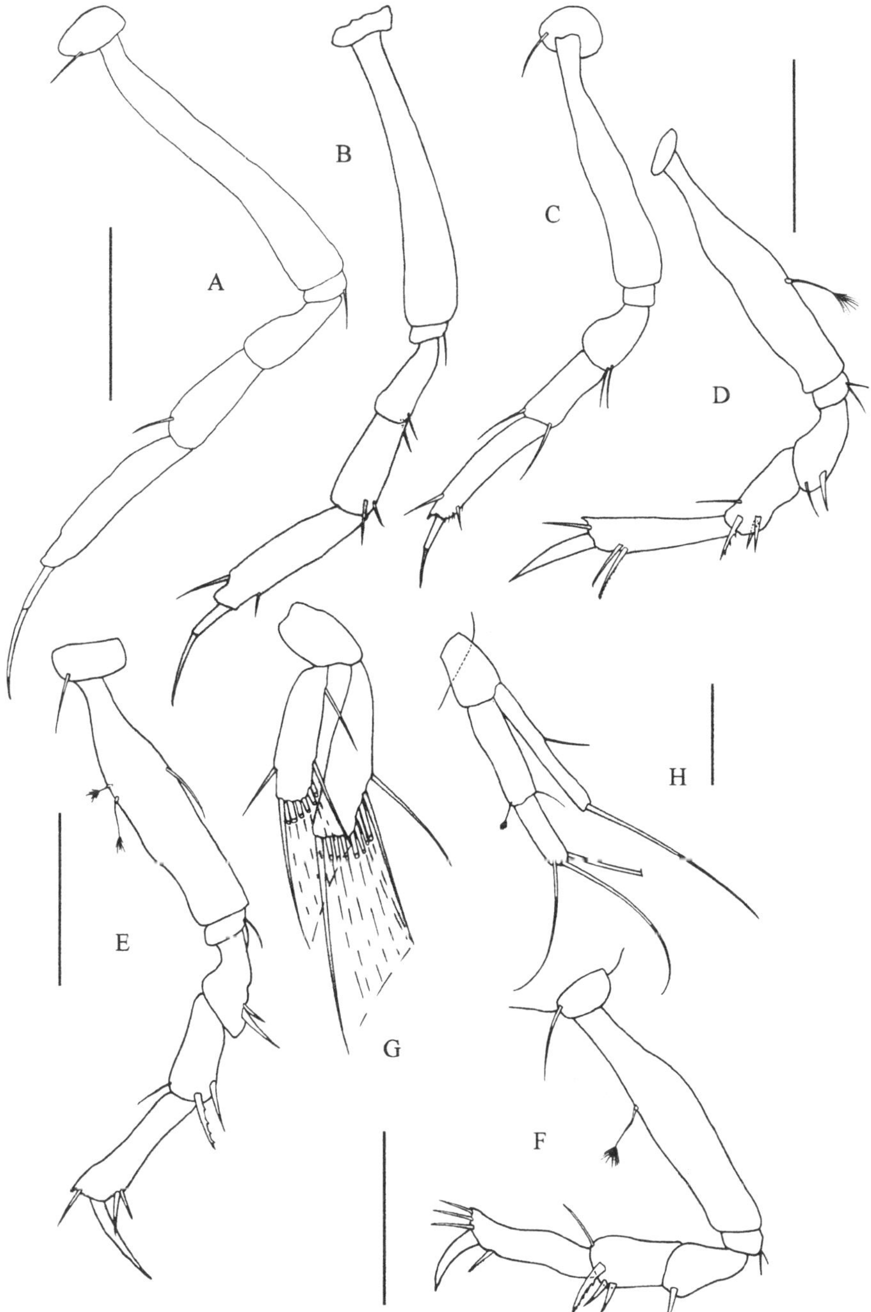

Fig. 63. *Meromonakantha gilrowei* sp. nov., female; A, pereopod 1; B, pereopod 2; C, pereopod 3; D, pereopod 4; E, pereopod 5; F, pereopod 6; G, pleopod; H, uropod. Scale bars equal 0.1 mm.

Pereopod 2 (fig. 63B) as pereopod 1 except: merus with two setae distally. Carpus 0.67 times as long as propodus, with two setae distally. Propodus 0.5 times as long as basis, with one seta distally on each margin.

Pereopod 3 (fig. 63C) as pereopod 2 except: ischium naked. Propodus with small setules distally.

Pereopod 4 (fig. 63D) as pereopod 2 except: basis with setulose seta ventrally. Ischium with two setae. Merus with one spiniform and one simple seta. Carpus with one simple seta dorso-distally and three spiniform setae ventro-distally, of which one is serrated. Propodus with one simple seta dorso-distally, a dorsal spine, and one spiniform and one serrated spiniform seta ventrally. Dactylus and unguis together shorter than propodus and fused, but not into a claw.

Pereopod 5 (fig. 63E) as pereopod 4 except: basis with two setulose setae dorsally and one simple seta ventrally.

Pereopod 6 (fig. 63F) coxa with one simple seta. Basis longer than the three succeeding articles combined, with one setulose seta dorsally. Ischium with two setae ventrally. Merus as long as carpus, widening distally, and with one spiniform seta distally. Carpus 0.67 times as long as propodus, with one simple seta, two spiniform setae, and one serrated spiniform seta distally. Propodus shorter than 0.5 times basis, with three setae dorso-distally, a dorsal spine, and one seta ventrally. Dactylus and unguis together longer than propodus and not fused to a claw.

Pleopod (fig. 63G) basal article naked. Exopod with one outer and nine inner simple setae, large gap between proximal inner seta and others. Endopod with ten simple inner setae.

Uropod (fig. 63H) basal article shorter than first endopod article, naked. Endopod longer than pleotelson; article 1 with one simple and one setulated seta distally; article 2 with three setae distally. Exopod uni-articulate, almost as long as endopod, with one seta medially and one seta distally.

Distribution. — A rare species only occurring in the northwestern part of the Gulf. Bathymetric range narrow: 1320–1875 m. See map 41.

Remarks. — This species has a number of characters in common with *Insociabilitanais ventrospinatus* (see Familiae indeterminatae), however, the following characters prevent its incorporation in *Insociabilitanais*: (1) antennule with 3 articles, (2) all pereopods with coxae, and (3) no pereonite spine. This species also has several characters in common with *Paratyphlotanais* but cannot be included in that genus according to the amended diagnosis of Bird (2004a), because of: (1) no pereonite spine, (2) carapace not clearly of conical shape, (3) antennular articles 1 and 3 not of equal length, (4) pereopods 4–6 with unguis not bifid.

However, the similarity between these genera and the overlapping characters suggest that we are still far from seeing a resolution of the systematics of this group of taxa.

Genus **Typhlotanais** G. O. Sars, 1882

Tanais Lilljeborg, 1864, (in part), Inbjudingsskrift, University of Uppsala: 12.
Typhlotanais G. O. Sars, 1882, Archiv for Matematik og Naturvidenskab, **7**: 52.

Diagnosis. — (Modified after Lang, 1970b.) Female. Body elongate, at least 4.5 times as long as wide. Carapace as long as wide or longer. Eyes and eyelobes absent. Antennule with three articles. Antenna with six articles. Molar process broad, with several terminal denticles. Maxillule with eight spiniform terminal setae. Maxillipedal bases partially fused, endites not fused. Chelipeds attached via sclerite, propodus slender (including fixed finger at least 1.5 times as long as wide). Pereopods 1–3 with coxae. Pereopods 4–6 without coxae; basis stouter than in pereopods 1–3; dactylus and unguis fused or partly fused. Pleopods present, with simple or plumose setae. Uropods biramous; endo- and exopod with 1 or 2 articles.

Male (where known). Antennule with six articles. Mouthparts reduced.

Gender of generic name. — Masculine.

Remarks. — See genus *Meromonakantha*.

Species currently assigned to this genus. — *T. aequiremis* (Lilljeborg, 1864); *T. angularis* Kudinova-Pasternak, 1966; *T. assimilis* G. O. Sars, 1882; *T. brachyurus* Beddard, 1886; *T. brevicornis* (Lilljeborg, 1864); *T. compactus* Kudinova-Pasternak, 1966; *T. cornutus* (G. O. Sars, 1879); *T. crassus* Dojiri & Sieg, 1997; *T. dubius* Tzareva, 1982; *T. elegans* Kudinova-Pasternak, 1978; *T. eximius* Hansen, 1913; *T. filatovae* Kudinova-Pasternak, 1975; *T. finmarchicus* G. O. Sars, 1882; *T. gracilipes* Hansen, 1913; *T. grandis* Hansen, 1913; *T. greenwichensis* Shiino, 1970; *T. gruzovi* Tzareva, 1982; *T. inaequipes* Hansen, 1913; *T. inermis* Hansen, 1913; *T. irregularis* Hansen, 1913; *T. kerguelenensis* Beddard, 1886; *T. kussakini* Kudinova-Pasternak, 1970; *T. longicephala* Kudinova-Pasternak, 1970; *T. longimanus* Dollfus, 1897; *T. longisetosus* Kudinova-Pasternak, 1990; *T. longus* Kudinova-Pasternak, 1975; *T. magnificus* Kudinova-Pasternak, 1970; *T. magnus* Kudinova-Pasternak, 1990; *T. mananensis* Wallace, 1919; *T. messinensis* G. O. Sars, 1882; *T. microcheles*

G. O. Sars, 1882; *T. mucronatus* Hansen, 1913; *T. parangularis* Kudinova-Pasternak, 1975; *T. parvus* Sieg, 1986; *T. peculiaris* Lang, 1968; *T. penicillatus* G. O. Sars, 1882; *T. plebejus* Hansen, 1913; *T. plicatus* Kudinova-Pasternak, 1993; *T. proctagon* Tattersall, 1904; *T. profundus* Hansen, 1913; *T. pulcher* Hansen, 1913; *T. rectus* Kudinova-Pasternak, 1966; *T. richardi* Dollfus, 1897; *T. rotundirostris* Lang, 1970; *T. sandersi* Kudinova-Pasternak, 1985; *T. setosus* Kudinova-Pasternak, 1966; *T. simplex* Kudinova-Pasternak, 1984; *T. solidus* Hansen, 1913; *T. spinicauda* Hansen, 1913; *T. spinipes* Kudinova-Pasternak, 1982; *T. spiniventris* Dollfus, 1897; *T. tenuicornis* G. O. Sars, 1882; *T. tenuimanus* (Lilljeborg, 1864); *T. trispinosus* Hansen, 1913; *T. variabilis* Hansen, 1913; *T. williamsae* Dojiri & Sieg, 1997.

Typhlotanais cf. **trispinosus** Hansen, 1913

Typhlotanais trispinosus Hansen, 1913, Danish Ingolf Expedition, **3**: 49.

Material examined. — 1 non-ovigerous ♀, NGoMCSS Sta. E1-5. 2 non-ovigerous ♀♀, NGoMCSS Sta. E1-6. 1 non-ovigerous ♀, NGoMCSS Sta. E1C-2. 1 non-ovigerous ♀, NGoMCSS Sta. E2-6. 1 non-ovigerous ♀, NGoMCSS Sta. E2A-1.

Diagnosis. — (After Hansen, 1913.) Female. Carapace as long as lengths of first two pereonites combined. Antenna article 2 with one or two ventral spines and article 3 with three ventral spines. Uropods short (as long as pleotelson or shorter), endo- and exopod uni-articulate.

Male. Unknown.

Distribution. — Davis Strait. Bathymetric range: 2626 m. In the Gulf of Mexico only found in a cluster off the Florida Panhandle. Bathymetric range: 350–622 m. See map 42.

Remarks. — *Typhlotanais trispinosus* was reported from the NGoMCSS programme conducted during the 1980's. The material from the NGoMCSS study is currently located at the Gulf Coast Research Laboratory, Mississippi, and was examined by the present author. However, the material, as with that of *Paranarthrura insignis*, seems to have crystallized and does not allow for reliable identification. This species does have the three spines on the antenna as also found in *T. trispinosus* and *T. tenuicornis* G. O. Sars, 1882 (from Norway) but species assignment is uncertain, since Hansen's (1913) description is fragmentary and since the type locality is in the Davis Strait (Hansen, 1913). As with *P. insignis*, it is unlikely that this Polar species is encountered in warm waters as those of the Gulf of Mexico.

Family PSEUDOTANAIDAE Sieg, 1976

Tanaidae Lang, 1949, Arkiv för Zoologi, **42**(18): 5.
Paratanaidae – Lang, 1967, (in part), Arkiv för Zoologi, **42**(19): 343.
Pseudotanaidae Sieg, 1976, Zeitschrift für Zoologische Systematik und Evolutionsforschung, **14**: 185.

Diagnosis. — (Modified after Larsen & Wilson, 2002.) Female. Medium or rigid body calcification. No plates in carapace. Eye lobes present or absent, with or without visual elements. Pereon with six free pereonites. Pereonite 1 reduced. Pleon with five free pleonites, as wide as pereon, no articulated setae on pleonites. Antennule with 3–4 articles. Antenna with six articles, article 3 without spiniform setae dorsally. Molar process broad or pointed. Labium with one pair of lobes, medial processes absent. Maxillipedal bases and endites fused, or only bases fused, endites narrower than basis and with three, one, or without flat setae. Cheliped attachment via sclerite. Marsupium (where known) with one or possibly four (Bird & Holdich, 1989c: 234) pairs of oostegites. Pereopod coxae present, dactylus and unguis either fused to a hook, or not. Pleopods well developed, or with simple setae only, or rudimentary. Uropodal endopod with two articles, exopod with 1–2 articles.

Male (where known). Without functional mouthparts. Carapace shorter than that of females. Antennule with 6–7 articles and several aesthetascs. Cheliped only moderately larger than in female. Pleopod plumose setae generally longer than in female.

Remarks. — Several new species belonging to this family (genera *Pseudotanais*, *Cryptocope*, and *Parapseudotanais*) were discovered during the DGoMB project. These new species are not described, since this family is currently the focus of a Ph.D. dissertation (J. McLelland, pers. comm.), and for the same reason only a key to the genera of the Pseudotanaidae in the Gulf of Mexico is given herein.

KEY TO THE DEEP-SEA GENERA OF PSEUDOTANAIDAE IN THE GULF OF MEXICO

1. Pereopod 4–6 unguis reduced and serrated .... *Iungentitanais* Sieg, 1976
– Pereopod 4–6 unguis not reduced or serrated 2
2. Maxillipedal endites not fused .... *Cryptocope* G. O. Sars, 1882
– Maxillipedal endites fused 3

3. Pereonites 4 and 5 narrower than other pereonites .... *Parapseudotanais* Bird & Holdich, 1989
– Pereonites 4 and 5 as wide as other pereonites .... *Pseudotanais* G. O. Sars, 1882

Subfamily CRYPTOCOPINAE Sieg, 1977

Cryptocopinae Sieg, 1977, Mitteilungen aus dem Zoologischen Museum Berlin, **53**(1): 11.

Diagnosis. — (Modified after Larsen & Wilson, 2002.) Female. Rigid body calcification. Eyes or eye lobes present. Antennule with four articles. Molar process of mandible broad or pointed. Maxillipedal endites not fused, either naked or with three flat setae. Marsupium (where known) with one or possibly four (Bird & Holdich, 1989c: 234) pairs of oostegites. Pereopods 4–6 with dactylus and unguis not fused to a hook. Pleopods well developed, or with simple setae only, or rudimentary.

Male. Unknown.

Type genus. — *Cryptocope* G. O. Sars, 1882.

Genera included. — *Cryptocope* G. O. Sars, 1882; *Cryptocopoides* Sieg, 1976; *Curtichelia* Kudinova-Pasternak, 1987; *Iungentitanais* Sieg, 1976; *Latitanais* Kudinova-Pasternak, 1987; *Paraiungentitanais* Sieg, 1977.

Remarks. — This subfamily is represented in the Gulf by *Iungentitanais primitivus* Sieg, 1977 and a new deep-sea species currently under description (J. McLelland, pers. comm.).

Genus **Cryptocope** G. O. Sars, 1882

Cryptocope G. O. Sars, 1882, Archiv for Matematik og Naturvidenskab, **7**: 49.

Diagnosis. — (After Sieg, 1977.) Female. Pereonites 1–3 not reduced. Pereonites 4 and 5 not elongate. Antennule with four articles. Mandibular molar pointed. Maxillule endite with nine spiniform terminal setae. Cheliped attached via large (larger than basis) sclerite. Pereopods 4–6 with dactylus and unguis not fused, unguis not serrated. Pleopods reduced and without setae. Uropods biramous.

Male. Unknown.

Type species. — *Tanais abbreviatus* G. O. Sars, 1868.

Gender of generic name. — Feminine.

Remarks. — This genus is inadequately defined. It probably does not belong in the Pseudotanaidae at all, but is more likely related to the Leptognathiidae. One species belonging to this genus was found in the Gulf of Mexico, but is yet to be formally described (J. McLelland, pers. comm.).

Species currently assigned to this genus. — *C. abbreviata* (G. O. Sars, 1868); *C. vitjazi* Kudinova-Pasternak, 1982.

Genus **Iungentitanais** Sieg, 1977

Mortensenia Sieg, 1973, Ein Beitrag zum natürlichen System der Dikonophora Lang: 46.
Iungentitanais Sieg, 1977, Mitteilungen aus dem Zoologischen Museum Berlin, **53**(1): 11.

Diagnosis. — (Modified after Sieg, 1977; Sieg & Heard, 1988.) Female. Pereonites 1–3 not reduced. Pereonites 4 and 5 not elongate. Antennule with four articles. Maxillule endite with five spiniform terminal setae. Mandibular molar not pointed, with ring of denticles distally. Cheliped long and slender (carpus and propodus approx. 5 times as long as wide). Pereopods 4–6 with unguis small and serrated. Pleopods well developed. Uropods biramous and therami both bi-articulate.

Male. Unknown.

Type species. — *Mortensenia primitiva* Sieg, 1973.

Gender of generic name. — Masculine.

Species currently assigned to this genus. — *I. primitivus* (Sieg, 1973)

Iungentitanais primitivus (Sieg, 1973)

Mortensenia primitiva Sieg, 1973, Ein Beitrag zum natürlichen System der Dikonophora Lang: 173.
Iungentitanais primitivus – Sieg, 1977, Mitteilungen aus dem Zoologischen Museum Berlin, **53**(1): 11.

Diagnosis. — Monotypic genus, no species diagnosis given.

Distribution. — Type locality off St. Thomas. In the Gulf the species is found off the Florida Panhandle and in the Straits of Florida. Bathymetric range: 18–33 m (likely found at greater depths at well). See map 43.

Remarks. — *Iungentitanais primitivus* is recorded as a shallow water species, but there seem to be some doubts about the bathymetric range. Sieg & Heard (1988: 57, fig. 12) placed a record on their distribution map (around

28°N 84°W) where the depth should be at least 100 m. It is possible that this species will occasionally be found in deeper samples.

Subfamily PSEUDOTANAINAE Sieg, 1976

Pseudotanaidae Sieg, 1976, Zeitschrift für Zoologische Systematik und Evolutionsforschung, **14**: 185.
Pseudotanainae – Sieg, 1977, Mitteilungen aus dem Zoologischen Museum Berlin, **53**(1): 26.

Diagnosis. — (Modified from Bird & Holdich, 1989c; Larsen & Wilson, 2002.) Female. Weak body calcification. Eyes absent or present (*Pseudotanais mortenseni*). Antennule with three articles. Molar process of mandible pointed. Maxillipedal endites fused, either naked or with one flat seta. Cheliped fixed finger and dactylus elongate. Marsupium (where known) with one pair of oostegites. Pereopods 4–6 with dactylus and unguis fused to hook. Pleopods well developed or with simple setae only.

Male. See remarks under genus *Pseudotanais*.

Type genus. — *Pseudotanais* G. O. Sars, 1882.

Genera included. — *Mystriocentrus* Bird & Holdich, 1989; *Parapseudotanais* Bird & Holdich, 1989; *Pseudotanais* G. O. Sars, 1882.

Remarks. — The characters of this subfamily are stable but none of them are exclusively found in the subfamily. Members of Pseudotanainae can easily be separated from Cryptocopinae, the other subfamily in the family, by the presence of fused maxillipedal endites.

Genus **Parapseudotanais** Bird & Holdich, 1989

Parapseudotanais Bird & Holdich, 1989, Zoological Journal of the Linnean Society, **97**: 281.

Diagnosis. — (Modified after Bird & Holdich, 1989c.) Female. Pereonites 1–3 reduced. Pereonites 4 and 5 elongate. Antennule with three articles; distal article elongate, and setae of this article with simple apex. Antenna with six articles; articles 2 and 3 with spiniform setae dorsally. Mandibles with crenulated distal margin; right mandibular lacinia mobilis reduced to a tiny, spiniform process fused to body of mandible; left mandibular lacinia mobilis as large and wide as incisor, discrete; molar pointed. Cheliped fixed finger with two setae ventrally. Pereopod 1 longer and more elongate that pereopods 2 and 3, without any spiniform setae. Pereopods 2 and 3 have carpus with blade-like setae ventrally. Pereopods 4–6 have propodus with long (longer than dactylus/unguis) spiniform seta dorsally. Pleopods well developed. Uropods biramous.

Male. Unknown.

Type species. — *Parapseudotanais abyssalis* Bird & Holdich, 1989.

Gender of generic name. — Bird & Holdich (1989c) erected this genus but did not assign a gender to the generic name. However, as the name is derived from *Tanais*, which is masculine, the gender is presumed to be masculine as well.

Remarks. — One species belonging to this genus was found in the Gulf of Mexico, but is yet to be formally described (J. McLelland, pers. comm.).

Species currently assigned to this genus. — *P. abyssalis* Bird & Holdich, 1989.

Genus **Pseudotanais** G. O. Sars, 1882

Pseudotanais G. O. Sars, 1882, Archiv for Matematik og Naturvidenskab, **7**: 46.
Paratanais Hatch, 1947, University of Washington Publications in Biology, **10**(5): 161.

Diagnosis. — (Modified after Bird & Holdich, 1989c.) Female. Pereonite 1 reduced. Pereonites 4 and 5 of normal appearance (not elongate). Antennular article 3 with setae with complex tips. Antenna with six articles; as long as antennule; articles 2 and 3 with spiniform setae dorsally. Mandibular molar pointed or of intermediate width. Cheliped fixed finger usually with one seta ventrally. Pereopod 1 longer and more elongate that pereopods 2 and 3, without any spiniform setae. Pereopods 2 and 3 have carpus with blade-like setae ventrally. Pereopods 4–6 have propodus with long (longer than dactylus/unguis) spiniform seta dorsally. Pleopods present or absent. Uropods biramous.

Male. See remarks.

Type species. — *Tanais forcipatus* Lilljeborg, 1864.

Gender of generic name. — Masculine.

Remarks. — The characters of this genus are stable. The genus is subdivided into two subgenera, separated by the presence or absence of setae with complex tips on antennular article 3. There seem to be some discrepancies concerning the male in this genus. Bird & Holdich (1989c: 245) record males with functional mouthparts, thicker antennule, and thicker pleon than in female, whereas Sieg (1977: 9) claims that the males are of the non-feeding type and have 'leptognathiid'-like appearances. This issue is not resolved but it is possible that both types of males exist.

Species currently assigned to this genus. — *P. abyssi* Hansen, 1913; *P. affinis* Hansen, 1886; *P. baresnauti* Bird, 1999; *P. californiensis* Dojiri & Sieg, 1997; *P. colonus* Bird & Holdich, 1989; *P. corollatus* Bird & Holdich, 1989; *P. denticulatus* Bird & Holdich, 1989; *P. falcicula* Bird & Holdich, 1989; *P. forcipatus* (Lilljeborg, 1864); *P. gaussi* Vanhöffen, 1914; *P. gerlachi* Sieg, 1977; *P. guillei* Shiino, 1978; *P. jonesi* Sieg, 1977; *P. lilljeborgi* G. O. Sars, 1882; *P. longipes* Hansen, 1913; *P. longisetosus* Sieg, 1977; *P. longispinus* Bird & Holdich, 1989; *P. macrochelis* G. O. Sars, 1882; *P. makrothrix* Dojiri & Sieg, 1997; *P. malayensis* Sieg, 1977; *P. mediterraneus* G. O. Sars, 1882; *P. mexikolpos* Sieg & Heard, 1988; *P. mortenseni* Sieg, 1977; *P. nordenskioldi* Sieg, 1977; *P. oculatus* Hansen, 1913; *P. scalpellum* Bird & Holdich, 1989; *P. siegi* Kudinova-Pasternak, 1985; *P. similis* Sieg, 1977; *P. spatula* Bird & Holdich, 1989; *P. spicatus* Bird & Holdich, 1989; *P. unicus* Sieg, 1977; *P. vitjazi* Kudinova-Pasternak, 1966; *P. vulsella* Bird & Holdich, 1989.

Pseudotanais mexikolpos Sieg & Heard, 1988

Pseudotanais mexikolpos Sieg & Heard, 1988, Proceedings of the Biological Society of Washington, **101**: 39.

Diagnosis. — (Modified after Sieg & Heard, 1988.) Female. Carapace short, only hardly longer than wide. Eyes present. Pereopod 3 only hardly shorter than pereopod 4. Pereopods 2–6 with spiniform setae modified into a curled leaf. Uropod endo- and exopod bi-articulate.

Male. Unknown.

Distribution. — This species is rare and has only been recorded from a hyper-saline seep (East Flower Garden Bank) and from Stetson Bank in the northwestern Gulf of Mexico. Bathymetric range: 72–213 m. See map 44.

Remarks. — *Pseudotanais mexikolpos* is closely similar to *P. mediterraneus* G. O. Sars, 1882 and *P. unicus* Sieg, 1977, but can be separated from these species by the shorter carapace, longer pereopod 3, and longer setae on the propodi of pereopods 4–6 (Sieg & Heard, 1988). This species was previously reported as inhabiting shallow waters < 100 m (Sieg & Heard, 1988) but was found deeper during this study (213 m). The identification of this species is somewhat dubious, though, owing to the poor condition of the material.

Pseudotanais mortenseni Sieg, 1977

Pseudotanais mortenseni Sieg, 1977, Mitteilungen aus dem Zoologischen Museum in Berlin, **53**: 42.

Diagnosis. — (Modified after Sieg, 1977.) Female. Carapace as long as wide. Eyes present. Pereopod 3 as long as pereopod 4. Pereopods 4–6 with long spiniform setae, but these not modified into a curled leaf. Uropod endo- and exopod bi-articulate.

Male. Unknown.

Distribution. — This species is rare and has only been recorded off the southwest coast of Florida. Bathymetric range: 18–215 m. See map 45.

Remarks. — This species was previously reported as inhabiting shallow waters but without depth assignments (Sieg & Heard, 1988), yet was found deeper during this study. The identification of this species is, however, even more dubious than that of *P. mexikolpos*, owing to the condition of the material. This species can be separated from *P. mexikolpos* by the lack of modified spiniform setae on pereopods 2–6.

Family TANAELLIDAE Larsen & Wilson, 2002

Paratanaidae – Lang, 1968, Galathea Report, **9**: 161.
Leptognathiidae – Sieg, 1986, Biology of the Antarctic Seas, **45**: 72.
Anarthruridae – Heard, 2002, Contributions to the study of east Pacific Crustaceans, **1**: 376.
Tanaellidae Larsen & Wilson, 2002, Journal of Zoological Systematics and Evolutionary Research, **40**(4): 218.

Diagnosis. — (Modified after Larsen & Wilson, 2002.) Female. Heavy body calcification. No plates in carapace. Eyes and eye-lobes absent. Pereon with six free pereonites. Pereonite 1 not reduced. Pleon with five free pleonites, as wide as or narrower (*Arthrura*) than pereon, no articulated setae on pleonites. Antennule with four articles. Antenna with six articles (or seven with fusion line), article 3 without spiniform setae dorsally. Molar process of mandible broad or pointed. Labium with one pair of lobes, medial spiniform setae absent. Maxillipedal bases not fused (*Arthrura*) or fused, endites not fused, narrower than basis and with one or no flat seta or denticle. Cheliped attachment via sclerite. Marsupium (where known) with four pairs of oostegites. Pereopods 1–3 with coxae. Pereopods 4–6 mostly without coxae, dactylus and unguis not fused to a hook. Pleopods well developed, reduced or absent. Uropodal endopod with 1–2 articles, exopod reduced and fused with basal article.

Male (where known) sexual dimorphism low. Antennule thicker than that of female but with same number of articles. Functional mouthparts retained. Pleopods well developed or with simple setae only.

Type genus. — *Tanaella* Norman & Stebbing, 1886.

Genera included. — *Araphura* Bird & Holdich, 1984; *Araphuroides* Sieg, 1986; *Arthrura* Kudinova-Pasternak, 1966; *Tanaella* Norman & Stebbing, 1886.

Remarks. — The ratio of the lengths of dactylus and unguis seems to make a most stable character within the genera of this family, which is hence suitable for generic discrimination.

KEY TO THE GENERA OF THE TANAELLIDAE IN THE GULF OF MEXICO

1. Uropodal exopod process not shaped as a spine. Uropodal endopods curved inwards and shaped like a pincer .... *Tanaella* Norman & Stebbing, 1886
– Uropodal exopod process shaped as an acute spine. Uropodal endopods not shaped as a pincer 2
2. Body slender (9–13 times as long as wide). Cheliped basis with wide distal part and a narrow, elongate proximal extension .... *Araphura* Bird & Holdich, 1984
– Body less than 9 times as long as wide. Cheliped basis without narrow proximal extension .... *Araphuroides* Sieg, 1986

Genus **Araphura** Bird & Holdich, 1984

Leptognathia Lilljeborg, 1864, Uppsala Universitets Årsskrifter, **5**: 22.
Araphura Bird & Holdich, 1984, Zoologica Scripta, **13**: 295.

Diagnosis. — (Modified after Bird & Holdich, 1984; Sieg & Dojiri, 1989.) Female. Body slender (9–13 times as long as wide). Pleon short (less than 0.2 times body length), often wider than pereon. Eyes and eye-lobes absent. Antennule with four articles, basal part frequently covered by carapace. Antenna with 6–7 articles (with or without fusion line). Mandible molar of intermediate width. Maxillipedal bases fused; endites not fused, with one large, globe-shaped seta or process on distal margin, disto-lateral corners often with process. Cheliped basis with wide distal part and a narrow, elongate basal part attached to large elongate sclerite. Pereopods 1–3 with dactylus shorter than unguis. Pereopods 4–6 with dactylus longer than unguis. Pleopods small but well developed, with plumose setae. Uropodal endopod with two articles, exopod fused to basal article and shaped like a spine.

Male. Unknown.

Type species. — *Leptognathia brevimanus* Lilljeborg, 1864.

Gender of generic name. — Feminine.

Remarks. — The characters appear stable but in two species the characters could not be ascertained: *A. filiformis* (Lilljeborg, 1864) is poorly described and for *A. elongata* (Shiino, 1970) the mouthparts have never been described or illustrated. The pseudobiramous uropod is a good, stable charater but is shared with the closely allied genus *Araphuroides*. The length of the exopod process is essential for species identification in *Araphura*. A new and potentially diagnostic generic character is the characteristic shape of the cheliped basis and its attachment to the carapace. The cheliped basis has a wide distal part and a narrow, elongate basal part on top of which a large, elongate sclerite attaches the cheliped to the carapace. This character has been verified in the new species described below, in *A. extensa* Larsen, 2003; *A. higginsi* Sieg & Dojiri, 1989; *A. elongata* Shiino, 1970; *A. parabrevimanus* Lang, 1968; *A. filiformis* (Lilljeborg, 1864); *A. brevimanus* (Lilljeborg, 1864); and *A. joubinensis* Sieg & Dojiri, 1989. The description of *A. rectifrons* Kudinova-Pasternak, 1973, does not allow to verify this condition.

Sieg (1986a) removed *A. brevispina* from *Araphura* and assigned it to his new genus *Araphuroides* on the basis of body shape and the structure of the mandibular molar. However, neither the molar nor the body shape of *A. brevispina* entirely fit Sieg's (1986a) diagnosis of *Araphuroides*. The status of *A. brevispina* is, therefore, not obvious, and possibly represents a link between *Araphura* and *Araphuroides*.

As regards the original spelling of Lang (1968) of *A. parabrevimanus*, i.e., *Leptognathia parabrevimanu*, without terminal "s", Sieg & Dojiri (1989) have argued that would be in discordance with the (1985) Code, Art. 32c, i. They concluded the species name had to be changed to *parabrevimanus*. Their opinion is followed in this work, although their reference to Art. 32c, i, may be unjustified. For the moment, the matter is thus regarded to be in need of further investigation.

Species currently assigned to this genus. — *A. breviaria* Dojiri & Sieg, 1997; *A. brevimanus* (Lilljeborg, 1864); *A. curticauda* sp. nov.; *A. cuspirostris* Dojiri & Sieg, 1997; *A. elongata* (Shiino, 1970); *A. extensa* Larsen, 2003; *A. filiformis* (Lilljeborg, 1864); *A. higginsi* Sieg & Dojiri, 1989; *A. joubinensis* Sieg & Dojiri, 1989; *A. parabrevimanus* (Lang, 1968); *A. rectifrons* (Kudinova-Pasternak, 1973); *A. spinithenari* sp. nov.

Araphura curticauda sp. nov.
(figs. 64, 65)

Material examined. — Holotype, non-ovigerous ♀ (USNM 1068903), body length 1.95 mm, DGoMB Sta. MT3-1. Paratypes: 1 non-ovigerous ♀ (USNM 1068904), body length 1.45 mm, DGoMB Sta. MT3-2. 1 non-ovigerous ♀ (USNM 1068905), DGoMB Sta. MT3-1 (dissected).

Additional material. — 2 non-ovigerous ♀♀, 4 mancas, DGoMB Sta. MT3-2. 1 ovigerous ♀, 5 non-ovigerous ♀♀, DGoMB Sta. MT3-4.

Diagnosis. — Female. Body elongate (about 9 times as long as wide). Carapace longer than pereonite 2. Pereonite 6 longer than wide. Pleotelson shorter than the length of two free pleonites combined. Cheliped propodus without dorso-distal process. Pereopods 4–6 as long as pereopods 1–3. Pleopods well developed, with plumose setae. Uropodal exopod longer than 0.33 times but shorter than 0.5 times first endopod article.

Male. Unknown.

Etymology. — Named after the comparatively short pleotelson and uropodal exopod process (Latin, curti + cauda = short + tail). The name is a noun in apposition to the generic name.

Type locality. — DGoMB Sta. MT3-1, 28°13'2246"N 89°29'7679"W, depth 983 m.

Description. — Adult female. Body (fig. 64A) about 9 times as long as wide.

Carapace lateral margins straight. Almost twice as long as wide. Shorter than the lengths of pereonites 1 and 2 combined but longer than pereonite 2. Eyes and eye-lobes absent. Covers basal part of antennules.

Pereon lateral margins straight. Pereonite 1 wider than long. Pereonites 2–6 longer than wide.

Pleon short (including pleotelson about 0.2 times total body length). Pleotelson shorter than the lengths of two free pleonites combined, apex blunt and covered by dorsal plate.

Antennule (fig. 64B) shorter than carapace, with four articles. Article 1 shorter than rest of antennule, with one simple and one setulose seta distally. Article 2 longer than article 3, with two simple and one setulose setae distally. Article 3 shorter than article 4, with one simple seta distally. Article 4 with four simple setae and one aesthetasc (broken in figure) distally.

Antenna (fig. 64C) with five articles; 0.75 times as long as antennule. Articles 1 and 2 wider than other articles, both with one simple seta dorsally. Article 3 without trace of fusion line, with three simple setae distally. Article 4

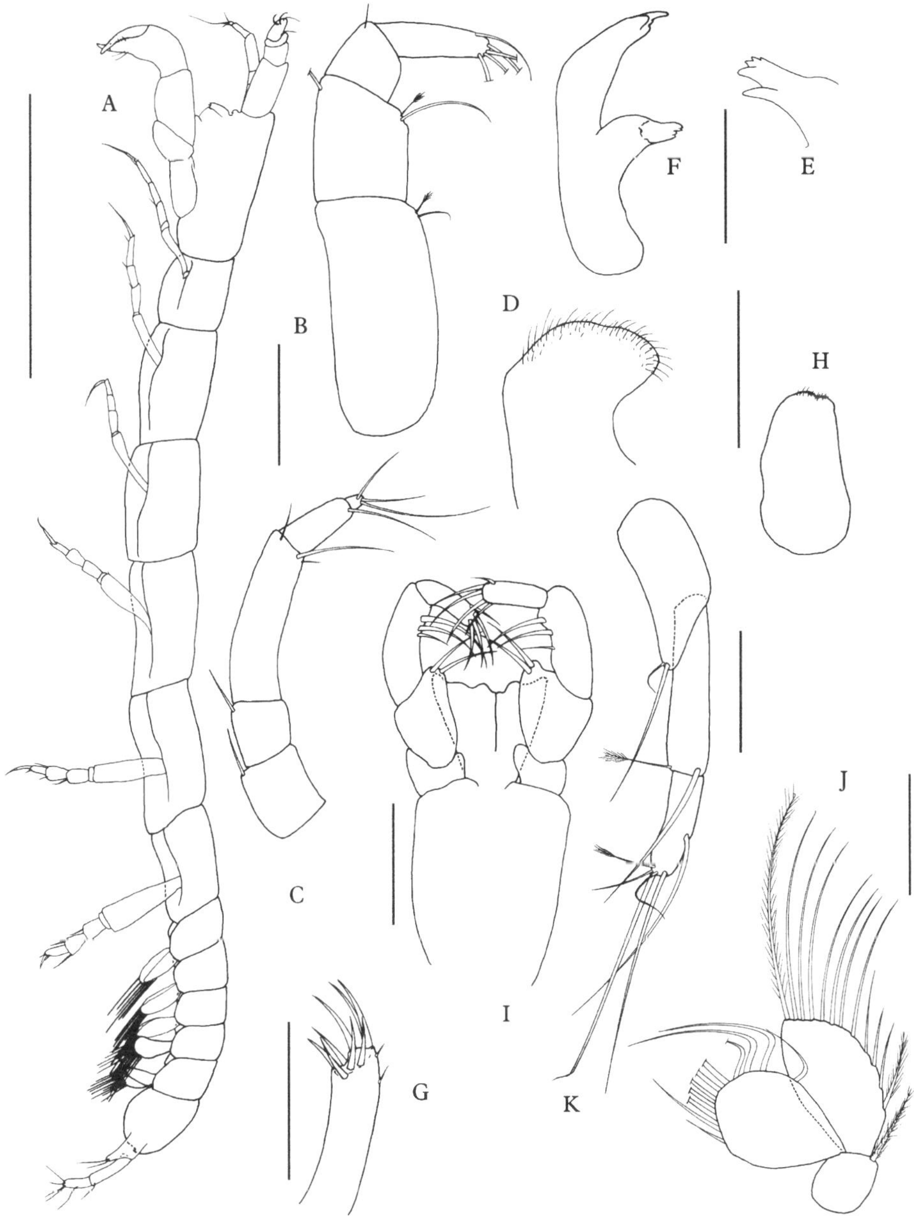

Fig. 64. *Araphura curticauda* sp. nov., female; A, holotype, lateral view; B, antennule; C, antenna; D, labrum; E, left mandible; F, right mandible; G, maxillule; H, maxilla; I, maxilliped; J, pleopod; K, uropod. Scale bars equal: A, 0.5 mm; B-C, 0.1 mm; D-K, 0.05 mm.

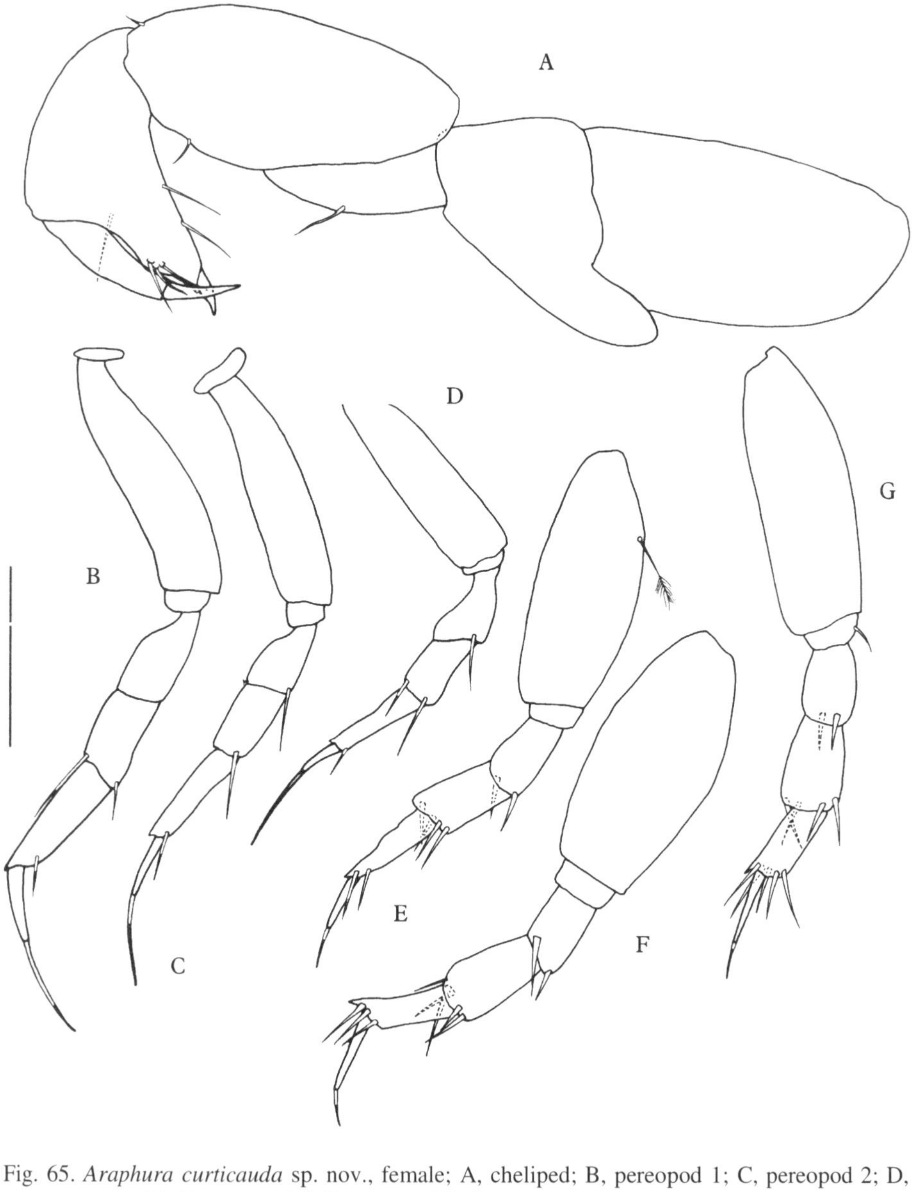

Fig. 65. *Araphura curticauda* sp. nov., female; A, cheliped; B, pereopod 1; C, pereopod 2; D, pereopod 3; E, pereopod 4; F, pereopod 5; G, pereopod 6. Scale bar equals 0.1 mm.

shorter than article 3, with one simple seta distally. Article 5 minute, with three setae distally.

Mouthparts. Labrum (fig. 64D) with finely setose apex. Mandibular molar tapering distally, with several denticles distally. Left mandible (fig. 64E) lacinia mobilis shorter than incisor, with one denticle; incisor broad, with three small and one large separate denticles. Right mandible (fig. 64F) incisor narrow and pointed, divided into two large and one tiny distal denticles. Labium lost during dissection. Maxillule (fig. 64G) endite with eight spiniform setae distally and two simple setae; palp lost during dissection. Maxilla (fig. 64H) elongate, distally with fine setules. Maxilliped (fig. 64I) endites with one, apparently naked, outer process and with one large inner denticle; palp article 1 naked, article 2 with two inner setae, article 3 with three setae on inner margin, article 4 with five setae. Epignath not recovered.

Cheliped (fig. 65A) basis with narrow proximal part and wide distal part, attached to long, elongate sclerite, naked, shorter than carpus. Merus with one seta ventrally. Carpus longer than propodus including fixed finger, with one simple seta on each margin. Propodus without dorso-distal process, with one seta at dactylus insertion. Fixed finger with two setae ventrally and three on inner margin as well as two denticles distally, apex pointed. Dactylus as long as fixed finger, apex pointed.

Pereopod 1 (fig. 65B) coxa small and naked. Basis longer than the three succeeding articles combined, naked. Ischium naked. Merus longer than carpus, widening, naked. Carpus shorter than 0.5 times basis, with one stiff seta distally on each margin. Propodus with one stiff seta distally and a dorsal spine. Dactylus and unguis together longer than propodus and not fused, dactylus shorter than unguis.

Pereopod 2 (fig. 65C) as pereopod 1 except: merus and carpus each with one stiff seta distally.

Pereopod 3 (fig. 65D) as pereopod 1 except: merus with one stiff seta distally.

Pereopod 4 (fig. 65E) no coxa present. Basis longer than the three succeeding articles combined, with one setulose seta medio-ventrally. Ischium naked. Merus shorter than carpus, widening distally and with two stiff setae distally. Carpus shorter than 0.5 times basis, with four stiff setae distally. Propodus longer than carpus, with two stiff setae distally and a dorsal spine. Dactylus and unguis together as long as propodus, not fused, dactylus longer than unguis.

Pereopod 5 (fig. 65F) as pereopod 4 except: basis naked. Carpus with five stiff setae distally. Propodus longer than carpus, with three stiff setae distally and a dorsal spine.

Pereopod 6 (fig. 65G) as pereopod 4 except: basis naked. Ischium with one simple seta. Propodus with five setae distally and a dorsal spine.

Pleopods (fig. 64J) small but well developed. Basal article with one circum-plumose seta. Endopod and exopod of similar size. Endopod with one outer and ten inner plumose setae. Exopod with 12 plumose setae.

Uropod (fig. 64K) longer than pleotelson. Basal article naked and 0.5 times as long as first endopod article. Endopod article 1 with one simple and one setulose setae distally; article 2 shorter than article 1, with one long seta sub-distally, two long setae distally, two short setae distally, and one setulose seta distally. Exopod process about 0.33 times as long as first endopod article, with one long and one short simple setae distally.

Distribution. — This is a very rare species, only found in the northern part of the Gulf. Bathymetric range narrow: 983–987 m. See map 46.

Remarks. — This species is very similar to A. *spinithenari* sp. nov. but differs in having a more slender cheliped with pointed dactylus and fixed finger, without dorsal cheliped propodus process, a shorter pleotelson and uropodal exopod process, and a considerably longer pereonite 6. Antenna with five articles and no sign of fusion line. *Araphura curticauda* can be separated from *A. higginsi* and *A. brevimanus* by the pleotelson being longer than wide; from *A. parabrevimanus*, *Araphura cuspirostris* Dojiri & Sieg, 1997, and *Araphura breviaria* Dojiri & Sieg, 1997 by the uropodal exopod process being shorter than 0.5 times the first endopod article; from *A. rectifrons* by the pleotelson being shorter than two free pleonites combined, and by the antenna article 4 showing no fusion line; from *A. elongata*, *A. extensa*, *A. joubinensis*, and *A. filiformis* by the uropodal exopod process being longer than 0.33 times as long as the first endopod article.

Araphura extensa Larsen, 2003

Araphura extensa Larsen, 2003, Journal of Crustacean Biology, **23**: 789–792.

Diagnosis. — (Modified after Larsen, 2003b.) Female. Body extremely elongate (length/width ratio = 13 : 1). Carapace short (as long as pereonite 2). Pereonite 6 longer than wide. Pleotelson shorter than the lengths of three free pleonites combined. Cheliped merus long (more than 0.5 times as long as carpus), propodus without dorso-distal process. Pereopods 4–6 as long as pereopods 1–3. Pleopods well developed, with plumose setae. Uropodal exopod tiny (less than 0.33 times as long as first endopod article).

Male. Unknown.

Distribution. — This species was found in only one sample from a cold seep: GB 425, Sta. BL7, 27°33'60"N 92°32'33"W. Bathymetric range: 570 m. See map 47.

Araphura higginsi Sieg & Dojiri, 1989

Araphura higginsi Sieg & Dojiri, 1989, Zoologica Scripta, **18**:128.

Diagnosis. — (Modified after Sieg & Dojiri, 1989.) Female. Body elongate (length/width ratio = 11 : 1). Carapace long (longer than pereonite 2). Pereonite 6 shorter than wide. Pleotelson shorter than the lengths of three free pleonites combined. Cheliped merus long (more than 0.5 times as long as carpus). Cheliped propodus with three blunt processes. Pereopods 4–6 as long as pereopods 1–3. Pleopods well developed, with plumose setae. Uropodal exopod long (more than 0.5 times as long as first endopod article).

Male. Unknown.

Distribution. — Atlantic coast of the United States from South Carolina to Fort Pierce, Florida. Only one Gulf record, off Tampa. Bathymetric range: 22–145 m. See map 48.

Remarks. — This species has been reported from the Atlantic coast off Florida but only once from the Gulf of Mexico proper. In the Atlantic, *Araphura higginsi* has only been taken from relatively shallow water (< 50 m) while the Gulf record is somewhat deeper, 145.5 m (Sieg & Dojiri, 1989).

Araphura spinithenari sp. nov.
(figs. 66–68)

Material examined. — Holotype, non-ovigerous ♀ (USNM 1068906), body length 1.75 mm, DGoMB Sta. W2-1. Paratypes: 1 non-ovigerous ♀ (USNM 1068907), body length 2.5 mm, DGoMB Sta. RW2-1. 1 non-ovigerous ♀ (USNM 1068908), DGoMB Sta. W2-1 (dissected).

Additional material. — 1 non-ovigerous ♀, DGoMB Sta. W2-4. 1 non-ovigerous ♀ (dissected), DGoMB Sta. S42-3. 4 mancas, DGoMB Sta. C7-2. 2 mancas, DGoMB Sta. MT3-2. 8 non-ovigerous ♀♀, 6 mancas, DGoMB Sta. MT2-2. 1 non-ovigerous ♀, DGoMB Sta. C12-1. 3 non-ovigerous ♀♀, 2 mancas, DGoMB Sta. MT3-1. 100+ non-ovigerous ♀♀ and mancas, DGoMB Sta. MT1-3.

Diagnosis. — Female. Body elongate (about 9 times as long as wide). Carapace longer than pereonites 1 and 2 combined. Pereonite 6 wider than long.

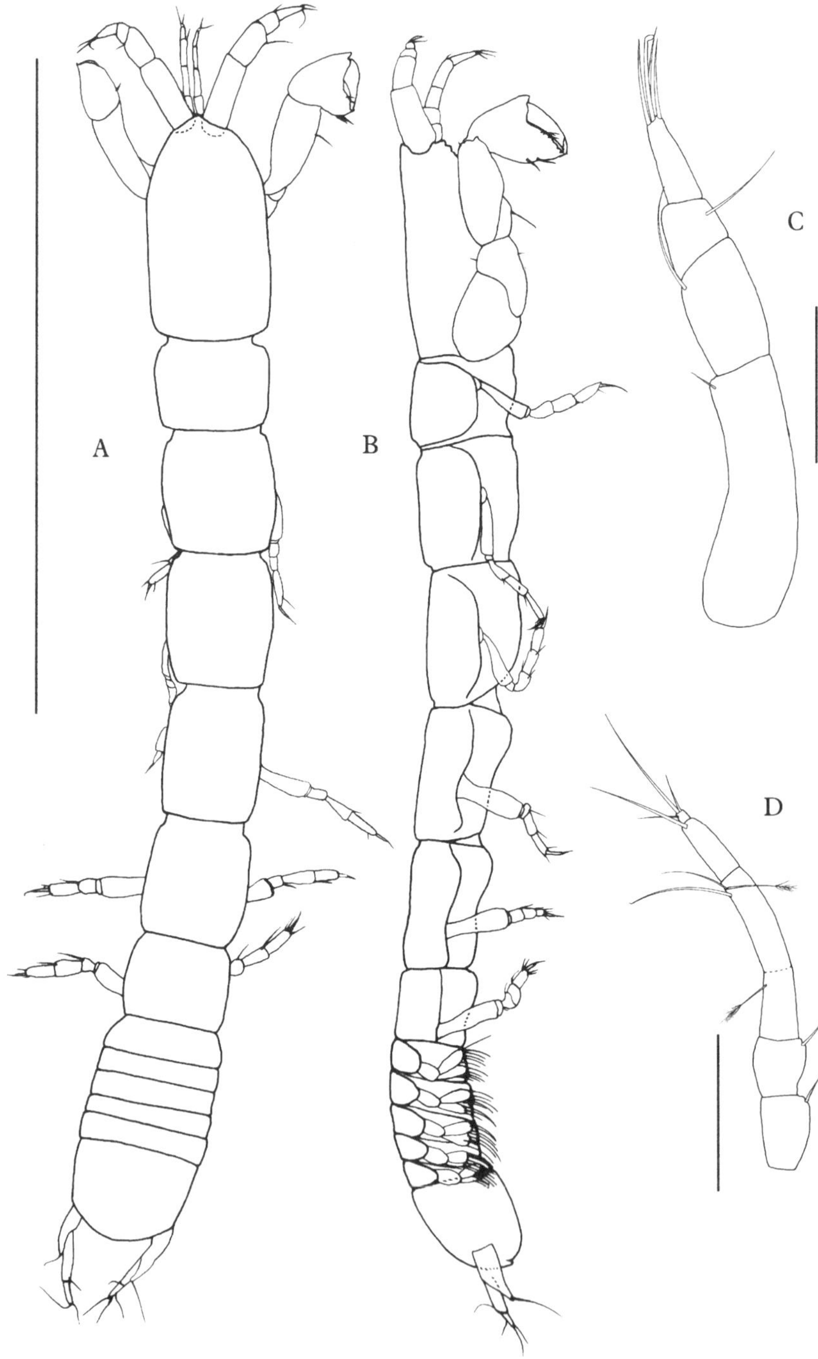

Fig. 66. *Araphura spinithenari* sp. nov., female; A, holotype, dorsal view; B, holotype, lateral view; C, antennule; D, antenna. Scale bars equal: A, 1.0 mm; C-D, 0.1 mm.

Pleotelson as long as lengths of four free pleonites combined. Cheliped propodus with dorso-distal process. Pereopods 4–6 longer than pereopods 1–3. Pleopods well developed, with plumose setae. Uropodal exopod 0.5 times as long as first endopod article.

Male. Unknown.

Etymology. — Named after the spiniform process on the cheliped propodus in combination with the Latinized Greek word for hand or palm (thenar), in the genitive singular.

Type locality. — DGoMB Sta. W2-1, 27°24'8008"N 93°20'2579"W, depth 625 m.

Description. — Adult female. Body (fig. 66A, B) about 9 times as long as wide.

Carapace with lateral margins straight. Almost twice as long as wide. Longer than the lengths of pereonites 1 and 2 combined. Eyes and eye-lobes absent. Covers basal part of antennules.

Pereon with lateral margins straight. Pereonites 1 and 6 wider than long. Pereonites 2–5 longer than wide.

Pleon short (including pleotelson about 0.2 times total body length). Pleonite 1 apparently larger than other pleonites. Pleotelson as long as lengths of four free pleonites combined, apex blunt and covered by a dorsal plate.

Antennule (fig. 66C) shorter than carapace, with four articles. Article 1 as long as rest of antennule, with one simple seta distally. Article 2 longer than article 3, with one simple seta distally. Article 3 shorter than article 4, with one simple seta distally. Article 4 with three thick simple setae and one aesthetasc distally.

Antenna (fig. 66D) with five articles; 0.75 times as long as antennule. Articles 1 and 2 wider than other articles, both with one seta dorsally. Article 3 with fusion line; first pseudoarticle with one setulose seta distally, second pseudoarticle with one setulose and two simple setae distally. Article 4 shorter than article 3, with one simple seta distally. Article 5 minute, with three setae distally.

Mouthparts. Labrum (fig. 67A) finely setose and ovoid. Mandibular molar tapering distally, with several denticles distally. Left mandible (fig. 67C) lacinia mobilis shorter than incisor, with two denticles; incisor broad, with three separate denticles. Right mandible (fig. 67B) incisor broad, with one large and three small denticles distally. Labium (fig. 67E) as wide as maxillipedal endites, outer corners setose. Maxillule (fig. 67D) endite with seven

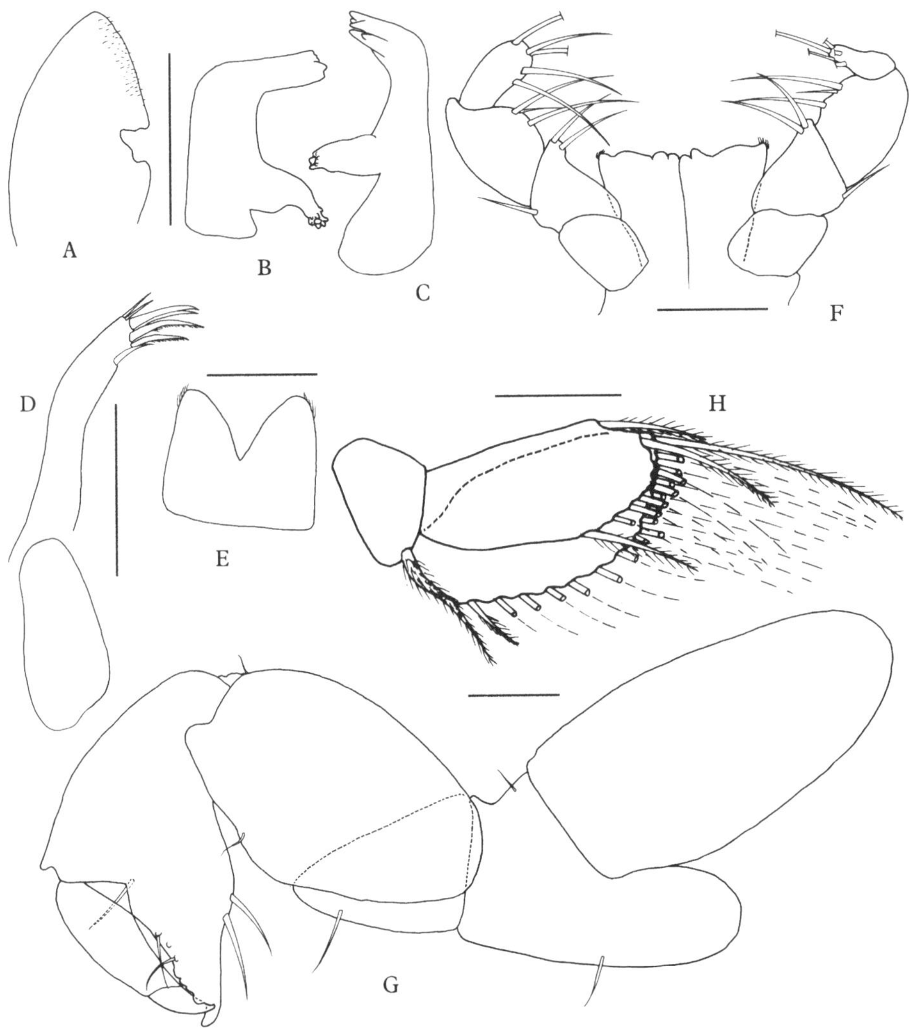

Fig. 67. *Araphura spinithenari* sp. nov., female; A, labrum; B, right mandible; C, left mandible; D, maxillule and maxilla; E, labium; F, maxilliped; G, cheliped; H, pleopod. Scale bars equal: A-F, 0.05 mm; G-H, 0.1 mm.

spiniform setae distally of which two are pinnate and two are small; palp lost during dissection. Maxilla (fig. 67D) elongate. Maxilliped (fig. 67F) endites with one outer setose process and two inner denticles; palp article 1 naked, article 2 with two inner and one outer setae, article 3 with three setae on inner margin, article 4 only 0.5 times as wide as article 3, with five setae. Epignath not recovered.

Cheliped (fig. 67G) basis with narrow proximal part and wide distal part, attached to long, elongate sclerite, with one simple seta on each margin, shorter than carpus. Merus with one seta ventrally. Carpus as long as propodus including fixed finger, with one simple seta on each margin. Propodus wide, with dorso-distal process, with one seta at dactylus insertion. Fixed finger with two stout setae ventrally and three on inner margin as well as four denticles distally, apex blunt. Dactylus as long as fixed finger, apex blunt.

Pereopod 1 (fig. 68A) coxa small and partly fused with the pereonite, with one seta. Basis longer than the three succeeding articles combined, with one proximal simple seta. Ischium with one simple seta. Merus shorter than carpus, widening distally and with one stiff seta disto-ventrally. Carpus shorter than 0.5 times basis, with one stiff seta distally on each margin. Propodus with two stiff setae distally and a dorsal spine. Dactylus and unguis together shorter than propodus and not fused. Dactylus shorter than unguis.

Pereopod 2 (fig. 68B) as pereopod 1 except: basis with two setulose setae dorso-medially. Merus with two stiff setae distally. Propodus with one stiff seta distally and a dorsal spine.

Pereopod 3 (fig. 68C) as pereopod 2 except: basis with one setulose seta dorso-medially. Propodus with a blunt dorsal spine.

Pereopod 4 (fig. 68D) longer than pereopods 1–3. No coxa present. Basis longer than the three succeeding articles combined, with two setulose setae ventro-medially. Ischium with one seta. Merus shorter than carpus, widening distally and with two stiff setae distally. Carpus shorter than 0.5 times basis, with four stiff setae distally. Propodus longer than carpus, with three stiff setae distally, of which one is setulose and a dorsal spine. Dactylus and unguis together longer than propodus, not fused. Dactylus longer than unguis.

Pereopod 5 (fig. 68E) as pereopod 4 except: basis and ischium naked. Carpus with five stiff setae distally.

Pereopod 6 (fig. 68F) as pereopod 4 except: basis naked. Merus with three stiff setae distally. Carpus with four stiff setae distally. Propodus with four setae distally and a dorsal spine.

Pleopods (fig. 67H) small but well developed. Basal article with one circum-plumose seta. Endo- and exopod of similar size. Endopod with one outer and seven inner plumose setae. Exopod with 16 plumose setae.

Uropod (fig. 68G) as long as pleotelson. Basal article naked and shorter than 0.5 times first endopod article. Endopod article 1 with one simple and two setose setae distally; article 2 shorter than article 1, with four long setae distally.

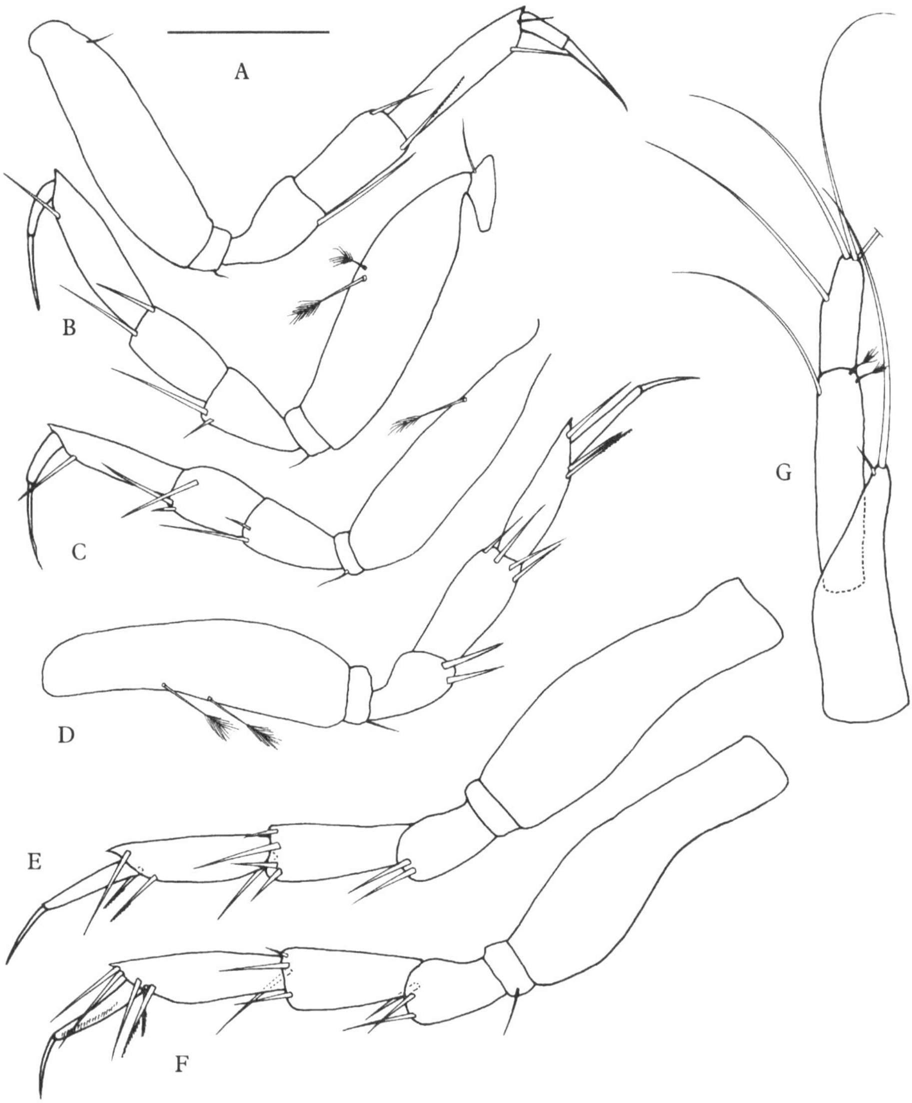

Fig. 68. *Araphura spinithenari* sp. nov., female; A, pereopod 1; B, pereopod 2; C, pereopod 3; D, pereopod 4; E, pereopod 5; F, pereopod 6; G, uropod. Scale bar equals 0.1 mm.

Exopod process about 0.5 times as long as first endopod article, with two simple setae distally.

Distribution. — Only found in the northern part of the Gulf. Bathymetric range wide: 482–2922 m. See map 49.

Remarks. — The dorso-distal process on the cheliped propodus separates this species from all other species of *Araphura*.

KEY TO THE SPECIES OF *ARAPHURA*, FEMALES (GLOBAL)

1. Cheliped propodus with dorsal pointed process at dactylus insertion .... *A. spinithenari* sp. nov.
– Cheliped propodus without dorsal pointed process at dactylus insertion 2
2. Uropodal exopodal process shorter than 0.33 times first endopod article 3
– Uropodal exopodal process longer than 0.33 times first endopod article 6
3. Pereonite 2 elongate (1.2–1.3 times as long as wide) 4
– Pereonite 2 as wide as long, or wider .... *A. joubinensis* Sieg & Dojiri, 1989
4. Carapace more than twice as long as wide 5
– Carapace less than 1.5 times as long as wide .... *A. elongata* (Shiino, 1970)
5. Pereonite 6 as wide as long, or wider .... *A. filiformis* (Lilljeborg, 1864)
– Pereonite 6 twice as long as wide .... *A. extensa* Larsen, 2003
6. Pleotelson apex with conical process .... *A. cuspirostris* Dojiri & Sieg, 1997
– Pleotelson apex without conical process 7
7. Pleotelson as wide as long, or wider 8
– Pleotelson longer than wide 10
8. Pereonites 2–4 clearly longer than pereonite 5 .... *A. higginsi* Sieg & Dojiri, 1989
– Pereonites 2–5 more or less subequal 9
9. Pleotelson nearly twice as wide as long .... *A. breviaria* Dojiri & Sieg, 1997
– Pleotelson nearly as long as wide .... *A. brevimanus* (Lilljeborg, 1864)
10. Uropodal exopod process more than 0.5 times as long as first endopod article .... *A. parabrevimanus* (Lang, 1968)
– Uropodal exopod process 0.5 times as long as first endopod article or less 11
11. Pleotelson as long as lengths of four free pleonites combined. Antenna article 4 with clear fusion line .... *A. rectifrons* (Kudinova-Pasternak, 1973)
– Pleotelson as long as lengths of only two free pleonites combined. Antenna article 4 without trace of fusion line .... *A. curticauda* sp. nov.

Genus **Araphuroides** Sieg, 1986

Araphura Bird & Holdich, 1984, [pro parte], Zoologica Scripta, **13**: 295.
Araphuroides Sieg, 1986, Biology of the Antarctic Seas, **45**: 127.

Diagnosis. — (Modified after Sieg, 1986b; Sieg & Dojiri, 1989.) Female. Body less than 9 times as long as wide. Pleon short (less than 0.25 times as long as body). Eyes and eye-lobes absent. Antennule with four articles. Antenna with six articles and article 3 with or without fusion line. Mandibular molars relatively well developed, always with ring of apical spines or denticles. Maxillipedal bases fused; endites not fused, with process on distal margin. Cheliped attached via sclerite. Cheliped basis without narrow proximal extension. Pereopods with coxae. Pereopods 1–3 with propodus longer than dactylus/unguis combined and dactylus shorter than unguis. Pereopods 4–6 with propodus shorter than dactylus/unguis combined and dactylus longer than unguis. Pleopods present or absent (*A. bombus* sp. nov.). Uropods stout and exopod fused to a pointed process on basal article, as in *Paranarthrura* and *Araphura*, endopod composed of two pseudoarticles and not curved inwards like in *Tanaella*.

Male. In general as the female. Lateral shield more pronounced. Antennule stouter than in female, with four articles. Fully functional mouthparts present, almost identical to those of female. Pleopods present in male and loosely coalesced with each other.

Type species. — *Araphuroides breviremis* Sieg, 1986.

Gender of generic name. — Masculine.

Remarks. — The species of this genus do possess a coxa on pereopods 4–6, whereas the other genera in the family do not.

Species currently assigned to this genus. — *A. bombus* sp. nov.; *A. brevispina* (Bird & Holdich, 1984); *A. parabreviremis* Sieg, 1986; *A. septentrionalis* Sieg & Dojiri, 1989.

Araphuroides bombus sp. nov.
(figs. 69–73)

Material examined. — Holotype, non-ovigerous ♀ (USNM 1068909), body length 2.7 mm, DGoMB Sta. S35-1. Paratypes: 1 ♂ (USNM 1068910), body length 2.3 mm, same locality. 1 non-ovigerous ♀ (USNM 1068913), body length 2.7 mm, DGoMB Sta. MT2-2. 1 ♂ (USNM 1068911), DGoMB Sta. S35-1 (dissected). 1 non-ovigerous ♀ (USNM 1068912), same locality (dissected).

Additional material. — 2 non-ovigerous ♀♀, 4 mancas, DGoMB Sta. MT2-2. 1 ♂, 1 ovigerous ♀, 2 non-ovigerous ♀♀, 1 manca, DGoMB Sta. MT2-1. 2 non-ovigerous ♀♀, 1 manca, DGoMB Sta. S35-2.

Diagnosis. — Female. Body stout, about 5 times as long as wide, with heavy calcification. Pereonites strongly curved in dorsal view. Pereopods 1–3 with long (as long as propodus) carpal bipinnate spiniform seta.

Male. Body shorter than in female, lateral shield more pronounced. Antennule stouter and shorter than in female. Functional mouthparts present. Pleopods present.

Etymology. — Named after the association of the stout body shape with a bumblebee of the genus *Bombus*, as a noun in apposition.

Type locality. — DGoMB Sta. S35-1, 29°20'0500"N 87°03'3758"W, depth 658 m.

Description. — Adult female. Body (fig. 69A, B) 5 times as long as wide.

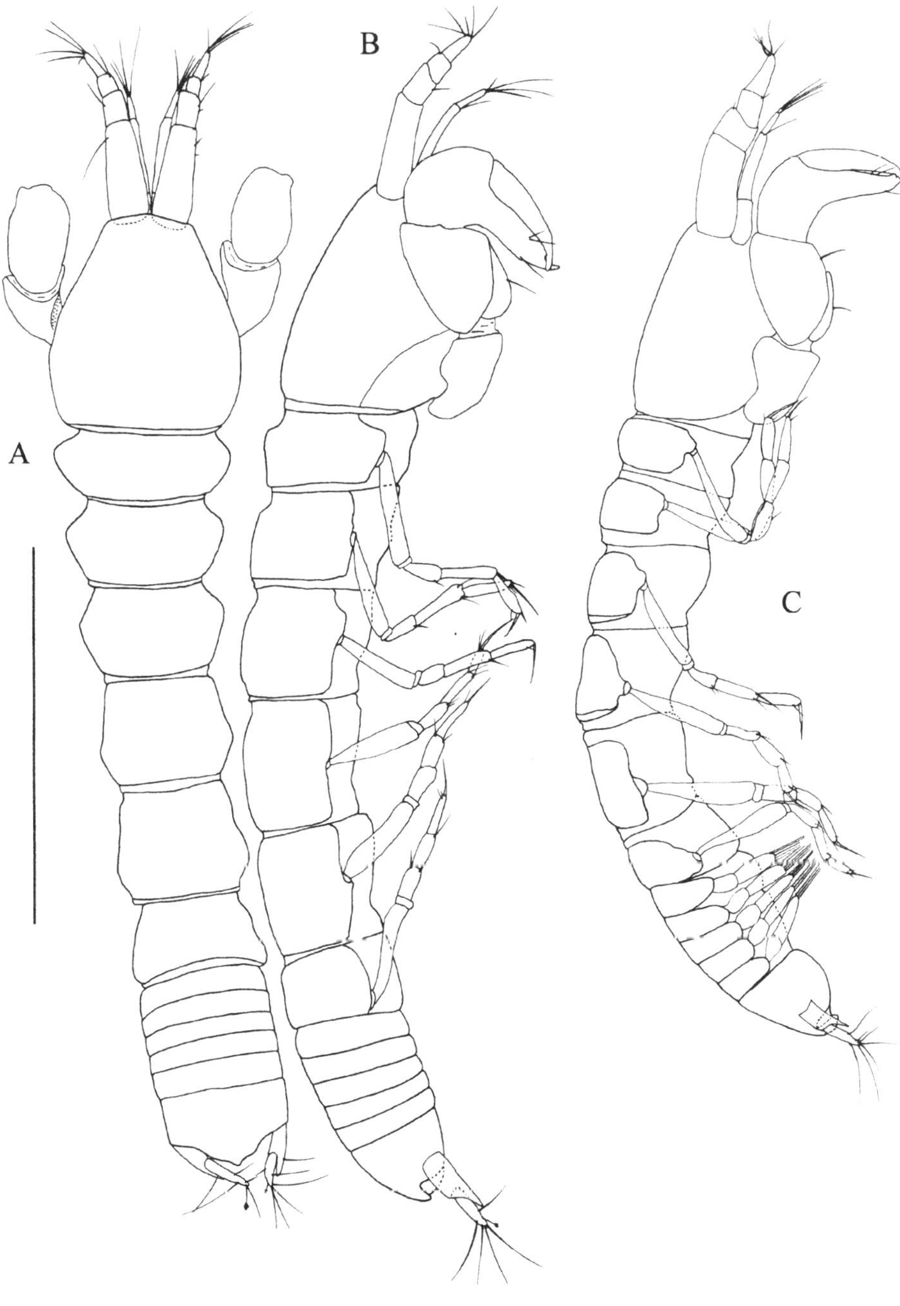

Fig. 69. *Araphuroides bombus* sp. nov.; A, female, holotype, dorsal view; B, female holotype, lateral view; C, male paratype, lateral view. Scale bar equals 1.0 mm.

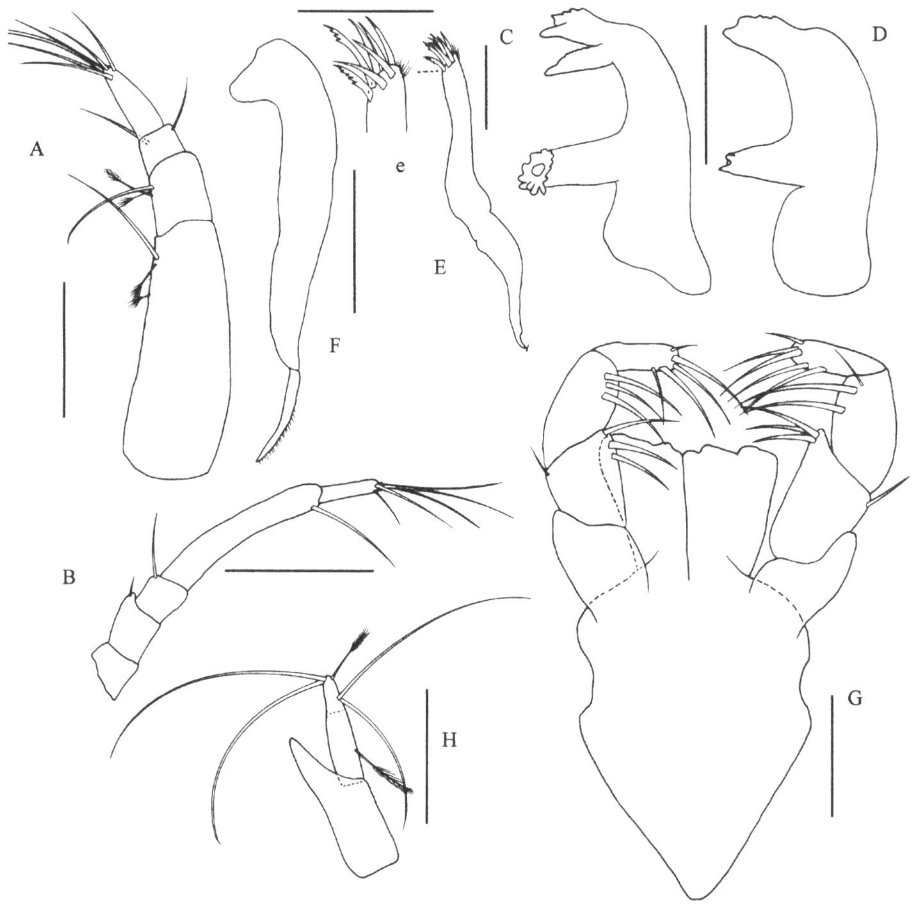

Fig. 70. *Araphuroides bombus* sp. nov., female; A, antennule; B, antenna; C, left mandible; D, right mandible; E, maxillule; e, maxillule terminal seta; F, epignath; G, maxilliped; H, uropod. Scale bars equal: A-B, H, 0.1 mm; C-G, 0.05 mm.

Carapace longer than the lengths of pereonites 1 and 2 combined. Eyes and eye-lobes absent.

Pereon with pereonites wider than long.

Pleon short (including pleotelson only 0.2 times total body length). All pleonites subequal. Pleotelson as long as lengths of three free pleonites combined, lateral edges straight, apex pointed and covered by a dorsal plate.

Antennule (fig. 70A) shorter than carapace, with four articles. Article 1 longer than rest of antennule, with two setulose and one simple setae distally. Article 2 longer than article 3, with one simple and two setulose setae distally. Article 3 shorter than other articles, with two simple setae distally. Article 4 longer than article 2, with five long setae and one aesthetasc distally.

Antenna (fig. 70B) with six articles; 0.75 times as long as antennule. Article 1 naked. Article 2 longer than article 3, with one simple seta dorso-distally. Article 3 longer than article 1, with one simple seta dorso-distally. Article 4 longer than other articles, with one simple seta distally. Article 5 longer than article 2, with two simple setae distally. Article 6 minute, with four long setae distally.

Mouthparts. Labrum lost during dissection. Mandibular molar relatively broad and longer than incisor, with small distal spines. Left mandible (fig. 70C) lacinia mobilis shorter than incisor; incisor broad, with five denticles. Right mandible (fig. 70D) incisor broad, with five denticles. Labium lost during dissection. Maxillule (fig. 70E) endite with seven spiniform setae distally of which two are multifurcate (fig. 70e); palp as long as endite. Maxilla not recovered. Maxilliped (fig. 70G) endites with one medial and one outer process on distal margin, almost as wide as basis; palp article 1 naked, article 2 with three setae on inner margin and one on outer margin, article 3 with three setae on inner margin, article 4 only 0.5 times as wide as article 3, with five setae. Epignath (fig. 70F) broad and longer than maxillule, terminal seta setulated.

Cheliped (fig. 71A) basis divided by prominent sclerite, shorter than carpus and without trace of pseudocoxa. Merus prominent and naked. Carpus shorter than propodus including fixed finger, with two setae ventrally, ventral margin only 0.5 times as long as that of propodus. Propodus shorter than fixed finger. Fixed finger with two setae of dissimilar length ventrally and two on inner margin (possibly an artifact, see under male cheliped). Dactylus as long as fixed finger.

Pereopod 1 (fig. 71B) coxa rounded. Basis as long as the three succeeding articles combined, with one setulose seta. Ischium with one simple seta. Merus only 0.5 times as long as carpus, widening distally, and with one stiff pinnate seta distally. Carpus more than 0.5 times as long as basis, with three thick, stiff pinnate and one simple setae distally. Propodus as long as carpus, with two setae of dissimilar length distally and a dorsal spine. Dactylus and unguis together shorter than propodus and not fused, apex spatulate.

Pereopod 2 (fig. 71C) as pereopod 1 except: basis with two setulose setae. Propodus with one stiff pinnate and one simple seta distally.

Pereopod 3 (fig. 71D) as pereopod 1 except: basis with one simple and one setulose seta proximally. Propodus with two setae distally.

Pereopod 4 (fig. 71E) not wider than pereopods 1–3. Coxa naked. Basis as long as the three succeeding articles combined, with one long setulose seta.

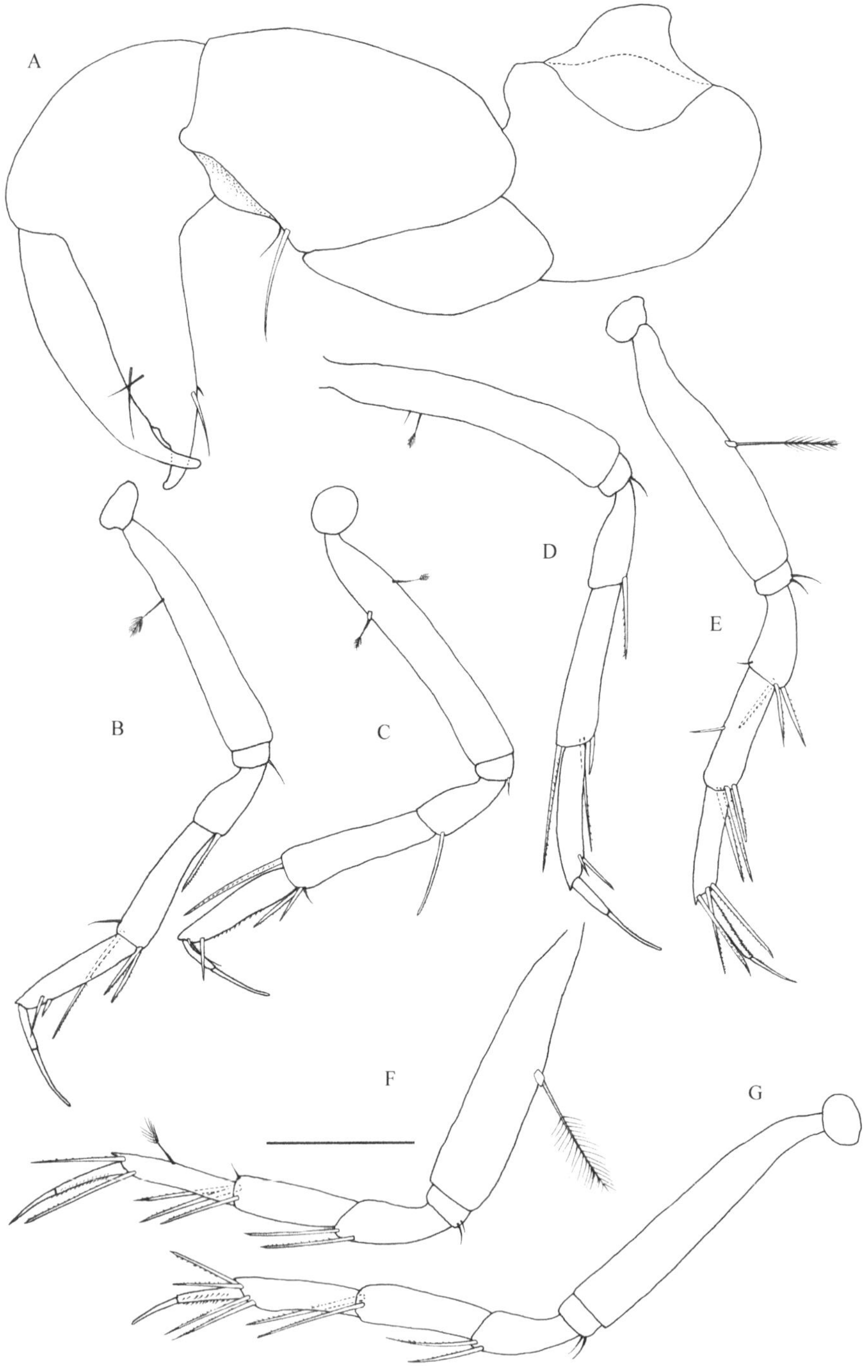

Fig. 71. *Araphuroides bombus* sp. nov., female; A, cheliped; B, pereopod 1; C, pereopod 2; D, pereopod 3; E, pereopod 4; F, pereopod 5; G, pereopod 6. Scale bar equals 0.1 mm.

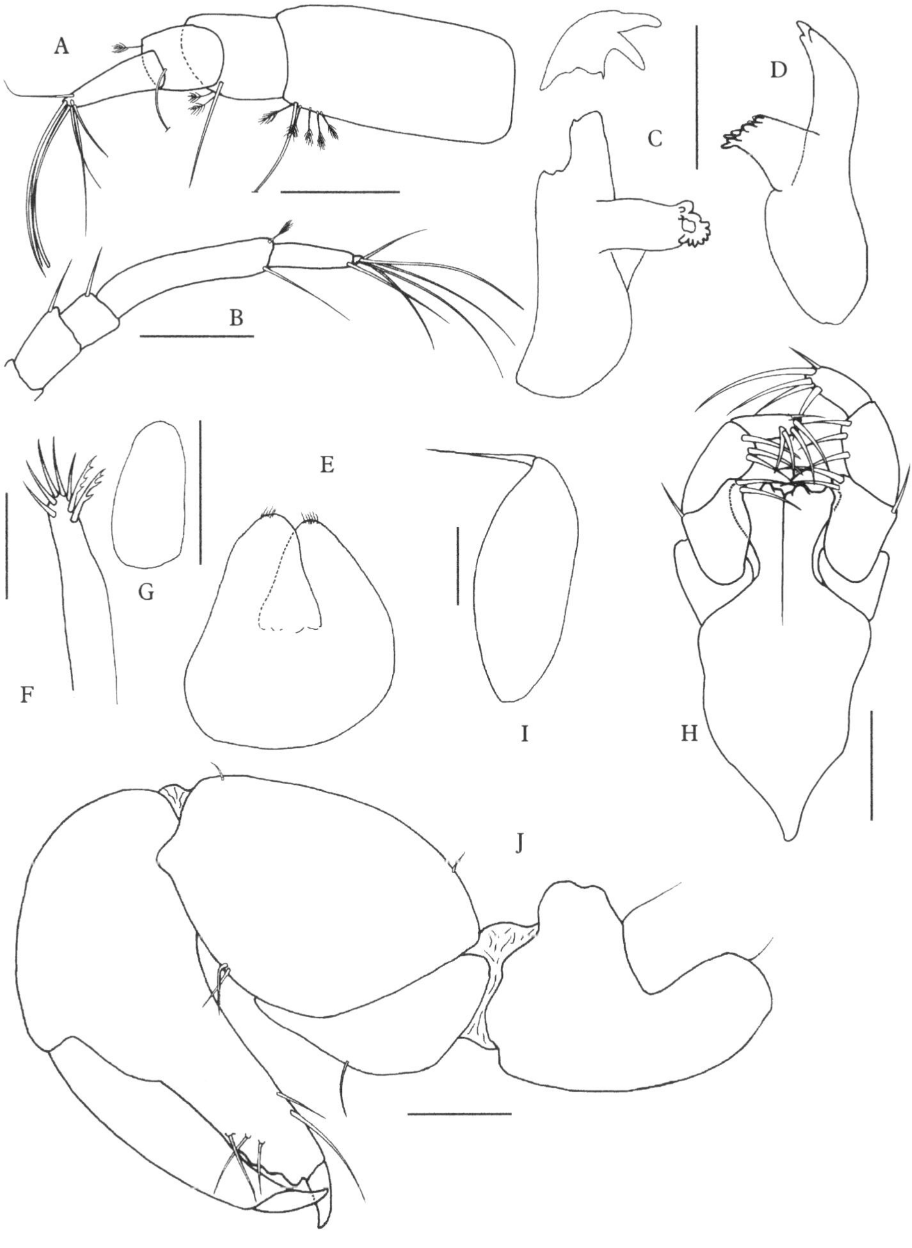

Fig. 72. *Araphuroides bombus* sp. nov., male; A, antennule; B, antenna; C, left mandible; D, right mandible; E, labium; F, maxillule; G, maxilla; H, maxilliped; I, epignath; J, cheliped. Scale bars equal: A-B, J, 0.1 mm; C-I, 0.1 mm.

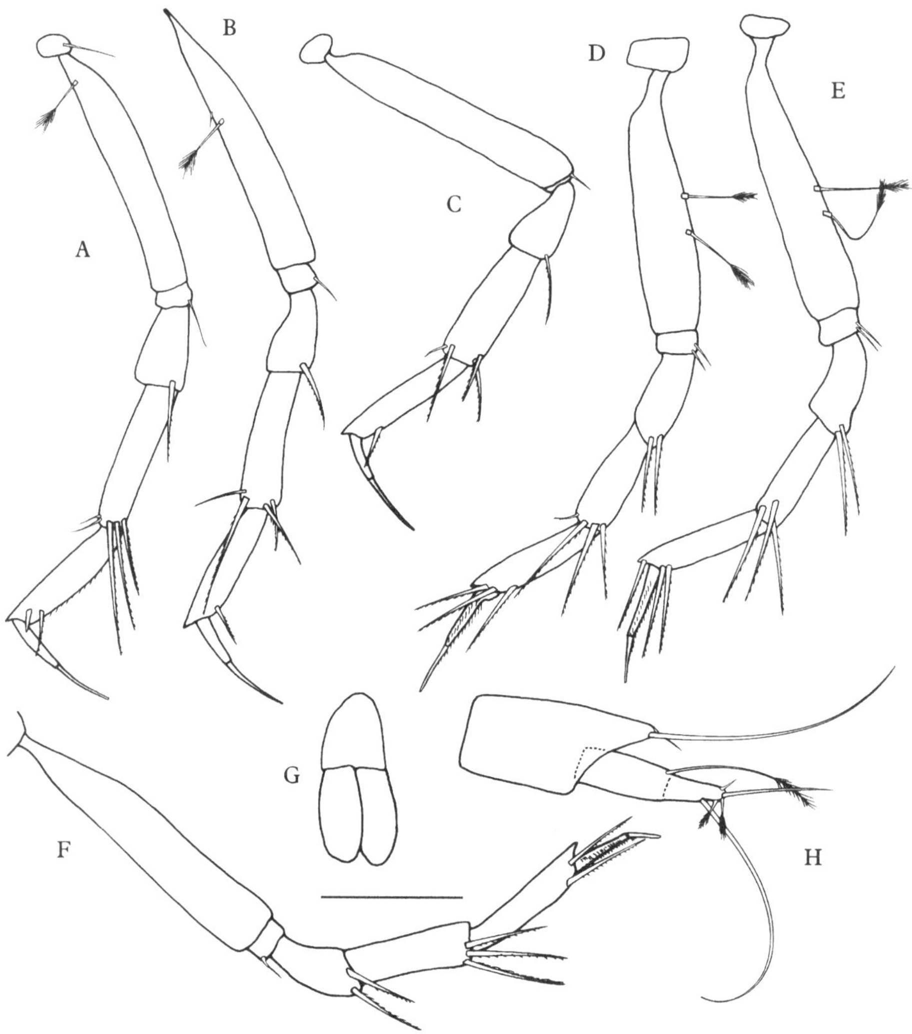

Fig. 73. *Araphuroides bombus* sp. nov., male; A, pereopod 1; B, pereopod 2; C, pereopod 3; D, pereopod 4; E, pereopod 5; F, pereopod 6; G, pleopods; H, uropod. Scale bar equals 0.1 mm.

Ischium with two setae. Merus more than 0.5 times as long as carpus, widening distally and with three stiff pinnate and one simple setae distally. Carpus shorter than 0.5 times basis, with three stiff pinnate setae distally and one simple seta medially. Propodus shorter than carpus, with three stiff pinnate setae distally and a dorsal spine. Dactylus and unguis together longer than propodus, not fused, and with a row of small ventral spines.

Pereopod 5 (fig. 71F) as pereopod 4 except: merus with two stiff pinnate setae distally. Propodus with two stiff pinnate setae distally and one setulose seta dorso-medially.

Pereopod 6 (fig. 71G) as pereopod 5 except: basis naked. Carpus with two stiff pinnate setae distally. Propodus with four stiff pinnate setae distally and a dorsal spine.

Pleopods not present in female.

Uropod (fig. 70H) longer than pleotelson. Basal article naked, as long as endopod. Endopod divided into two pseudoarticles; first pseudoarticle with one setulose seta medially; second pseudoarticle with four long simple and one setulose setae. Exopod process more than 0.5 times as long as endopod, spiniform.

Male, where differing from female. Body from adult male paratype; appendages from dissected juvenile male. Body (fig. 69C) shorter than in female.

Pereon with pereonites of same proportions as in female but lateral shield shorter and more pronounced.

All pleonites with pleopods and lateral shield.

Antennule (fig. 72A) stouter and shorter than in female.

Antenna (fig. 72B) as in female.

Mouthparts (fig. 72C-I). Functional, almost identical to those of female except: labium (fig. 72E) with blunt and sparsely setose apex. Epignath (fig. 72I) rather wide compared to that of female, terminal seta apparently naked.

Cheliped (fig. 72J) fixed finger with three setae and three small denticles on inner margin.

Pereopods (fig. 73A-F) identical to those of the female, given the odd seta.

Pleopods (fig. 73G) biramous, rami of equal size. Juvenile pleopods without setae; adult pleopods with plumose setae.

Uropod (fig. 73H) basal article somewhat stouter than in female.

Distribution. — Only found in the northern part of the Gulf. Bathymetric range narrow: 658–677 m. See map 50.

Remarks. — This species has a number of features in common with *Tanaella* but cannot be assigned to this genus owing to its uropodal exopod process.

KEY TO THE SPECIES OF *ARAPHUROIDES*, FEMALES (GLOBAL)

1. Pereopods 1–3 with one long (as long as propodus) carpal bipinnate spiniform seta. Antenna without trace of fusion line .... *A. bombus* sp. nov.

– Pereopods 1–3 without long (as long as propodus) carpal bipinnate spiniform seta. Antenna with fusion line 2
2. Pleotelson longer than wide .... *A. brevispina* (Bird & Holdich, 1984)
– Pleotelson wider than long 3
3. Antenna article 4 (both pseudoarticles) four times as long as article 5. Pereopods 4 and 5 have carpus with four spiniform and one simple setae .... *A. parabreviremis* Sieg, 1986
– Antenna article 4 (both pseudoarticles) twice as long as article 5. Pereopods 4 and 5 have carpus with three spiniform and one simple setae .... *A. septentrionalis* Sieg & Dojiri, 1989

Genus **Tanaella** Norman & Stebbing, 1886.

Tanaella Norman & Stebbing, 1886, Transactions of the Zoological Society of London, **12**: 117–118.
Leptognathia – Lang, 1968, Galathea Report, **9**: 65.

Diagnosis. — (Modified from Larsen & Heard, 2004a.) Female. Body less than 8 times as long as wide. Pleon short (about 0.25 times body length). Eyes and eye-lobes absent. Antennule with four articles. Antenna with five articles, and article 3 with or without fusion line. Mandibular molars well developed, of width intermediate between those of *Typhlotanais* and *Leptognathia*, and always with ring of small apical spines. Maxillipedal bases fused; endites not fused, with process or globe-shaped seta on distal margin. Cheliped attached via sclerite. Cheliped basis with narrow proximal extension. Pereopods 1–3 with propodus longer than dactylus and unguis combined. Pereopods 4–6 propodus shorter than dactylus and unguis combined. Pleopods present or absent. Uropods stout and uniramous (although minute exopod process can be observed in some species), shaped as a pincer and consisting of one or two articles.

Male. In general as female. Antennule stouter than in female, usually with four articles but sometimes with five (*T. tuberculata* Kudinova-Pasternak, 1989). Fully functional mouthparts present, almost identical to those of the female. Pleopods always present in male.

Type species. — *Tanaella unguicillata* Norman & Stebbing, 1886.

Gender of generic name. — Feminine.

Remarks. — The width of the molar, while stable within the genus, is not a strong character, as it depends on the researcher's assessment of what is 'intermediate' width (i.e., between those of *Typhlotanais* and *Leptognathia*). This is also subject to variation within a taxon. Other generic characters are stable. For species recognition, the following characters are valid: (1) uropod articulation and length, (2) carapace length/width ratio, (3) pleotelson shape, (4) denticulations of the fixed finger of the cheliped, and (5) process of the dactylus of the cheliped. A key to this genus is given in Larsen & Heard (2004a).

Species currently assigned to this genus. — *T. eltaninae* Guerrero-Kommritz & Błażewicz-Paszkowycz, 2004; *T. forcifera* (Lang, 1968); *T. kimi* Guerrero-Kommritz & Błażewicz-Paszkowycz, 2004; *T. mclellandi* Larsen & Heard, 2004; *T. paraforcifera* (Lang, 1968); *T. profunda* Guerrero-Kommritz & Błażewicz-Paszkowycz, 2004; *T. prolixcauda* Larsen & Heard, 2004; *T. propinqua* Dojiri & Sieg, 1997 [originally published as *T. propinquus* [sic]]; *T. rotundicephala* Sieg, 1986; *T. tuberculata* Kudinova-Pasternak, 1989; *T. unguicillata* Norman & Stebbing, 1886; *T. unisetosa* Sieg, 1986.

Tanaella mclellandi Larsen & Heard, 2004

Tanaella mclellandi Larsen & Heard, 2004, Journal of Natural History, London, **38**: 552–559.

Diagnosis. — (After Larsen & Heard, 2004a.) Female. Antennular article 2 about 0.5 times as long as article 1. Cheliped fixed finger without pronounced serration. Cheliped dactylus without large depression and processes. Maxillipedal endites with process. Pleopods absent. Uropods shorter than pleotelson.

Male. Antennular article 2 more than 0.5 times as long as article 1. Cheliped fixed finger with pronounced proximal serration. Cheliped dactylus with proximal depression and two processes. Pleopods present on all pleonites, armed with simple setae.

Distribution. — This species has only been recorded in the Gulf of Mexico but seems to be quite common at depths below 400 m. Bathymetric range: 213–2920 m. See map 51.

Tanaella prolixcauda Larsen & Heard, 2004

Tanaella prolixcauda Larsen & Heard, 2004, Journal of Natural History, London, **38**: 566–570.

Diagnosis. — (Modified after Larsen & Heard, 2004a.) Male. Antennular article 2 shorter than 0.5 times article 1. Cheliped fixed finger with pronounced proximal serration. Cheliped dactylus without proximal depression or processes. Pleopods present on all pleonites, all with plumose setae. Uropods longer than pleotelson and last two pleonites combined.

Female. Unknown.

Distribution. — This species is rare and has only been recorded in the Gulf of Mexico. Bathymetric range: 550–2030 m. See map 52.

Remarks. — The sexual assignment of this species is considered uncertain, owing to the lack of specimens with oostegites (Larsen & Heard, 2004a).

FAMILAE INDETERMINATAE

A large number of paratanaoidean genera could not be assigned to family by Larsen & Wilson (2002) in their phylogenetic analysis, owing to a combination of poor descriptions, lost type material, a lack of female specimens, and, most importantly, a high degree of homoplasy. The genera listed below should be perceived as a repository, and **not** as belonging to a monophyletic group.

Genera included. — *Andrognathia* Sieg, 1983; *Allodaposia* Sieg, 1986; *Anarthrura* G. O. Sars, 1882; *Androtanais* Sieg, 1976; *Armatognathia* Kudinova-Pasternak, 1987; *Armaturatanais* n. gen.; *Bifida* Sieg & Zibrowius, 1988; *Chauliopleona* Dojiri & Sieg, 1997; *Coalecerotanais* Larsen, 2003; *Dimorphognathia* Sieg, 1986; *Expina* Lang, 1968; *Insociabilitanais* n. gen.; *Leptognathioides* Bird & Holdich, 1984; *Mimicarhaphura* Sieg, 1986 [sic]; *Monstrotanais* Kudinova-Pasternak, 1981; *Nototanoides* Sieg & Heard, 1985; *Paraleptognathia* Kudinova-Pasternak, 1981; *Parafilitanais* Kudinova-Pasternak, 1989; *Pseudoarthrura* n. gen.; *Pseudomacrinella* Kudinova-Pasternak, 1990; *Pseudoparatanais* Sieg, 1973; *Robustognathia* Kudinova-Pasternak, 1989; *Robustochelia* Kudinova-Pasternak, 1983; *Safaritanais* Kudinova-Pasternak, 1987; *Salemia* Lang, 1971; *Scoloura* Sieg & Dojiri, 1991; *Siphonolabrum* Lang, 1972; *Spinitanaopsis* nomen novum; *Stenotanais* Bird & Holdich, 1984; *Tanabnormia* Guțu, 1986; *Tanaopsis* G. O. Sars, 1896.

KEY TO THE GENERA CURRENTLY PLACED AS FAMILAE INDETERMINATAE, THAT OCCUR IN THE GULF OF MEXICO, ADULT FEMALES ONLY

1. Pleon completely fused with 6th pereonite. Only five pairs of pereopods present .... *Coalecerotanais* Larsen, 2003
– Pleon not fused with 6th pereonite. Six pairs of pereopods present 2
2. Last pleonite with posteriorly directed spine .... *Chauliopleona* Dojiri & Sieg, 1997
– Last pleonite without posteriorly directed spine 3
3. Maxillule without spiniform terminal setae .... *Expina* Lang, 1968
– Maxillule with spiniform terminal setae 4
4. Maxillipedal endites partly fused .... *Nototanoides* Sieg & Heard, 1985
– Maxillipedal endites separate 5

5. Pereopods 1–3 with basis wider than basis of pereopods 4–6 .... *Stenotanais* Bird & Holdich, 1984
– Pereopods 1–3 with basis narrower than basis of pereopods 4–6 6
6. Uropods uniramous .... *Parafilitanais* Kudinova-Pasternak, 1989
– Uropods biramous 7
7. Body dorso-ventrally flattened. Pereopods and pleopods attached ventrally inside the lateral shield. Uropods not longer than 25% of pleotelson length .... *Armaturatanais* n. gen.
– Body cylindrical. Pereopods and pleopods attached laterally. Uropods longer than 50% of pleotelson length 8
8. Mandibles reduced, no molar present .... *Spinitanaopsis* nomen novum
– Mandibles not reduced, molar present 9
9. Chelipeds with propodus massive, larger than carpus .... *Robustochelia* Kudinova-Pasternak, 1983
– Chelipeds with propodus normal, as large as carpus or smaller 10
10. Pereonite 1 with strong, anteriorly directed ventral spine .... *Insociabilitanais* n. gen.
– Pereonite 1 without anteriorly directed ventral spine .... *Pseudoarthrura* n. gen.

Genus **Armaturatanais** n. gen.

Diagnosis. — Female. Body dorso-ventrally flattened. Cuticle heavily calcified. Pereopods and pleopods attached ventrally inside the lateral shield. Pleotelson acorn-shaped in lateral view. Antennule with four articles. Antenna with six articles and fusion line on article 4. Mandibles small; molar fairly broad, with flat distal crushing area surrounded by small denticles. Maxilliped endites without denticles or flat setae. Pereopods with coxa; dactylus and unguis not fused to a claw; pereopod 1–3 unguis longer than dactylus; pereopod 4–6 dactylus longer than unguis. Pleopods with simple setae only. Uropods biramous, endopod with two articles; exopod uni-articulate.

Male. Pleon hardly larger than in female. Antenna article 3 shorter than that of female and without fusion line. Functional mouthparts retained. All pleonites with pleopods bearing plumose setae; pleopods larger than those of females.

Etymology. — Named after the 'tank' like appearance of the body (Latin, "armatura" = armour).

Gender of generic name. — Masculine.

Type species. — *Armaturatanais minipodus* sp. nov.

Remarks. — The ventral attachment of the pereopods and uropods is similar to that of *Isopodidus*, but the body shape is radically different (much shorter and with fused pleon in *Isopodidus*). The cuticle is heavily calcified like in the Agathotanaidae, but here the body shape is even more radically deviant. The shape of the body and particularly the acorn-shaped pleotelson are similar to the situation in the Tanaellidae, but the ventral attachment of the appendages and the biramous uropods conflict with that family diagnosis as well. This character combination makes the genus unique and currently unassignable to any known family. Unfortunately, most specimens of the type species are in poor condition.

Species currently assigned to this genus. — *A. minipodus* sp. nov.

Armaturatanais minipodus sp. nov.
(figs. 74–78)

Material examined. — Holotype, non-ovigerous ♀ (USNM 1068914), body length 1.6 mm, DGoMB Sta. MT3-2. Paratypes: 1 non-ovigerous ♀ (USNM 1068915), body length 0.9 mm, same locality. 1 ♂ (USNM 1068916), body length 1.7 mm, DGoMB Sta. MT2-2. 1 non-ovigerous ♀ (USNM 1068917), DGoMB Sta. MT3-2 (dissected). 1 ♂ (USNM 1068918), same locality (dissected).

Additional material. — 2 mancas, DGoMB Sta. RW5-1. 3 ♂♂, 1 manca, DGoMB Sta. MT2-2. 3 ♂♂, 1 manca, DGoMB Sta. RW5-2. 2 ♂♂, 1 manca, DGoMB Sta. WC12-2. 1 non-ovigerous ♀, DGoMB Sta. S43-1.

Diagnosis. — Monotypic genus, no species diagnosis given.

Etymology. — Named after the small pereopods, relative to the chelipeds (from latinized Greek, mini + podous = small + footed). The name is an adjective agreeing in gender with the generic name.

Type locality. — DGoMB Sta. MT3-2, 28°13'4659"N 89°30'3941"W, depth 982 m.

Description. — Adult female. Body (fig. 74A, B) dorso-ventrally flattened, 5.5 times as long as wide.

Carapace as long as lengths of pereonites 1 and 2 combined. Eyes and eye-lobes absent.

Pereonites all wider than long.

Pleon short (including pleotelson only 0.25 times total body length). All pleonites subequal; lateral edge of dorsal shield extends down to ventral surface. Pleotelson as long as all free pleonites combined, apex acorn-shaped, and with uropods extending from a depression.

Antennule (fig. 74C) about as long as carapace. Article 1 shorter than rest of antennule combined, with one simple seta distally. Article 2 more than 0.5 times as long as article 1, partly covering article 3 on inner margin, with one

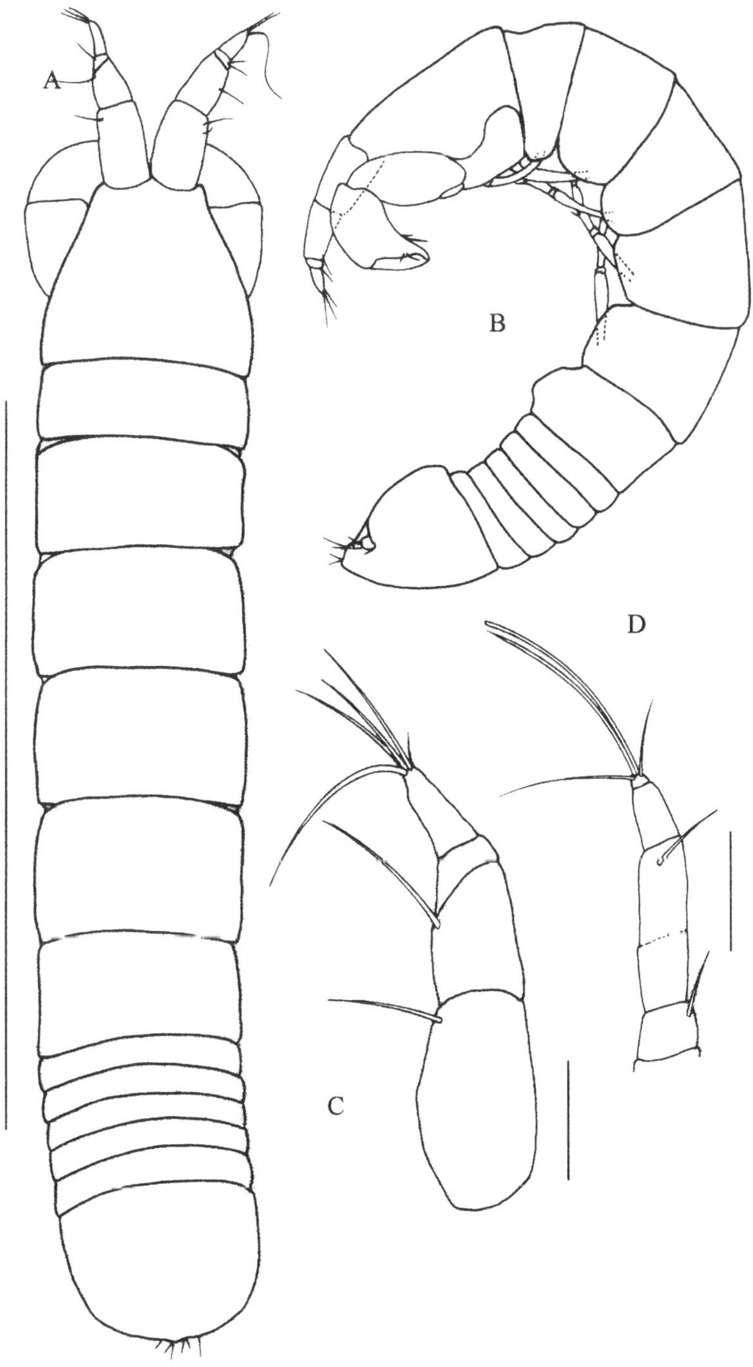

Fig. 74. *Armaturatanais minipodus* gen. et sp. nov., female; A, holotype, dorsal view; B, paratype, lateral view; C, antennule; D, antenna. Scale bars equal: A, B, 0.5 mm; C-D, 0.1 mm.

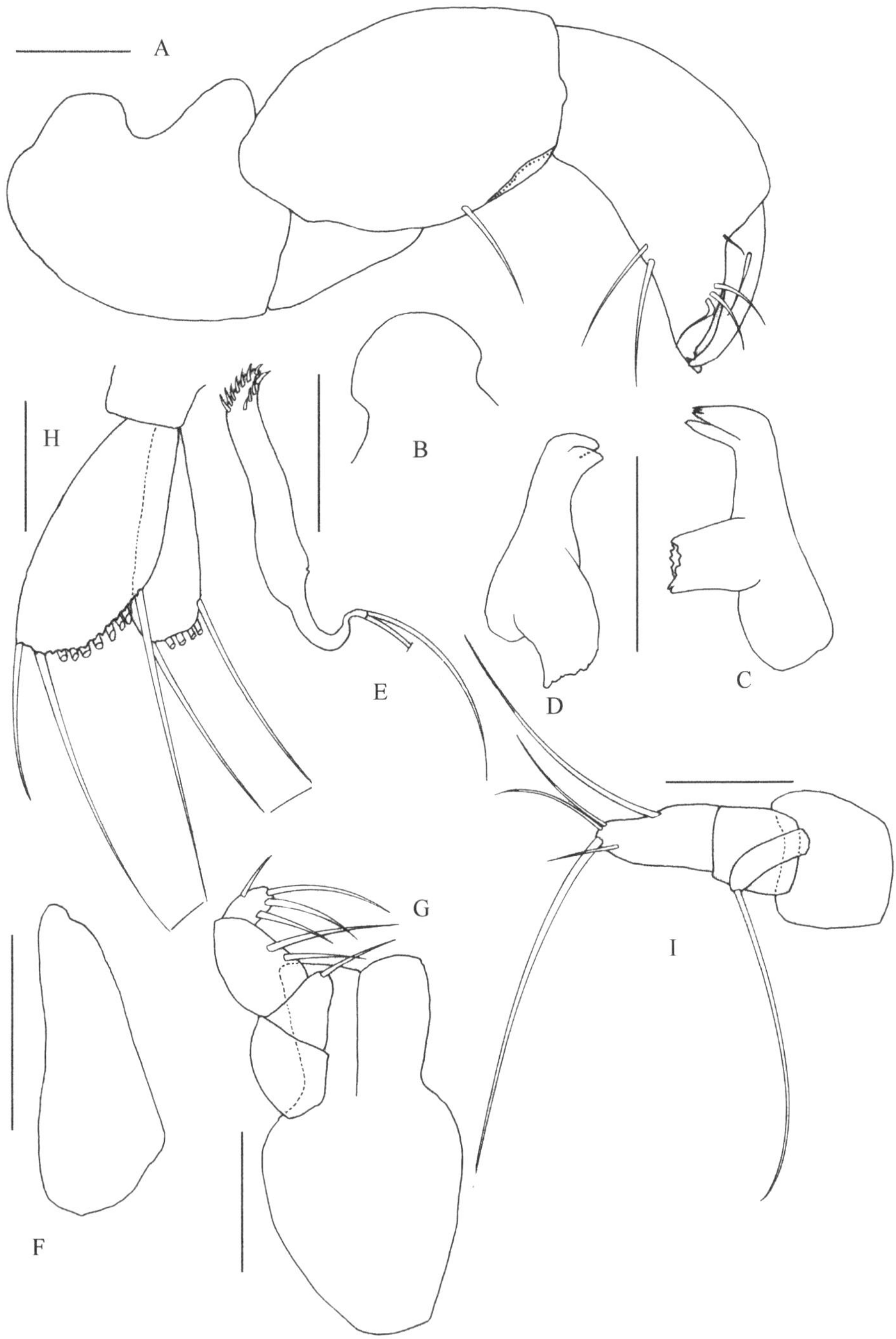

Fig. 75. *Armaturatanais minipodus* gen. et sp. nov., female; A, cheliped; B, labrum; C, left mandible; D, right mandible; E, maxillule; F, maxilla; G, maxilliped; H, pleopod; I, uropod. Scale bars equal: A, H-I, 0.1 mm; B-G, 0.05 mm.

simple seta distally. Article 3 shorter than article 4, naked. Article 4 shorter than article 2, with five simple setae distally.

Antenna (fig. 74D) almost as long as antennule. Article 1 (not illustrated) hardly wider than following articles, with one simple seta dorso-distally. Article 2 shorter than article 4, with one simple seta dorsally. Article 3 longer than other articles and with medial fusion line, with one simple seta distally. Article 4 as long as proximal part of article 3, naked. Article 5 minute, with three setae and one aesthetasc distally.

Mouthparts. Labrum (fig. 75B) naked and hood-shaped. Mandibular molar process broad and as long as incisor, with terminal ring of several small denticles. Left mandible (fig. 75C) lacinia mobilis as long as incisor; incisor with three pointed denticles. Right mandible (fig. 75D) incisor broad, with two large processes. Labium not recovered. Maxillule (fig. 75E) endite with 11 short, spiniform setae distally; palp shorter than 0.33 times length of endite, with two terminal setae. Maxilla (fig. 75F) elongate. Maxilliped (fig. 75G) endites naked and narrower than basis; palp article 1 naked, article 2 with one seta on inner margin, article 3 with two setae on inner margin, article 4 with three setae on inner margin and one on outer margin. Epignath not recovered.

Cheliped (fig. 75A) basis unequally divided by sclerite, shorter than carpus. Merus naked. Carpus longer than propodus including fixed finger, with one seta ventrally. Propodus with one seta between dactylus and fixed finger. Fixed finger with two setae ventrally and three on inner margin, with one small denticle on inner margin. Dactylus as long as fixed finger, with one prominent proximal seta.

Pereopods 1 and 2 (fig. 76A, B) with coxa naked. Basis as long as the three succeeding articles combined, naked. Ischium with one seta ventrally. Merus longer than carpus, widening distally and with one seta ventrally. Carpus 0.5 times as long as propodus, with one spiniform seta dorso-distally. Propodus more than 0.5 times as long as basis, with one spiniform seta distally and a blunt dorsal spine. Dactylus and unguis combined shorter than propodus and not fused to a claw.

Pereopod 3 (fig. 76C) as pereopod 2 except: no visible coxa.

Pereopod 4 (fig. 76D) as pereopod 3 except: ischium with two prominent setae. Merus with two spiniform setae. Carpus with one simple and two spiniform setae distally. Propodus with two simple and one spiniform setae distally and a dorsal spine. Dactylus and unguis combined longer than propodus.

Pereopod 5 (fig. 76E) as pereopod 4 except: ischium with one prominent seta. Carpus with four setae distally. Propodus with three simple setae distally.

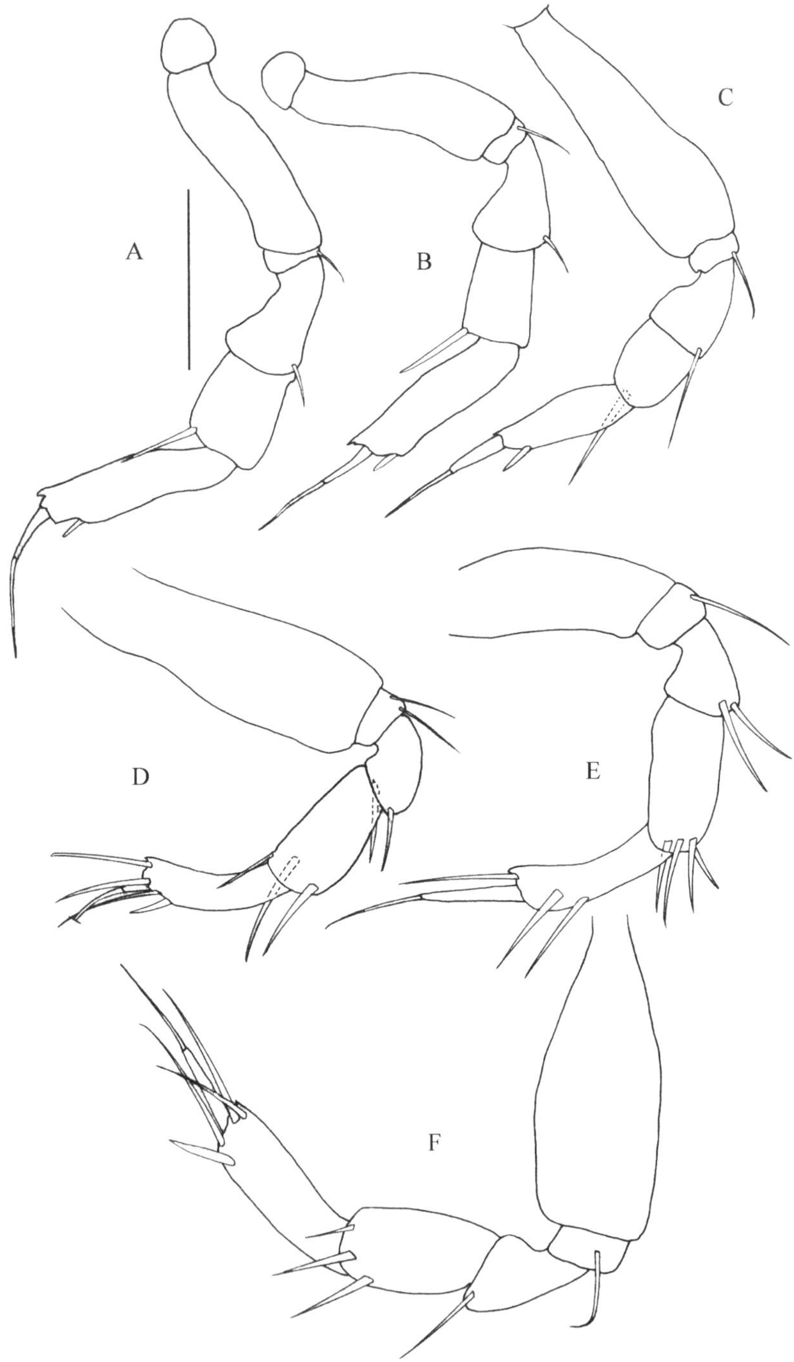

Fig. 76. *Armaturatanais minipodus* gen. et sp. nov., female; A, pereopod 1; B, pereopod 2; C, pereopod 3; D, pereopod 4; E, pereopod 5; F, pereopod 6. Scale bar equals 0.1 mm.

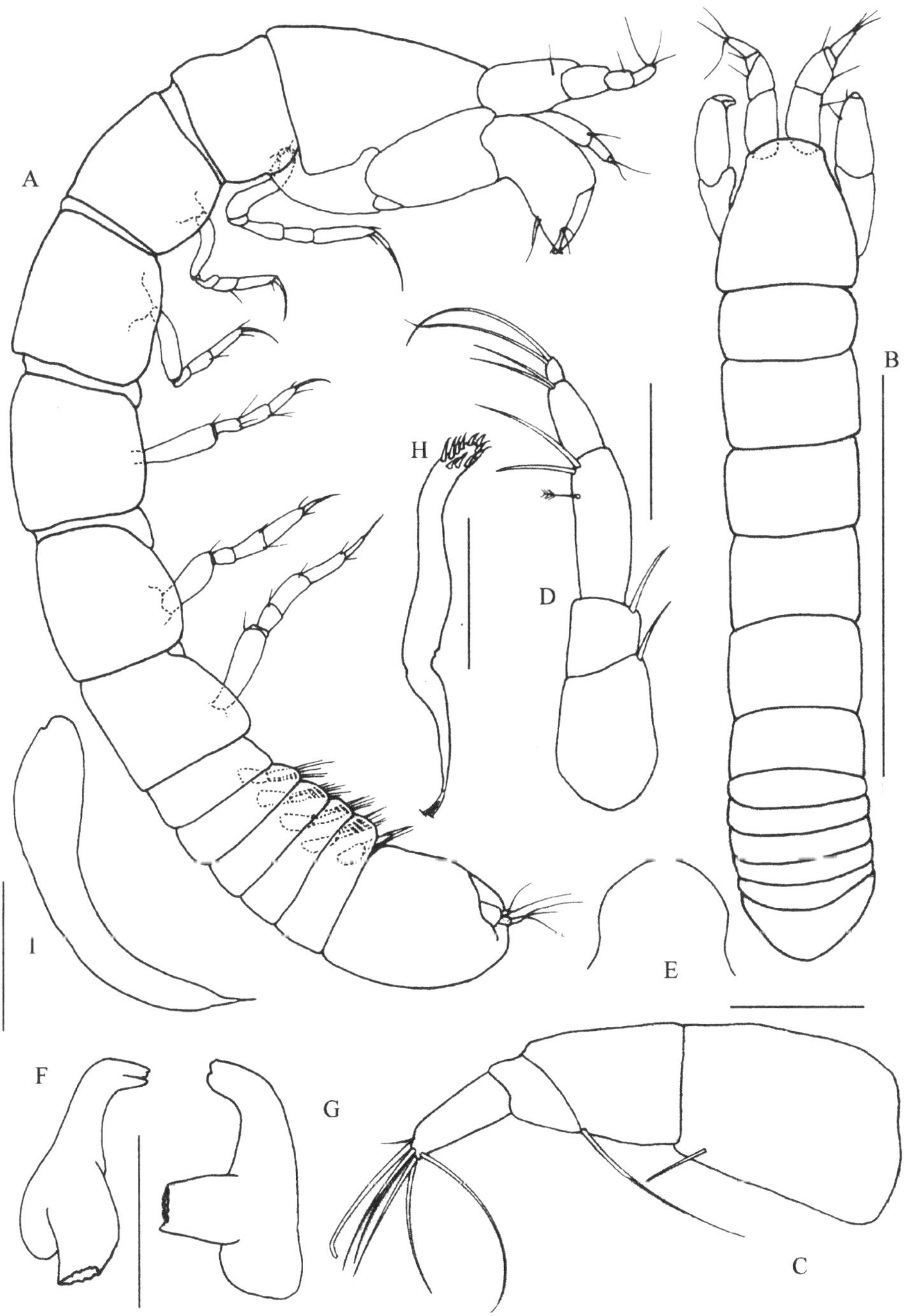

Fig. 77. *Armaturatanais minipodus* gen. et sp. nov., male; A, paratype, lateral view; B, paratype, dorsal view; C, antennule; D, antenna; E, labrum; F, left mandible; G, right mandible; H, maxillule; I, epignath. Scale bars equal: A, B, 0.5 mm; C-D, 0.1 mm; E-I, 0.05 mm.

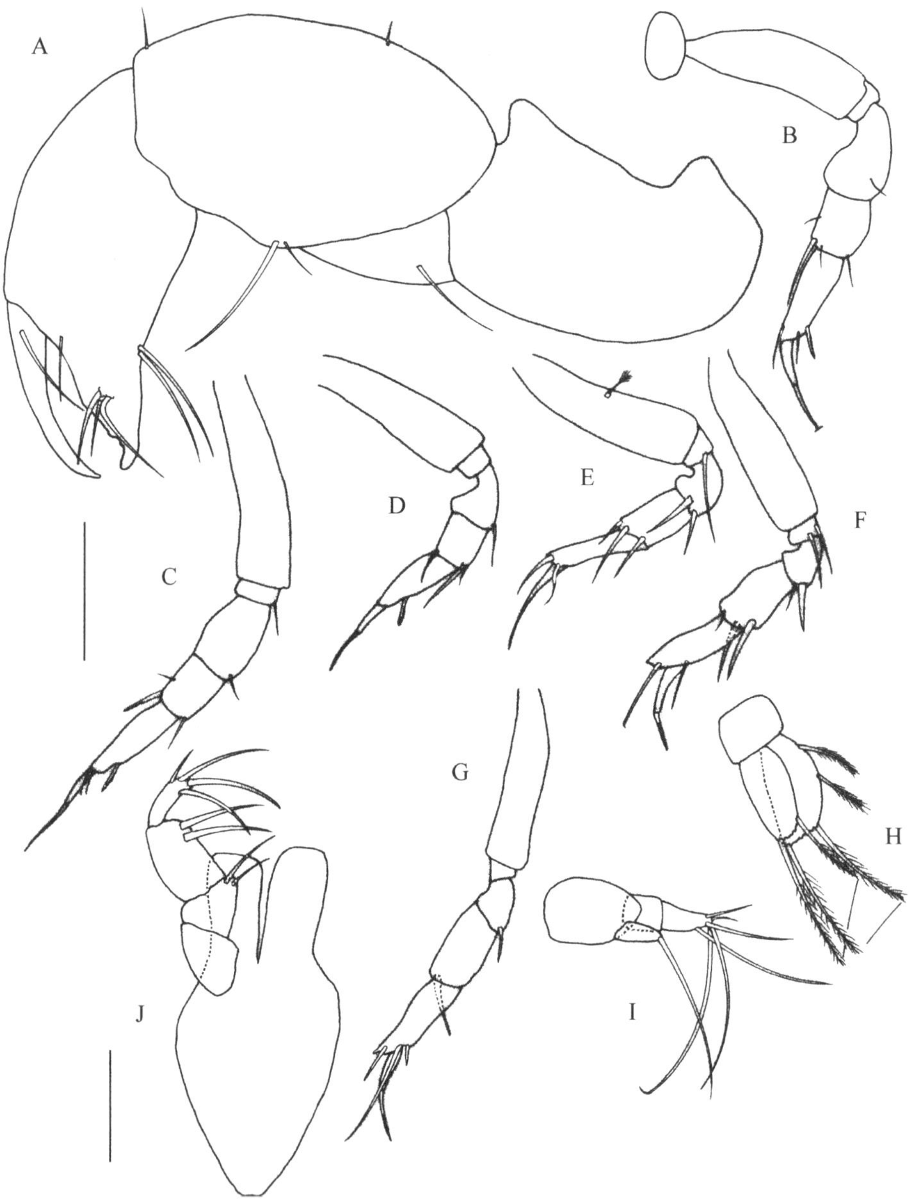

Fig. 78. *Armaturatanais minipodus* gen. et sp. nov., male; A, cheliped; B, pereopod 1; C, pereopod 2; D, pereopod 3; E, pereopod 4; F, pereopod 5; G, pereopod 6; H, pleopod; I, uropod; J, maxilliped. Scale bars equal: A-I, 0.1 mm; J, 0.05 mm.

Pereopod 6 (fig. 76F) as pereopod 5 except: merus with one simple seta. Carpus with three setae distally. Propodus with three simple setae distally and one spiniform seta ventrally.

Pleopods (fig. 75H) biramous, small. Endopod with ten simple setae. Exopod with six simple setae.

Uropod (fig. 75I) about 0.25 times as long as pleotelson. Basal article with dorsal process, naked. Endopod with two articles; article 1 naked; article 2 with five setae distally. Exopod as long as first endopod article, with one simple seta distally.

Male, where differing from female. Body (fig. 77A, B) pleon hardly longer and deeper than in female. Pleotelson shorter than the lengths of all free pleonites combined.

Antennule (fig. 77C) about 1.25 times as wide as that of female. Article 4 with five setae and a prominent aesthetasc distally.

Antenna (fig. 77D) shorter than that of female, article 3 without fusion line.

Mouthparts (figs. 77E-I, 78J) as in female except: epignath (fig. 77I) elongate, with naked apical seta.

Cheliped (fig. 78A) as in female.

Pereopods (fig. 78B-G) as in female except: pereopod 2 (fig. 78C) coxa not visible.

Pleopods (fig. 78H) with plumose setae present.

Uropod (fig. 78I) longer than in female.

Distribution. — Rare in the northern part of the Gulf. Bathymetric range: 366–1620 m. See map 53.

Remarks. — The female of this species has small pleopods (frequently not visible in lateral view). The ventral attachment of the pereopods and uropods are unusual and only found in a few apparently unrelated species. The length of the pereopod dactylus/unguis relative to that of the propodus is similar to that of *Tanaella*, but the structure of the uropods is different.

Genus **Chauliopleona** Dojiri & Sieg, 1997

Leptognathia group a, subdivision γ Hansen, 1913, Danish Ingolf Expedition, **3**: 65.
Chauliopleona Dojiri & Sieg, 1997, Taxonomic atlas of the benthic fauna of the Santa Maria Basin and western Santa Barbara Channel, **11**, The Crustacea (2, The Isopoda, Cumacea and Tanaidacea) (1–5): 231.

Diagnosis. — (Modified from Dojiri & Sieg, 1997; Guerrero-Kommritz, 2005.) Female. Eyes and eye-lobes absent. Pleon short (including pleotelson only 0.2 times total body length). Pleonite 5 with conspicuous, posteriorly directed spine, reaching under the pleotelson. Antennule with four articles. Antenna with 5–7 articles (occasionally with fusion line). Mandibular molar pointed. Maxillule with 7–9 spiniform terminal setae, some of which are often setulose. Maxilla elongate and often with distal setules. Carapace almost completely surrounding the cephalothorax and including the ventral part of the cheliped attachment point. Cheliped attached via lateral sclerite. Pereopods 1–3 with coxa. Pereopod 1 merus with one long, spiniform seta reaching carpus. Pereopods 4–6 without coxa; dactylus and unguis not fused to a claw, propodus often serrated, dactylus with two parallel rows of spinules. Pleopods present and biramous, albeit often reduced. Uropod biramous, both endo- and exopod bi-articulate; exopod shorter than first endopod article.

Male. Antennule with five articles, considerably wider than that of female.

Type species. — *Chauliopleona dentata* Dojiri & Sieg, 1997.

Gender of generic name. — Feminine.

Remarks. — Most diagnostic characters appear stable at this time. The characters separating the species include: (1) antennule length/width ratio, (2) length of cheliped carpus relative to that of propodus including fixed finger, (3) dactylus denticulations, and (4) pereopod 1 armament. The number of antennal articles is a dubious character, owing to the frequent presence of a fusion line and to the difficulties in determining whether the small proximal 'article' is really an article, or part of the cephalothorax itself, or muscle-tissue torn out from the cephalothorax during dissection.

This genus is easily recognizable by the ventral, posteriorly directed spine on pleonite 5 that protrudes under the pleotelson. A key to the genus is given by Guerrero-Kommritz (2005).

Species currently assigned to this genus. — *C. armata* (Hansen, 1913); *C. amdrupii* (Hansen, 1913); *C. dentata* Dojiri & Sieg, 1997; *C. faini* sp. nov.; *C. hastata* (Hansen, 1913).

Chauliopleona faini sp. nov.
(figs. 79, 80)

Material examined. — Holotype, non-ovigerous ♀ (USNM 1068919), body length 2.45 mm, DGoMB Sta. W3-1. Paratypes: 1 non-ovigerous ♀ (USNM 1068920), body length 2.4 mm, DGoMB Sta. RW6-3. 1 non-ovigerous ♀ (USNM 1068921), DGoMB Sta. S36-1 (dissected).

Additional material. — 3 mancas, DGoMB, Sta. NB4-1. 1 non-ovigerous ♀, 2 mancas, DGoMB Sta. S42-2. 1 non-ovigerous ♀, DGoMB Sta. MT4-4.

Diagnosis. — Female. Antennule stout (length/width ratio 3.7). Cheliped carpus longer than propodus including fixed finger; dactylus without proximal dorsal crenulation. Pereopod 1 propodus without serration/setules. Pleonite 5 ventral posteriorly-directed spine not protruding beyond pleotelson apex.

Male. Unknown.

Etymology. — Named after our DGoMB colleague, the polychaete scientist Dr. Guinn Fain Hubbard. The name is thus to be conceived a noun in the genitive singular.

Type locality. — DGoMB Sta. W3-1, 27°10'3711"N 93°19'3081"W, depth 860 m.

Description. — (Body from holotype, appendages from other adult female.) Body (fig. 79A, B) elongate, 6.5 times as long as wide.

Carapace about as long as lengths of pereonites 1 and 2 combined, covering basal part of antennules. Eyes and eye-lobes absent.

Pereon with all pereonites wider than long. Pereonites longer than dorsal shield and exposing margins of weakly calcified areas between the individual pereonites.

Pleon short (including pleotelson only 0.2 times total body length). All pleonites subequal. Pleonite 5 with ventral posteriorly-directed spine not protruding beyond pleotelson apex. Pleotelson shorter than the lengths of two free pleonites combined, apex rounded, and covered by a dorsal plate.

Antennule (fig. 79C) shorter than carapace. Stout (length/width ratio 3.7). Article 1 as long as rest of antennule, with one simple and two setulose setae distally. Article 2 shorter than 0.5 times article 1, with one simple and three setulated setae distally. Article 3 shorter than article 4, with two simple setae distally. Article 4 hardly shorter than article 2, with four long setae and one tiny seta distally.

Antenna (fig. 79D) 0.75 times as long as antennule. Article 1 wider than following articles, naked. Article 2 longer than article 5, with one seta distally. Article 3 longer than other articles, with clear fusion line and two simple and two setulose setae distally. Article 4 as long as article 1, naked. Article 5 smallest, with five simple setae distally.

Mouthparts. Labrum (fig. 79E) with flat and finely setose apex. Mandibular molar process pointed and longer than incisor, with small distal spines. Left mandible (fig. 79F) lacinia mobilis longer than incisor; incisor narrow and blunt, with two denticles. Right mandible not recovered. Labium not recovered. Maxillule (fig. 79G) endite with nine spiniform setae distally of which at least three are serrated; palp not recovered. Maxilla (fig. 79I) remarkably long and with outer distal margin finely setose. Maxilliped (fig. 79H) basis with one

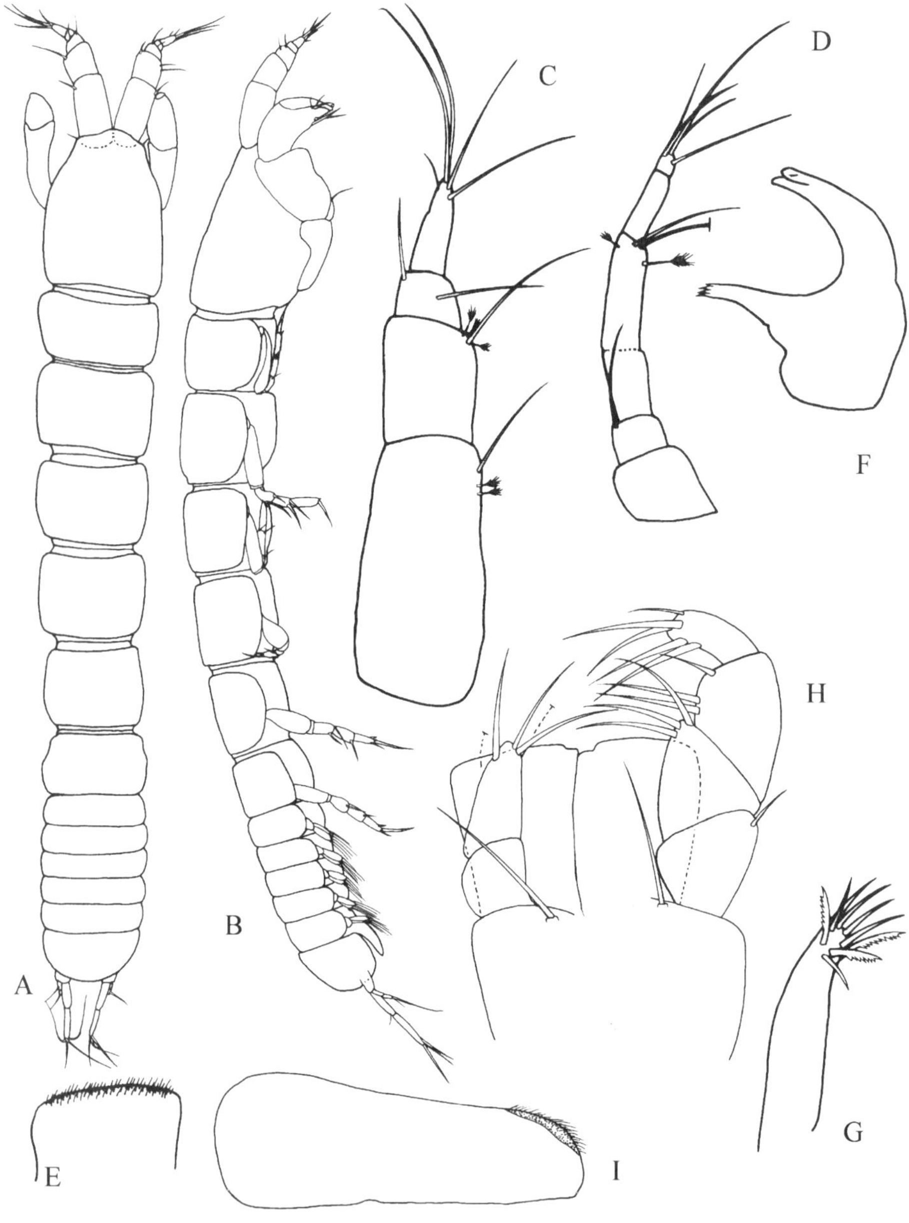

Fig. 79. *Chauliopleona faini* sp. nov., female; A, holotype, dorsal view; B, same, lateral view; C, antennule; D, antenna; E, labrum; F, left mandible; G, maxillule; H, maxilliped; I, maxilla. Scale bars equal: A, B, 1.0 mm; C-D, 0.1 mm; E-I, 0.05mm.

Fig. 80. *Chauliopleona faini* sp. nov., female; A, cheliped; B, pereopod 1; C, pereopod 2; D, pereopod 3; E, pereopod 4; F, pereopod 5; G, pereopod 6; H, pleopod; I, uropod. Scale bar equals 0.1 mm.

long distal seta at palp insertion; endites with small medial depression on distal margin but otherwise completely naked, almost as wide as basis; palp article 1 with one seta on outer margin, article 2 with distal expansion and three setae on inner margin, article 3 with three setae on inner margin, article 4 only 0.5 times as wide as article 3, with five distal setae. Epignath not recovered.

Cheliped (fig. 80A) basis unequally divided by prominent sclerite, shorter than carpus. Merus triangular, with one seta ventrally. Carpus longer than propodus including fixed finger, with one seta ventro-medially and one seta dorso-distally. Propodus with one seta near dactylus insertion. Fixed finger with two setae ventrally and three on inner margin, with four denticles on inner margin. Dactylus as long as fixed finger, naked.

Pereopod 1 (fig. 80B) coxa partly fused with body, with one seta. Basis longer than the three succeeding articles combined, with one seta dorsomedially. Ischium naked. Merus hardly shorter than carpus, widening distally and with one long spiniform seta distally. Carpus more than 0.5 times as long as propodus, with two long spiniform setae distally and one tiny spine. Propodus more than 0.5 times as long as basis, with one spiniform seta distally and a dorsal spine. Dactylus and unguis combined shorter than propodus, and not fused to a claw.

Pereopod 2 (fig. 80C) as pereopod 1 except: coxa naked. Seta on basis setulose. Carpus with three spiniform setae distally and a row of small ventral setules. Propodus with row of small ventral setules, one spiniform, and one simple setae distally, and a dorsal spine.

Pereopod 3 (fig. 80D) as pereopod 1 except: basis naked. Ischium with one seta. Carpus with three setae distally.

Pereopod 4 (fig. 80E) no visible coxa. Basis stouter than those of pereopods 1–3, with one medial and one simple setae distally. Ischium with one seta. Merus with two setae distally. Carpus with four setae distally. Propodus with two simple setae distally and a dorsal spine. Dactylus and unguis combined longer than propodus, and dactylus with pronounced serration.

Pereopod 5 (fig. 80F) as pereopod 4 except: basis with two setulated setae proximally and one simple seta distally. Merus with one seta distally. Carpus with three setae distally. Propodus with three setae and a dorsal spine distally.

Pereopod 6 (fig. 80G) as pereopod 4 except: basis naked. Ischium naked. Merus with two setae distally. Propodus with five setae and a dorsal spine distally.

Pleopods (fig. 80H), all pairs similar. Endopod with eight simple setae. Exopod with five simple setae.

Uropod (fig. 80I) longer than pleotelson and last pleonite combined. Basal article short (as long as exopod). Endopod with two articles; article 1 with one setulated and one simple seta distally; article 2 longer than article 1, with four very long and one short setae distally. Exopod with two articles, shorter than first endopod article; article 1 with one seta distally; article 2 with two unequal setae distally.

Distribution. — A rare species in the northern part of the Gulf. Bathymetric range: 768–3000 m. See map 54.

Remarks. — This species is unique in its lack of serration along the ventral margin of the propodus of pereopod 1, the medio-dorsal process of the basis of pereopod 1, and the proximo-ventral process on the basis of pereopods 2 and 3. Another character combination that can be used in separating this species is the lack of a carpal shield and the lack of crenulations on the cheliped dactylus.

Genus **Coalecerotanais** Larsen, 2003

Coalecerotanais Larsen, 2003, Journal of Crustacean Biology, **23**: 777–781.

Diagnosis. — (After Larsen, 2003b.) Sex undetermined but presumably female (see Remarks). Body consisting of five pereonites and pleotelson only, pereonite 6 and all pleonites fused with pleotelson. Eyes and eye-lobes absent. Antennule with four articles. Antenna with six articles. Mandibles almost rudimentary, no chitinization and no identifiable molar. Maxillule with six spiniform setae. Maxilliped palp conspicuously large (as long as cheliped basis) while endites remarkably small (not longer than first palp article). Cheliped carpus and propodus elongate. Only five pairs of pereopods present, pereopod 6 absent. Pereopods 1–3 with coxa; dactylus/unguis combined shorter than propodus. Pereopods 4–5 without coxa; basis thicker than pereopods 1–3; dactylus/unguis fused to a claw. Pleopods and lateral edge of dorsal shield absent. Uropods uniramous; endopod uni-articulate.

Male. Unknown.

Type species. — *Coalecerotanais inflatus* Larsen, 2003.

Gender of generic name. — Masculine.

Remarks. — This genus is remarkable in several respects. The lack of pereopod 6 is unique among adult tanaidaceans. Fused pleonites occur in a number of genera and fusion with the pleotelson occurs both in some deep-sea (*Anarthrura*, *Isopodidus*) and some shallow-water genera (*Pseudozeuxo*, *Curtipleon*). Fusion of all pleonites, the pleotelson, and the 6th pereonite has only been recorded from one species, *Metagathotanais insulcatus*, and then only in the female. *Metagathotanais insulcatus*, however, has retained pereopod 6 (Bird & Holdich, 1988).

A number of characters of this genus corresponds with those of the Agathotanaidae to which also *Metagathotanais* belongs. These are: (1) a reduced mandible molar, (2) uniramous uropods, (3) a uni-articulate uropodal endopod, (4) ventral cheliped attachment, and (5) the dorsal shape of the pereonites. However, a number of other characters conflicts with the Agathotanaidae diagnosis, as given by Larsen & Wilson (2002). Those are: (1) body calcification not rigid (in fact almost weak, but this could be an artifact of fixation) and (2) dactylus/unguis of the last pereopods fused to a claw.

The reduced state of the mandibles is puzzling, possibly reflecting a special dietary adaptation. Since this genus (species) has only been taken from a cold seep habitat with large sheets of bacteria covering the substrate, it is here speculated that *Coalecerotanais* might feed upon these bacteria. An alternative hypothesis is that the animal carries chemoautotrophic symbionts.

Species currently assigned to this genus. — *C. inflatus* Larsen, 2003.

Coalecerotanais inflatus Larsen, 2003

Coalecerotanais inflatus Larsen, 2003, Journal of Crustacean Biology, **23**: 778–781.

Diagnosis. — Monotypic genus, no species diagnosis given.

Distribution. — This species was only found in one sample from a cold seep: GB 425, Sta. BL7, 27°33'60"N 92°32'33"W. Bathymetric range: 570 m. See map 55.

Genus **Spinitanaopsis** nomen novum

Crurispina Larsen, 2003, Journal of Crustacean Biology, **23**: 781.

Diagnosis. — (Modified after Larsen, 2003b.) Female. Body somites not fused. Eyes and eye-lobes absent. Antennule with four articles. Antenna with six articles; articles 2 and 3 without spiniform setae dorsally. Mandibles reduced. Maxillule with seven spiniform terminal setae. Maxillipedal endites and bases completely naked; bases fused but endites not fused. Chelipeds attached via sclerite; propodus separated from fixed finger by pronounced ridge; fixed finger without setae ventrally and with only two setae on inner margin; dactylus with conspicuous proximal setae, and distal spine on unguis. Pereopod 1 longer than other pereopods. Pereopods 1–3 with coxae, dactylus/unguis not fused. Pereopods 4–6 without coxae, basis thicker than pereopods 1–3, merus and carpus with very large, cone-shaped spiniform setae, dactylus/unguis not fused. Pleopods present and well developed. Uropods biramous; exopod uni-articulate and partly fused with basal article; endopod with two articles.

Male. Unknown.

Etymology. — The new name (from Latin, spinosus + *Tanaopsis* = spiny + tanaid-like), has been chosen because it resembles the meaning of the old name (spiny legs) and agrees with it in gender.

Type species. — *Crurispina insolituchelia* Larsen, 2003.

Gender of generic name. — Feminine.

Remarks. — The genus was described under the name *Crurispina* Larsen, 2003b but as this name is preoccupied by a Brachiopoda taxon from the middle Devonian (Goldman & Mitchell, 1990), this genus is here renamed *Spinitanaopsis*.

The cheliped is unusual by the distinct separation (not articulation) of the propodus and fixed finger, having two setae ventrally on the propodus above the fixed finger and only two setae on the inner margin, and the dactylus with a spine on the unguis. Pereopods 4–6 resemble those of the shallow water families Paratanaidae and Leptocheliidae, but without the dactylus and unguis being fused.

The reduced mandibles suggest a specialized feeding behaviour (see *Coalecerotanais*).

Species currently assigned to this genus. — *S. insolituchelia* (Larsen, 2003).

Spinitanaopsis insolituchelia (Larsen, 2003)

Crurispina insolituchelia Larsen, 2003, Journal of Crustacean Biology, **23**: 781–784.

Diagnosis. — Monotypic genus, no species diagnosis given.

Distribution. — This species was only found in one sample from a cold seep: GB 425, Sta. BL7, 27°33'60"N 92°32'33"W. Bathymetric range: 570 m. See map 56.

Genus **Expina** Lang, 1968

Expina Lang, 1968, Galathea Report, **9**: 188.

Diagnosis. — (After Lang, 1968.) Female. Body somites not fused. Eyes and eye-lobes absent. Antennule with four articles. Antenna with six articles;

articles 2 and 3 without spiniform setae dorsally. Mandibles with reduced, membranous molar. Maxillule without any spiniform terminal setae. Maxillipedal endites and basis completely naked; bases fused, but endites not fused. Chelipeds attached via sclerite. Pereopods all with coxae. Pereopods 4–6 with basis not thicker than those of pereopods 1–3, dactylus/unguis not fused, but unguis bifurcate. Pleopods absent. Uropods biramous; endopod with two articles; exopod uni-articulate and as long as first endopod article.

Male. Antennule thicker and longer than that of female. Pleopods present but reduced.

Type species. — *Expina typica* Lang, 1968.

Gender of generic name. — Feminine.

Remarks. — The genus is unique among the Tanaidacea by lacking any spiniform setae on the maxillule. This genus has affinities with the family Colletteidae but cannot be assigned to any known family with certainty (Larsen & Wilson, 2002).

Species currently assigned to this genus. — *E. typica* Lang, 1968.

Expina typica Lang, 1968

Strongylura antarctica Vanhöffen, 1914, Deutsche Südpolar Expedition 1901–1903, **15**: 479–481.
Expina typica Lang, 1968, Galathea Report, **9**: 188–192.

Diagnosis. — Monotypic genus, no species diagnosis given.

Distribution. — This species is among the most widespread species of Tanaidacea and has been recorded from Iceland to Antarctica and from Panama to New Zealand (Sieg, 1979; Thurston et al., 1987). Bathymetric range: 1687–4895 m. See map 57.

Remarks. — The unusually wide distribution range of this species is suggested to be a consequence of its parasitic (or at least symbiotic, see Biology and Distribution) relationship with holothurians (Thurston et al., 1987).

Genus **Insociabilitanais** n. gen.

Diagnosis. — Female. Body almost completely cylindrical. Pereonite 1 with strong, anteriorly directed ventral spine (possibly a hyposphenian). Antennule with four articles; 0.5 times as long as carapace. Antenna with five articles. Molar process of mandible broad, with several terminal denticles. Maxillule with seven spiniform terminal setae of which at least one is pinnate. Maxillipedal bases fused; endites not fused. Chelipeds ventrally attached via sclerite (see remarks). Pereopods long (longer than pleon). Pereopods 1–3 with coxae. Pereopods 4–6 without coxae, not stouter than pereopods 1–3; dactylus and unguis not fused to a claw. Pleopods absent. Uropods biramous, longer than pleotelson; endopod with two articles; exopod with one article.

Male. Essentially as female. Antennule 1.33 times as thick as that of female. Functional mouthparts present. Pleopods present, with simple setae only; all pairs closely held together in a cone-shaped manner.

Etymology. — Named after the apparent difficulties associating this genus with a family (Latin, insociabilis = unsociable).

Type species. — *Insociabilitanais ventrospinatus* sp. nov.

Gender of generic name. — Masculine.

Remarks. — This genus has a number of conflicting characters. Some characters agree with the Agathotanaidae: (1) the cuticle is heavily chitinized and even pitted like in *Agathotanais*, (2) the male pleopods are closely held together in a cone-shaped manner, like in all male Agathotanaidae, although not fused. However, the mandibular molar is rather broad and the maxillipedal endites are rectangular, which features are both inconsistent with the Agathotanaidae, and point more towards an affinity with the Colletteidae. The uropods and general morphology of the mouthparts, cheliped, and pereopods appear much like *Paratyphlotanais* Kudinova-Pasternak, 1978, but the four-articulate antennule and the condition of the pleopods is not consistent with such a relationship. The condition of the cheliped attachment as ventral, but via a sclerite, is consistent with *Isopodidus*, while just about every other character is not.

The antennule of both sexes is very short (0.5 times as long as the carapace) and very stout. All specimens of this species were found with a large, anteriorly directed ventral hyposphenian on pereonite 1. Hyposphenians are normally not good characters, as they vary with ontogeny (Lang, 1953b). However, all specimens from the Gulf, whether male, female, or manca had the very same ventral hyposphenian. This may, therefore, be a 'real' ventral spine, and thus a valid character rather than a hyposphenian.

Species currently assigned to this genus. — *I. ventrospinatus* sp. nov.

Insociabilitanais ventrospinatus sp. nov. (figs. 81–84)

Material examined. — Holotype, non-ovigerous ♀ (USNM 1068922), body length 1.85 mm, DGoMB Sta. MT3-2. Paratypes: 1 non-ovigerous ♀, body length 1.2 mm, 1 ♂, body length 1.65 mm (USNM 1068923), same locality. 1 non-ovigerous ♀ (USNM 1068924), same locality (dissected). 1 ♂ (USNM 1068925), same locality (dissected).

Additional material. — 2 non-ovigerous ♀♀, 3 mancas, DGoMB Sta. MT3-1. 1 ♂, DGoMB Sta. RW5-3. 1 ♂, 2 mancas, DGoMB Sta. S42-3. 6 non-ovigerous ♀♀, 4 ♂♂, 3 mancas, DGoMB Sta. S35-2. 2 non-ovigerous ♀♀, 1 ♂, DGoMB Sta. MT3-4. 2 non-ovigerous ♀♀, 2 ♂♂, 2 mancas, DGoMB Sta. MT3-1. 1 non-ovigerous ♀, 2 mancas, DGoMB Sta. S39-4. 1 non-ovigerous ♀, DGoMB Sta. WC12-4.

Diagnosis. — Monotypic genus, no species diagnosis given.

Etymology. — Named after the ventral spine on the first pereonite (Latin, ventro- + spinatus = belly-side + spined). The name is a compound word to be treated as an adjective, agreeing in gender with the generic name.

Type locality. — DGoMB Sta. MT3-2, 28°13'1533"N 89°29'7860"W, depth 987 m.

Description. — Adult female. Body (fig. 81A) elongate, eight times as long as wide.

Carapace longer than the lengths of pereonites 1 and 2 combined. Eyes and eye-lobes absent.

Pereon with pereonites 1 and 6 wider than long. Pereonite 2 as wide as long. Pereonites 3 and 4 longer than wide.

Pleon short (including pleotelson only 0.2 times total body length). All pleonites subequal; no lateral edge of dorsal shield visible in female. Pleotelson shorter than the lengths of three free pleonites combined, apex acorn-shaped and covered by a dorsal plate.

Antennule (fig. 81B) about 0.5 times as long as carapace. Article 1 as long as rest of antennule, with one bipinnate and three simple setae distally. Article 2 less than 0.5 times as long as article 1, with one simple seta distally. Article 3 shorter than article 4, with one simple seta distally. Article 4 shorter than article 2, with four setae distally.

Antenna (fig. 81C) 0.67 times as long as antennule. Article 1 not wider than following articles, naked. Article 2 shorter than article 1, with one simple seta dorso-distally. Article 3 longer than other articles, with two simple setae distally. Article 4 as long as article 2, with one simple seta dorso-distally. Article 5 minute, with three setae distally.

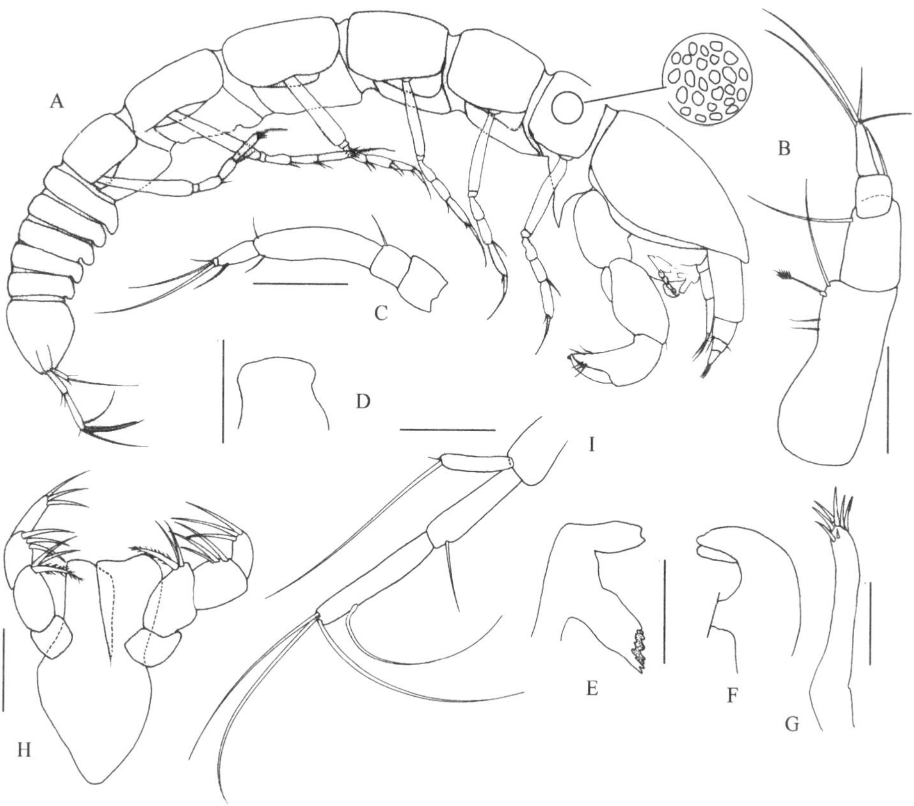

Fig. 81. *Insociabilitanais ventrospinatus* gen. et sp. nov., female; A, holotype, lateral view; B, antennule; C, antenna; D, labrum; E, right mandible; F, left mandible; G, maxillule; H, maxilliped; I, uropod. Scale bars equal: A, 1.0 mm; B-C, 0.1 mm; D-I, 0.05 mm.

Mouthparts. Labrum (fig. 81D) naked and hood-shaped. Mandibular molar process broad and longer than incisor, with terminal ring of several small denticles. Left mandible (fig. 81F) lacinia mobilis as long as incisor; incisor broad, with one denticle. Right mandible (fig. 81E) incisor broad, with two denticles. Labium not recovered. Maxillule (fig. 81G) endite with seven spiniform setae distally, of which at least one is pinnate; palp not recovered. Maxilla not recovered. Maxilliped (fig. 81H) endites with one denticle laterally on distal margin, almost as wide as basis; palp article naked, articles 2–4 each with three setae on inner margin. Epignath not recovered.

Cheliped (fig. 82A) basis unequally divided by small sclerite, shorter than carpus. Merus naked. Carpus shorter than propodus including fixed finger, with two setae ventrally and one seta dorsally. Propodus naked. Fixed finger

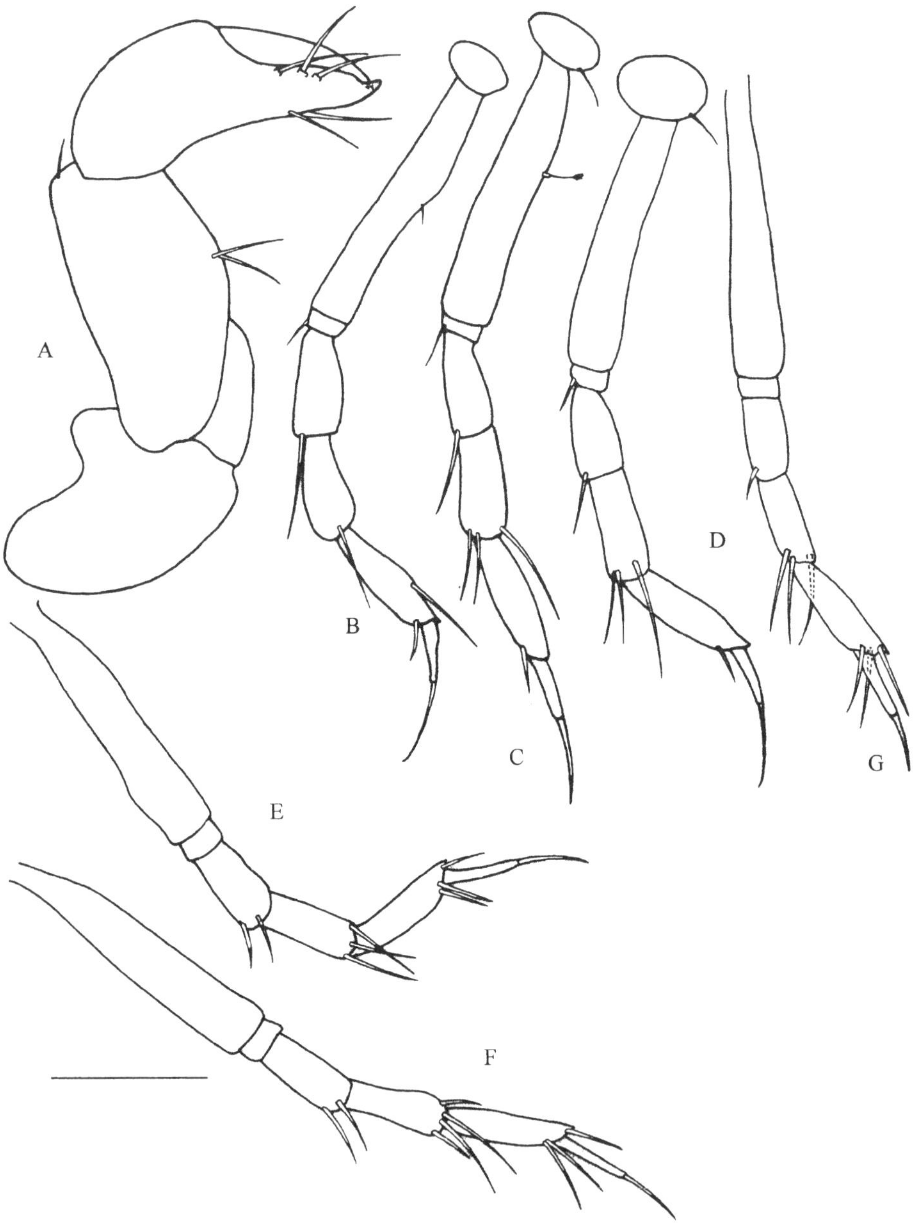

Fig. 82. *Insociabilitanais ventrospinatus* gen. et sp. nov., female; A, cheliped; B, pereopod 1; C, pereopod 2; D, pereopod 3; E, pereopod 4; F, pereopod 5; G, pereopod 6. Scale bar equals 0.1 mm.

with two setae ventrally and three setae on inner margin, without visible denticles on inner margin. Dactylus as long as fixed finger.

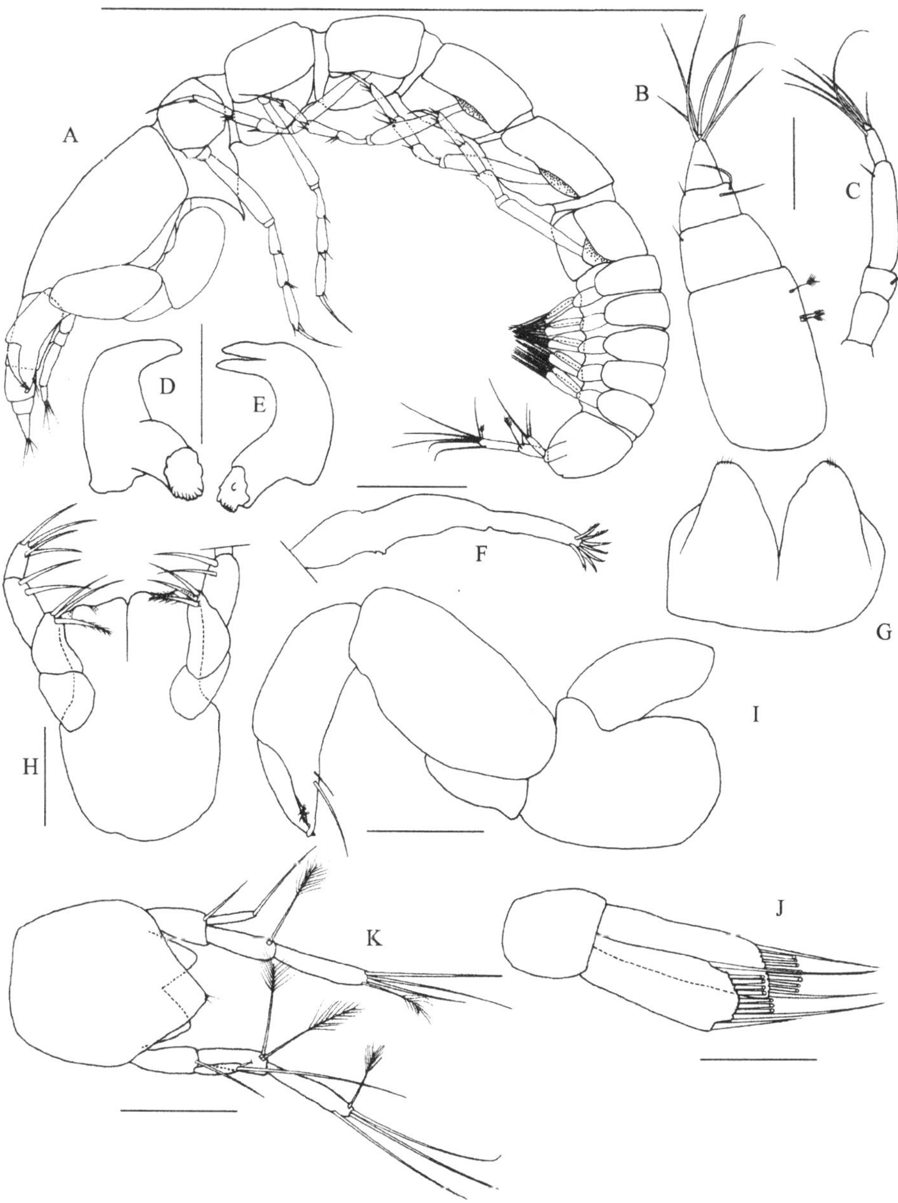

Fig. 83. *Insociabilitanais ventrospinatus* gen. et sp. nov., male; A, paratype, lateral view; B, antennule; C, antenna; D, right mandible; E, left mandible; F, maxillule; G, labium; H, maxilliped; I, cheliped; J, pleopod; K, pleotelson and uropods. Scale bars equal: A, 1.0 mm; B-C, 0.1 mm; D-K, 0.05 mm.

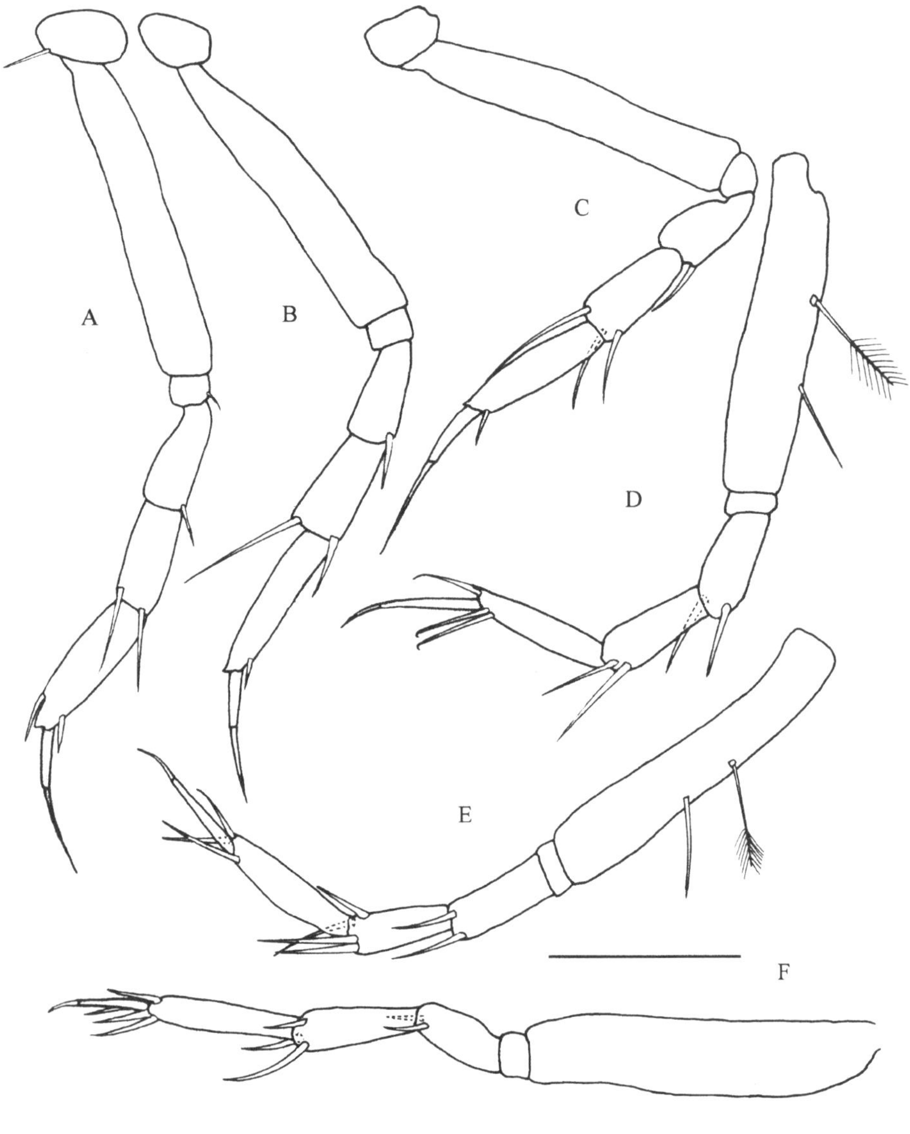

Fig. 84. *Insociabilitanais ventrospinatus* gen. et sp. nov., male; A, pereopod 1; B, pereopod 2; C, pereopod 3; D, pereopod 4; E, pereopod 5; F, pereopod 6. Scale bar equals 0.1 mm.

Pereopod 1 (fig. 82B) coxa naked. Basis longer than the three succeeding articles combined, with one seta dorsally. Ischium with one seta ventrally. Merus as long as carpus, widening distally and with one seta distally. Carpus 0.67 times as long as propodus, with one seta distally. Propodus shorter than 0.5 times as long as basis, with one seta distally on each margin and a dorsal spine. Dactylus and unguis combined longer than propodus, and not fused to a claw.

Pereopod 2 (fig. 82C) as pereopod 1 except: coxa with one seta. Basis with one bipinnate seta dorsally. Carpus with three setae distally. Propodus without the dorsal seta.

Pereopod 3 (fig. 82D) as pereopod 2 except: basis naked.

Pereopod 4 (fig. 82E) as pereopod 3 except: no coxa. Ischium naked. Propodus with three simple setae distally and a dorsal spine. Dactylus and unguis combined as long as propodus, not fused to a claw.

Pereopod 5 (fig. 82F) as pereopod 4 except: merus with two setae distally. Propodus with three setae distally and a dorsal spine. Dactylus and unguis combined longer than propodus.

Pereopod 6 (fig. 82G) as pereopod 5 except: carpus with three setae distally.

Pleopods not present in female.

Uropod (fig. 81I) basal article as long as exopod, naked. Endopod with two articles, longer than pleotelson; article 1 with one simple seta subdistally; article 2 with one seta subdistally and three setae distally. Exopod shorter than first endopod article, with two unequal simple setae distally.

Male, where differing from female. Body (fig. 83A) as female, except pleon with clear lateral shield.

Antennule (fig. 83B) about 1.33 times as wide as that of female. Article 1 longer than rest of antennule, with three bipinnate setae distally. Article 4 shorter than article 2, with five setae and prominent aesthetasc distally.

Antenna (fig. 83C) as in female.

Mouthparts (fig. 83D-H) as in female except: left mandible (fig. 83E) lacinia mobilis as wide as, and longer than, incisor; incisor without obvious denticles. Right mandible (fig. 83D) incisor broad, with one denticle. Labium (fig. 83F) with broad basis and sparse setation distally. Maxillule (fig. 83G) endite with seven spiniform setae distally of which at least two are pinnate and one is bifurcate; palp not recovered.

Cheliped (fig. 83I) as that of female except: carpus apparently naked (possibly an artifact).

Pereopods (fig. 84A-F) as in female except: pereopods 5 (fig. 81E) and 6 (fig. 81F) with heavier armament on distal articles.

Pleopods (fig. 83J) held together (but not fused) in a cone. Basal article naked. Endopod with nine simple setae. Exopod with ten simple setae.

Uropod (fig. 83K) endopod articles with several prominent, bipinnate setae distally.

Distribution. — Rare in the northern part of the Gulf. Bathymetric range: 667–3000 m. See map 58.

Remarks. — The male of this species conforms well to the 'typical' deep-sea tanaid male characteristics mentioned by Larsen (2002a).

Genus **Nototanoides** Sieg & Heard, 1985

Nototanoides Sieg & Heard, 1985, Gulf Research Reports, **8**: 51.

Diagnosis. — (Modified after Sieg & Heard, 1985.) Female. Carapace with well developed eyes. Pleon with five free pleonites. Antennule with three articles. Antenna with six articles. Mandibular molar well developed. Labium consisting of one pair of lobes without lateral and medial processes. Maxillule with nine spiniform terminal setae. Maxillipedal bases fused distally, endites partly fused, with two denticles distally, and narrower than basis. Chelipeds attached laterally by sidepiece. Pereopods 1–3 with coxae. Pereopods 4–6 without coxae; dactylus and unguis fused to a claw. Pleopods present and well developed. Uropodal endopod and exopod consisting of two articles each.

Male. Antennule with four articles, article 3 minute and covered by aesthetascs. Mouthparts reduced. Chelipeds enlarged.

Type species. — *Nototanoides trifurcatus* Sieg & Heard, 1985.

Gender of generic name. — Masculine.

Remarks. — The character combination of: (1) uropodal endopod biarticulate, (2) presence of eyes, (3) the partly fused maxillipedal endites, (4) fused dactylus/unguis of pereopod 4–6, (5) tri-articulate antennule (female), and (6) enlarged chelipeds (male); makes this genus unique.

Species currently assigned to this genus. — *N. trifurcatus* Sieg & Heard, 1985.

Nototanoides trifurcatus Sieg & Heard, 1985

Nototanoides trifurcatus Sieg & Heard, 1985, Gulf Research Reports, **8**: 52–61.

Diagnosis. — Monotypic genus, no species diagnosis given.

Distribution. — Species widespread in the northern part of the Gulf of Mexico but in populations with apparently patchy distributions. Bathymetric range: 19–190 m. See map 59.

Remarks. — This species is mostly a shallow water inhabitant, but has been taken as deep as 190 m.

Genus **Parafilitanais** Kudinova-Pasternak, 1989

Parafilitanais Kudinova-Pasternak, 1989, Zoologicheskii Zhurnal, **68**: 34.

Diagnosis. — (Modified after Larsen, 2002a.) Female. Body elongate and cylindrical. Antennule with four articles. Antenna with 5–6 articles. Mandibular molar pointed. Labium consisting of one pair of lobes, without lateral and medial processes. Maxillipedal bases fused distally; endites not fused, narrower than basis, and with 1 minute distal denticle. Chelipeds attached ventrally. Pereopods 1–3 with coxae; dactylus and unguis combined shorter than propodus. Pereopods 4–6 without coxae and dactylus and unguis combined as long as, or longer than, propodus, and not fused into a claw. Pleopods absent. Uropods long and uniramous, endopod consisting of two articles.

Male. Body shorter than in female. Functional mouthparts present in adult male, resembling but not identical to those of female. Antennule only marginally thicker than in female. Pleonites with reduced pleopods, with simple setae (juvenile male pleopods without setae).

Type species. — *Parafilitanais caudatus* Kudinova-Pasternak, 1989.

Gender of generic name. — Masculine.

Remarks. — All the diagnostic characters for this genus seem to be stable. The maxillipedal bases fusion is, however, a character that should not be given too much weight, owing to the difficulties in observing whether or not the bases are completely fused or not. Furthermore, it is doubtful if this character has been identified correctly throughout tanaidacean systematics, and is thus weak to begin with. A key to the genus is given by Larsen (2002a).

Species currently assigned to this genus. — *P. caudatus* Kudinova-Pasternak, 1989; *P. mexicanus* Larsen, 2002; *P. similis* Kudinova-Pasternak, 1990.

Parafilitanais mexicanus Larsen, 2002

Parafilitanais mexicana Larsen, 2002 [sic], Gulf and Caribbean Research, **14**: 54–57.

Diagnosis. — Female. Body not tapering distally. Carapace shorter than the lengths of two pereonites combined. Maxillule with two bifurcate terminal setae. Cheliped dactylus hardly longer than fixed finger. Pleotelson apex not acute.

Male. As female, except for thicker antennules and cheliped dactylus not longer than fixed finger.

Distribution. — *Parafilitanais mexicanus* is only recorded from the northern part of the Gulf. Bathymetric range: 625–2030 m. See map 60.

Genus **Pseudoarthrura** n. gen.

Diagnosis. — Female. Cuticle heavily calcified. Pereon with six pereonites. Pleon narrower than pereon, but not narrower than pleotelson, with five free pleonites. Antennule with four articles Antenna with six articles. Molar process of mandible well developed. Maxillipedal basis partially fused, endites not fused, narrower than basis, and without flat setae. Cheliped attachment via sclerite. Marsupium with four pairs of oostegites. Pereopod coxae present. Pereopods 4–6 with dactylus and unguis not fused. Pleopods absent. Uropodal endopod with two articles, and not fused with basal article; exopod minute.

Male. Functional mouthparts retained. Antennule thicker than that of female. Pleopods present but coalesced, and with simple setae only. Pleon always as wide as pereon or wider.

Type species. — *Pseudoarthrura heardi* sp. nov.

Etymology. — Named after the superficial similarity with *Arthrura* (latinized Greek, pseudo = fake).

Gender of generic name. — Feminine.

Remarks. — This genus closely resembles *Arthrura* Kudinova-Pasternak, 1966 in most respects, except that the uropodal exopod is not fused with the basal article and the maxillipedal basis is fused. The uropod feature is indeed hard to identify in *Pseudoarthrura* and, regarding the maxilliped, Kudinova-Pasternak's (1966b: 259, fig. 2g) illustration is possibly erroneous (as the five-articulate palp surely is a mistake). However, it was not possible to get hold of the type or other material of *Arthrura*. Therefore, it was not possible to confirm the observation of Kudinova-Pasternak (1966b), and a new genus is thus erected. *Pseudoarthrura* and *Anarthrura* with their well developed mandibles and agathotanaid-like body shape, heavy calicification, and coalesced male pleopods, might be the link between the Tanaellidae and Agathotanaidae. The body shape, chelipeds, pereopods, and uropods also somewhat remind of those of *Paranarthrurella* Kudinova-Pasternak, 1965, although the mouthparts appear dissimilar from those as illustrated by Kudinova-Pasternak (1965: 89), but these are not described in the text.

Species currently assigned to this genus. — *P. heardi* sp. nov.

Pseudoarthrura heardi sp. nov.
(figs. 85–87)

Material examined. — Holotype, non-ovigerous ♀ (USNM 1068926), body length 1.8 mm, DGoMB Sta. C4-2. Paratypes: 1 ♂ (USNM 1068927), body length 1.5 mm, DGoMB Sta. C4-1. 1 non-ovigerous ♀ (USNM 1068928), body length 1.9 mm, DGoMB Sta. C4-2. 1 non-ovigerous ♀ (USNM 1068929), same locality (dissected).

Additional material. — 2 non-ovigerous ♀♀, DGoMB Sta. S40-1. 2 non-ovigerous ♀♀, 1 manca, DGoMB Sta. C4-1. 2 non-ovigerous ♀♀, DGoMB Sta. RW3-1. 4 non-ovigerous ♀♀, DGoMB Sta. S40-2. 1 non-ovigerous ♀, 1 manca, DGoMB Sta. S41-1. 1 non-ovigerous ♀, DGoMB S42-1. 1 manca, DGoMB Sta. S37-2. 1 non-ovigerous ♀, DGoMB Sta. RW6-3.

Diagnosis. — Monotypic genus, no species diagnosis given.

Etymology. — Named after Prof. Richard Heard, Gulf Coast Research Laboratory, in recognition of his many years of work with the Tanaidacea. The name is a noun in the genitive singular.

Type locality. — DGoMB Sta. C4-2, 27°27'6103"N 89°46'7720"W, depth 1452 m.

Description. — Adult female. Body (fig. 85A, B) about eight times as long as wide. Lateral shoulders inconspicuous.

Carapace with smooth lateral edges in dorsal view. Longer than wide. Shorter than the lengths of pereonites 1 and 2 combined. Eye-lobes absent.

Pereon with pereonites gently curved. Pereonite 1 wider than long. Pereonites 3, 4, and 5 longer than wide. Pereonites 2 and 6 as long as wide.

Pleon short (including pleotelson only 0.15 times total body length). Pleonite 1 apparently larger than other pleonites, no visible lateral edges of dorsal shield. Pleotelson longer than the lengths of three free pleonites combined, acorn-shaped, apex blunt, and covered by a dorsal plate.

Antennule (fig. 86A) shorter than carapace, with four articles. Article 1 longer than rest of antennule, with one simple distal and two setulose setae medially. Article 2 longer than article 3, with one setulose and two simple setae distally. Article 3 as long as article 4, with two simple setae distally. Article 4 with three simple setae and one aesthetasc distally.

Antenna (fig. 86B) with six articles. Almost as long as antennule. Article 1 longer than article 3, naked. Article 2 longer than article 1, naked. Article 3 shorter than article 5, naked. Article 4 longer than each of the other articles, with two simple setae distally. Article 5 shorter than article 2, with two simple setae distally. Article 6 minute, with three long setae distally.

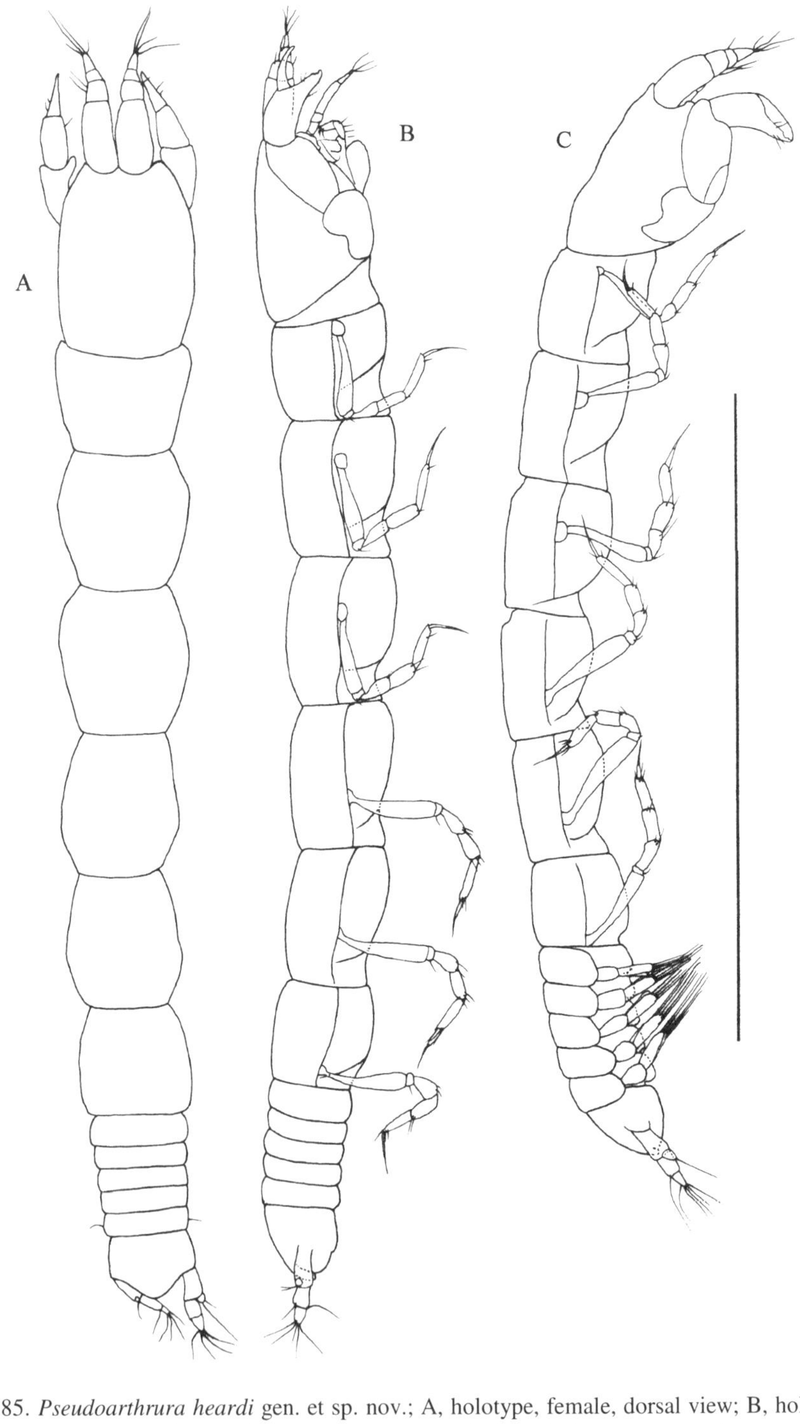

Fig. 85. *Pseudoarthrura heardi* gen. et sp. nov.; A, holotype, female, dorsal view; B, holotype, female, lateral view; C, paratype, male, lateral view. Scale bar equals 1.0 mm.

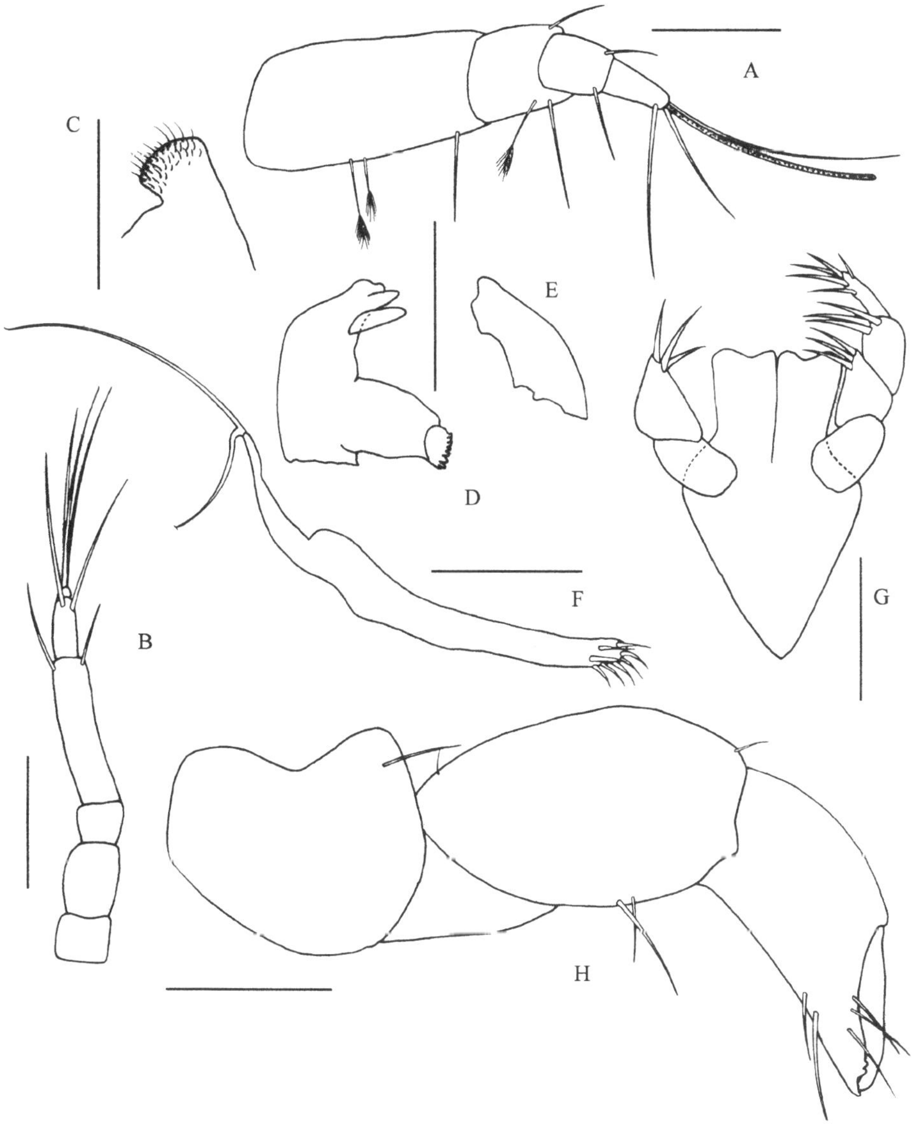

Fig. 86. *Pseudoarthrura heardi* gen. et sp. nov., female; A, antennule; B, antenna; C, labrum; D, left mandible; E, right mandible, incisor; F, maxillule; G, maxilliped; H, cheliped. Scale bars equal: A-B, H, 0.1 mm; C-G, 0.05 mm.

Mouthparts. Labrum (fig. 86C) finely setose and hood-shaped. Mandibular molar well developed and with several denticles distally. Left mandible (fig. 86D) lacinia mobilis longer than incisor; incisor narrow, with two separate

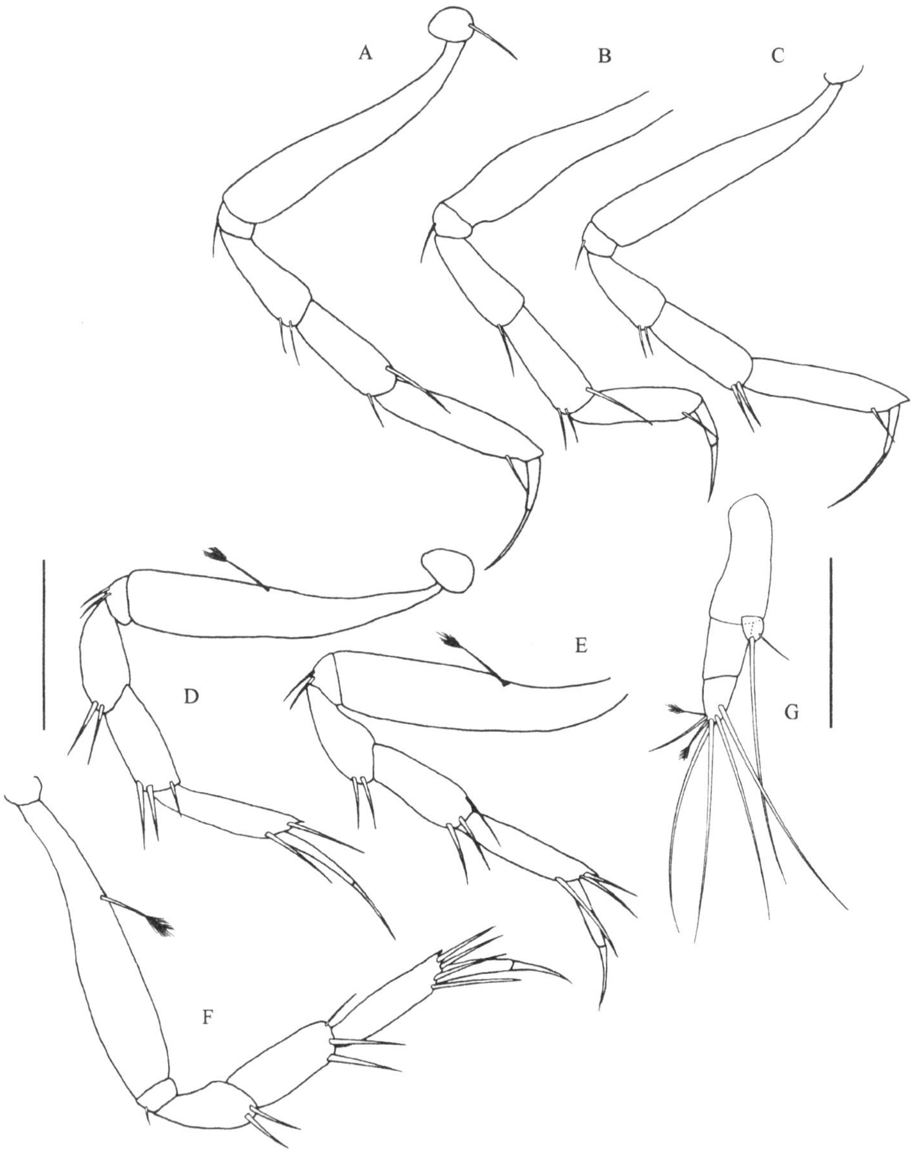

Fig. 87. *Pseudoarthrura heardi* gen. et sp. nov., female; A, pereopod 1; B, pereopod 2; C, pereopod 3; D, pereopod 4; E, pereopod 5; F, pereopod 6; G, uropod. Scale bars equal 0.1 mm.

denticles. Right mandible (fig. 86E) incisor broad and blunt. Maxillule (fig. 86F) endite with eight spiniform setae distally; palp only 0.5 times as long as endite, with two setae of unequal length. Maxilla not recovered. Maxilliped (fig. 86G) endites naked, with medial depression; palp article 1 naked, articles

2 and 3 each with three setae on inner margin, article 4 only 0.5 times as wide as article 3, with five setae. Epignath not recovered.

Cheliped (fig. 86H) basis shorter than carpus, with one seta dorsally. Merus naked. Carpus longer than propodus including fixed finger, with two ventral and one dorsal setae. Propodus slender and naked. Fixed finger with two setae ventrally and three on inner margin, as well as two denticles distally. Dactylus as long as fixed finger.

Pereopod 1 (fig. 87A) coxa spherical, with one seta. Basis longer than the three succeeding articles combined, naked. Ischium with one simple seta. Merus shorter than carpus, widening distally, and with two setae ventro-distally. Carpus shorter than 0.5 times basis, with two setae distally. Propodus with one seta distally but no dorsal spine. Dactylus and unguis combined shorter than propodus, and not fused.

Pereopod 2 (fig. 87B) as pereopod 1 except: merus with one seta distally. Carpus with three setae distally.

Pereopod 3 (fig. 87C) as pereopod 1 except: propodus with blunt dorsal spine.

Pereopod 4 (fig. 87D) coxa spherical. Basis longer than the three succeeding articles combined, with one setulose seta medially. Ischium with two setae. Merus as long as carpus, widening distally, and with two setae distally. Carpus shorter than 0.5 times basis, with three setae distally. Propodus longer than carpus, with two setae distally and a dorsal spine. Dactylus and unguis combined longer than propodus, not fused.

Pereopod 5 (fig. 87E) as pereopod 4 except: propodus with three setae distally and a dorsal spine.

Pereopod 6 (fig. 87F) as pereopod 5 except: ischium with one simple seta. Propodus with four setae distally and a dorsal spine.

Pleopods not present in female.

Uropod (fig. 87G) shorter than pleotelson. Basal article naked and longer than endopod. Endopod article 1 naked; article 2 shorter than article 1, with four long, one short simple, and two setulose setae distally. Exopod shorter than 0.5 times first endopod article, not fused with basal article.

Male, not dissected. Body as in fig. 85C. Antennule thicker than in female. Functional mouthparts retained. Pleopods present on all pleonites, coalesced, with simple setae only.

Distribution. — Only found in the northern part of the Gulf. Bathymetric range wide and deep: 767–3000 m. See map 61.

Remarks. — See generic remarks.

Genus **Robustochelia** Kudinova-Pasternak, 1983

Leptognathia – Kudinova-Pasternak, 1972, Trudy Instituta Okeanologii, Akademiya Nauk SSSR, **86**: 341–381.
Robustochelia Kudinova-Pasternak, 1983, Zoologicheskii Zhurnal, **62**: 1171.

Diagnosis. — (Modified after Kudinova-Pasternak, 1983.) Female. Antennule with four articles. Antenna with 5–6 articles. Mandibular molar pointed, or tapering into small crushing area. Labium consisting of one pair of lobes with sparse lateral setulation, without medial processes. Maxillule with seven terminal setae. Maxillipedal bases not fused, distally with lateral setulation, endites not fused; with inner setae distally and/or with or without two denticles distally, narrower than basis. Chelipeds attached by ventral sidepiece, carpus and propodus massive, dactylus and fixed finger shorter than [proximal part of] propodus, and heavily chitinized. Pereopods 1–3 with coxae, dactylus and unguis combined shorter than, or as long as, propodus, unguis shorter than dactylus. Pereopods 4–6 without coxae, dactylus and unguis not fused to a claw, together longer than propodus, unguis shorter than dactylus. Pleopods present or absent. Uropods small (shorter than pleotelson). Exopod uniarticulate, endopod consisting of two articles.

Male. Unknown.

Type species. — *Robustochelia longa* Kudinova-Pasternak, 1983.

Gender of generic name. — Feminine.

Remarks. — All the characters and in particular the characteristic shape and chitinization of the cheliped seem to be stable in the genus.

This genus is closely related to *Leptognathia* and shares characters of the antennae and uropods. However, Kudinova-Pasternak (1983) was probably correct to establish a new genus on the basis of the body shape, the massive chelipeds, and the pereopods. The situation about the mouthparts needs to be resolved when more material becomes available.

Species currently assigned to this genus. — *R. angusticephala* Kudinova-Pasternak, 1986; *R. longa* Kudinova-Pasternak, 1983; *R. robusta* (Kudinova-Pasternak, 1970); *R. solida* sp. nov.

Robustochelia solida sp. nov.
(figs. 88, 89)

Material examined. — Holotype, non-ovigerous ♀ (USNM 1068930), body length 2.9 mm, DGoMB Sta. MT5-4. 1 non-ovigerous ♀ (USNM 1068931), DGoMB Sta. S40-2 (dissected).

Diagnosis. — Female. Carapace shorter than the lengths of pereonites 1 and 2 combined. Mandibular molar tapering into a small crushing area. Maxillipedal endites without medial seta but with two denticles distally. Pleopods absent.

Male. Unknown.

Etymology. — Named after the dense appearance of the chelipeds (Latin, solidus = dense). The name thus is an adjective agreeing in gender with the generic name.

Type locality. — DGoMB Sta. MT5-4, 27°20'1910"N 88°39'5716"W, depth 2263 m.

Description. — Adult female. Body (fig. 88A) about 12 times as long as wide (owing to the curved nature of the holotype, body morphometric calculations are based on the lateral view).

Carapace tapering in anterior direction. More than 1.5 times as long as wide. Shorter than the lengths of pereonites 1 and 2 combined. Eye-lobes absent.

Pereon lateral margins almost straight. Pereonite 1 wider than long. Pereonites 2–5 longer than wide.

Pleon short (including pleotelson about 0.2 times total body length). Pleotelson longer than the lengths of three free pleonites combined, apex blunt and covered by dorsal plate.

Antennule (fig. 88B) 0.67 times as long as carapace, with four articles. Article 1 shorter than rest of antennule, with seta proximally and ventral process. Article 2 longer than article 3, naked. Article 3 shorter than article 4, with one simple seta distally. Article 4 longer than article 2, with three thick and three short simple setae and one aesthetasc distally.

Antenna (fig. 88B) with six articles (see Remarks). About 0.8 times as long as antennule. Article 1 (see Remarks) wider than long, naked, not wider than other articles. Article 2 as long as article 3, with one simple seta dorsally. Article 3 shorter than article 5, with one simple seta dorsally. Article 4 longer than other articles, with one setulose seta medially, one simple and two setulose setae distally. Article 5 longer than article 2, with two simple setae distally. Articlc 6 minute, with four setae distally.

Mouthparts. Labrum (fig. 88C) finely setose, with flat apex. Mandibular molar tapering, but ending in a small crushing area ringed by dorso-lateral denticles and a ventral row of spines. Left mandible (fig. 88D) lacinia mobilis longer than, and as wide as, incisor, shaped as one spine; incisor broad, with four dorsal denticles. Right mandible (fig. 88E, broken) incisor with medial

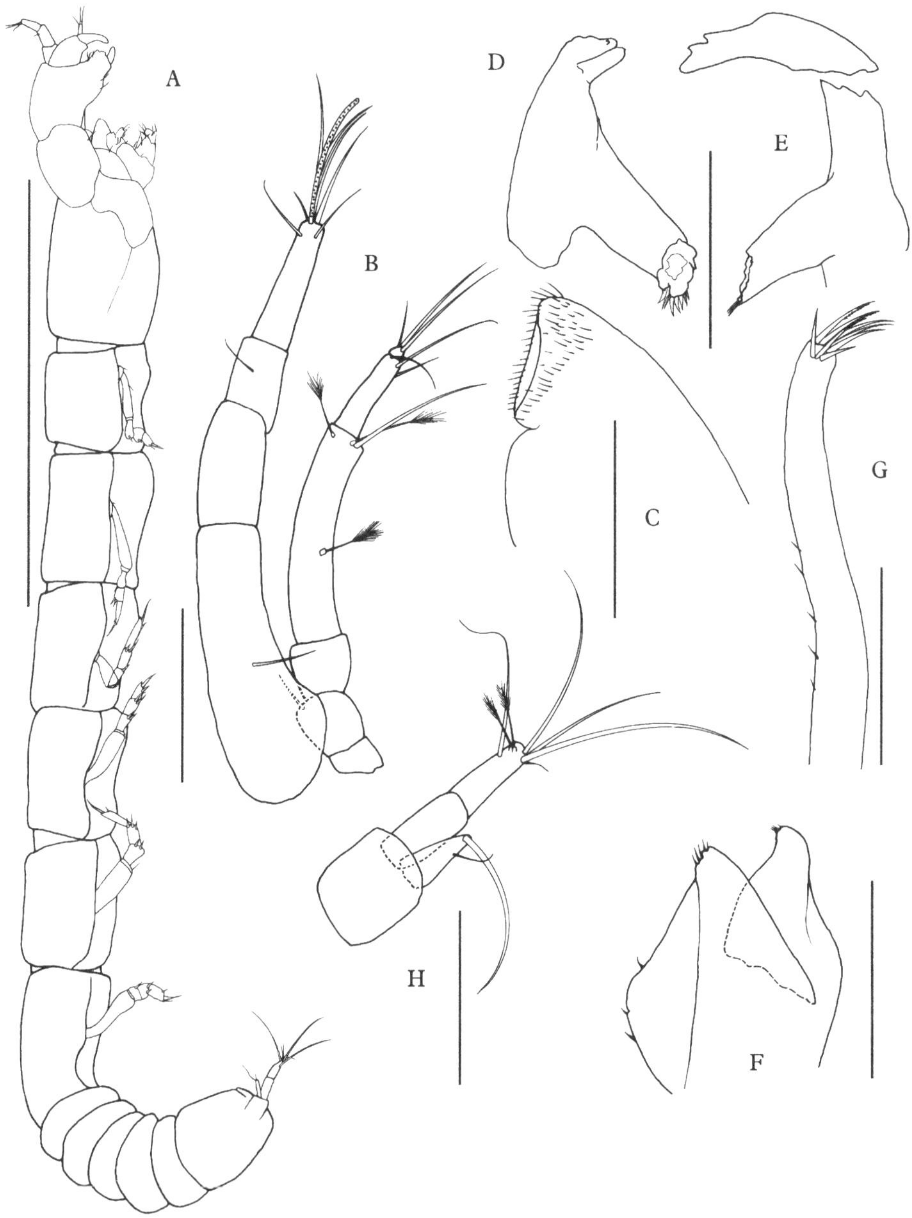

Fig. 88. *Robustochelia solida* sp. nov.; A, holotype, female, lateral view; B, antennae; C, labrum; D, left mandible; E, right mandible; F, labium; G, maxillule; H, uropod. Scale bars equal: A, 1.0 mm; B-C, 0.1 mm; D-H, 0.05 mm.

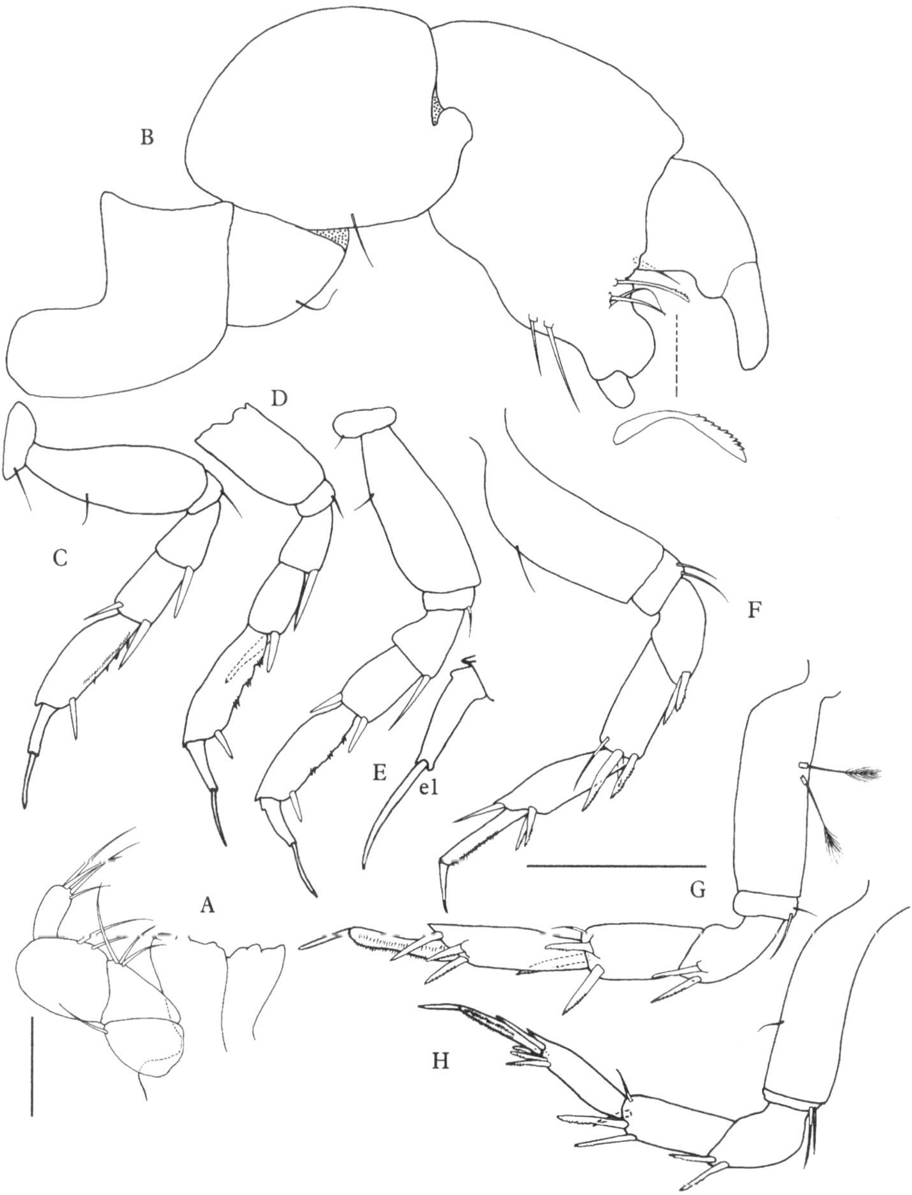

Fig. 89. *Robustochelia solida* sp. nov., female; A, maxilliped; B, cheliped; C, pereopod 1; D, pereopod 2; E, pereopod 3; e1, dactylus/unguis of same, enlargement; F, pereopod 4; G, pereopod 5; H, pereopod 6. Scale bars equal: A, 0.05 mm; B-H, 0.1 mm.

depression and dorsal denticle. Labium (fig. 88F) as wide as maxillipedal endites, outer corners setulated, widest medially. Maxillule (fig. 88G) endite with seven spiniform setae distally of which two are serrated; palp lost during dissection. Maxilla not recovered. Maxilliped (fig. 89A) endites with two denticles distally; palp article 1 with one outer seta, article 2 with three inner setae, article 3 with two long and one short setae on inner margin, article 4 with 4–5 setae. Epignath not recovered.

Cheliped (fig. 89B) basis shorter and narrower than carpus, attached to ventral sclerite, naked. Merus with one simple seta ventrally. Carpus shorter than propodus including fixed finger, with one simple seta ventrally and prominent distal process. Propodus massive, with one seta at dactylus insertion. Fixed finger with two setae ventrally and three on inner margin of which one is spatulate and serrated, with one large distal process. Dactylus longer than fixed finger, with medial process.

Pereopod 1 (fig. 89C) coxa with one simple seta. Basis longer than the three succeeding articles combined, with one dorso-medial seta. Ischium with one seta. Merus as long as carpus, widening distally, and with one spiniform seta ventro-distally. Carpus shorter than 0.5 times basis, with two spiniform setae distally. Propodus with proximal row of spinules, four single spines ventrally, one spiniform seta ventro-distally, and a dorsal spine. Dactylus and unguis combined about as long as propodus, not fused, dactylus shorter than unguis, and unguis attached in a socket-like structure on dactylus.

Pereopod 2 (fig. 89D) as pereopod 1 except: propodus with paired ventral spines.

Pereopod 3 (fig. 89E) as pereopod 2 except: dactylus (fig. 89e1) with blunt process dorsally.

Pereopod 4 (fig. 89F) has no visible coxa. Basis about as long as the three succeeding articles combined, with one seta dorso-proximally. Ischium with two simple setae. Merus with two serrated spiniform setae ventro-distally. Carpus with three serrated spiniform and one simple setae ventro-distally. Propodus longer than carpus, with two serrated spiniform setae ventro-distally, one stiff seta dorso-distally, and a dorsal spine. Dactylus and unguis combined longer than propodus, not fused, with paired rows of spinules. Unguis shorter than dactylus and not inserted into a socket.

Pereopod 5 (fig. 89G) as pereopod 4 except: basis with two setulated setae ventro-medially.

Pereopod 6 (fig. 89H) as pereopod 4 except: basis naked. Propodus with three spiniform setae ventro-distally, one stiff seta and one spine dorsally.

Pleopods not present in female.

Uropod (fig. 88H) about 0.65 times as long as pleotelson. Basal article naked and about as long as first endopod article. Endopod article 1 longer than article 2, naked; article 2 with four long simple, one short simple, and two setulose setae distally. Exopod hardly shorter than first endopod article, cone-shaped, with two dissimilar setae distally.

Distribution. — A very rare species, only found in the northern part of the Gulf. Bathymetric range: 2263–2975 m. See map 62.

Remarks. — This species shows a lot of characters remarkably similar with *Robustochelia longa* Kudinova-Pasternak, 1983, particularly in the body shape, structures of the cheliped, uropods, and pereopods. However, the radically different mouthparts (mandibles and maxillipedal endites) suggest that *R. solida* sp. nov. is not even to be placed in the same genus as *R. longa*. Unfortunately, the type material of *R. longa*, as well as all slides, appear to have been lost (M. Błażewicz-Paszkowycz, pers. comm.) so no resolution to this problem could be attained. As for *R. angusticephala* Kudinova-Pasternak, 1986, this is only incompletely described and nothing is known about the mouthparts. *R. solida* is only tentatively placed in *Robustochelia* and, as reflected in the diagnosis, this genus needs revision when more material becomes available.

KEY TO THE SPECIES OF *ROBUSTOCHELIA*, FEMALES ONLY (GLOBAL)

1. Carapace shorter than the lengths of pereonites 1 and 2 combined 3
– Carapace longer than the lengths of pereonites 1 and 2 combined 2
2. Body elongate (length/width ratio > 8.5) ....
.... *R. angusticephala* Kudinova-Pasternak, 1986
– Body not elongate (length/width ratio < 7) ....
.... *R. robusta* (Kudinova-Pasternak, 1970)
3. Mandibles pointed. Maxillipedal endites with medial seta but without denticles. Pleopods present .... *R. longa* Kudinova-Pasternak, 1983
– Mandibular molar tapering into a small crushing area. Maxillipedal endites without medial seta but with two denticles distally. Pleopods absent .... *R. solida* sp. nov.

Genus **Stenotanais** Bird & Holdich, 1984

Stenotanais Bird & Holdich, 1984, Zoologica Scripta, **13**: 298.

Diagnosis. — (Modified from Bird & Holdich, 1984.) Female. Body elongate (11.5–13 times as long as wide). Pereonites with lateral edges straight. Pereonites longer than wide. Pereonite 1 long (almost as long as carapace). Pleon short (about 0.15 times total body length). Antennule shorter than carapace, with four articles. Antenna with six articles. Mandibular molar pointed, with 2–4 terminal spines.

Maxillule with 8–9 spiniform terminal setae. Maxillipedal endites with two rounded processes and spiniform lateral corners. Cheliped carpus prominent and with carpal shield. Chela robust. Fixed finger with very robust spiniform setae ventrally. Pereopods 1–3 without coxae; basis wider than on pereopods 4–6; merus and carpus with robust spiniform setae. Pereopods 4–6 without coxae; dactylus and unguis not fused; dactylus with rows of small spines. Pleopods short and broad, with plumose or simple setae. Uropods prominent and folded back under themselves; endopod with two articles; exopod wider than endopod, with 1–2 articles, armed with one or two heavy, articulated setae distally.

Male. Unknown.

Type species. — *Stenotanais crassiseta* Bird & Holdich, 1984.

Gender of generic name. — Masculine.

Remarks. — The characters of the uropod setae and the folding, the cheliped carpus/fixed finger/dactylus, the wide basis of pereopods 1–3, the maxillipedal endites, and the elongate pereon are remarkably stable in this genus.

This genus cannot be mistaken for any other 'leptognathiid'. Particularly the 'reversed' character of the thickness of the pereopod basis is interesting. Normally, the basis is wider in pereopods 4–6 than in pereopods 1–3, or otherwise they are more or less identical. In *Stenotanais* this condition is reversed, and the anterior pereopods have a wider basis than the posterior ones. The reason for this 'reversal' cannot be determined at this stage. A good character for immediate genus identification without the use of a compound microscope, is the similar appearance of antennae and uropods.

Species currently assigned to this genus. — *S. crassiseta* Bird & Holdich, 1984; *S. hamicauda* Bird & Holdich, 1984; *S. macrodactylus* sp. nov.

Stenotanais macrodactylus sp. nov.
(figs. 90–92)

Material examined. — Holotype, non-ovigerous ♀ (USNM 1068932), body length 1.7 mm, DGoMB Sta. NB4-4. Paratypes: 1 non-ovigerous ♀ (USNM 1068933), body length 1.6 mm, DGoMB Sta. RW5-2. 1 non-ovigerous ♀ (USNM 1068934), DGoMB Sta. NB4-4 (dissected).

Additional material. — 1 non-ovigerous ♀, DGoMB Sta. WC12-2. 1 non-ovigerous ♀, DGoMB Sta. WC12-4. 1 ovigerous ♀, 2 non-ovigerous ♀♀, 1 manca, DGoMB Sta. NB2-2. 3 non-ovigerous ♀♀, DGoMB Sta. S37-3. 3 non-ovigerous ♀♀, DGoMB Sta. B1–2. 2 non-ovigerous ♀♀, DGoMB Sta. RW5-2. 2 non-ovigerous ♀♀, DGoMB Sta. NB4-2. 1 ♀, DGoMB Sta. RW3-1. 3 non-ovigerous ♀♀, DGoMB Sta. MT5-2. 3 non-ovigerous ♀♀, DGoMB Sta. NB5-2. 1 non-ovigerous ♀, DGoMB Sta. C7-4. 1 non-ovigerous ♀, DGoMB Sta. C4-4.

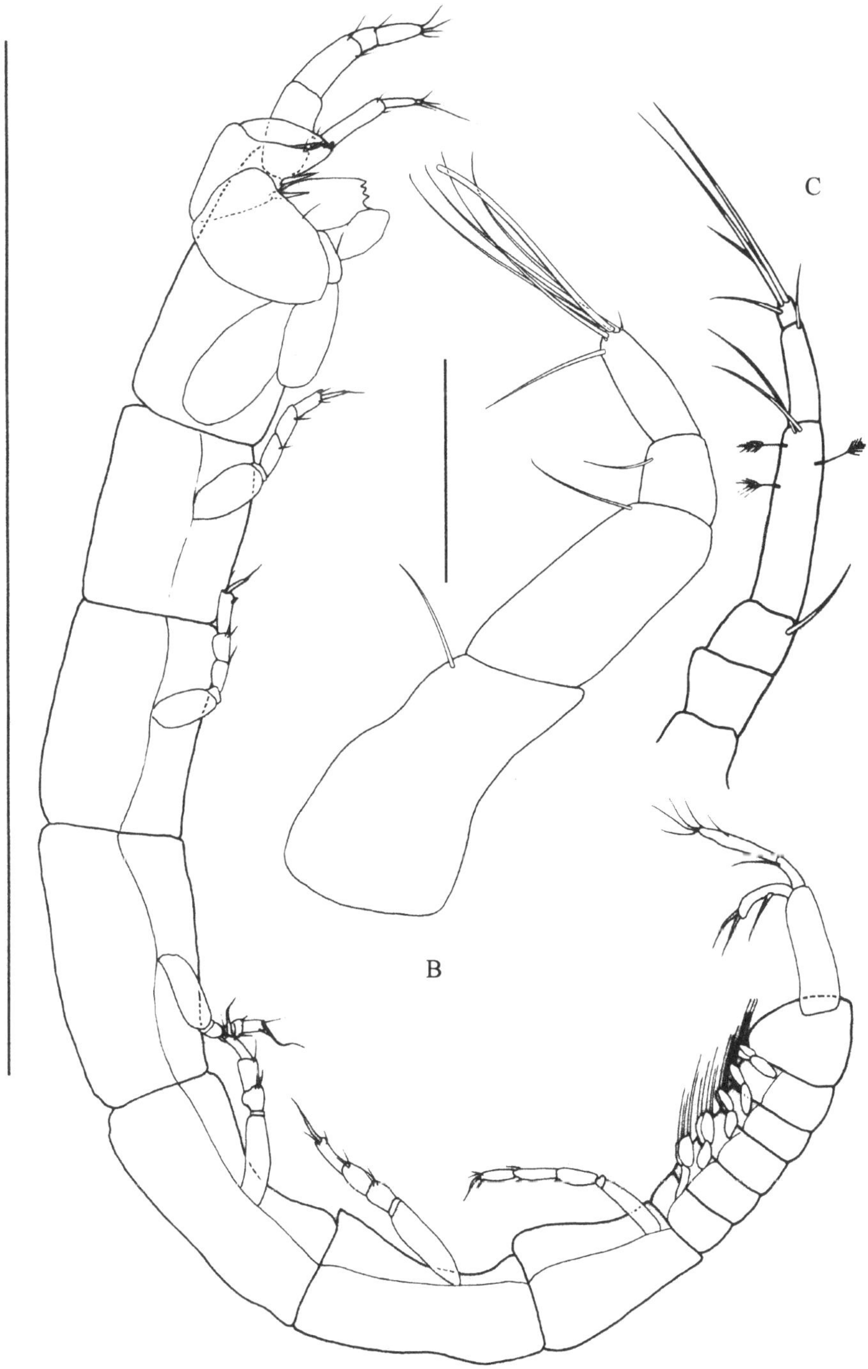

Fig. 90. *Stenotanais macrodactylus* sp. nov., female; A, holotype, lateral view; B, antennule; C, antenna. Scale bars equal: A, 1.0 mm; B-C, 0.1 mm.

Diagnosis. — Female. Uropod basal article longer than pleotelson and about as long as rest of uropod; exopod with two specialized articulated setae.

Male. Unknown.

Etymology. — Named after the heavy cheliped dactylus of this species (Greek, macro + dactylos = big + finger). The name thus is a noun in apposition with the generic name.

Type locality. — DGoMB Sta. NB4-4, 26°14'9712"N 92°23'5933"W, depth 2050 m.

Description. — Adult female. Body (fig. 90A) slightly dorsally flattened, very elongate, 12 times as long as wide.

Carapace elongate. Hardly longer than pereonite 2. Not wider than pereonites.

Pereon with pereonites all longer than wide. Pereonite 1 hardly shorter than pereonite 2. Pereonite 2 hardly shorter than pereonite 5. Pereonite 3 as long as pereonite 4. Pereonite 5 hardly shorter than pereonite 3. Pereonite 6 shortest.

Antennule (fig. 90B) hardly shorter than carapace. Article 1 barely longer than second article, with one simple seta distally. Article 2 longer than articles 3 and 4 combined, with one simple seta distally. Article 3 shorter than 0.5 times article 4, with one simple seta distally. Article 4 about 0.5 times as long as article 1, with six setae and one aesthetasc distally.

Antenna (fig. 90C) 0.75 times as long as antennule. Article 1 not wider than following articles, naked. Article 2 longer than article 1, naked. Article 3 as long as article 2, with one simple seta distally. Article 4 longer than other articles, with two simple and three small setulated setae distally. Article 5 longer than article 3, with one simple seta distally. Article 6 smallest, decreasing in width at midlength, with one simple seta at midlength and three simple setae distally.

Mouthparts. Labrum (fig. 91A) elongate, with ventral setules. Mandibular molar pointed, with 2–3 terminal spines. Left mandible (fig. 91B) incisor broad with two denticles; lacinia mobilis broad and separated from incisor by wide gap. Right mandible (fig. 91C) incisor broad with dorsal process. Labium (fig. 91D) lobes naked, wider at basis. Maxillule (fig. 91E) endite with eight slender spiniform setae distally, inner margin with setules; palp lost during dissection. Maxilla (fig. 91F) stout. Maxilliped (fig. 91G) basis with one long seta at palp insertion; endites almost as wide as basis, with two globular processes and corner with spiniform process; palp article 1 naked, article 2 with two simple inner setae, article 3 with three inner setae, article 4 with four inner and one outer setae. Epignath not recovered.

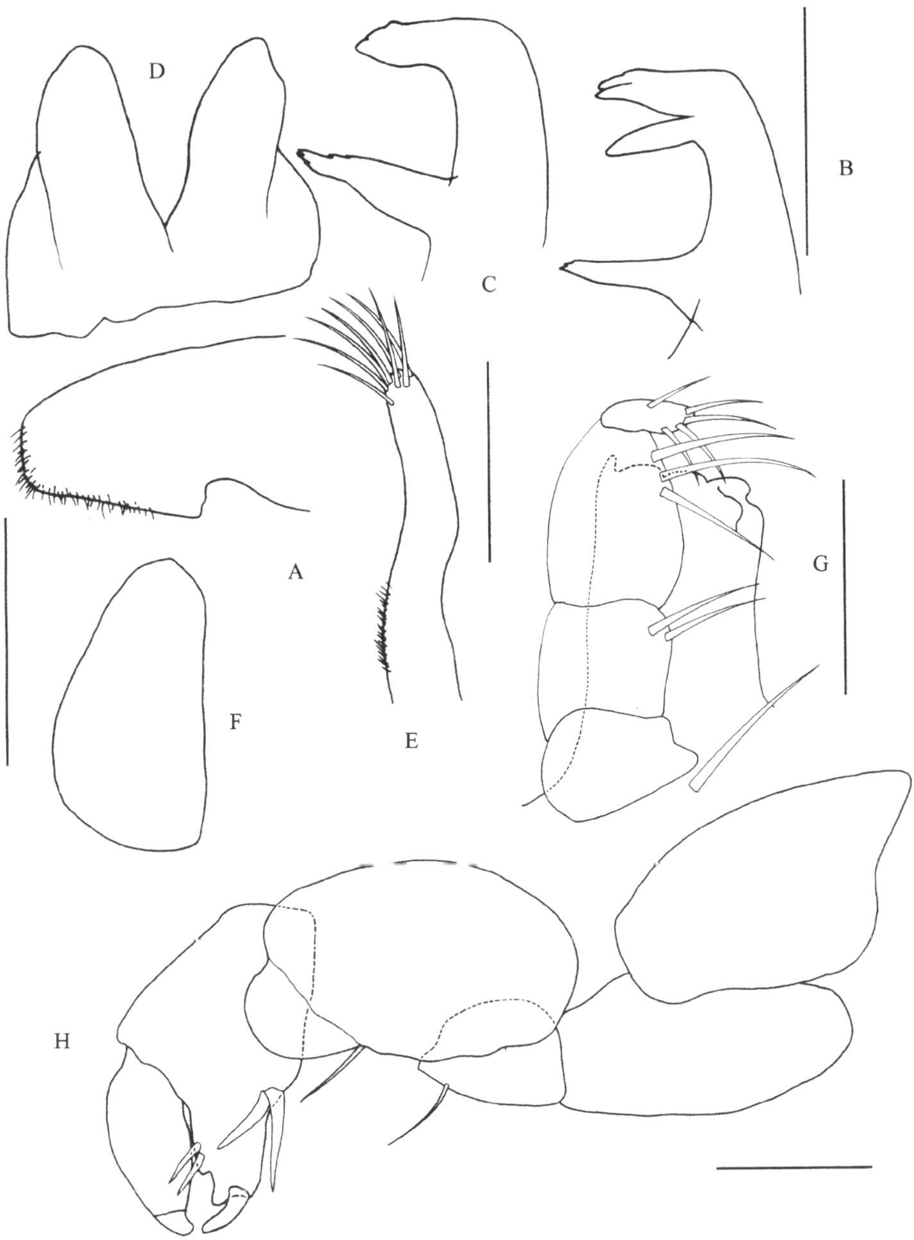

Fig. 91. *Stenotanais macrodactylus* sp. nov., female; A, labrum; B, left mandible; C, right mandible; D, labium; E, maxillule; F, maxilla; G, maxilliped; H, cheliped. Scale bars equal: A-G, 0.05 mm; H, 0.1 mm.

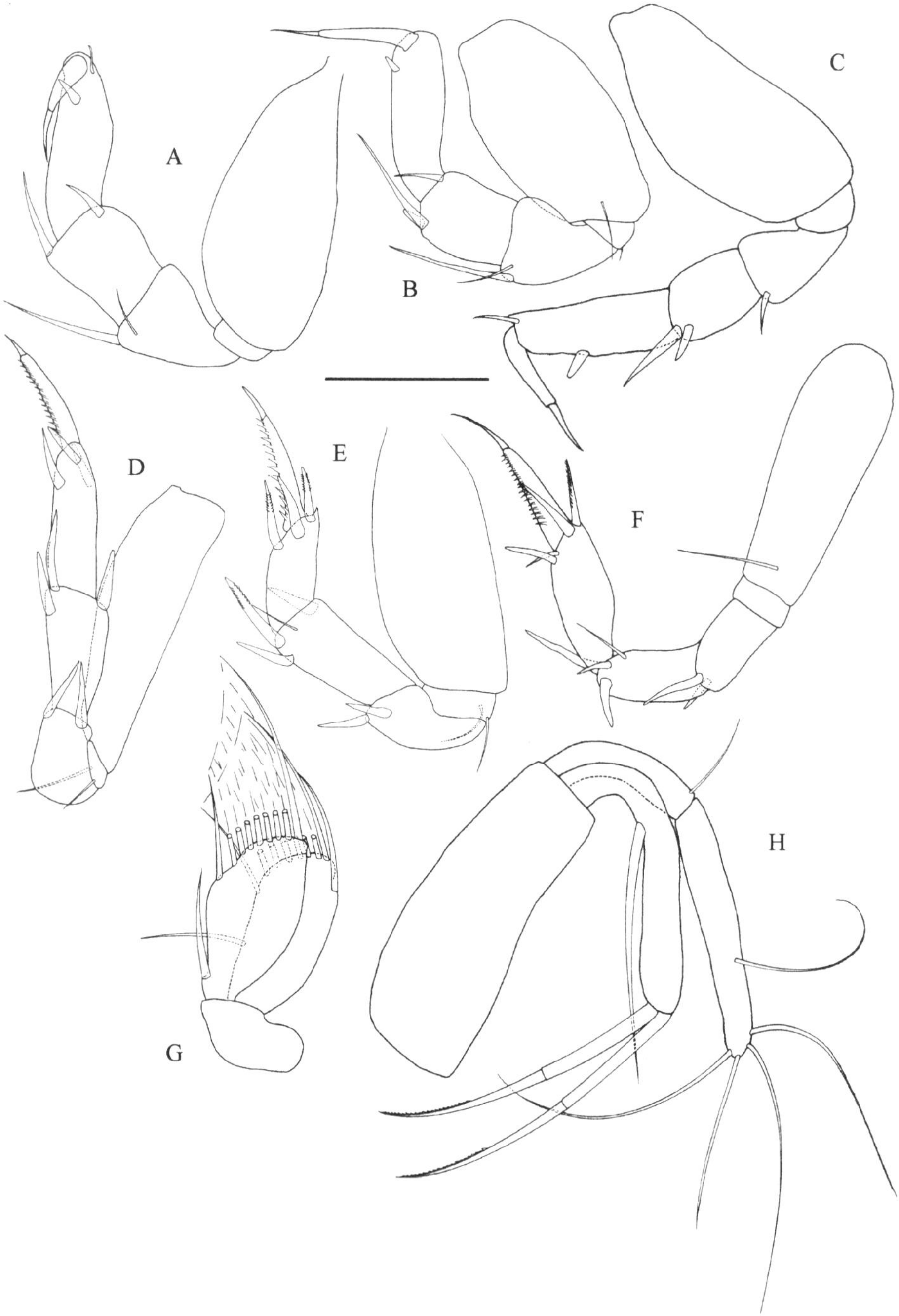

Fig. 92. *Stenotanais macrodactylus* sp. nov., female; A, pereopod 1; B, pereopod 2; C, pereopod 3; D, pereopod 4; E, pereopod 5; F, pereopod 6; G, pleopod; H, uropod. Scale bar equals 0.1 mm.

Cheliped (fig. 91H) basis slender (0.5 times as wide as carpus), attached to cephalothorax by large (larger than basis) elongate sclerite. Merus with one seta ventrally. Carpus longer than propodus including fixed finger, with one seta ventrally and large carpal shield. Propodus robust and naked. Fixed finger with two robust spiniform setae ventrally and two spiniform setae on inner margin, also with two prominent processes on inner margin. Dactylus very robust, as wide and as long as fixed finger, naked.

Pereopod 1 (fig. 92A) without coxa. Basis wider than those of pereopods 4–6, naked. Ischium naked. Merus with one long robust seta and one simple seta distally. Carpus as long as merus, with two spiniform setae distally. Propodus with one spiniform seta ventrally and one simple seta and a spine dorsally (not shown in figure, owing to propodus orientation). Dactylus and unguis combined shorter than propodus and not fused.

Pereopod 2 (fig. 92B) as pereopod 1 except: basis and ischium with one seta distally. Carpus with one short and one long robust setae, and one spiniform seta distally. Propodus without dorsal seta and apparently without dorsal spine.

Pereopod 3 (fig. 92C) as pereopod 1 except: merus with one seta distally.

Pereopod 4 (fig. 92D) basis narrower than on pereopods 1–3, naked. Ischium with two setae. Merus with two spiniform setae distally. Carpus with three spiniform setae distally. Propodus with three spiniform setae distally. Dactylus and unguis combined as long as propodus and not fused. Dactylus with row of conspicuous spines.

Pereopod 5 (fig. 92E) as pereopod 4 except: carpus with three heavy spiniform setae of which one is bi-serrate, and one simple seta distally. Propodus with three heavy spiniform setae of which two are bi-serrate, and a dorsal spine. Dactylus and unguis combined longer than propodus.

Pereopod 6 (fig. 92F) as pereopod 5 except: basis with one seta distally. Ischium naked. Merus with two spiniform setae distally. Propodus with one simple and three heavy spiniform setae of which one is bi-serrate, and a dorsal spine.

Pleopods (fig. 92G) short and wide. Basal article naked. Endopod with nine simple setae and one very robust proximal seta. Exopod with six simple setae and one robust outer seta.

Uropods (fig. 92H) long (as long as antennule) and folded back on themselves. Basal article as long as exopod and three times as wide, naked. Endopod with two partly fused articles; article 1 about 0.5 times as long as article 2, with one seta distally; article 2 with one seta at midlength and four setae distally. Exopod somewhat wider than endopod, with one specialized seta at midlength and two articulated, specialized setae distally.

Distribution. — Only found in the northern part of the Gulf. Bathymetric range: 1320–2387 m. See map 63.

Remarks. — This species can be recognized from the other species of *Stenotanais* by the long basal article of the uropod (longer than the pleotelson).

Key to the Species of *Stenotanais*, Females only (global)

1. Uropod basal article longer than pleotelson .... *S. macrodactylus* sp. nov.
– Uropod basal article shorter than pleotelson 2
2. Uropodal exopod bi-articulate, reaching only to about end of first endopod article. Endopod with one wide, flattened specialized seta distally. Exopod without specialized setae .... *S. crassiseta* Bird & Holdich, 1984
– Uropodal exopod uni-articulate, reaching at least to 0.75 of entire endopod length. Endopod without specialized setae. Exopod with two specialized flattened setae .... *S. hamicauda* Bird & Holdich, 1984

Filiform species

A large number of filiform species were found in the deep-sea samples from the Gulf of Mexico, and are known from several other localities as well. Such specimens may easily escape detection owing to their small size. Even more often, they are never worked upon by taxonomists, because of the difficulties involved in the dissection of such tiny, elongate specimens. Also, the thread-like body shape seems to take fixation very badly.

These specimens were only worked up in those, unfortunately rare, cases were the material was in usable condition (*Filitanais filiformis*, *Collettea elongata*, *Leptognathiella vermiformis*). Several specimens (species?) found in this study were not treated, as the material was in a very poor condition, which would not allow identification with the desired degree of confidence.

The filiform body shape does not appear to be restricted to one natural group, and thus several different evolutionary lineages seem to have adapted to the interstitial niche by developing a 'worm-like' habitus. Most of the filiform tanaidaceans found in the Gulf seem to be colletteids, but an anarthrurid type is reported from the North Atlantic (Bird, 2004b) and a leptognathiid component is known from Australia (K. Larsen, unpubl. data). It is expected that this filiform interstitial fauna exists almost everywhere, but has yet to be recorded and investigated. It is possible, that a significant proportion of the total tanaidacean diversity is still hidden in this 'filiformis group'.

TANAIDACEAN DISTRIBUTION

The distribution patterns displayed by the Peracarida, which lack a free-swimming larval stage, have been thoroughly discussed by many authors (Barnard, 1961, 1962; McCain, 1968; Hessler, 1970; Jones & Sanders, 1972; Laubitz, 1972; Menzies et al., 1973; Highsmith, 1985; Wilson & Hessler, 1987; Rex et al., 1997; Bamber, 1998). The general consensus is that most peracarid species have relatively limited distributions.

When extensive distributions are reported, the following explanations have been suggested: (1) either rafting (Highsmith, 1985; Alverez-Cadena et al., 1996; Knowles & Bell, 1998), or transport by other organisms, such as sea turtles (Caine, 1986; Peck, 1993); (2) human intervention, either by transport in ballast water (Jones, 1991), or as fouling organisms on ship hulls (Hoese, 1972; El-Nassry, 1973; Ghobashy et al., 1980; Shoukr et al., 1991); (3) a high degree of individual mobility, or else the ability to stay suspended in a moving water column (Mauchline, 1980).

The tube-dwelling Tanaidomorpha behave almost as sedentary organisms. Females apparently only rarely leave their tube voluntarily, and even the mancas leave the parental tube only in order to build their own tubes in the immediate surroundings, and frequently even in physical association with the tube of the parent (Bückle-Ramirez, 1965; Johnson & Attramadal, 1982a, b). Because of this sedentary life-style and limited swimming capacity (see under Locomotion), it has been assumed that tanaidaceans in general, and the tube-dwelling Tanaidomorpha in particular, have limited distributional ranges, even when compared with other peracarids (Sieg, 1983b).

However, a number of species have been reported as cosmopolitan, for example *Sinelobus stanfordi* (cf. Miyadi, 1938; Gardiner, 1975a) and *Expina typica* (cf. Sieg, 1979; Thurston et al., 1987). Other species are reported with a bipolar (antitropical) distribution, for example *Leptognathia breviremis* (cf. Schmidt & Brandt, 2001a). The many widely reported species (see tables IV and V, maps 64 and 65) suggest a very effective dispersal power that does not conform well to what is expected from a semi-sedentary peracarid.

Data collected during this study are not in good accordance with either view. The distribution patterns found have to be evaluated at both species and genus level, and be separated into a deep-sea and a shallow-water component, before general conclusions can be drawn.

TABLE IV
List of cosmopolitan species in the Tanaidacea

Species	References
Expina typica Lang, 1968	Sieg, 1979
Collettea minima (Hansen, 1913)	Kudinova-Pasternak, 1978; this study
Collettea cylindrata (G. O. Sars, 1882)	Kudinova-Pasternak, 1970, 1973; Larsen, 1999a, 2000a
Leptochelia dubia (Krøyer, 1842)	Sieg, 1986b
Leptognathia brevimanus (Lilljeborg, 1864)	Sieg, 1986b
Paraleptognathia gracilis (Krøyer, 1842)	Sieg, 1986b
Paraleptognathia longiremis (Lilljeborg, 1864)	Sieg, 1986b
Neotanais affinis Wolff, 1956	Gardiner, 1975b; Sieg, 1986b
Neotanais hadalis Wolff, 1956	Gardiner, 1975b
Mirandotanais vorax Kussakin & Tzareva, 1974	Kussakin & Tzareva, 1974; Sieg, 1983c; Schmidt & Brandt, 2001a; this study
Agathotanais ingolfi Hansen, 1913	Larsen, 1999a
Sinelobus stanfordi (Richardson, 1901)	Gardiner, 1975a

TABLE V
List of known bipolar (antitropic) species in the Tanaidacea

Species	References
Apseudes spectabilis Studer, 1883	Lang, 1968; Shiino, 1970
Cryptocopoides arcticus (Hansen, 1886)	Kudinova-Pasternak, 1978
Tanaissus lilljeborgi Hansen, 1913	Sieg, 1986b
Meromonakantha macrocephala Hansen, 1913	Sieg, 1986b
Paranarthrura insignis (Stebbing, 1891)	Sieg, 1986b
Pseudotanais abyssi (Hansen, 1913)	Sieg, 1986b
Paraleptognathia gracilis (Krøyer, 1842)	Sieg, 1986b
Leptognathia breviremis (Lilljeborg, 1864)	Sieg, 1986b; Schmidt & Brandt, 2001a

Species level distributions in shallow waters

Most tropical shallow-water species display rather narrow distribution patterns (Sieg, 1983a, b). This is perhaps not unexpected. According to Rapoport's Rule (Rapoport, 1982), narrow distributions are more common among tropical species than among cold-climate species (Grant, 1984; Bot et al., 1989; Stevens, 1989, 1992; Palumbi & Wilson, 1990; Palumbi, 1992; Gage & May, 1993). Outside the tropics, many shallow-water tanaids display wide distributions.

Rafting or human intervention

When presented with extensive distributions, one can argue in most cases for dispersal by rafting or human intervention. Several species are known to cling on to, or building tubes on, floating objects, ship hulls, and large marine organisms (Sieg, 1980a; Highsmith, 1985; Caine, 1986; Shoukr et al., 1991). Introductions via ballast water have been confirmed for at least one species, *Tanais dulongii* (cf. Jones, 1991) but may account for the transport of several species, yet to be documented.

High mobility or transport by near-shore currents

Tanaidaceans are not great swimmers but as noted above (see Locomotion) many shallow-water species are capable of leaping into suspension if necessary. Tanaidaceans have been identified in some plankton surveys (Stephensen, 1913, 1915; Alldredge & King, 1980; McWilliam et al., 1981) whereas other surveys have revealed a conspicuous lack of tanaidaceans (Hansen, 1895; Hobson & Chess, 1976, 1978; Valentin, 1978; Sieg, 1982c). They are apparently also a rare component of the night-time surface plankton (Hansen, 1985). Alvarez-Cadena et al. (1996) suggested that the occurrence of tanaidaceans in plankton surveys is mainly due to contamination from the substrate, or floating seaweed. During this study it was observed, however, that in intertidal habitats some tanaidaceans (e.g., *Hargeria rapax*) were represented in the night-time plankton.

Thus, in shallow water it is possible that tanaidaceans could be transported via currents, and this may contribute to the distribution of shallow-water tanaidaceans.

Species level distributions in the deep sea

The distribution of the deep-water Peracarida is of particular interest, since there is little chance of rafting or human intervention, which thereby excludes these factors as means for dispersal.

Strangely enough, the deep-sea species of tanaidaceans appear to have far more extensive distributions than most shallow-water species, despite the limited sampling effort in the deep sea, which constitutes 67% of the surface of the earth. Cosmopolitan and bipolar (or antitropical) distributions are reported for some species (see chart 1, tables IV and V, maps 64 and 65). The reasons for these wide distributions are not obvious, since most of the deep-water species lack even the modest swimming capacity of the shallow-water forms.

High mobility or transport by oceanic currents

Most, if not all, deep-sea tanaidomorphans are tube-dwellers like their shallow-water counterparts, but while both sexes of most shallow-water taxa possess pleopods, most females of deep-sea taxa do not have functional pleopods. Males of the deep-sea species do possess pleopods but in many (widespread) deep-sea taxa, these are reduced or even rudimentary. Thus, most deep-sea species lack the capacity to swim or stay suspended long enough to be transported by currents. It should be mentioned, however, that some Tanaidacea have been found in one deep-sea plankton survey (Kudinova-Pasternak, 1970). The deep-sea taxa do not display any morphological adaptations to increase mobility.

Biotic factors affecting dispersal

Several factors, both biotic and abiotic, may influence species distributions. Regarding biotic factors as population structure, inbreeding, sediment type (or food supply associated with sediment type), and predation, we have only limited information.

Inbreeding. — The Tanaidacea, with their limited potential for dispersal and low population densities in the deep sea, could be susceptible to inbreeding (Sanders, 1977; review by Knowlton & Jackson, 1993) but may be able to reduce loss of heterozygosity by the "stepping stone" model (Felsenstein, 1975; Slatkin, 1985). In some species, the male's swimming capacity can contribute to a relatively fast colonization of vacant substrata (Bird & Holdich, 1989b). While males are unknown for many genera, some males are well adapted for active swimming, e.g., in some deep-sea genera such as *Leptognathia*, *Peraeospinosus*, and *Pseudotanais* (cf. Sieg, 1977; Wilson, 1987). In most deep-sea genera males are either not known, or in others, such as *Agathotanais*, *Paragathotanais*, *Collettea*, *Paranarthrura*, and *Parafilitanais* (cf. Bird & Holdich, 1989a; Larsen, 1999a, b, 2000b, 2002a), the males have reduced or non-functional pleopods and thus no swimming capacity. Also, even if male mobility contributes to gene flow, male tanaidaceans cannot make stable colonies alone, since protandry is unknown in the Tanaidacea, while protogyny does occur (Sieg, 1984; Larsen, 2001).

Sediment type and seasonality. — Sediment type and seasonality are not really biotic factors, but the food supply associated with it is. Sediment type and seasonality are well known to have a marked influence on the distribution of benthic organisms (Gray, 1974; Rhoads, 1974; Haedrich et al., 1975; Etter & Grassle, 1992; Snelgrove & Butman, 1994; Rex et al., 1997). Whether or not tanaid distribution patterns are affected by these factors and to what extent, remains to be examined. No obvious correlation between deep-sea taxonomic groups and sediment type/size was noted during this study. Food supply in the deep sea experiences seasonal pulses (Lauerman & Kaufmann, 1998) but it is not known to what extent these influences would affect distribution patterns, if any.

Predation and mortality. — Intensive predation or mortality caused by disturbance will cause a population to be composed of predominantly young individuals (Grassle & Sanders, 1973). A particularly large proportion of young individuals were neither found in the studies of Neotanaidomorpha (cf. Gardiner, 1975b; Wolff, 1977) nor in the other suborders examined during this study. Predation thus seems unlikely to be a factor that limits tanaid distribution in the deep sea. The rate of mancas versus adults in the shallow areas is much higher, and predation is a much more important factor in this environment, as also confirmed by direct observations (Menge, 1976; Kneib, 1985, 1988).

Abiotic factors affecting dispersal

Abiotic factors in the deep sea include physical parameters such as topography, currents, temperature, salinity, and hydrostatic pressure. Most of these factors are generally very stable in the deep sea. Indeed, the classical "time stability hypothesis" (Sanders, 1969) maintains that the high diversity that occurs in the deep sea is there, owing to the extreme physical stability of this environment over long periods of time. However, some abiotic factors could still limit dispersal in several ways.

Temperature and salinity. — Temperature and salinity are very stable in the deep sea relative to the conditions in shallow waters. Minor variations in salinity are normally not an obstacle for the distribution of benthic peracarids, whereas temperature may be (Gerken & Watling, 1997; Larsen, 1998). Seasonal variations in salinity are observed to have little effect on population densities (Kneib, 1999).

Upwelling and benthic storms. — Upwelling may affect species richness and abundance (Rowe et al., 1982; Cosson et al., 1997) but is not necessarily a barrier for distribution or temporary small-scale colonization. The effects of strong near-bottom currents are not clear, but tanaid communities have been shown to be relatively stable, albeit at low levels, both in terms of species richness and relative abundance, in habitats disturbed by benthic storms (Reidenauer & Thistle, 1985). As a means of transportation, benthic storms may have an effect, but mortality of the relatively fragile tanaids is likely to be high during such a catastrophic event.

Pressure gradients. — Most clear-cut abiotic barriers in the deep sea involve pressure gradients associated with trenches or ridges. In order for any benthic organism to cross barriers such as a trench or a ridge, pressure tolerance is required. Most invertebrate species on continental slopes and rises have narrow, band-shaped distributions (Ekman, 1953; Hartman & Fauchald, 1971; Jones & Sanders, 1972; Grassle et al., 1975; Marshall, 1979; review by Rex, 1981; Rice et al., 1990) and this is particularly true for peracarids (Sanders & Hessler, 1969; Sanders, 1977; Wilson & Hessler, 1987; Poore et al., 1994). However, most species of the Tanaidacea appear to have extensive vertical ranges. Sieg (1986) reported that approximately 15% of all tanaid species have vertical distributions over around 6000 m. Species with large vertical ranges were also reported by Holdich & Bird (1985). Many species examined in detail during this study were also found to have surprisingly large bathymetric ranges, often in the order of 2000 m or more in the Gulf of Mexico and 4000 m or more in deeper environments (see charts 1 and 2).

Explanations for wide species distributions

Previous hypotheses, proposed to explain observed wide distributions (rafting, human intervention, current transport), cannot explain the tanaid distributions observed. The question remains how so many species with wide distribution records can have attained this, despite all biotic and abiotic barriers, if all those hypotheses fail. The data are insufficient to give a completely satisfactory explanation for the tanaid distribution pattern at present, but some plausible hypotheses can be posed.

Misidentification

Misidentification of species is the most plausible explanation for the observed, wide species-level distributions of deep-sea tanaids. The Tanaidomorpha (the bulk of the deep-sea species) are small, difficult to identify, and their systematics are very unstable (Larsen, 2001; Larsen & Wilson, 2002). Other widespread invertebrate species have often been split into several species when more extensive material became available (Wilson, 1983; Wilson & Hessler, 1987; Grassle et al., 1990). This occurs even with what were considered to be well-known bioindicator species (Lobel et al., 1990).

The presence of cryptic or sibling species complexes presents challenges to taxonomists (Grassle & Grassle, 1976; Palumbi & Benzies, 1991; Knowlton, 1993) and the knowledge of the existence of such cryptic species complexes among the Tanaidacea (cf. Gardiner, 1975b; Larsen, 2001) supports the misidentification hypothesis. In order to determine the extent of tanaidacean species distribution, one cannot rely completely on morphological characters. There are some species, known to be separate as a result of genetic examination, that cannot be conclusively identified by morphological observation and, since such cryptic species can even be sympatric (Larsen, 2001), the reported, wide species distributions thus seem even more questionable.

Pressure tolerance

Misidentification may thus explain wide distributions but may not be the reason for the reported extensive bathymetric range. If wide vertical distributions are reported in concert with wide horizontal distributions, then misidentification may indeed be a feasible explanation. However, when the same species is collected from considerably different depths at close localities (i.e., down a slope), misidentification becomes a less satisfactory explanation. Though not impossible, as seen in the polychaete *Chaetozone setosa* Malmgren, 1867, which was reported to have a vertical distribution of 4500 m but was later identified as a complex of at least three species (Christie, 1985).

One possible explanation for reported wide bathymetric distributions may reside in hydrostatic pressure tolerance. There are no physiological limitations that should exclude extensive vertical distributions. It has been shown that the enzymes of deep-sea organisms are not necessarily pressure-dependent (Siebenaller, 1978; Siebenaller & Somero, 1978). It is, therefore, possible that the level of adaptation necessary for pressure tolerance is low. Furthermore, a number of invertebrates undergo migrations over large pressure gradients either as eggs (Hofmann et al., 1992), larvae (Bouchet & Waren, 1979, 1985; Rex & Waren, 1982; Domansky, 1984; Lutz et al., 1984; Killingley & Rex, 1985), or adults (Sanders & Grassle, 1971; Grassle et al., 1975; Angel, 1997). Even diurnal migrations of morc than a 1000 m have been reported in cephalopods (Denton, 1974). Angel (1997) also states that hydrostatic pressure does not appear to be the sole reason for the exclusion of shallow-water taxa from the deep sea and vice-versa.

An individual tanaid would rarely experience a significant pressure change, owing to the limited mobility and sedentary life style of the tanaidomorphan tanaids (see mobility) and the vertical expansion of a population through generations would cause little if any adverse affects. The bathymetric ranges of the tanaidomorphans examined during this study indicate a substantial pressure tolerance. Extensive depth changes such as at trenches or ridges are thus negotiable for tanaidaceans and could, in turn, allow for wide horizontal distributions, as vertical barriers can be crossed in this manner.

Behavioural modifications

A behavioural change towards higher dispersal (by walking, not swimming) of the mancas is here hypothesized for the deep-sea tanaidomorphans, although no morphological modification for greater mobility could be found in the material examined during this study.

(1) Almost nothing is known about breeding by the deep-sea taxa but competition for food between siblings and their parent in a low-food environment, such as the abyssal plain, would require a certain minimum distance before settlement. Such (relatively) higher mobility and settling patterns are seen in several other deep-sea invertebrate taxa (Sokolova, 1959; Rowe & Menzies, 1969; Sanders & Grassle, 1971; Haedrich et al., 1975; Menge, 1976; Rex, 1977; Sanders, 1977; Wilson, 1981; Grassle, 1989).

(2) Organisms capable of feeding over large distances have an advantage in a low-food environment like the deep sea (Jumars & Fauchald, 1977), a strategy not conforming well to the almost sessile behaviour observed in shallow-water tanaidomorphans.

(3) The low dispersal capacity owing to a sedentary life style would also predispose tanaidaceans to high speciation, leading to high local endemism (Sanders, 1977; Scheltema, 1977, 1978; Valentine & Jablonski, 1983). The tanaidaceans have previously been suggested to have such a high local endemism (Sieg, 1983b; Holdich & Bird, 1986; Bamber, 1997; Bamber & Bird, 1997), but this is contradicted by the wide distributions reported. A higher mobility of deep-sea tanaidaceans may contribute to the explanation of why organisms with such low potential for gene flow have such an apparently low rate of identifiable speciation.

It is noteworthy that two of the most widely distributed deep-sea species of tanaidaceans, *Expina typica* and *Mirandotanais vorax*, may be parasitic on holothurians (Kussakin & Tzareva, 1974; Sieg, 1979; Thurston et al., 1987; this study). Holothurians themselves are reported to have wide distributions owing to their larval dispersal stage (Madsen, 1961; Hansen, 1967, 1975; Doyle, 1972). How this could affect the distribution of the tanaid is not clear, since a tanaidacean could not possibly parasitize a holothurian larva.

TABLE VI

List of presumed cosmopolitan genera in the Tanaidacea

Genus	References
Anarthrura G. O. Sars, 1882	This study
Apseudes Leach, 1814	Shiino, 1970
Heterotanais G. O. Sars, 1882	Bamber, 1998
Kalliapseudes Stebbing, 1910	Shiino, 1966; Bamber, 1993
Neotanais Beddard, 1886	Lang, 1956c; Gardiner, 1975b; Larsen, 1999a
Sinelobus Sieg, 1980	Gardiner, 1975a
Agathotanais Hansen, 1913	Larsen, 1999b
Carpoapseudes Lang, 1968	Larsen, 1999c
Collettea Lang, 1973	Larsen, 2000b
Leptochelia Dana, 1849	Lang, 1973; Larsen & Rayment, 2002
Leptognathia G. O. Sars, 1882	This study
Leviapseudes Sieg, 1983	This study
Paranarthrura Hansen, 1913	This study
Paratanais Dana, 1852	Larsen, 2000a
Synapseudes Miller, 1940	This study

Explanations for wide distributions at the generic level

If misidentification explains most widespread species distributions, it cannot explain the distribution patterns at the generic level, unless the genera themselves are artefacts (a consideration by no means rejected here). Several shallow-water genera and most deep-sea genera have been identified as having extensive distributions (see table VI). In fact, most deep-water genera seem to have representatives in all major oceans (Larsen, 1999a, b, 2000a, 2002a, this work) and while this may be true for a number of macrofaunal taxa, it is still unsettling to find almost completely identical species in the Tasman Sea and the Davis Strait.

Biological success

Biological success is another possible explanation for wide distributions in the deep sea. The sheer number (of both species and specimens), of deep-sea tanaidomorphans indicates that they are very successful in the deep sea and probably occupy well-defined microhabitats (Grassle & Maciolek, 1992; Gage, 1996) to accommodate their diversity.

Tanaidomorphans have some special adaptations to the deep-sea habitats. Most importantly they are small, especially the deep-sea species. Their small size would allow them to employ a special feeding strategy by selecting the most nutrient-rich particles from the detritus. Such a strategy is an advantage in a food-impoverished environment as the deep sea (Jumars et al., 1990).

Although the morphology of deep-sea genera from different oceans indicates a close relationship, the genetic distance may be (and probably is) significant, if distance prevents gene flow between populations. Future research, utilizing molecular techniques that compare members of the same genera from different oceanic regions, may reveal substantial genetic distances and identify some of these genera as unnatural groupings, resulting from morphological conservatism and hence plesiomorphic limitations.

Unfortunately, a number of logistic problems are inherent in the application of genetic studies in the deep sea. Deep-sea sampling is notoriously expensive and usually involves cooperation with specialists on many different groups. The problem with this is that sampling for taxonomic studies involving assignment of specimens for museum collections means that the material has to be fixed with formalin in order to be suitable for storage. Fixing with formalin renders the material useless for most molecular studies and, owing to the rapid decay of deep-sea material, fixing must occur within a relatively short time (within hours in warmer climates). These problems, combined with the problems of sorting individual groups of small Crustacea while at sea, makes standard deep-sea sampling unsuitable for obtaining material for molecular studies.

Conclusions

The reported wide distribution patterns of deep-sea Tanaidacea are here attributed to the following reasons: (1) at the species level, cryptic species have been misidentified; (2) at the generic level the conservative morphology of tanaidaceans has led to unnatural groupings (i.e., non-monophyletic taxa); (3) the Tanaidacea, or at least the tanaidomorphans, are highly successful in the deep sea; (4) the Tanaidacea have a relatively high pressure tolerance when given time to adapt; and (5) deep-sea tanaidaceans are under higher selection pressure to disperse than are shallow-water taxa.

TANAIDACEAN DISTRIBUTION PATTERNS IN THE GULF OF MEXICO

The distribution patterns of Tanaidacea in the Gulf of Mexico revealed during this study are unremarkable, and indicate that no biological barriers are present. The high diversity and endemism suggest that the fauna is well settled in a late state of succession.

Except for the apparently rare species that were found at only a few stations, most species showed a widespread distribution, conspicuously related only to sampling effort.

It is likely that most of the species encountered in this study will also be present in the southern part of the Gulf, but little sampling has been conducted in that area so far. Both the deep- and shallow-water tanaidacean genera and their numbers of species found in the Gulf of Mexico and the western Caribbean Sea, are listed in tables VII and VIII.

TABLE VII

List of shallow-water genera of Tanaidacea in the Gulf of Mexico/western Caribbean Sea

Taxon	Number of species	References
Ascumnella	1	Guţu & Heard, 2002
Atlantapseudes	1	Meyer & Heard, 1989
Apseudes	5	Richardson, 1903; Sieg, 1983a; Guţu, 1986; Hansknecht & Heard, 2001; Heard et al., 2004
Apseudomorpha	1	Heard et al., 2004
Araphura	1	Sieg & Dojiri, 1989
Bunakenia	2	Heard et al., 2004
Calozidion	3	Gardiner, 1973c; Guţu, 1984; Heard et al., 2004
Cycloapseudes	1	Gardiner, 1973c
Discapseudes	2	Băcescu & Guţu, 1975; Guţu & Heard, 2002
Halmyrapseudes	3	Băcescu & Guţu, 1974; Sieg et al., 1982
Hargeria	1	Harger, 1879
Heterotanais	1	Ogle, 1977
Hexapleomera	1	Moore, 1894
Iungentitanais	1	Kudinova-Pasternak, 1975; Sieg & Heard, 1988
Kalliapseudes	3	Ogle, 1977; Sieg, 1982c; Drumm, 2003
Leptochelia	4	Krøyer, 1842; Stebbing, 1896; Makkaveeva, 1968; Lang, 1973; Ogle, 1977; Heard et al., 2004
Metapseudes	1	Ogle, 1977
Nototanoides	1	Sieg & Heard, 1985

TABLE VII
(Continued)

Taxon	Number of species	References
Pagurapseudes	3	Bouvier, 1918; Guţu & Gomez, 1976; Ogle, 1977; McSweeny, 1982
Pakistanapseudes	1	Heard et al., 2004
Parapseudes	1	Makkaveeva, 1968; Ogle, 1977
Paratanais	1	Ogle, 1977
Protanaissus	1	Larsen & Heard, 2004b
Pseudoleptochelia	1	Lang, 1973; Heard et al., 2004
Pseudonototanais	1	Lang, 1973
Pseudotanais	1	Sieg, 1977
Psammokallioapseudes	2	Silva Brum, 1978
Saltipedis	2	Hansknecht et al., 2001
Sinelobus	1	Sieg, 1980a; this study
Sphyrapoides	2	Guţu & Iliffe, 1998; Guţu & Heard, 2002
Synapseudes	1	Gardiner, 1973c
Tanaissus	2	Wallace, 1919; Heard et al., 2004
Tanapseudes	1	Hansknecht et al., 2002
Teleotanais	1	Lang, 1956b
Typhlapseudes	1	Beddard, 1886
Zeuxo	2	Sieg, 1980a
Total	58	

TABLE VIII
List of deep-sea genera of Tanaidacea in the Gulf of Mexico/eastern Caribbean Sea

Genus	Number of species	References
Agathotanais	1	Lang, 1971a
Anarthruropsis	2	Lang, 1968; this study
Araphura	4	Larsen, 2003b; this study
Araphuroides	1	This study
Armaturatanais	1	This study
Apseudes	2	Guţu, 1986; Hansknecht & Heard, 2001
Atlantapseudes	1	Meyer & Heard, 1989
Bathyleptochelia	1	Larsen, 2003b
Carololangia	1	Gardiner, 1975b
Carpoapseudes	1	Lang, 1968
Caudalonga	1	This study
Chauliopleona	1	This study
Cetiopyge	1	Larsen & Heard, 2002
Coalecerotanais	1	Larsen, 2003b
Collettea	2	Larsen, 2002a; this study
Spinitanaopsis	1	Larsen, 2003b

TABLE VIII
(Continued)

Genus	Number of species	References
Cryptocope	2	Kudinova-Pasternak & Pasternak, 1978; J. McLelland, pers. comm.
Dactyloprion	1	Guţu, 2002
Expina	1	Lang, 1968
Filitanais	1	This study
Insociabilitanais	1	This study
Isopodidus	1	Larsen & Heard, 2002
Iungentitanais	1	Sieg & Heard, 1988
Kudinopasternakia	2	Viskup & Heard, 1989; Guţu & Heard, 2002
Leptognathia	2	This study
Leptognathiella	2	This study
Leptognathiid males not coupled with females	(9)	This study
Leviapseudes	3	Lang, 1968; Kudinova-Pasternak, 1966a; this study
Libanius	1	Lang, 1971b
Meromonakantha	1	This study
Mesotanais	2	Sieg & Heard, 1989
Neotanais	5	Wolff, 1956a, b; Gardiner, 1975b; Messing, 1977; Larsen & Hansknecht, 2003
Nototanoides	1	Sieg & Heard, 1985
Parafilitanais	1	Larsen, 2002a
Paragathotanais	7	Lang, 1971a; Kudinova-Pasternak, 1978; this study
Paranarthrura	5	Lang, 1971b; this study
Parapseudes	1	Ogle, 1977
Parapseudotanais	1	J. McLelland, pers. comm.
Paratyphlotanais	1	Kudinova-Pasternak & Pasternak, 1978
Pectinapseudes	2	Guţu, 2002
Pseudoarthrura	1	This study
Pseudotanais	8	Kudinova-Pasternak & Pasternak, 1978; Sieg & Heard, 1988; J. McLelland, pers. comm.; this study
Pseudosphyrapus	1	This study
Robustochelia	1	This study
Stenotanais	1	This study
Sphyrapoides	1	Guţu & Iliffe, 1998
Tanaella	2	Larsen & Heard, 2004a
Tanaopsis	1	Lang, 1967
Tumidochelia	1	Knight et al., 2003
Typhlotanais	2	Kudinova-Pasternak & Pasternak, 1978
Venusticrus	1	Gardiner, 1975b
Unidentified tanaidomorphan taxa	27	
Total	123	

MAPS 1–65

CHARTS 1–2

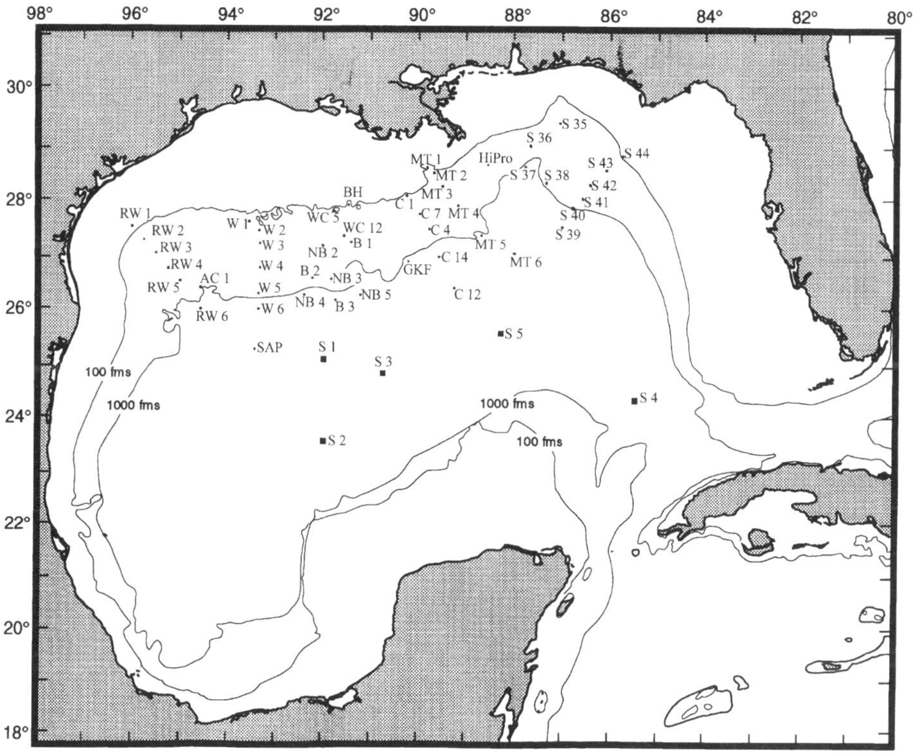

Map 1. DGoMB stations.

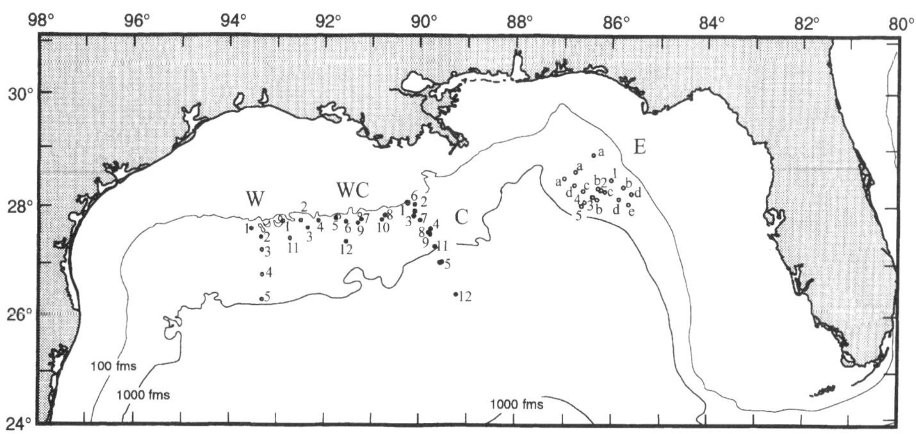

Map 2. NGoMCSS stations.

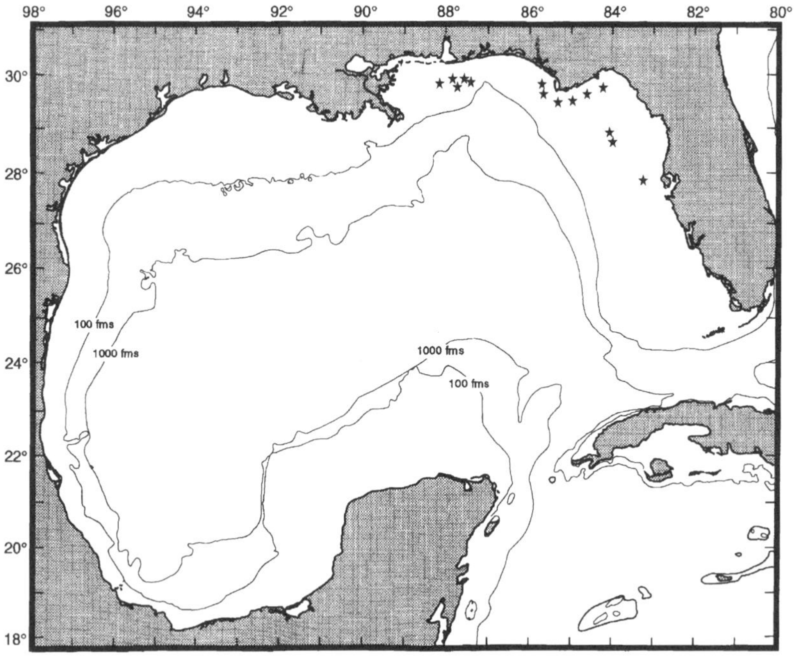

Map 3. Known distribution of *Apseudes olimpiae* Guţu, 1986.

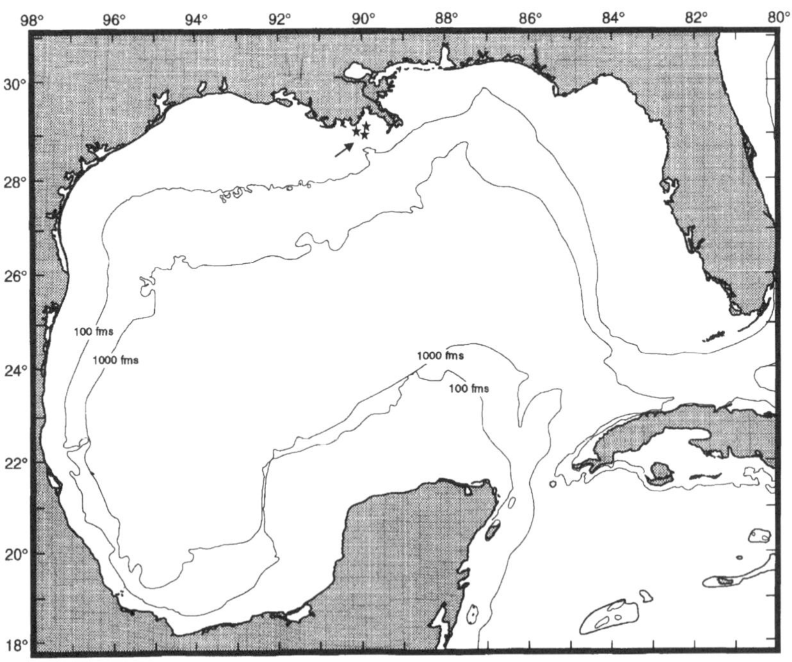

Map 4. Known distribution of *Apseudes* cf. *spinosus* M. Sars, 1858.

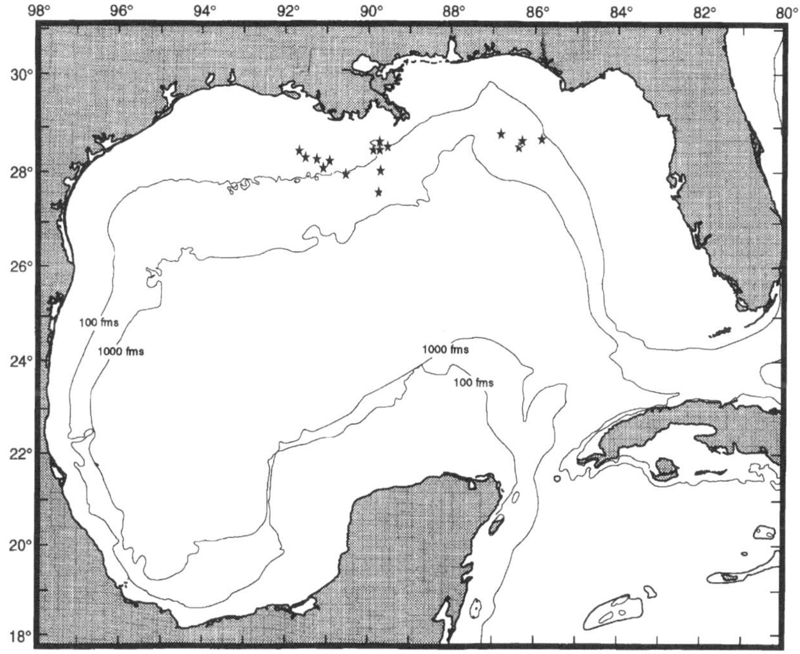

Map 5. Known distribution of *Atlantapseudes lindae* Meyer & Heard, 1989.

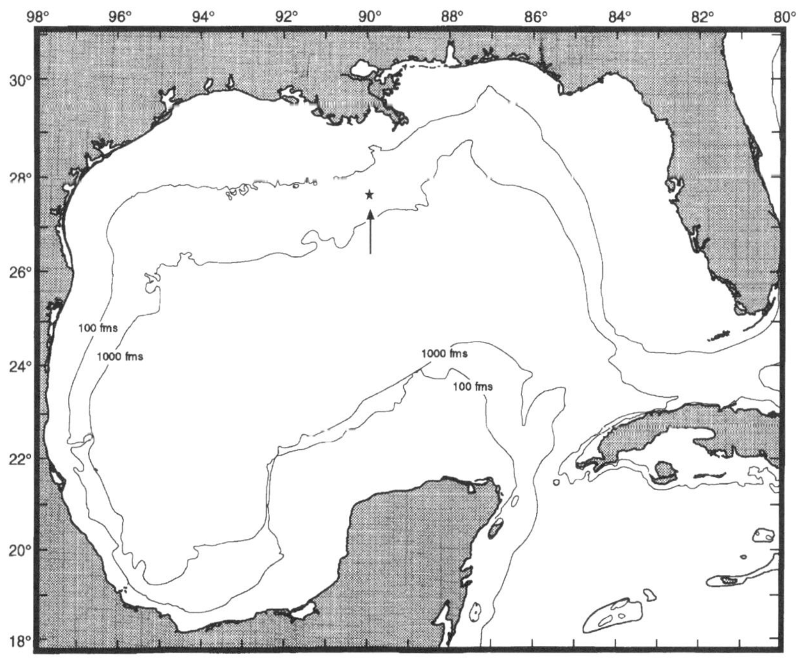

Map 6. Known distribution of *Dactyloprion anakanthes* Guţu, 2002(b).

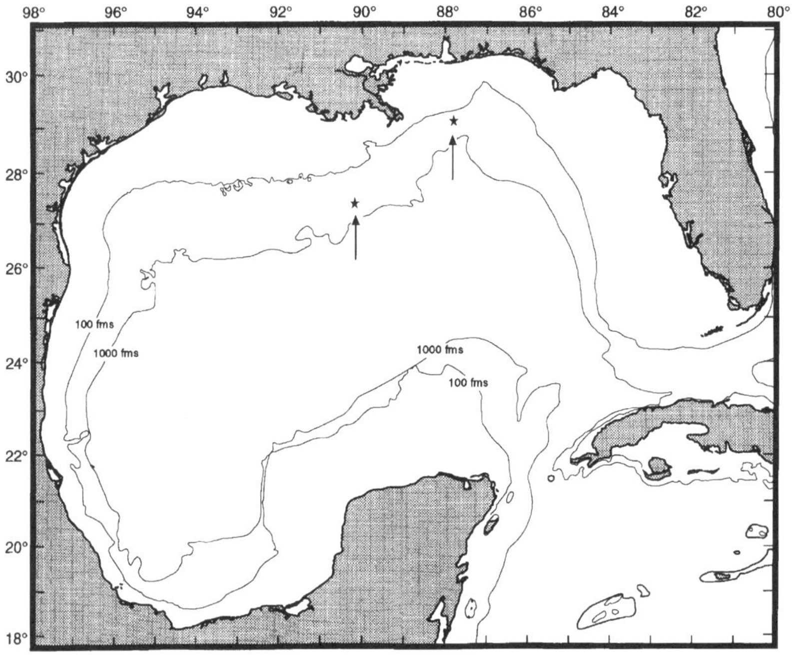

Map 7. Known distribution of *Pectinapseudes magnus* Guţu, 2002(b).

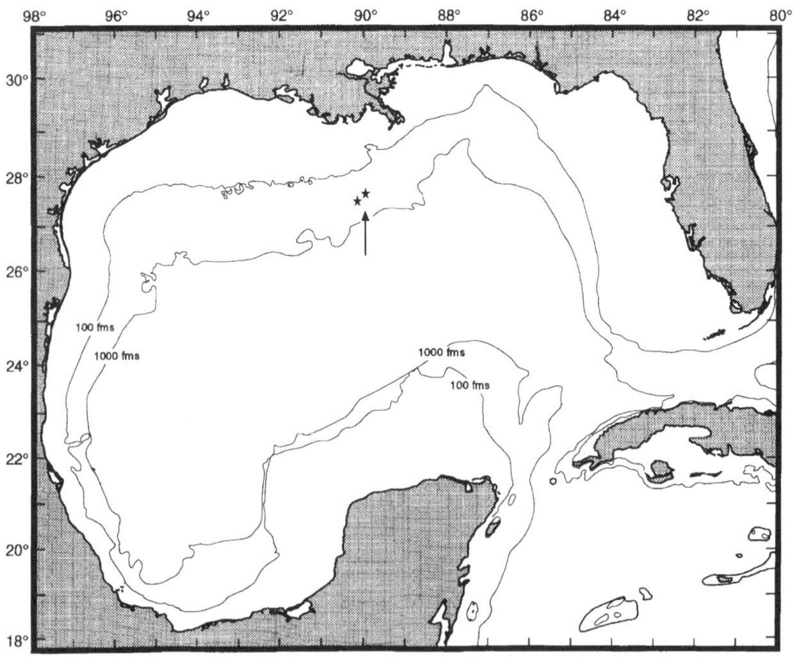

Map 8. Known distribution of *Pectinapseudes mesteri* Guţu, 2002(b).

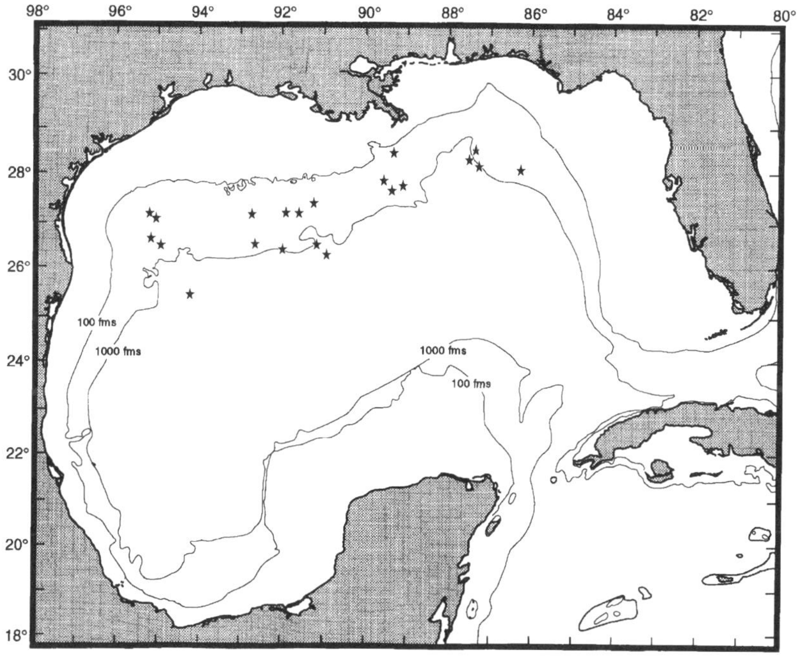

Map 9. Known distribution of *Leviapseudes bipartitus* sp. nov.

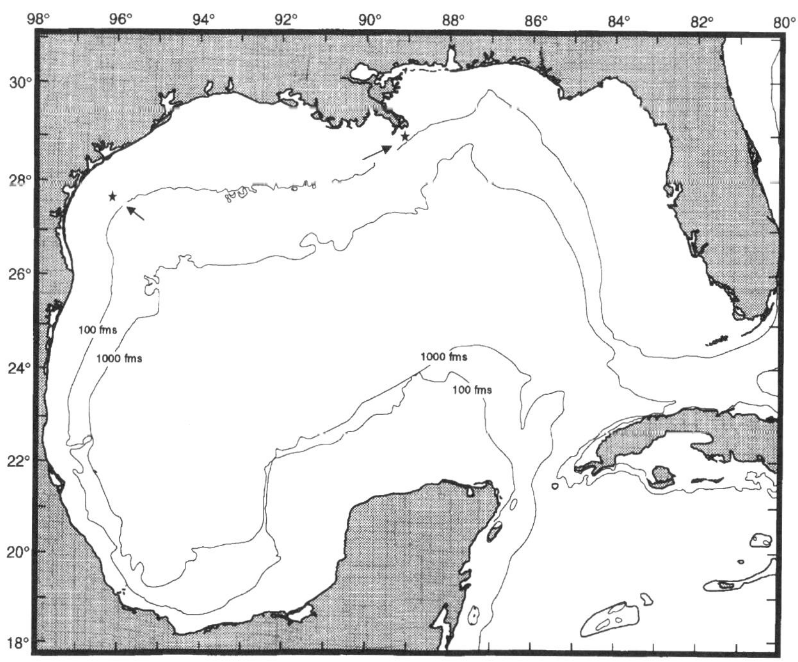

Map 10. Known distribution of *Pseudosphyrapus* sp. A.

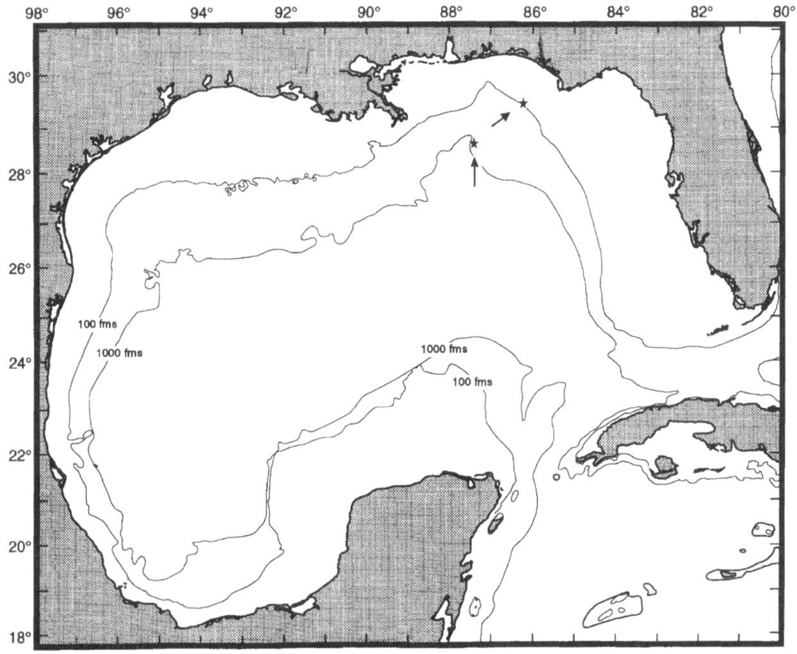

Map 11. Known distribution of *Kudinopasternakia bispinosa* Guţu & Heard, 2002.

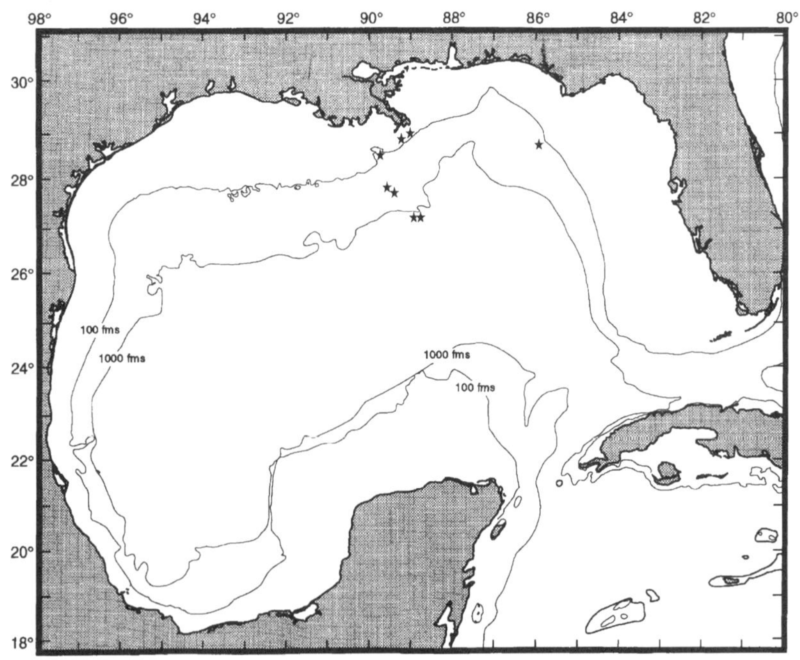

Map 12. Known distribution of *Kudinopasternakia siegi* Viskup & Heard, 1989.

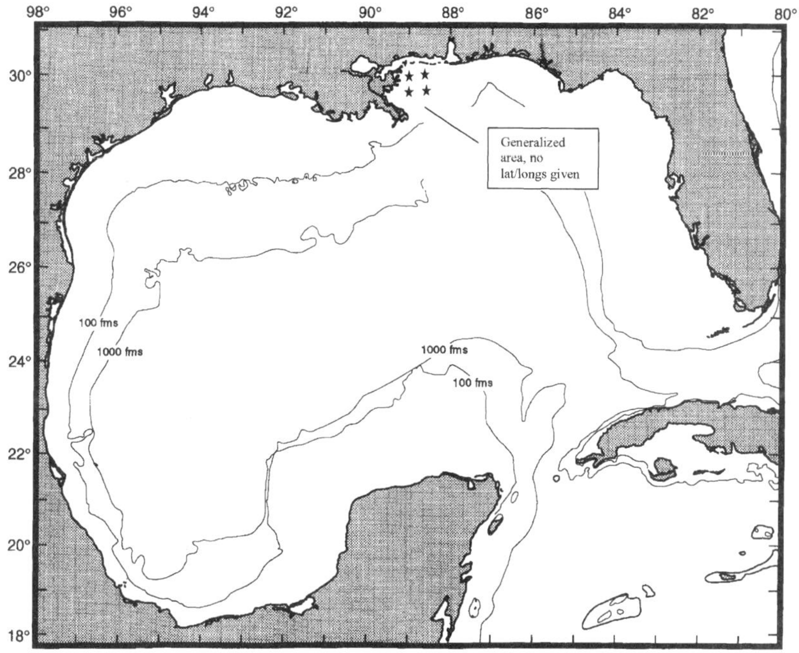

Map 13. Known distribution of *Parapseudes* cf. *latifrons* (Grube, 1864).

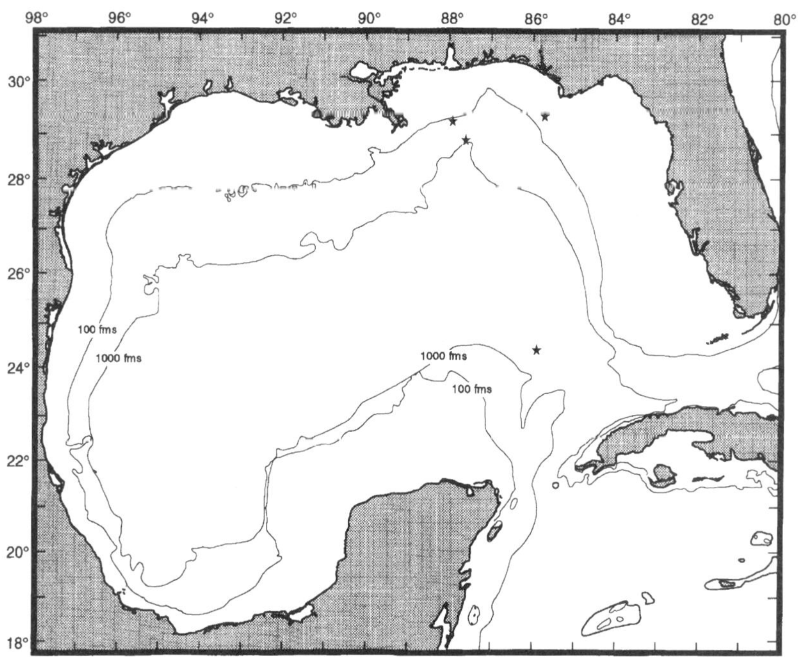

Map 14. Known distribution of *Neotanais armiger* Wolff, 1956(b).

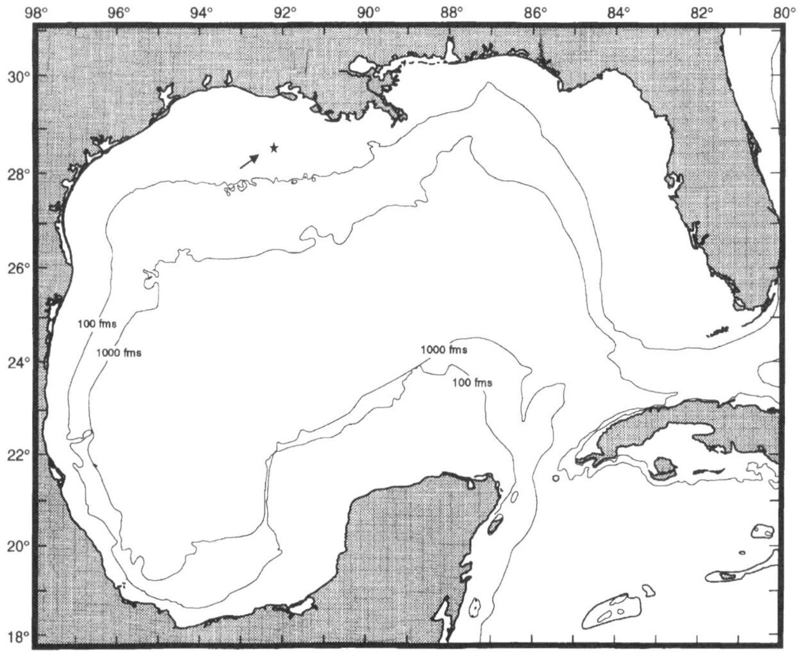

Map 15. Known distribution of *Neotanais curvimanus* Larsen & Hansknecht, 2003.

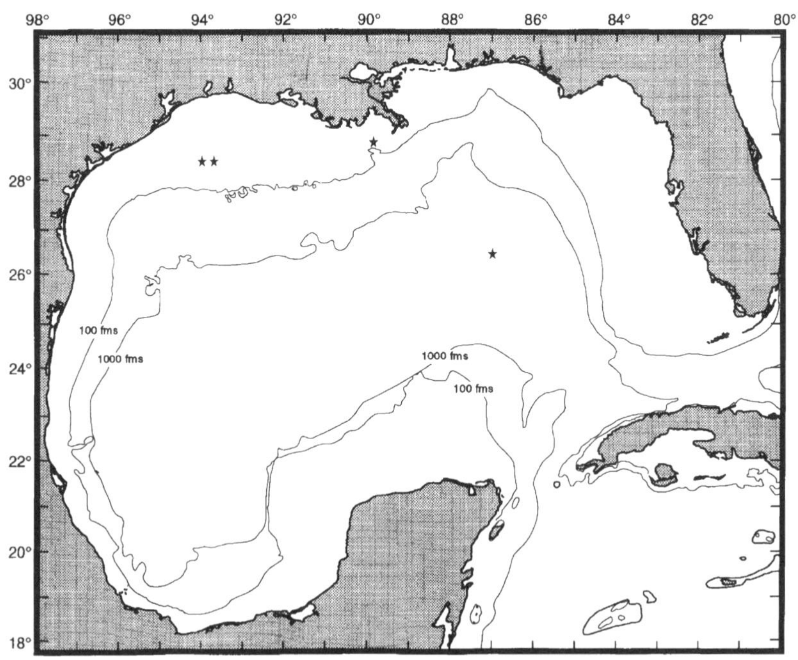

Map 16. Known distribution of *Neotanais minimus* Larsen & Hansknecht, 2003.

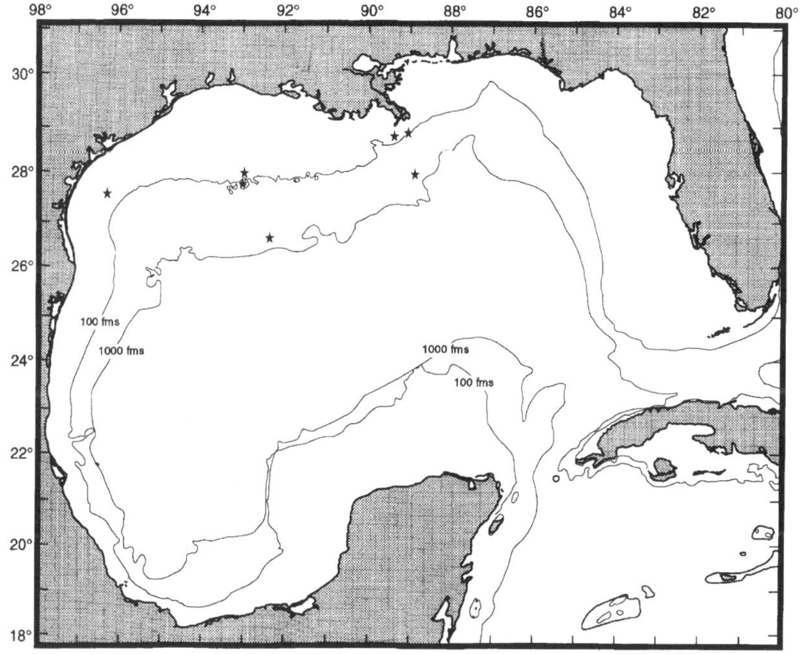

Map 17. Known distribution of *Agathotanais hanseni* Lang, 1971(b).

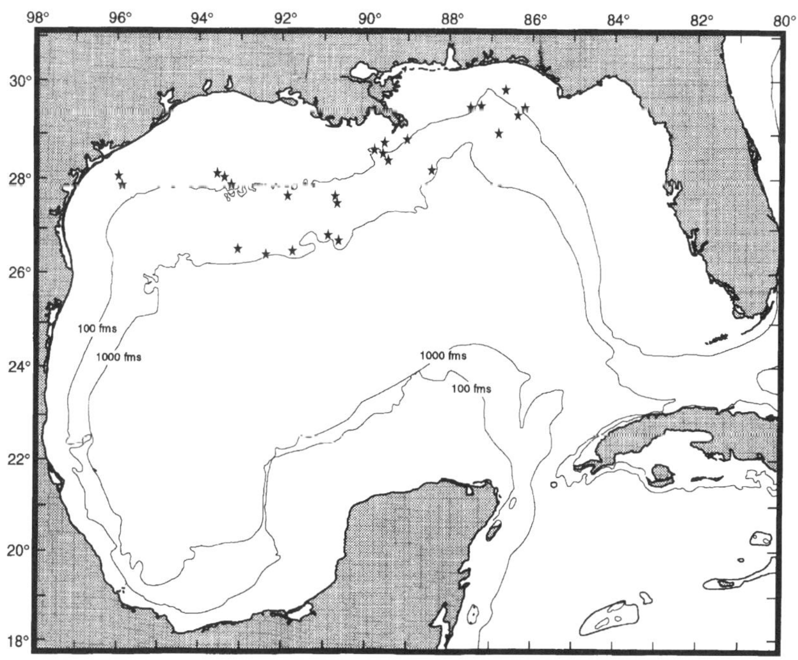

Map 18. Known distribution of *Paragathotanais medius* Larsen, 2002(a).

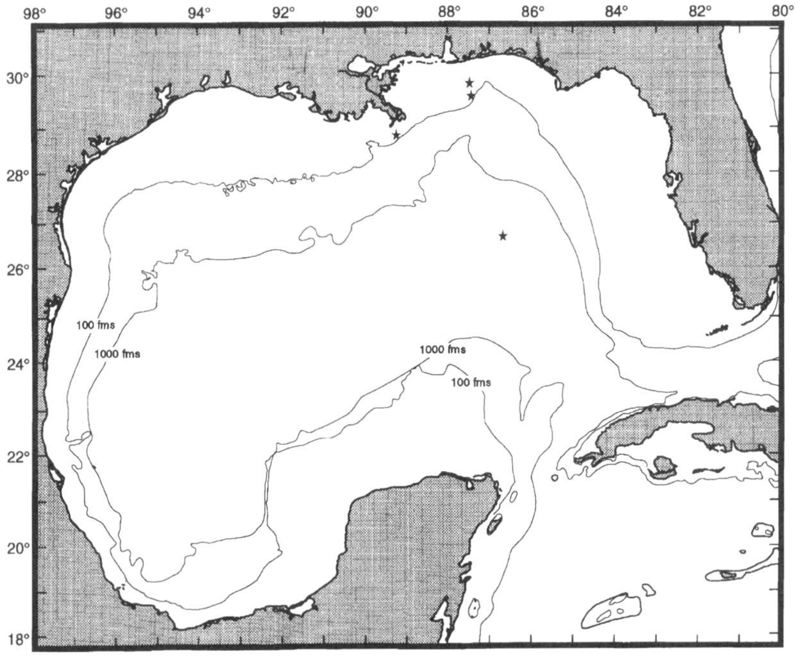

Map 19. Known distribution of *Paragathotanais spinosus* sp. nov.

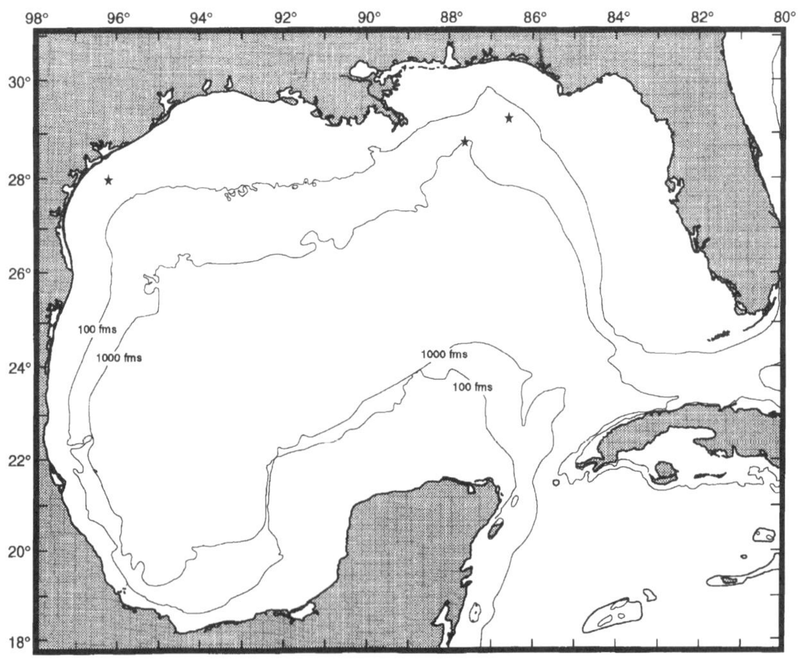

Map 20. Known distribution of *Paragathotanais typicus* Lang, 1971(b).

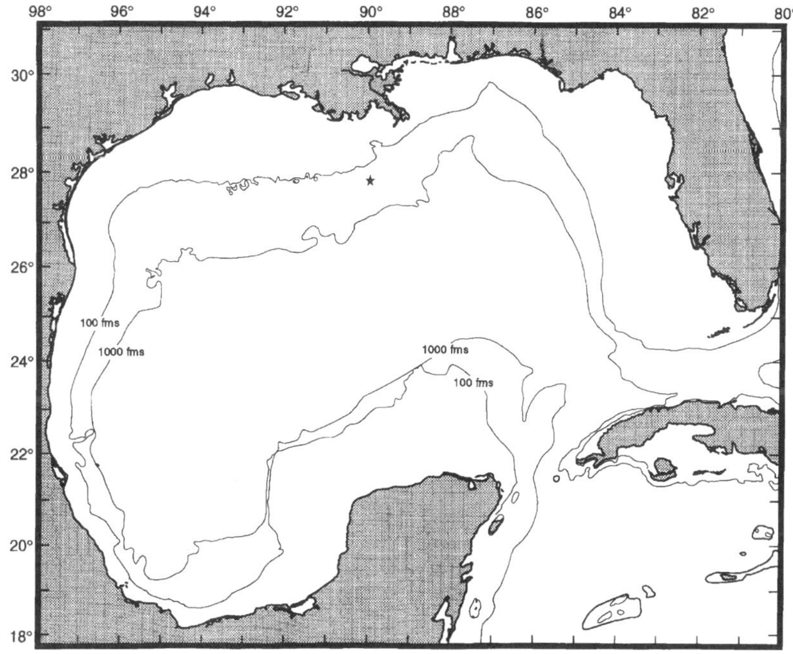

Map 21. Known distribution of *Paragathotanais* sp. A.

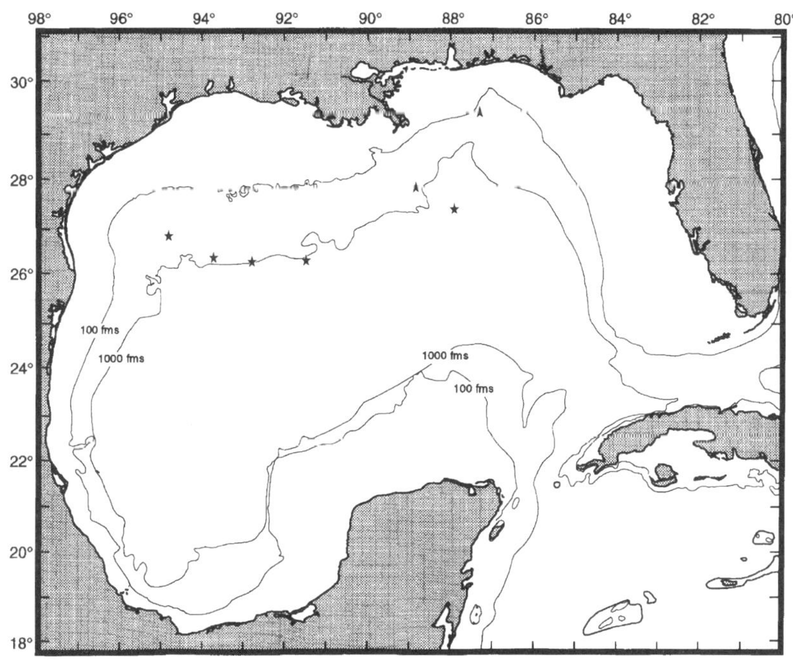

Map 22. Known distribution of *Paranarthrura bispinosa* sp. nov.

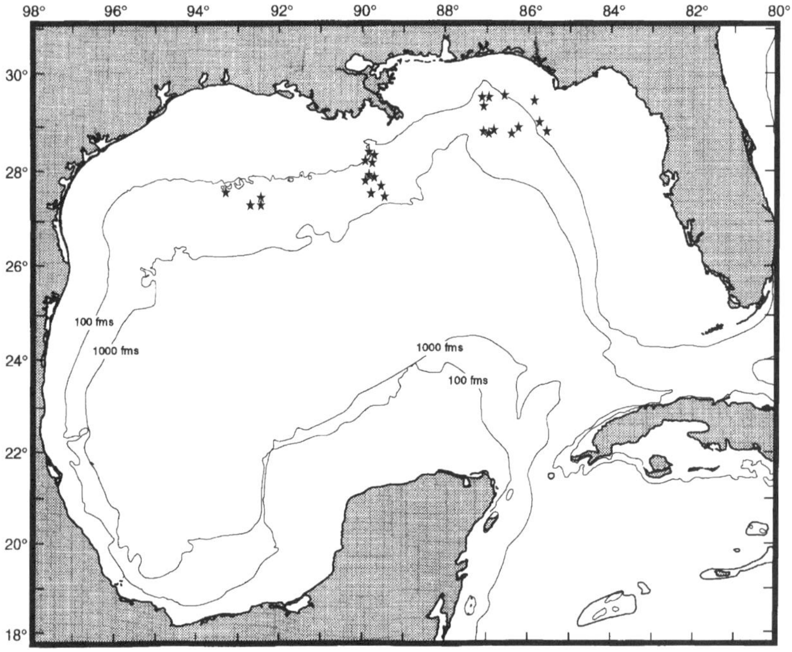

Map 23. Known distribution of *Paranarthrura* cf. *insignis* Hansen, 1913.

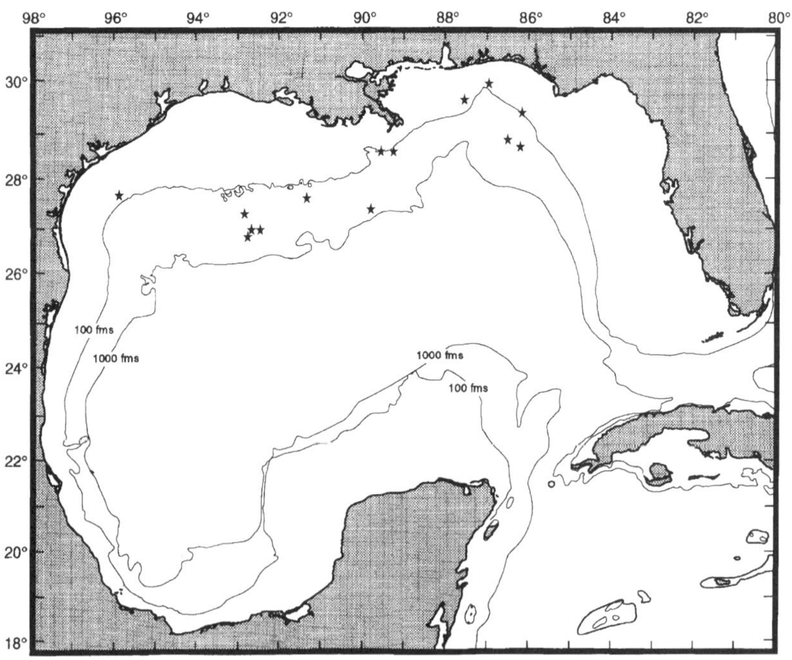

Map 24. Known distribution of *Paranarthrura tenuimanus* sp. nov.

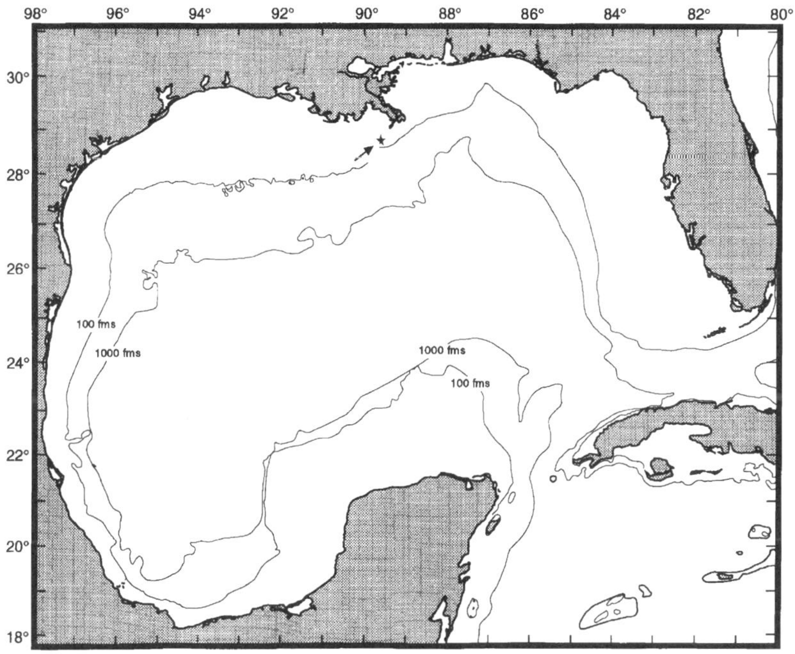

Map 25. Known distribution of *Paranarthrura* cf. *undulata* Lang, 1968.

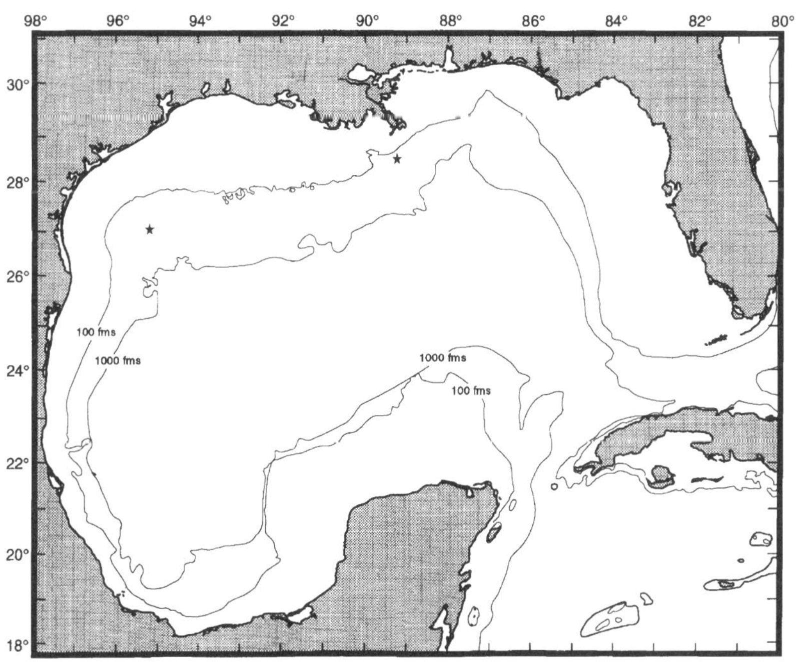

Map 26. Known distribution of *Anarthruropsis edentula* sp. nov.

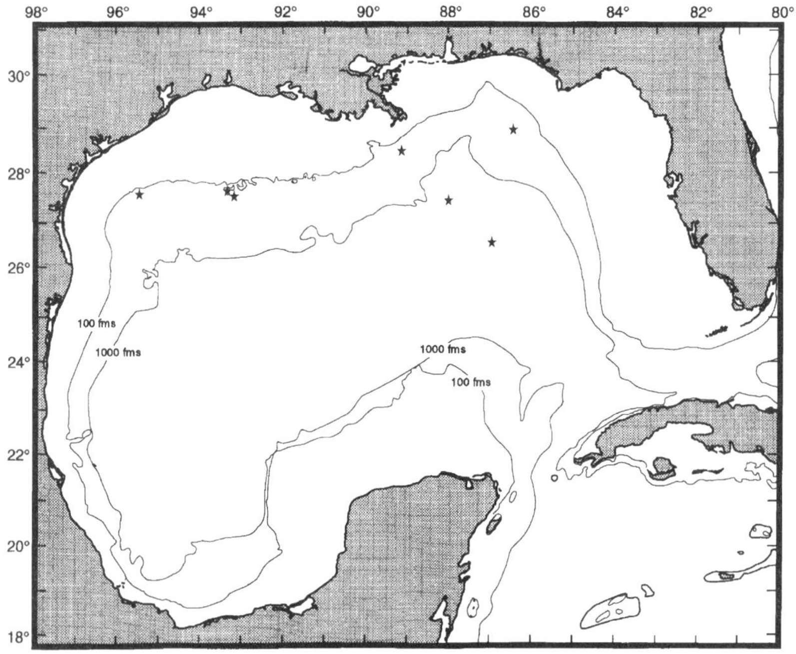

Map 27. Known distribution of *Caudalonga quatropleon* sp. nov.

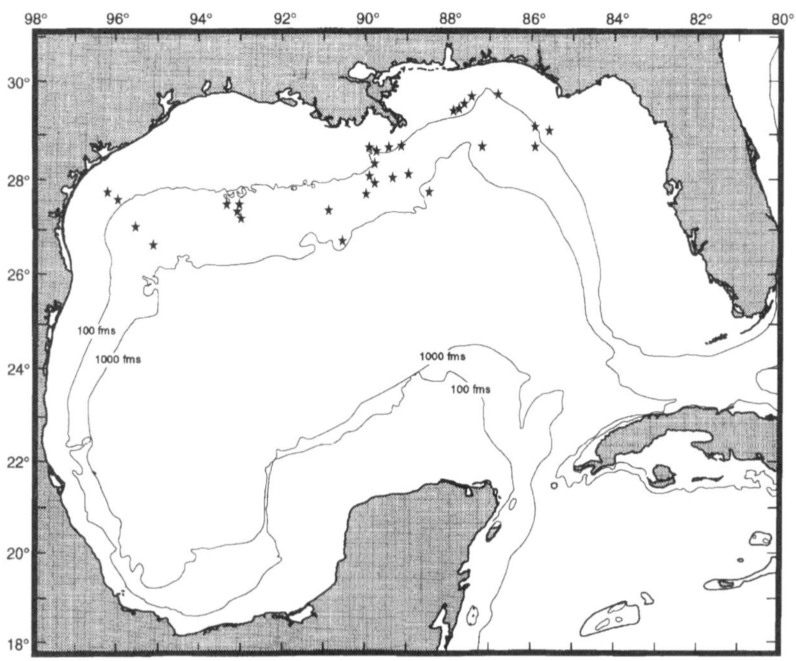

Map 28. Known distribution of *Cetiopyge mira* Larsen & Heard, 2002.

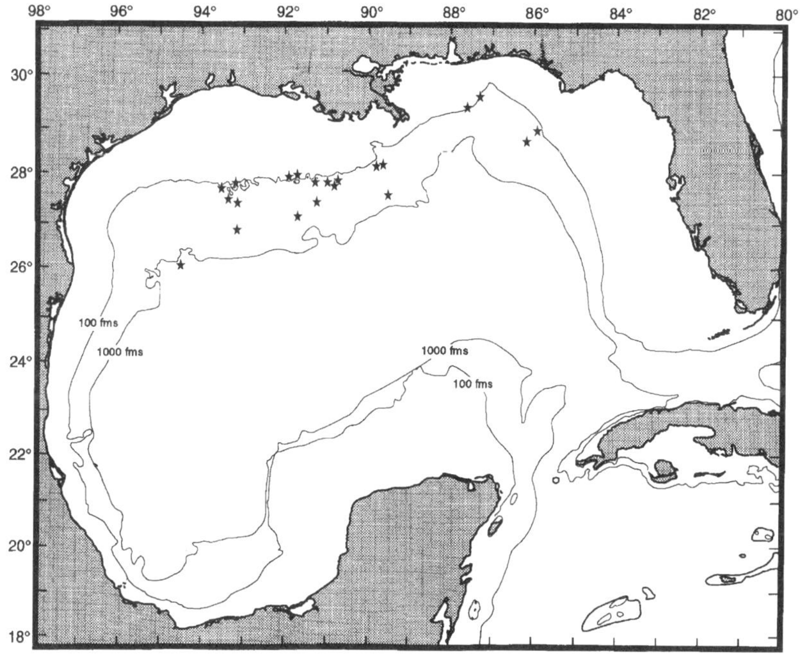

Map 29. Known distribution of *Collettea elongata* Larsen, 2002(a).

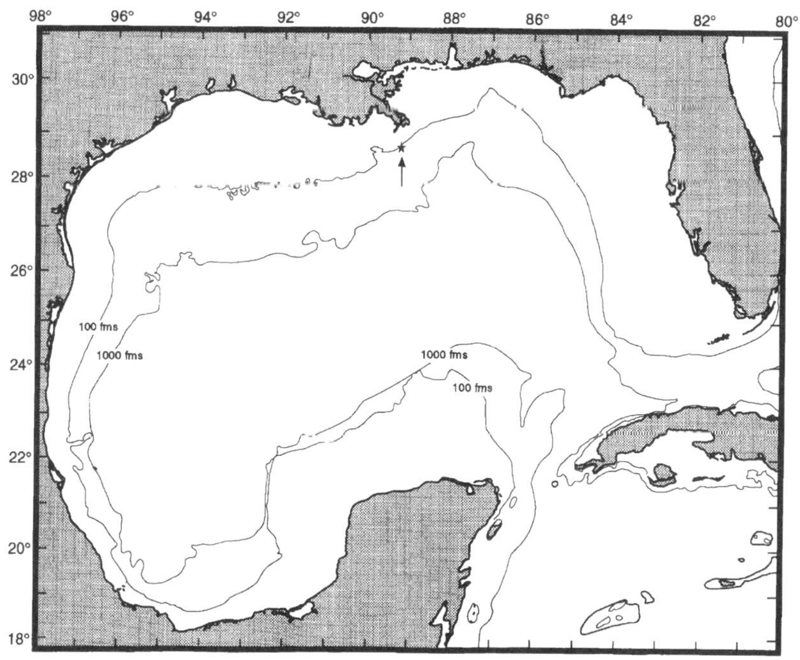

Map 30. Known distribution of *Collettea* sp. A.

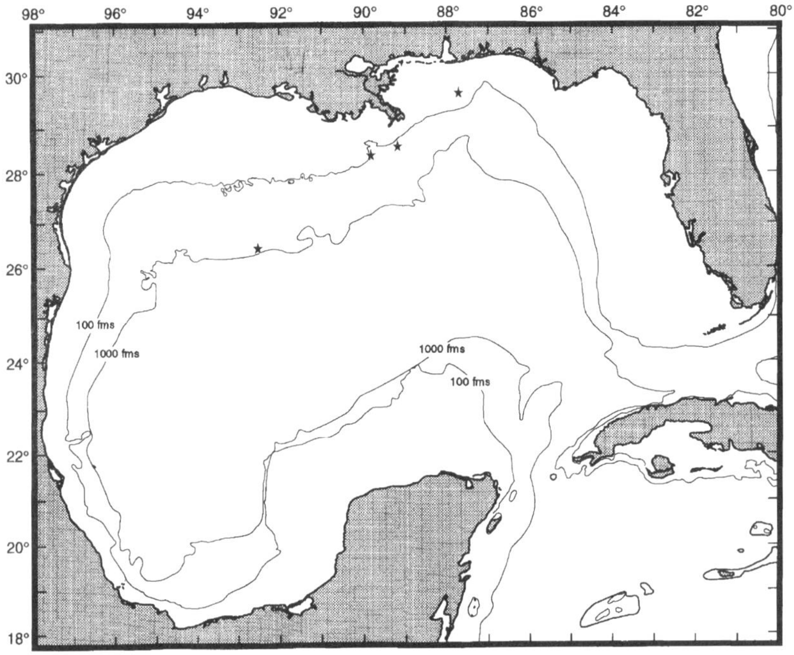

Map 31. Known distribution of *Filitanais filiformis* sp. nov.

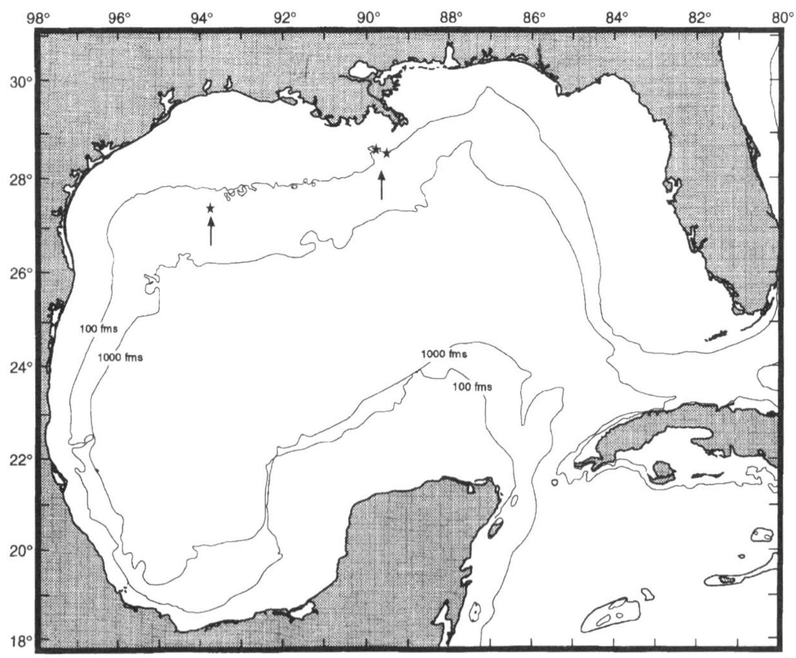

Map 32. Known distribution of *Isopodidus janum* Larsen & Heard, 2002.

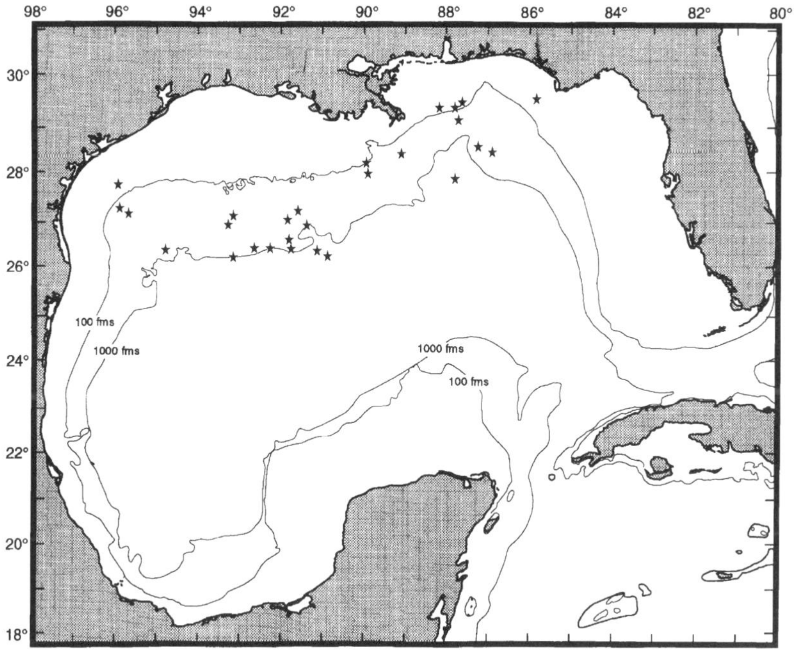

Map 33. Known distribution of *Leptognathiella occidentis* sp. nov.

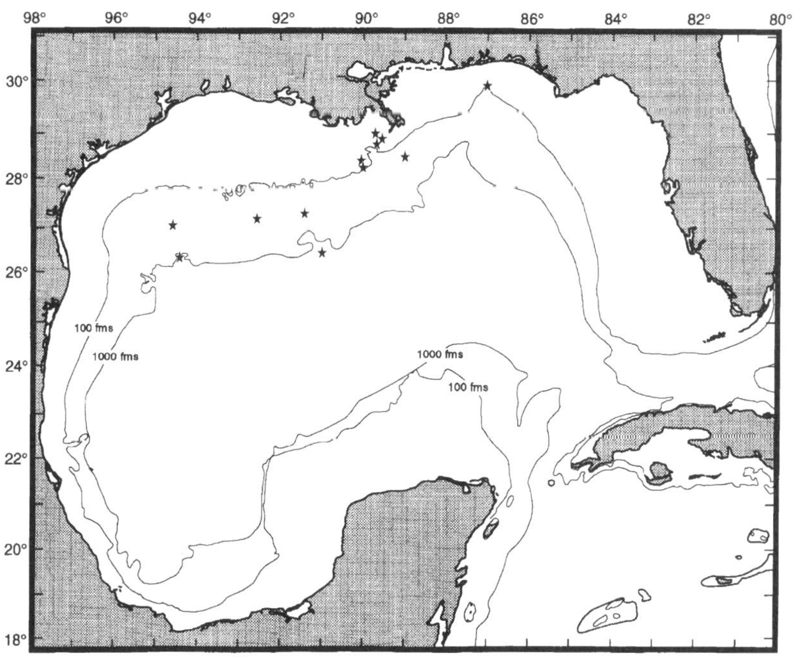

Map 34. Known distribution of *Leptognathiella vermiformis* sp. nov.

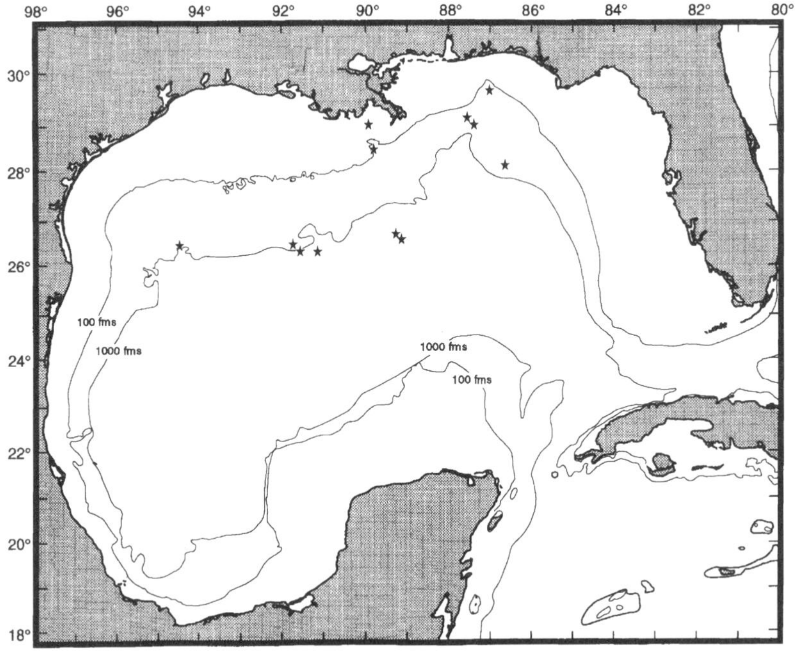

Map 35. Known distribution of *Tumidochelia randyi* Knight, Larsen & Heard, 2002.

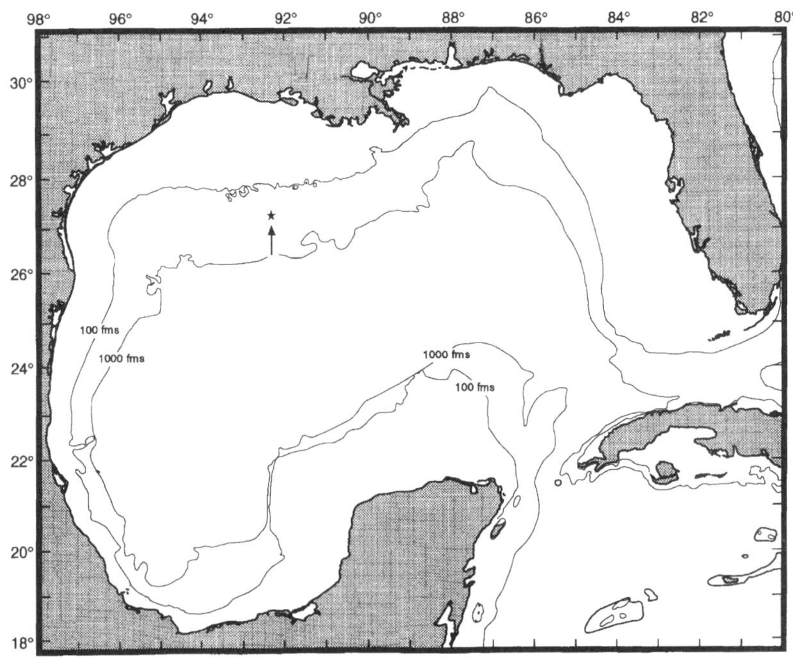

Map 36. Known distribution of *Bathyleptochelia oculata* Larsen, 2003(b).

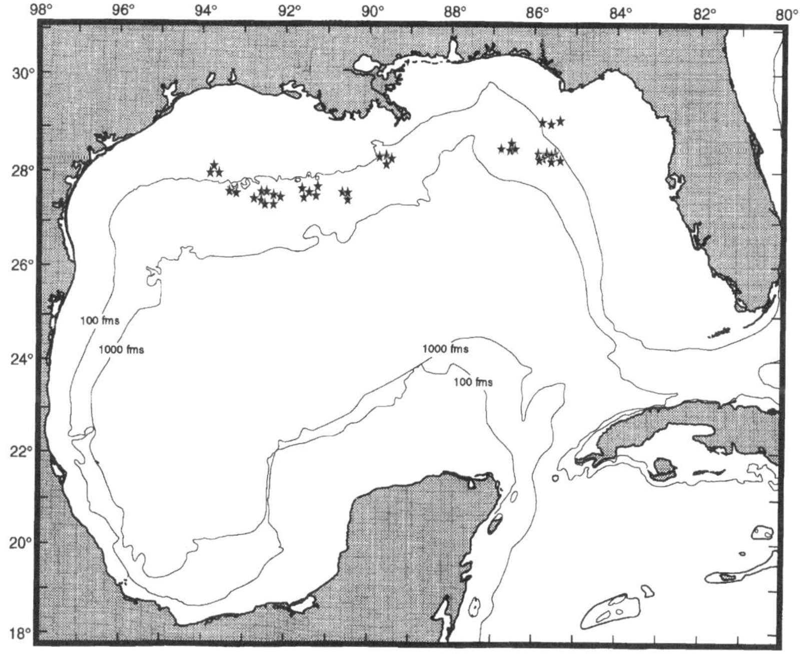

Map 37. Known distribution of *Mesotanais longisetosus* Sieg & Heard, 1989.

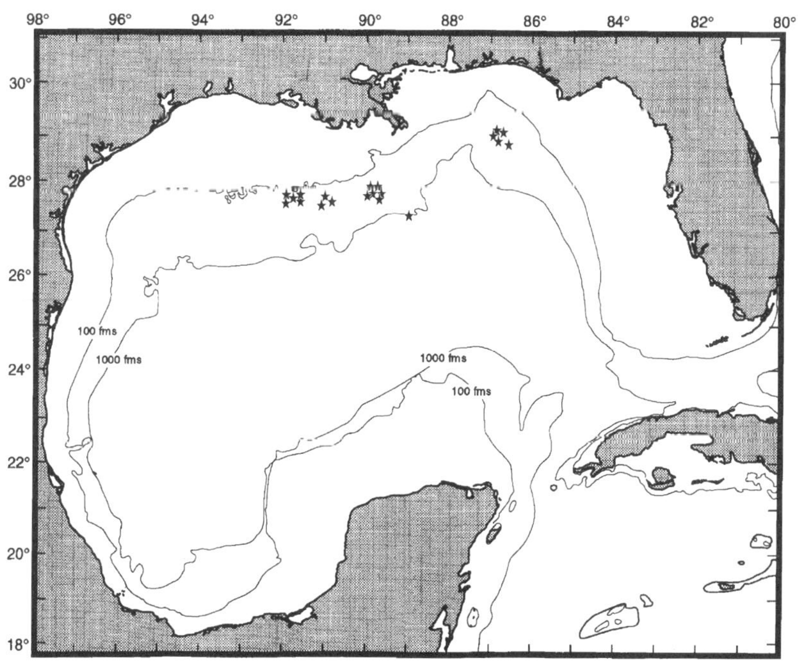

Map 38. Known distribution of *Mesotanais vadicola* Sieg & Heard, 1989.

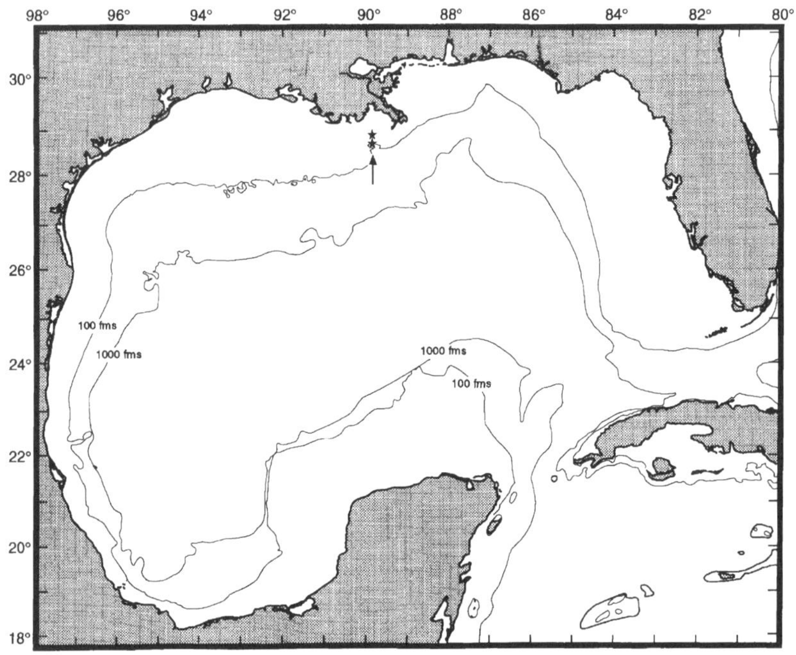

Map 39. Known distribution of *Leptognathia gyreae* sp. nov.

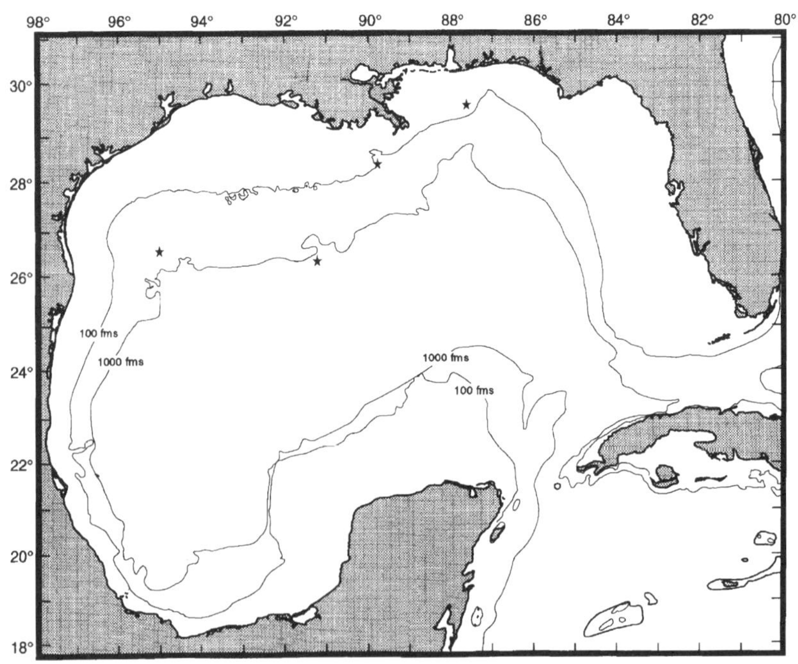

Map 40. Known distribution of *Leptognathia longimanus* sp. nov.

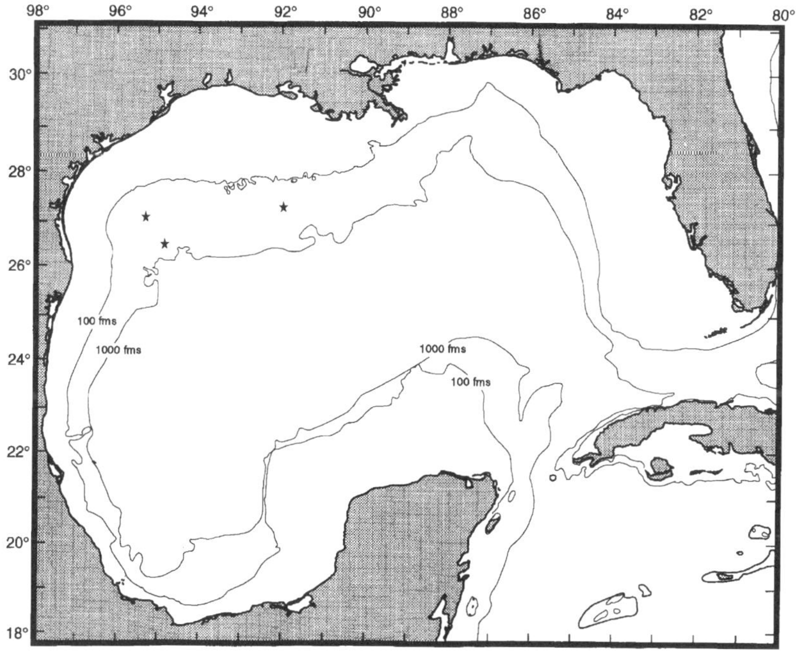

Map 41. Known distribution of *Meromonakantha gilrowei* sp. nov.

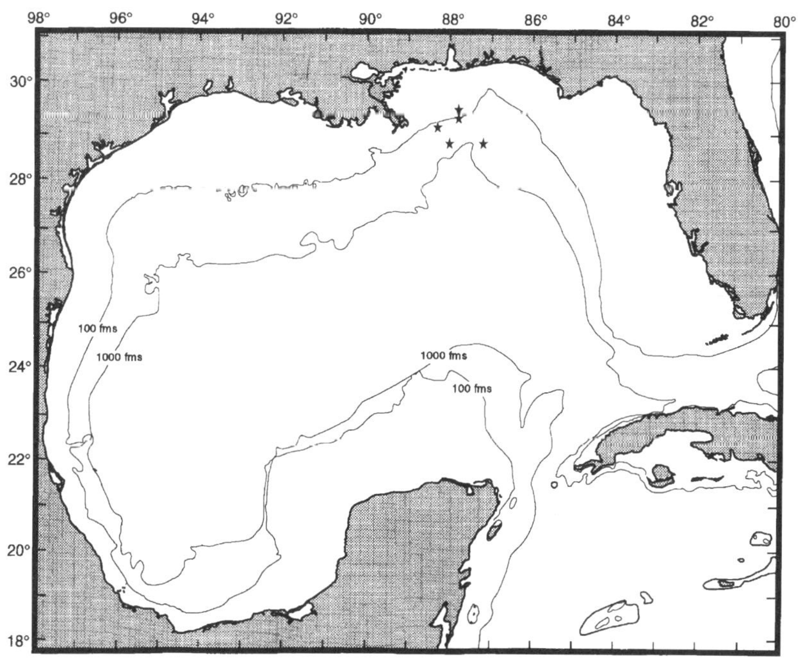

Map 42. Known distribution of *Typhlotanais* cf. *trispinosus* Hansen, 1913.

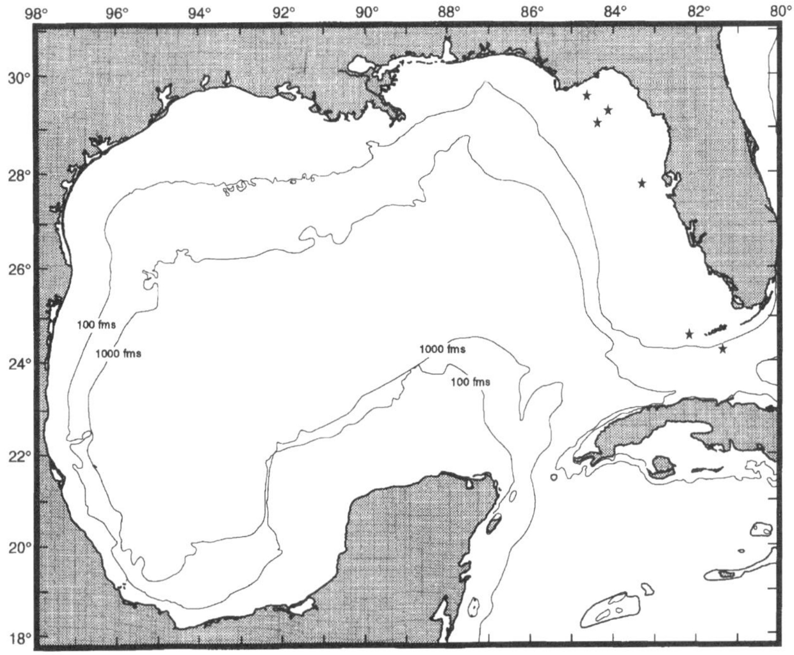

Map 43. Known distribution of *Iungentitanais primitivus* (Sieg, 1973).

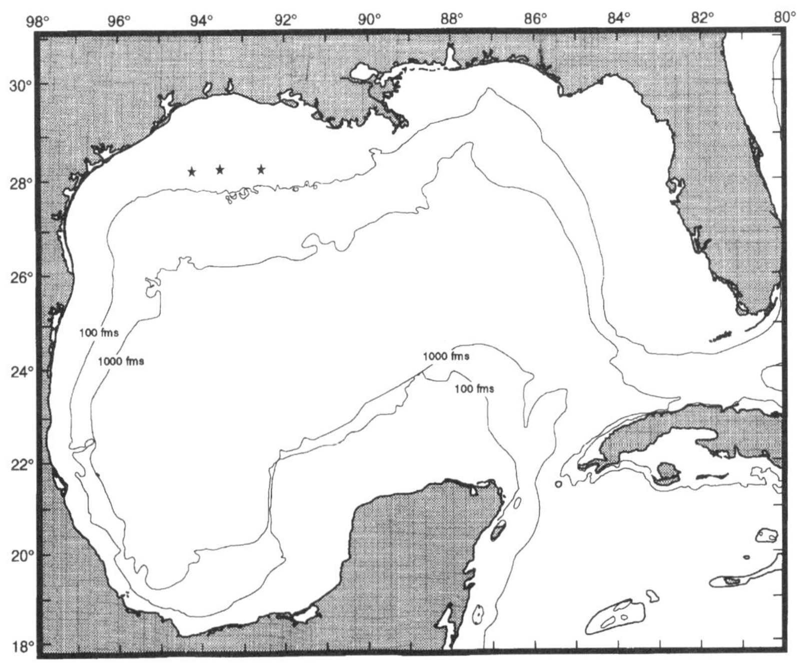

Map 44. Known distribution of *Pseudotanais mexikolpos* Sieg & Heard, 1988.

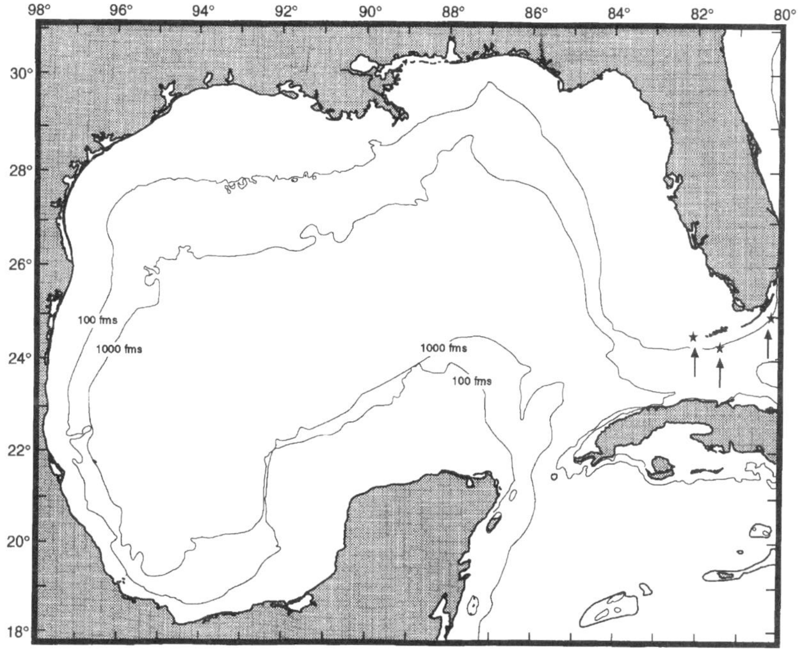

Map 45. Known distribution of *Pseudotanais mortenseni* Sieg, 1977.

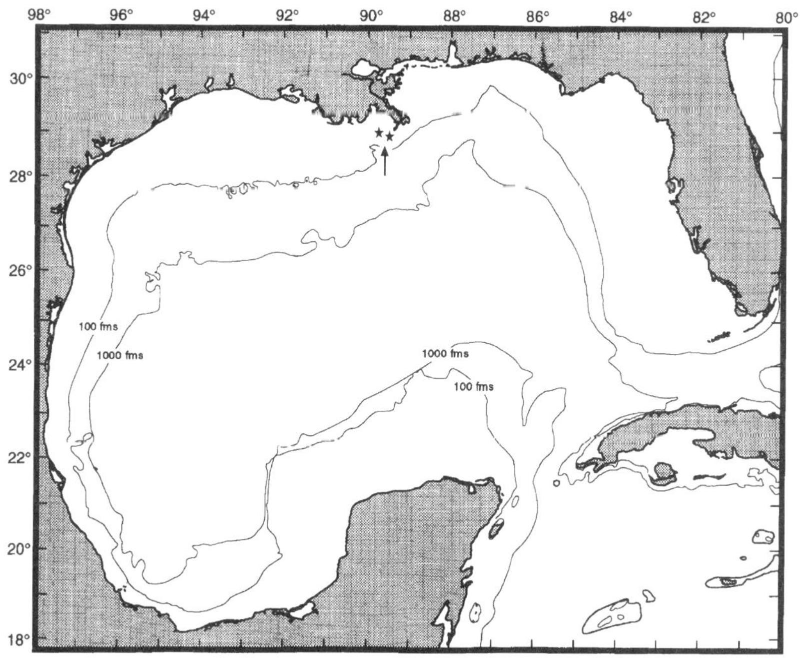

Map 46. Known distribution of *Araphura curticauda* sp. nov.

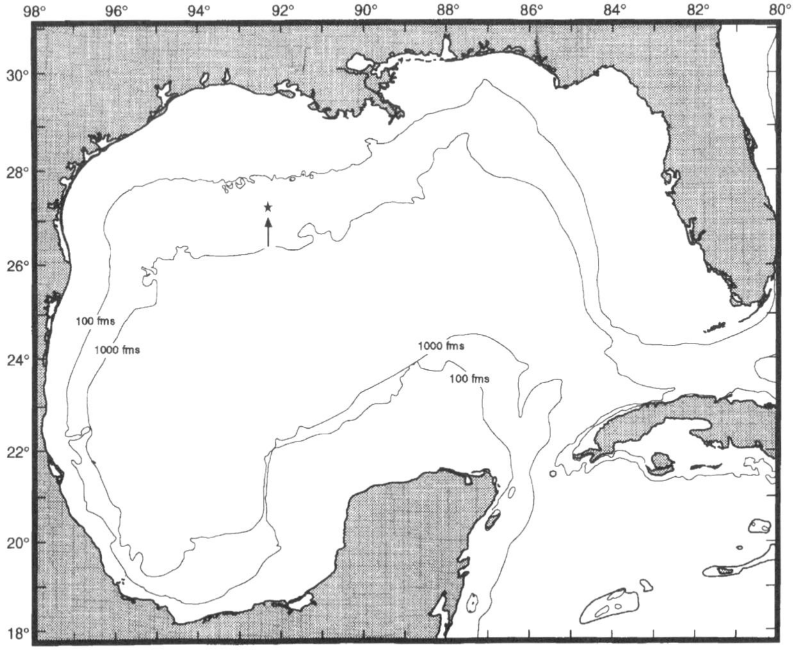

Map 47. Known distribution of *Araphura extensa* Larsen, 2003(b).

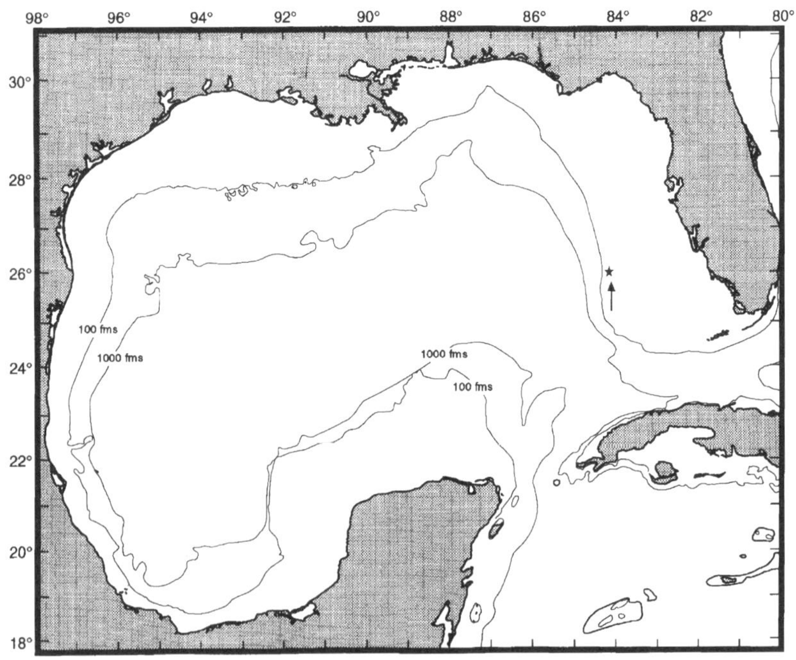

Map 48. Known distribution of *Araphura higginsi* Sieg & Dojiri, 1989.

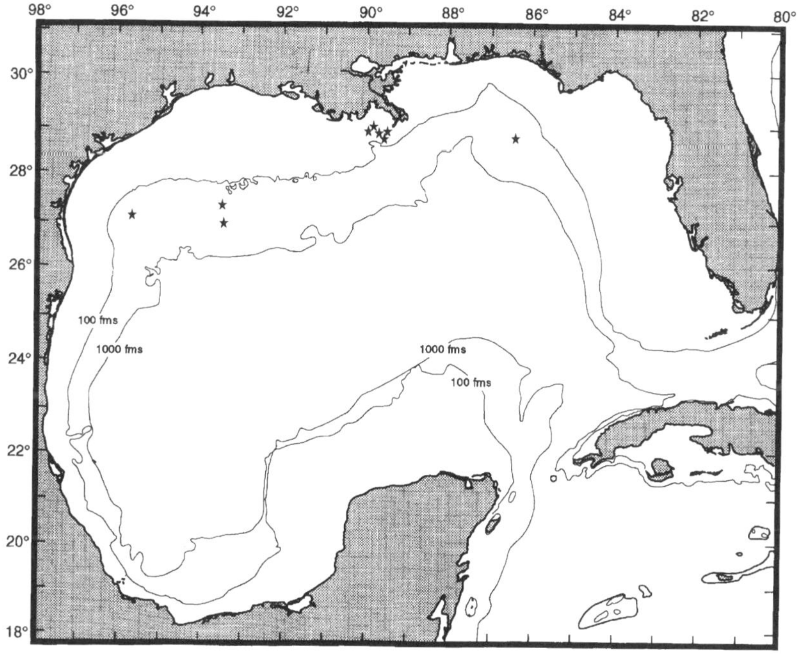

Map 49. Known distribution of *Araphura spinithenari* sp. nov.

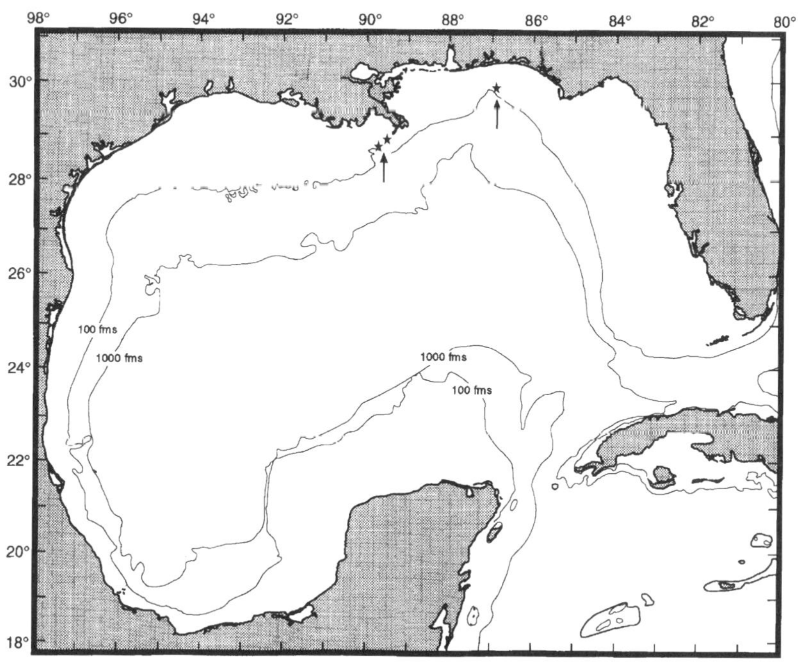

Map 50. Known distribution of *Araphuroides bombus* sp. nov.

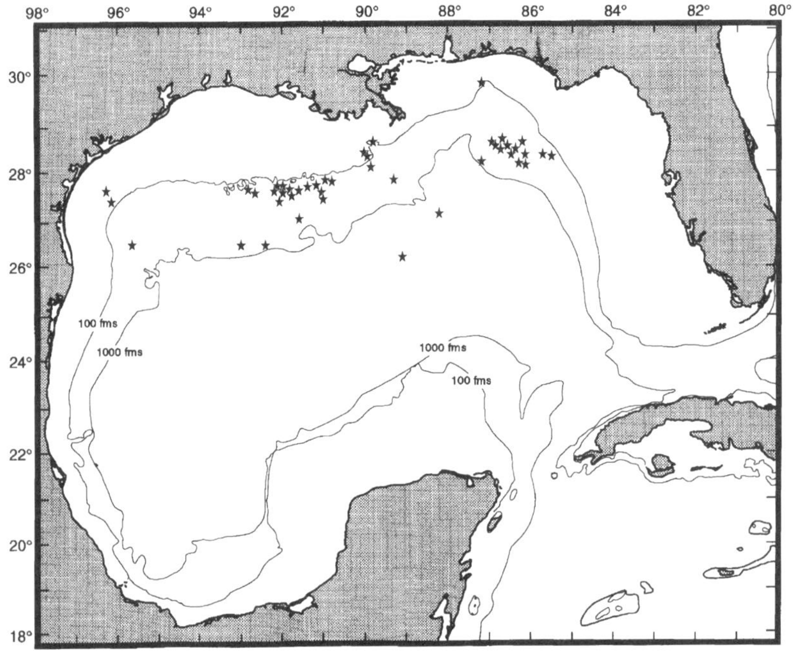

Map 51. Known distribution of *Tanaella mclellandi* Larsen & Heard, 2004(a).

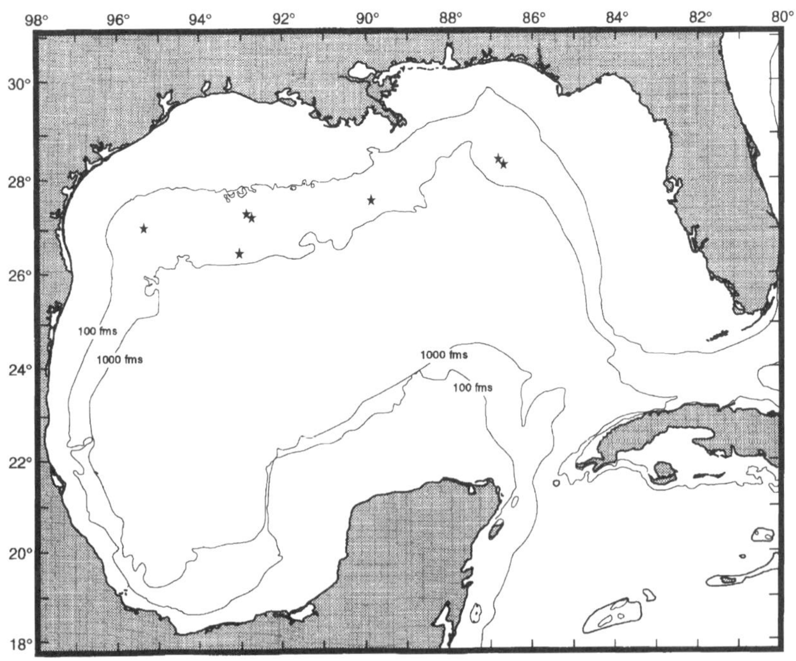

Map 52. Known distribution of *Tanaella prolixcauda* Larsen & Heard, 2004(a).

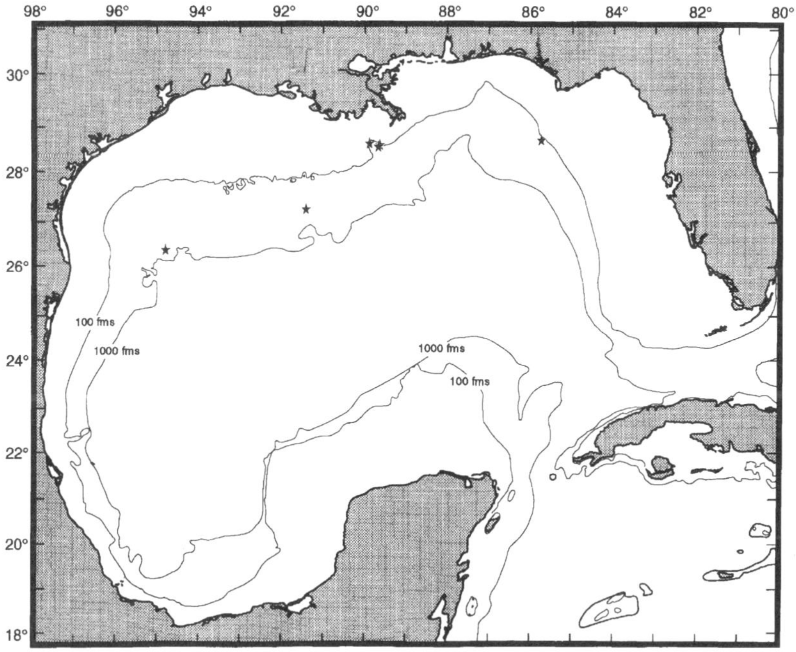

Map 53. Known distribution of *Armaturatanais minipodus* sp. nov.

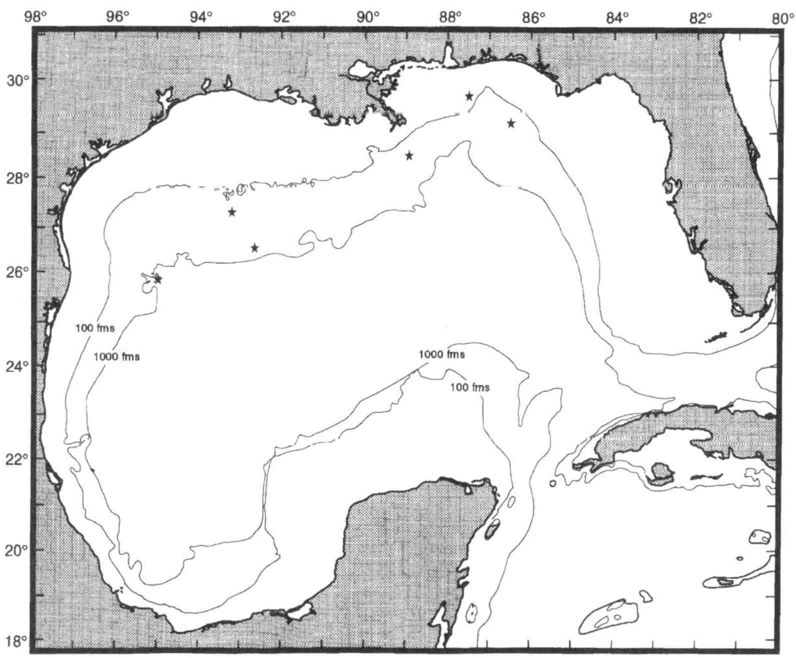

Map 54. Known distribution of *Chauliopleona faini* sp. nov.

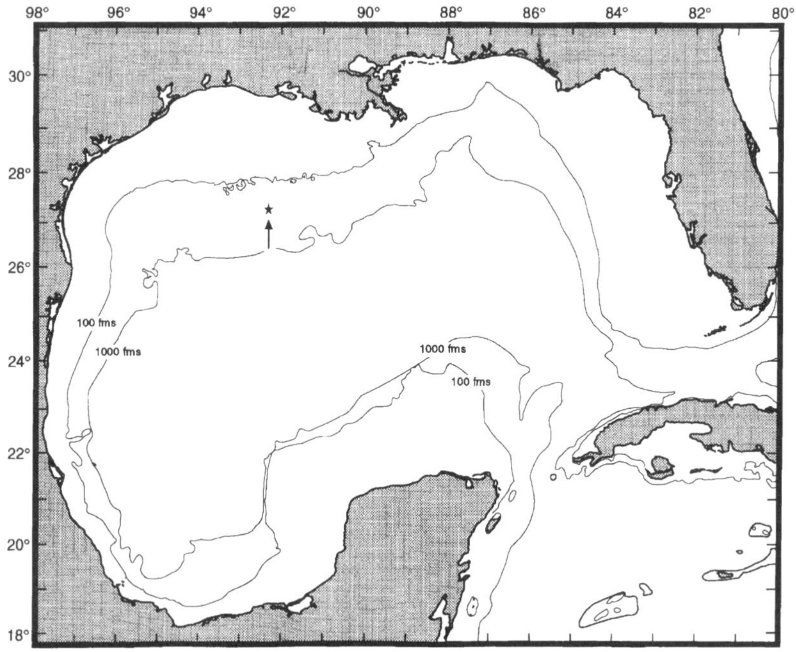

Map 55. Known distribution of *Coalecerotanais inflatus* Larsen, 2003(b).

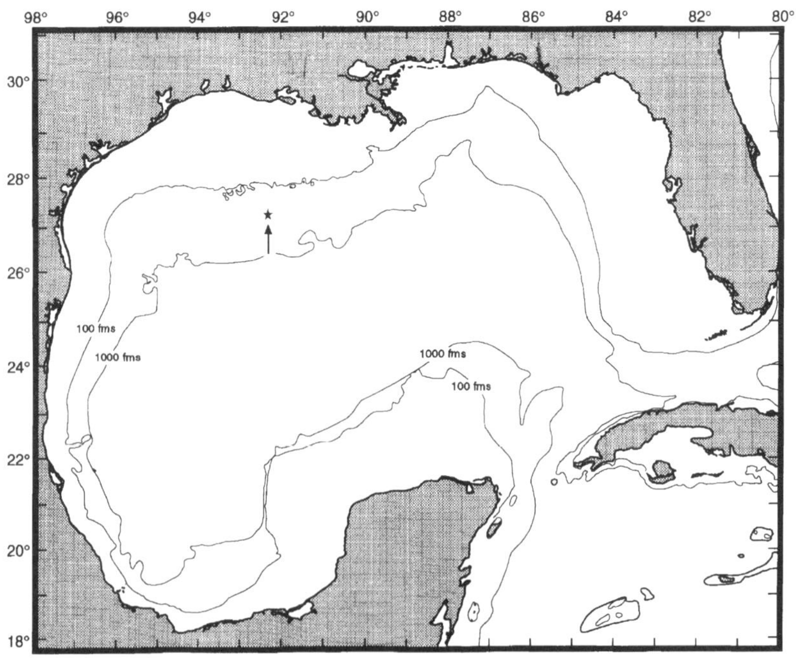

Map 56. Known distribution of *Spinitanaopsis insolituchelia* Larsen, 2003(b).

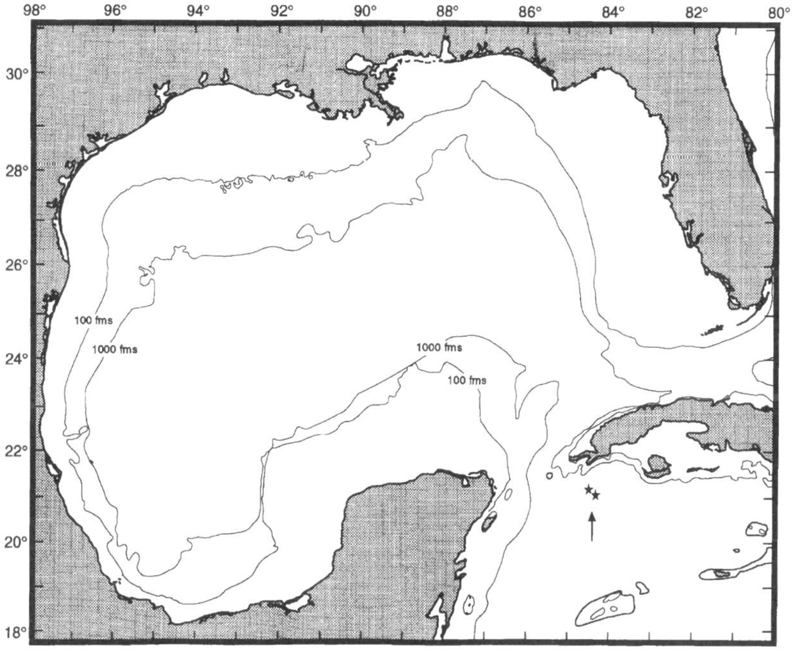

Map 57. Known distribution of *Expina typica* Lang, 1968.

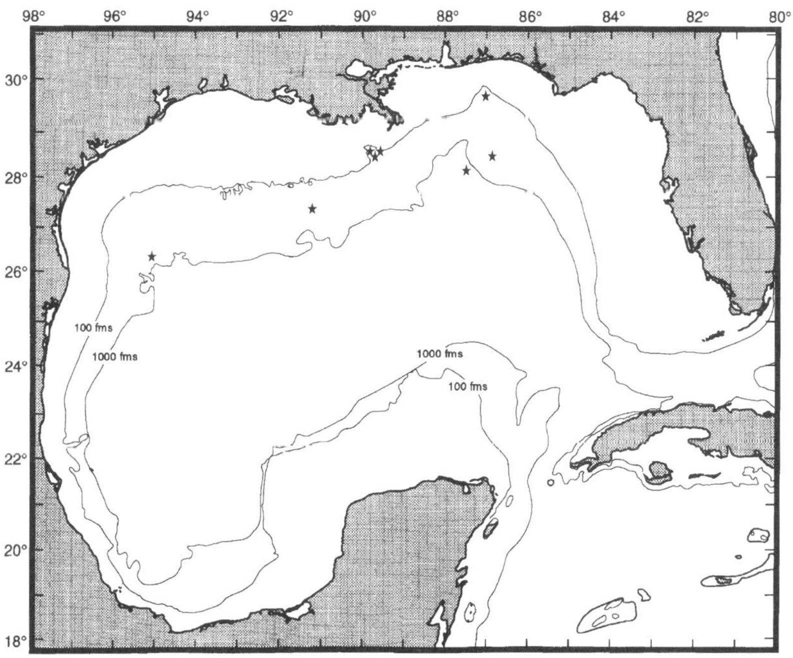

Map 58. Known distribution of *Insociabilitanais ventrospinatus* sp. nov.

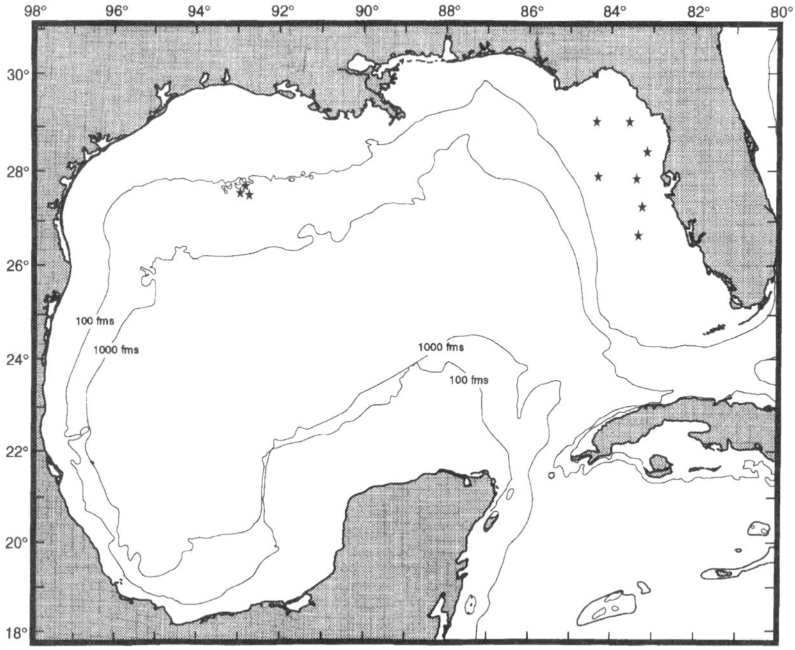

Map 59. Known distribution of *Nototanoides trifurcatus* Sieg & Heard, 1985.

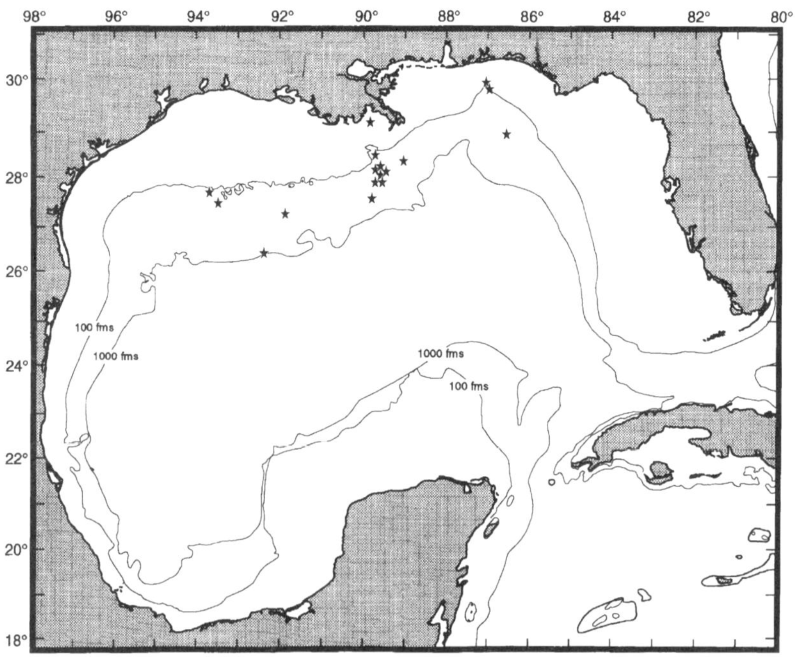

Map 60. Known distribution of *Parafilitanais mexicanus* Larsen, 2002(a).

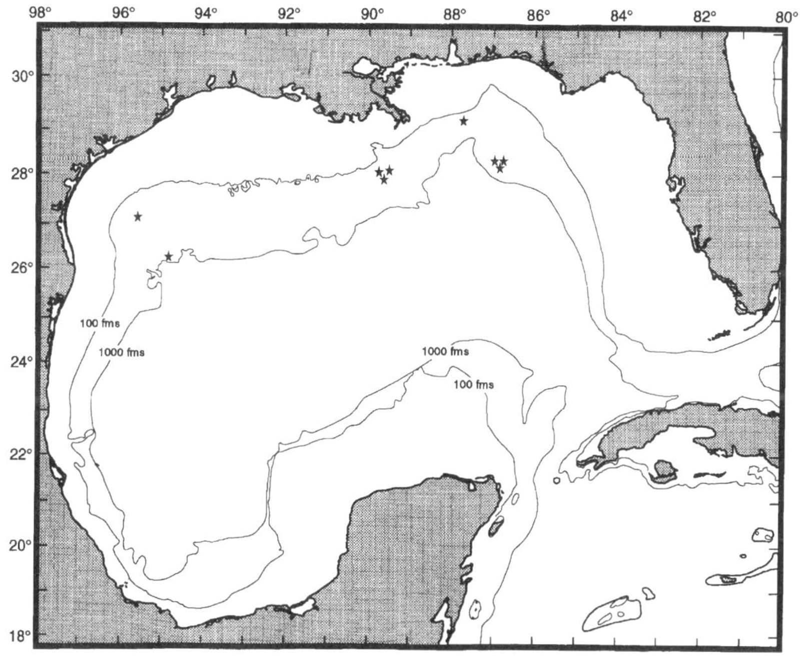

Map 61. Known distribution of *Pseudoarthrura heardi* sp. nov.

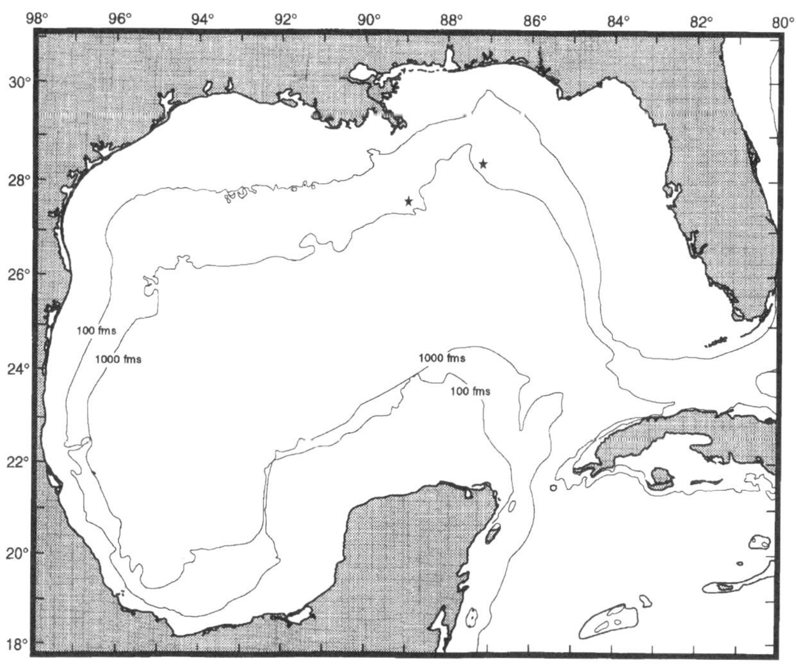

Map 62. Known distribution of *Robustochelia solida* sp. nov.

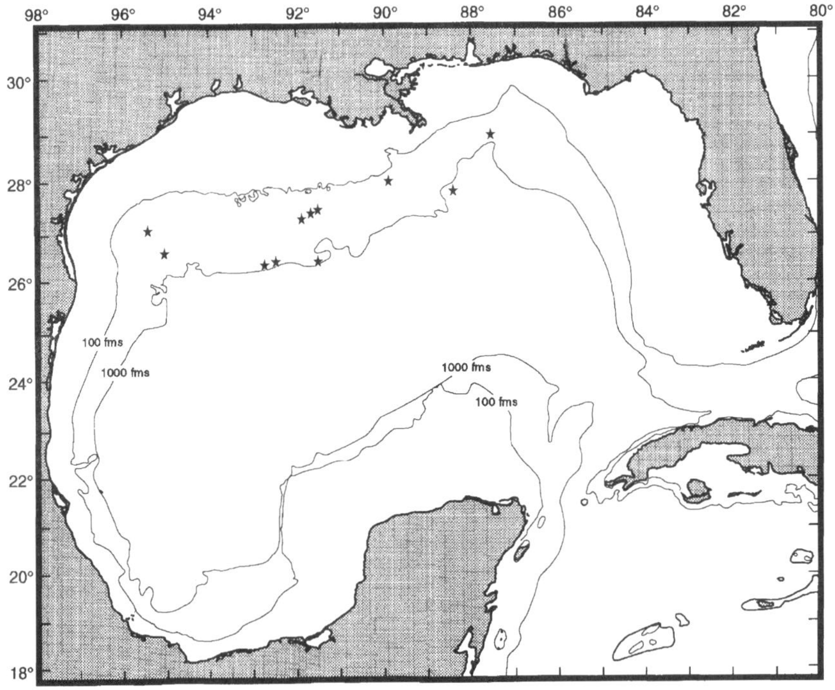

Map 63. Known distribution of *Stenotanais macrodactylus* sp. nov.

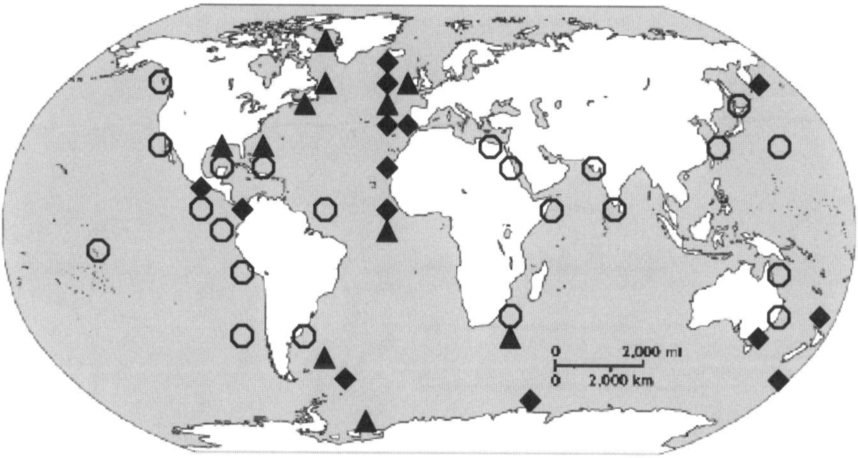

Map 64. Cosmopolitan distributions. ○, *Sinelobus stanfordi* (Richardson, 1901); ▲, *Neotanais affinis* Wolff, 1956; ◆, *Expina typica* Lang, 1968.

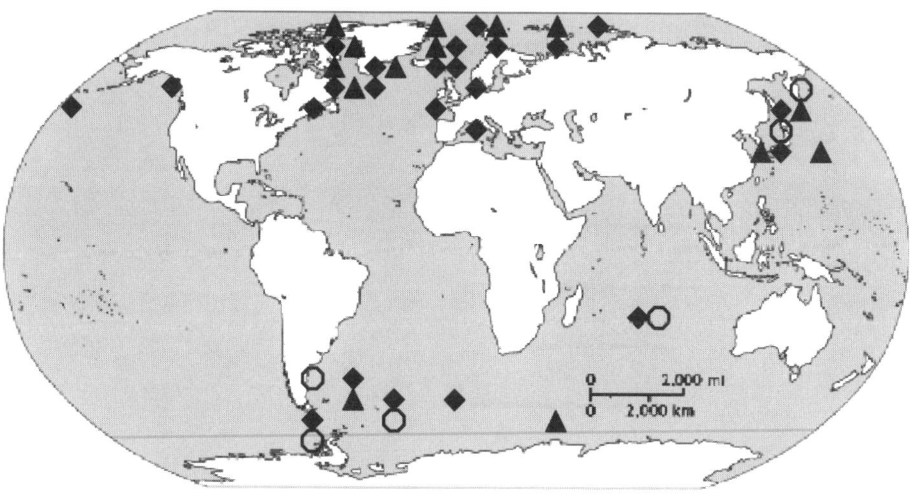

Map 65. Bipolar distributions. ▲, *Cryptocopoides arcticus* (Hansen, 1886); ◆, *Paraleptognathia gracilis* (Krøyer, 1842); ○, *Apseudes spectabilis* Studer, 1883.

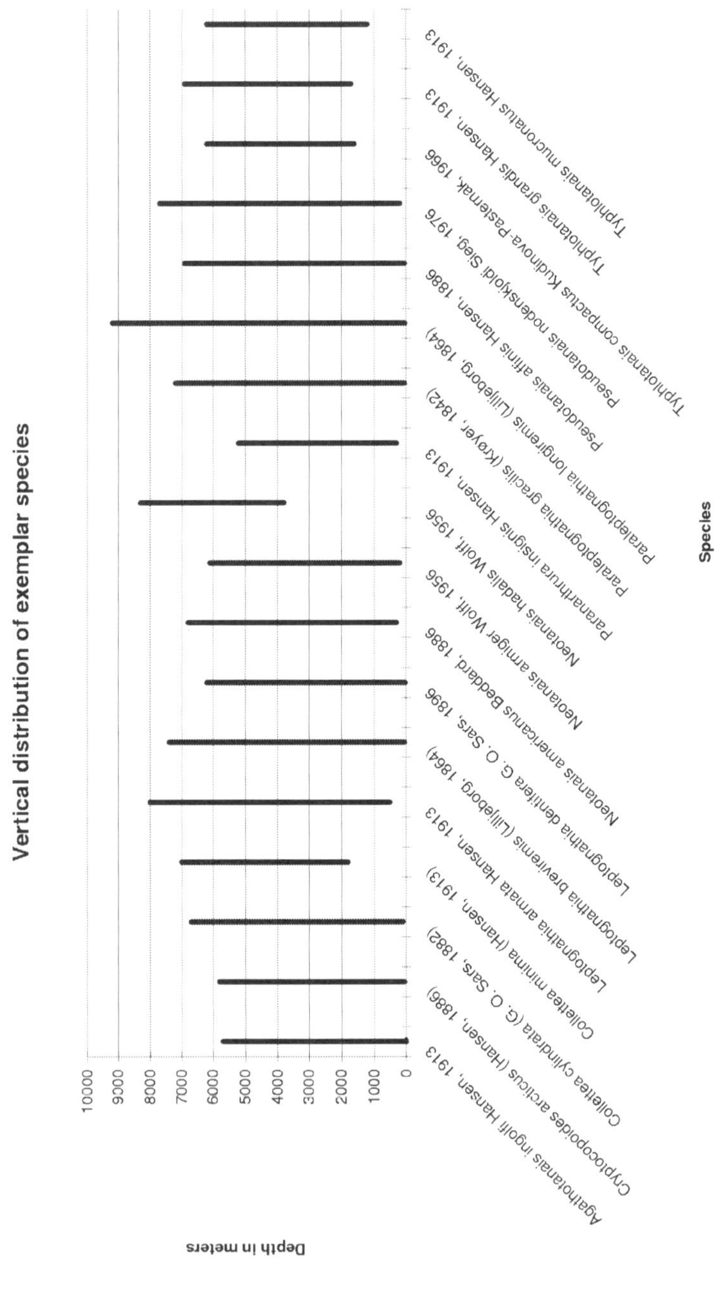

Chart 1. Vertical distribution ranges of exemplar species.

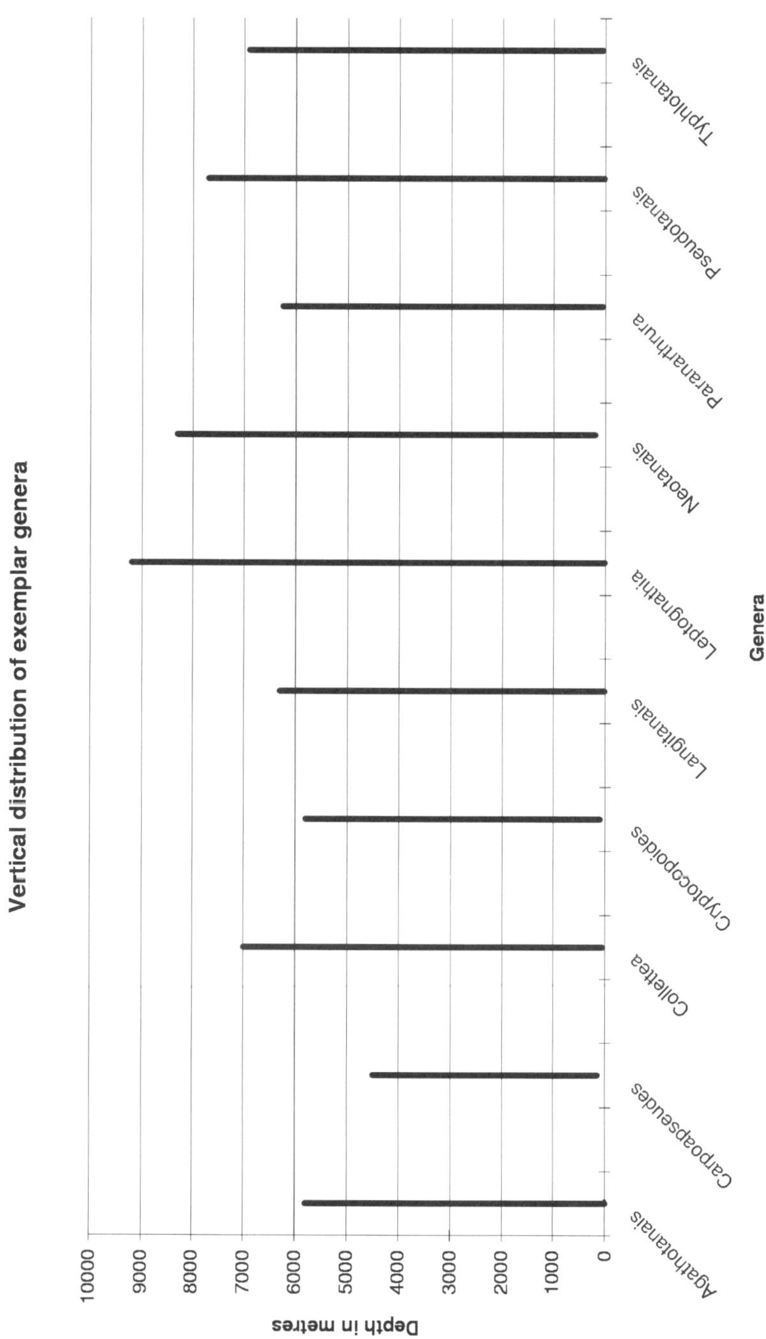

Chart 2. Vertical distribution ranges of exemplar genera, ranging from 4500 to > 9000 m depth.

ACKNOWLEDGEMENTS

The funding for this project was provided by the BP Group Environmental Technology Program Deep-Sea Biodiversity Fellowship, hosted by the Department of Marine Biology, Texas A&M University, Galveston, Texas.

Material used in this report was provided by Texas A&M University under a contract (1435-01-99-CT-30991) to them from the Minerals Management Service, Department of the Interior. I am greatly thankful to the many colleagues who have commented on parts of this monograph. In alphabetical order these are: Dr. Roger Bamber, The Natural History Museum, London; Dr. Graham Bird, Camborne; Dr. Magdalena Błażewicz-Paszkowycz, University of Łodz, Łodz; Dr. Jürgen Guerrero-Kommritz, Zoological Museum of Hamburg, Hamburg; Mr. Tom Hansknecht, Barry Vittor and Associates Inc., Mobile, Alabama; Dr. Richard Heard, Gulf Coast Research Laboratory, Ocean Springs, Mississippi; Ms. Julian Knight, Gulf Coast Research Laboratory, Ocean Springs, Mississippi; Mr. Jerry McLelland, Gulf Coast Research Laboratory, Ocean Springs, Mississippi. Thanks are extended to Dr. Mary Wicksten, Texas A&M University, College Station for proof reading of the manuscript and correction of the language.

Furthermore, I would like to acknowledge the extensive and very constructive work of the reviewers and the editor, which helped increase the quality of this study.

LITERATURE CITED

AFZELIUS, B. A., 1972. Sperm morphology and fertilization biology. In: R. A. BEATTY & S. GLUECKSOHN-WAELSCH, The genetics of the spermatozoon: 131–143. (Bogtrykkeriet Forum, Copenhagen).

ALLDREDGE, A. L. & J. M. KING, 1980. Effects of moonlight on the vertical migration patterns of demersal zooplankton. Journal of Experimental Marine Biology and Ecology, **44**: 133–156.

ALTENBACH, A. V. & M. SARTHEIN, 1989. Productivity record in benthic Foraminifera. In: W. H. BERGER, V. S. SMETACEK & G. WEFER, Productivity of the oceans: present and past: 255–279. (Wiley, New York).

ALVAREZ-CADENA, J. N., M. E. ISLAS-LANDEROS & E. SUÁREZ-MORALES, 1996. A preliminary zooplankton survey in a Mexican Caribbean eutrophic coastal lagoon. Bulletin of Marine Science, **58**(3): 694–708.

ANDERSSON, A., E. HALLBERG & J.-B. JOHNSON, 1978. The fine structure of the compound eye of *Tanais cavolinii* Milne-Edwards (Crustacea: Tanaidacea). Acta Zoologica, Stockholm, **59**: 49–55.

ANDERSON, G., R. W. HEARD & K. LARSEN, 2000 [and onwards]. Tanaidacea Homepage, http://tidepool.st.usm.edu/tanaids/

ANGEL, M. V., 1997. Spatial distribution of marine organisms: patterns and processes. In: R. J. A. ORMOND, J. D. GAGE & M. V. ANGEL, Marine biodiversity: causes and consequences: 59–109. (Cambridge University Press, Cambridge).

ARTNZ, W. E., T. BREY & V. A. GALLARDO, 1994. Antarctic zoobenthos. Oceanography and Marine Biology, an Annual Review, **32**: 241–304.

BACCETTI, B. & B. A. AFZELIUS, 1976. The biology of the sperm cell. In: A. WOLSKY, Monographs in developmental biology, **10**: 1–254. (Karger, Basel).

BĂCESCU, M., 1961. Contribution à la connaissance des Tanaidacés de la Mediterranée orientale, 1: les Apseudidae et Kalliapseudidae des côtes d'Israel. The Bulletin of the Research Council of Israel, (Zoology) **10**(b)(4): 137–170.

——, 1976a. Contribution to the knowledge of the family Pagurapseudidae (Crustacea–Tanaidacea) occurring in the infralittoral area of the west Indian Ocean (Tanzanian waters). Revue Roumaine de Biologie, (Biologie Animale) **21**(1): 3–11.

——, 1976b. Representatives of the family Synapseudidae (Crustacea, Tanaidacea) from the Tanzanian coral reefs: one new genus (*Curtipleon*) and three new species of *Synapseudes*. Travaux du Muséum d'Histoire Naturelle "Grigore Antipa", **17**: 51–63.

——, 1978. *Atlantapseudes nigrichela* n. g., sp. nov., Tanaïdacé nouveau capturé par le navire "Thalassa" dans les eaux Portugaises. Cahiers de Biologie Marine, **19**: 317–322.

——, 1981a. Contribution to the knowledge of the Monokonophora (Crustacea, Tanaidacea) of the eastern Australian coral reefs. Revue Roumaine de Biologie, (Biologie Animale) **26**(2): 111–120.

——, 1981b. Sur quelques Apseudidae (Crustacea-Tanaidacea) de la Méditerranée occidentale et de l'Adriatique, avec les diagnoses de deux taxons nouveaux: *Apseudes sicilianus* sp. nov. et *A. misarrai* sp. nov. Rapports et Proces–verbaux des Réunions, Commission Internationale pour l'Exploration Scientifique de la Mer Méditerranée, **27**(2): 215–218.

——, 1981c. *Carpoapseudes kudinovae* sp. nov. in the French waters of the Atlantic. Revue Roumaine de Biologie, (Biologie Animale) **26**(1): 9–14.

BĂCESCU, M. & M. GUȚU, 1974. *Halmyrapseudes cubanensis* n. g. sp. nov. and *H. bahamensis* sp. nov., brackish–water species of Tanaidacea (Crustacea). Travaux du Muséum d'Histoire Naturelle "Grigore Antipa", **15**: 91–101.

—— & ——, 1975. A new genus (*Discapseudes* n. g.) and three new species of Apseudidae (Crustacea, Tanaidacea) from the northeastern coast of South America. Zoologische Mededelingen, Leiden, **49**(11): 95–113.

BĂCESCU, M. & I. WILLIAMS, 1988. *Pectinapseudes carolinensis*, new gen., sp. nov. (Tanaidacea) discovered in continental slope waters of North and South Carolina (USA). Revue Roumaine de Biologie, (Biologie animale) **33**(1): 3–6.

BAILEY-BROCK, J. H., 1984. Ecology and tube–building of the polychaete *Diopatra leuckarti* Kinberg, 1865 (Onuphidae) in Hawaii: community structure and sediment stabilising properties. Zoological Journal of the Linnean Society, London, **80**: 191–199.

BALASUBRAMANIAN, T., C. B. LALITHAMBIKA-DEVI & M. KRISHNANKUTTY, 1979. Feeding behaviour and preying efficiency of *Metapenaeus dobsoni*. Indian Journal of Marine Science, **8**(3): 197–200.

BALDINGER, A. J. & M. F. GABLE, 1996. The occurrence of amphipods and other peracarid crustaceans in the rocky littoral zone of Bermuda. Polskie Archiwum Hydrobiologii, **42**(4): 431–439.

BAMBER, R. N., 1985. The autecology of *Cyathura carinata* (Crustacea: Isopoda) in a cooling water discharge lagoon. Journal of the Marine Biological Association of the United Kingdom, **65**: 181–194.

——, 1990. A new species of *Zeuxo* (Crustacea: Tanaidacea) from the French Atlantic coast. Journal of Natural History, London, **24**: 1587–1596.

——, 1993. A new species of *Kalliapseudes* (Crustacea: Tanaidacea: Kalliapseudidae) from Trinidad. Proceedings of the Biological Society of Washington, **106**: 122–130.

——, 1997. Peracarid crustaceans from Cape D'Aguilar and Hong Kong, II. Tanaidacea: Apseudomorpha. In: B. MORTON, The marine flora and fauna of Hong Kong and southern China, IV. Proceedings of the Eighth International Marine Biological workshop: The Marine Flora and Fauna of Hong Kong and Southern China. Hong Kong, 2–20 April 1995: 88–102. (Hong Kong University Press, Hong Kong).

——, 1998. Zoogeographic trends in some Hong Kong arthropods. In: B. MORTON, The marine flora and fauna of Hong Kong and southern China, IV. Proceedings of the Third International Conference on the Marine Biology of the South China Sea. Hong Kong, 28 October–1 November 1996: 92–112. (Hong Kong University Press, Hong Kong).

BAMBER, R. N. & G. J. BIRD, 1997. Peracarid crustaceans from Cape D'Aguilar and Hong Kong, III. Tanaidacea: Tanaidomorpha. In: B. MORTON, The marine flora and fauna of Hong Kong and southern China, IV. Proceedings of the Eighth International Marine Biological workshop: The Marine Flora and Fauna of Hong Kong and Southern China. Hong Kong, 2–20 April 1995: 104–142. (Hong Kong University Press, Hong Kong).

BAMBER, R. N., G. J. BIRD & S. ANGSUPANICH, 2003. Tanaidacea (Crustacea: Peracarida) from Thailand: new records and new species. Asian Marine Biology, **18**: 35–69.

BARNARD, J. L., 1961. Gammaridean Amphipoda from depths of 400 to 6000 meters. Galathea Report, **4**: 23–128.

——, 1962. South Atlantic abyssal amphipods collected by R. V. Vema. Vema Research Series, **1**: 1–78.

——, 1970. Benthic ecology of Bahia de San Quintin, Baja California. Smithsonian Contributions to Zoology, **44**: 1–60.

BARNARD, K. H., 1935. Report on the Amphipoda, Isopoda and Tanaidacea in the collection of the Indian Museum. Records of the Indian Museum, **37**: 279–319.

BATE, C. S. & J. O. WESTWOOD, 1866. A history of British sessile-eyed Crustacea, **2**(14): 117–154. (Jan van Voorst, London).

BEDDARD, F. E., 1886. Report on the Isopoda collected by HMS Challenger during the years 1873–1876. Second part. Challenger Report, (Zoology) **17**(1): 1–178.

BERREUR-BONNENFANT, J., 1967. Action de la glande androgène et du cerveau sur la gamétogenèse de Crustacés Péracarides. Archives de Zoologie Expérimentale et Générale, **108**: 521–558.

BIERBAUM, C. K. & E. L. WEGNER, 1993. Trapping of necrophagous crustaceans on the upper continental slope off South Carolina, U.S.A. Journal of Crustacean Biology, **13**(3): 601–608.

BIRD, G. J., 1999. A new species of *Pseudotanais* (Crustacea, Tanaidacea) from the cold seeps in the deep Caribbean, collected by the French submersible Nautile. Zoosystema, **21**(3): 445–451.

——, 2004a. The Tanaidacea (Crustacea, Peracarida) of the northeast Atlantic: the shelf and bathyal *Paratyphlotanais* of the 'Atlantic Margin'. Journal of Natural History, London, **38**(11): 1359–1384.

——, 2004b. Tanaidacea (Crustacea) of the northeast Atlantic: non–filiform species of Anarthruridae Lang from the Atlantic Margin. Zootaxa, **471**: 1–44.

BIRD, G. J. & R. N. BAMBER, 2000. Additions to the tanaidomorph tanaidaceans (Crustacea: Peracarida) of Hong Kong. In: B. MORTON, The marine flora and fauna of Hong Kong and southern China, IV. Proceedings of the Tenth International Marine Biological workshop: The Marine Flora and Fauna of Hong Kong and Southern China. Hong Kong, 2–26 April 1998, 2000: 66–104. (Hong Kong University Press, Hong Kong).

BIRD, G. J. & D. M. HOLDICH, 1984. New deep-sea leptognathiid tanaids (Crustacea, Tanaidacea) from the north–east Atlantic. Zoologica Scripta, **13**(4): 285–315.

—— & ——, 1985. A remarkable tubicolous tanaid (Crustacea: Tanaidacea) from the Rockall Trough. Journal of the Marine Biological Association of the United Kingdom, **65**: 563–572.

—— & ——, 1988. Deep-sea Tanaidacea (Crustacea) of the north–east Atlantic: the tribe Agathotanaini. Journal of Natural History, London, **22**: 1591–1621.

—— & ——, 1989a. Deep-sea Tanaidacea (Crustacea) of the north–east Atlantic: the genus *Paranarthrura* Hansen. Journal of Natural History, London, **23**: 137–167.

—— & ——, 1989b. Recolonisation of artificial sediment in the deep Bay of Biscay by tanaidaceans (Crustacea: Peracarida), with a description of a new species of *Pseudotanais*. Journal of the Marine Biological Association of the United Kingdom, **69**: 307–317.

—— & ——, 1989c. Tanaidacea (Crustacea) of the north–east Atlantic: the subfamily Pseudotanainae (Pseudotanaidae) and the family Nototanaidae. Zoological Journal of the Linnean Society, London, **97**: 233–298.

BLANC, H., 1884. Contribution à l'histoire naturelle des Asellotes Heteroptères. Observations faites sur la *Tanais oerstedii* Krøyer. Records de Zoologie Suisse, **1**: 189–258.

BŁAŻEWICZ, M. & K. JAŻDŻEWSKI, 1996. A contribution to the knowledge of Tanaidacea (Crustacea, Malacostraca) of Admiralty Bay, King George Island, Antarctica. Polish Polar Research, **17**: 213–220.

BŁAŻEWICZ-PASZKOWYCZ, M. & K. LARSEN, 2004. Three new deep-sea species of Apseudidae (Tanaidacea; Apseudomorpha) from the southern seas and Antarctica. Crustaceana, **77**: 467–498.

BŁAŻEWICZ-PASZKOWYCZ, M. & R. LIGOWSKI, 2002. Diatoms as food source indicator for some Antarctic Cumacea and Tanaidacea (Crustacea). Antarctic Science, **14**(1): 11–15.

BOLAND, P. N. & G. T. ROWE, 1991. Deep-sea benthic sampling with the GOMEX box corer. Limnology and Oceanography, **35**(5): 1015–1020.

BOLTT, R. E., 1969. The benthos of some South African lakes part II: the epifauna and infauna of the benthos of Lake Sibayi. Transactions of the Royal Society of South Africa, **38**(3): 249–269.

BOT, P. V. M., W. T. STAM, S. A. BOELE-BOS, C. VAN DEN HOEK & W. VAN DELDEN, 1989. Biogeographic and phylogenetic studies in three North Atlantic species of *Cladophora* (Cladophorales, Chlorophyta) using DNA-DNA hybridization. Phycologia, **28**: 159–168.

BOXSHALL, G. A. & R. J. LINCOLN, 1987. The lifecycle of Tantulocarida (Crustacea). Philosophical Transactions of the Royal Society of London, (B, Biological Sciences) **315**: 267–303.

BOUCHET, P. & A. WAREN, 1979. Planktotrophic larvae development in deep-water gastropods. Sarsia, **64**: 37–40.

— — & — —, 1985. Revision of the northeast Atlantic bathyal and abyssal Aclididae, Eulimnidae and Epitoniidae (Mollusca, Gastropoda). Bulletin Malacologia, (Supplement) **2**: 297–576.

BRIGHT, T. J., 1970. Food of deep-sea fishes. Contributions on the Biology of the Gulf of Mexico, **1**: 345–252.

BROWN, A. C., 1958. Report on the tanaidacean Crustacea of the Langebaan Lagoon and Saldanha Bay, on the west coast of South Africa. Annals and Magazine of Natural History, (13) **1**: 453–458.

BROWNE, D. W. & P. A. BERKMAN, 1992. Epizooic Foraminifera, tanaid and polychaete species association on Antarctic scallop shell. Antarctic Journal of the United States, **27**(5): 134–135.

BOUVIER, E. L., 1918. Sur une petite collection de Crustacés de Cuba offerte au Muséum par M. de Boury. Bulletin du Muséum National d'Histoire Naturelle, Paris, **24**: 6–15.

BÜCKLE-RAMIREZ, L. F., 1965. Unterschungen über die Biologie von *Heterotanais oerstedi* [sic] (Krøyer). Zeitschrift für Morphologie und Ökologie der Tiere, **55**: 711–782.

CAINE, E. A., 1986. Carapace epibionts of nesting loggerhead sea turtles: Atlantic coast of U.S.A. Journal of Experimental Marine Biology and Ecology, **95**: 15–26.

CHARNIAUX-COTTON, H., 1960. Sex determination. In: T. H. WATERMAN (ed.), The physiology of Crustacea, **4**: 411–447. (Academic Press, New York).

CHRISTIE, G., 1985. A comparative study of the reproductive cycles of three Northumberland populations of *Chaetozone setosa* (Polychaeta: Cirratulidae). Journal of the Marine Biological Association of the United Kingdom, **65**: 239–254.

CLAUS, C., 1884. Ueber *Apseudes latreillii* M-Edw. und die Tanaiden. Arbeiten aus dem Zoologischen Institut der Universität Wien, **5**: 319–332.

— —, 1887. Ueber den Organismus der Apseudiden. Anzeiger der Wissenschaftlichen Akademien aus Wien, (Mathematische und Naturwissenschaftliche Klasse) **24**: 156–161.

— —, 1888. Ueber *Apseudes latreillii* M-Edw. und die Tanaiden, II. Arbeiten aus dem Zoologischen Institut der Universität Wien, **7**: 139–220.

COSSON, N., M. SIBUET & J. GALERON, 1997. Community structure and spatial heterogeneity of the deep-sea macrofauna at three contrasting stations in the tropical northeast Atlantic. Deep-sea Research, **44**(2): 247–269.

COSTA, S., 1960. Note préliminaire sur l'éthologie alimentaire de deux Caprellides de la rade de Villefranche-sur-Mer. Travaux du Station Zoologique de Villefranche-sur-Mer, **19**(20): 103–105.

COTELLI, F. & C. L. L. DONIN, 1980. The spermatozoon of Peracarida II. The spermatozoon of Tanaidacea. Journal of Ultrastructure Research, **73**: 263–268.

COTELLI, F., M. FERRAGUTI, G. LAVAVECCHIA & C. L. L. DONIN, 1976. The spermatozoon of Peracarida I. The spermatozoon of terrestrial isopods. Journal of Ultrastructure Research, **55**: 378–390.

DANA, J. D., 1849. Conspectus Crustaceorum, Conspectus of the Crustacea of the U. S. Exploring Expedition. American Journal of Science, **2**: 424–428.

——, 1852. On the classification of the Crustacea Choristopoda or Tetrapoda. American Journal of Science, (2) **14**: 197–306.

DAWSON, C. E., 1966. Additions to the known marine fauna of Grand Isle, Louisiana. Proceedings of the Louisiana Academy of Science, **29**: 175–180.

DELILLE, D., L. D. GUIDI & J. SOYER, 1985. Nutrition of *Allotanais hirsutus* (Crustacea, Tanaidacea) at Kerguelen Islands. In: W. R. SIEGFRIED, P. R. CONDY & R. M. LAWS, Antarctic nutrient cycles and food webs: 378–380. (Springer-Verlag, Berlin).

DENNELL, R., 1937. On the feeding mechanism of *Apseudes talpa* and the evolution of the Peracarida. Transactions of the Royal Society of Edinburgh, **59**: 57–78.

DENTON, E. J., 1974. On buoyancy and the lives of modern and fossil cephalopods. Proceedings of the Royal Society of London, (B, Biological Sciences) **185**: 273–299.

DOJIRI, M. & J. SIEG, 1997. The Tanaidacea. In: J. A. BLAKE & P. H. SCOTT, Taxonomic atlas of the benthic fauna of the Santa Maria Basin and western Santa Barbara Channel, **11**, The Crustacea (2), The Isopoda, Cumacea and Tanaidacea, (1–5): 181–278. (Santa Barbara Museum of Natural History, Santa Barbara, California).

DOLLFUS, A., 1897. Note préliminaire sur les Tanaidae recueillis aux Açores pendant les campagnes de l'Hirondelle (1887–1888). Bulletin de la Société Zoologique de France, **21**: 207–215.

DOMANSKY, P. A., 1984. Giant larvae: prolonged planktonic larval phase in the asteroid *Luidia sarsi*. Marine Biology, Berlin, **80**: 189–195.

DOYLE, R. W., 1972. Genetic variation in *Ophiomusium lymani* (Echinodermata) populations in the deep sea. Deep-sea Research, **19**: 661–664.

DRUMM, D. T., 2003. *Kalliapseudes* (*Mesokalliapseudes*) *macsweenyi* a new species of Kalliapseudidae (Crustacea: Tanaidacea: Apseudomorpha) from mangroves in Florida. Zootaxa, **357**: 1–12.

——, 2005a. Comparative morphology of the mouthparts, chelipeds and foregut of two kalliapseudid apseudomorphans (Crustacea: Tanaidacea). Proceedings of the Academy of Natural Sciences of Philadelphia, **154**(1): 137–147.

——, 2005b. Comparison of feeding mechanisms, respiration and cleaning behaviour in two kalliapseudids, *Kalliapseudes macsweenyi* and *Psammokalliapseudes granulosus* (Peracarida: Tanaidacea). Gulf and Caribbean Research, **25**(2): 203–211.

ECKELBARGER, K. J., 1994. Ultrastructural features of gonads and gametes in deep-sea invertebrates. In: C. M. YOUNG & K. J. ECKLEBARGER (eds.), Reproduction, larval biology and recruitment of the deep-sea benthos: 256–301. (Columbia University Press, New York).

EL-NASSRY, M., 1973. Biological study on Crustacea associated with fouling growths in the eastern harbour of Alexandria: 1–348. (Unpubl. M.Sc. Thesis, Faculty of Science, Alexandria University, Alexandria).

EKMAN, S., 1953. Zoogeography of the sea: 1–417. (Sidgwick and Jackson, Ltd., London).

ESCOBAR-BRIONES, E., F. ALVAREZ & G. SALGADO-MALDONADO, 1999. *Discapseudes holthuisi* (Crustacea: Tanaidacae) as an intermediate host of *Caballerhynchus lamothei* (Acanthocephala: Cavisomidae). Journal of Parasitology, **85**(1): 134–137.

ETTER, R. J. & J. F. GRASSLE, 1992. Patterns of species diversity in the deep sea as a function of sediment particle size diversity. Nature, London, **360**: 576–578.

FAIN-MAUREL, M. A., J. F. REGER & P. CASSIER, 1975. Le gamète mâle des Schizopodes et ses analogies avec celui des autres Péracarides. Journal of Ultrastructure Research, **51**: 378–390.

FELLER, R. J., 1978. Predation on meiofauna established with immunological methods. American Zoologist, (Abstract) **18**(3): 662.

FELSENSTEIN, J., 1975. A pain in the torus: some difficulties with models of isolation with distances. American Naturalist, **109**(967): 359–368.

FORSMAN, B., 1938. Untersuchungen über die Cumaceen des Skagerraks. Zoologiska Bidrag från Uppsala, **18**: 1–162.

GAGE, J. D., 1996. Why are there so many species in the deep-sea sediments? Journal of Experimental Marine Biology and Ecology, **200**: 257–286.

GAGE, J. D. & G. G. COGHILL, 1977. Studies on the dispersion patterns of Scottish sea loch benthos from contiguous core transects. In: C. COULL, Ecology of marine benthos: 319–359. (University of South Carolina Press, Charleston).

GAGE, J. D. & M. MAY, 1993. A dip in the deep seas. Nature, London, **365**: 609–610.

GAMBLE, J. C., 1970. Effect of low dissolved oxygen concentrations on the ventilating rhythm of three tubicolous crustaceans, with special reference to the phenomena of intermittent ventilation. Marine Biology, Berlin, **6**: 121–127.

GAMO, S., 1984. A new remarkably giant tanaid, *Gigapseudes maximus* sp. nov. (Crustacea) from abyssal depths far off southeast of Mindanao, the Philippines. Scientific Reports of Yokohama Natural University Series, **11**(31): 1–12.

GARDINER, L. F., 1973a. A new species and genus of a new monokonophoran family (Crustacea: Tanaidacea), from southeastern Florida. Journal of Zoology, London, **169**: 237–253.

— —, 1973b. *Calozodion wadei*, a new genus and species of apseudid tanaidacean (Crustacea) from Jamaica. Journal of Natural History, London, **7**: 499–507.

— —, 1973c. New species of the genera *Synapseudes* and *Cyclopoapseudes* with notes on morphological variation, postmarsupial development and phylogenetic relationships within the family Metapseudidae (Crustacea: Tanaidacea). Zoological Journal of the Linnean Society, London, **53**: 25–58.

— —, 1975a. A fresh and brackish water tanaidacean *Tanais stanfordi* Richardson, 1901, from a hypersaline lake in the Galapagos Archipelago, with a report on West Indian specimens. Crustaceana, **29**: 127–140.

— —, 1975b. The systematics, postmarsupial development and ecology of the deep-sea family Neotanaidae (Crustacea: Tanaidacea). Smithsonian Contributions to Zoology, **170**: 1–265.

— —, 1977. Tanaidacea. In: S. H. HURLBERT, Biota acuática de Sudamerica Australis, **14**: 342. (State University of California at San Diego, San Diego, California).

GERKEN, S. & L. WATLING, 1997. Oceanographic relationships of the Cumacea from the Faeroe Islands. Abstract. The Crustacean Society 1997 Summer Meeting, Alabama, 21–24 May 1997. Journal of Crustacean Biology, **18**(5).

GERSTAECKER, A., 1883. Flohkrebse. In: A. GERSTAECKER, Bronns Klassen und Ordnungen des Tierreichs. Die Klassen und Ordnungen der Arthropoden, (2), Crustacea Malacostraca, **1**(5): 279–543.

GILLANDERS, B. M., 1995. Feeding ecology of the temperate marine fish *Achoerodus viridis* (Labridae). Size, seasonal and site specific differences. Marine and Freshwater Research, **46**(7): 1009–1020.

GHOBASHY, A. F., M. M. EL-KOMI & S. H. E. RAMADAN, 1980. Fouling in the Suez Canal. Proceedings of the 5th International Congress on Marine Correlated Fouling: 75–92.

GOLDMAN, D. & C. E. MITCHELL, 1990. Morphology, systematics and evolution of Middle Devonian Ambocoeliidae (Brachiopoda), western New York. Journal of Paleontology, **64**(1): 79–98.

GOODAY, A. J., 1984. Records of deep-sea rhizopod tests inhabited by metazoans in the north-east Atlantic. Sarsia, **69**: 45–53.

GOODAY, A. J., L. A. LEVIN, P. LINKE & T. HEEGER, 1992. The role of benthic Foraminifera in deep-sea foodwebs and carbon cycling. In: G. T. ROWE & V. PARIENTE, Deep-sea food chains and global carbon cycling: 63–91. (Kluwer Academic Publishers, Dordrecht).

GRANT, W. S., 1984. Biochemical population genetics of Atlantic herring, *Clupea harengus*. Copeia, **1984**: 357–364.

GRASSLE, J. F., 1989. Species richness in deep-sea communities. Trends in Ecology and Evolution, **4**: 12–15.

GRASSLE, J. F. & J. P. GRASSLE, 1976. Sibling species in marine pollution indicator *Capitella* (Polychaeta). Science, New York, **192**: 567–569.

GRASSLE, J. F. & N. J. MACIOLEK, 1992. Deep-sea species richness: regional and local diversity estimates from quantitative bottom samples. American Naturalist, **139**: 313–341.

GRASSLE, J. F., N. J. MACIOLEK & J. A. BLAKE, 1990. Are deep-sea communities resilient? In: G. M. WOODWELL, The earth in transition: 385–393. (Cambridge University Press, Cambridge).

GRASSLE, J. F. & H. L. SANDERS, 1973. Life histories and the role of disturbance. Deep-sea Research, **20**: 643–659.

GRASSLE, J. F., H. L. SANDERS, R. R. HESSLER, G. T. ROWE & T. MCLELLAN, 1975. Patterns and zonation: a study of the bathyal megafauna using the research submersible Alvin. Deep-sea Research, **22**: 457–481.

GRAY, J. S., 1974. Animal–sedimentation relationships. Oceanography and Marine Biology, an Annual Review, **12**: 223–261.

— —, 1994. Is deep-sea diversity really so high? Species diversity of the Norwegian continental shelf. Marine Ecology Progress Series, **112**: 205–207.

GREVE, L., 1965. A new epicaridean from western Norway, parasite on Tanaidacea. Sarsia, **20**: 15–19.

— —, 1967. On the tube building of some Tanaidacea. Sarsia, **29**: 295–298.

GRYGIER, M. J. & J. SIEG, 1990. *Microdajus* (Crustacea, Tantulocarida) parasitic on an Antarctic tanaidacean and a range extension of *M. langi* Greve. Journal of Natural History, London, **23**(5): 1495–1505.

GUERRERO-KOMMRITZ, J., 2005. Review of the genus *Chauliopleona* Dojiri & Sieg, 1997 (Crustacea, Peracarida) and description of three new species. Journal of Natural History, London, **39**(16): 1177–1210.

GUERRERO-KOMMRITZ, J. & R. W. HEARD, 2003. A new genus and species of deep-sea apseudomorph tanaidacean (Crustacea: Malacostraca: Peracarida) from the Angola Basin. Mitteilungen aus dem Hamburgischen Zoologischen Museum und Institut, **100**: 127–137.

GUERRERO-KOMMRITZ, J., A. SCHMIDT & A. BRANDT, 2002. *Paranarthrura* Hansen, 1913 (Crustacea: Tanaidacea) from the Angola Basin, description of *Paranarthrura angolensis* sp. nov. Zootaxa, **116**: 1–12.

GUŢU, M., 1972. Phylogenetic and systematic considerations upon the Monokonophora (Crustacea, Tanaidacea) with the suggestion of a new family and several new subfamilies. Revue Roumaine de Biologie, (Biologie Animale) **17**(5): 297–305.

— —, 1980a. On the status of the "groups" *Leiopus* and *Carpoapseudes* (Crustacea, Tanaidacea) and their systematic position. Travaux du Muséum d'Histoire Naturelle "Grigore Antipa", **22**: 385–392.

——, 1980b. *Pseudosphyrapus*, a new genus of a new family (Sphyrapidae) of Monokonophora (Crustacea, Tanaidacea). Travaux du Muséum d'Histoire Naturelle "Grigore Antipa", **22**: 393–400.

——, 1981. A new contribution to the systematics and phylogeny of the suborder Monokonophora (Crustacea, Tanaidacea). Travaux du Muséum d'Histoire Naturelle "Grigore Antipa", **23**: 81–108.

——, 1984. Contribution to the knowledge of the genus *Calozodion* (Crustacea, Tanaidacea). Travaux du Muséum National d'Histoire Naturelle "Grigore Antipa", **26**: 35–43.

——, 1986. Description of *Apseudes olimpiae* sp. nov. and of *Tanabnormia cornicauda* n. g., sp. nov. (Crustacea, Tanaidacea). Travaux du Muséum d'Histoire Naturelle "Grigore Antipa", **28**: 37–48.

——, 1989. Tanaidacea (Crustacea) collected by the "Benthédi" French Expedition (1977) in south–western Indian ocean. I. Travaux du Muséum d'Histoire Naturelle "Grigore Antipa", **30**: 135–160.

——, 1991. A few remarks on Sphyrapidae (Crustacea, Tanaidacea) and the description of a new genus *Kudinopasternakia*, belonging to this family. Travaux du Muséum d'Histoire Naturelle "Grigore Antipa", **31**: 341–348.

——, 1995. A new subfamily and three new genera of Apseudomorpha (Crustacea, Tanaidacea). Travaux du Muséum d'Histoire Naturelle "Grigore Antipa", **35**: 17–28.

——, 1996a. Tanaidaceans (Crustacea, Peracarida) from Brazil, with descriptions of new taxa and systematical remarks on some families. Travaux du Muséum d'Histoire Naturelle "Grigore Antipa", **35**: 23–133.

——, 1996b. The synoptic table and key to superspecific taxa of recent Apseudodomorpha (Crustacea, Tanaidacea). Travaux du Muséum d'Histoire Naturelle "Grigore Antipa", **35**: 135–146.

——, 1996c. The description of *Spinosapseudes* n. g. and amended diagnoses of two genera of Tanaidacea (Crustacea). Revue Roumaine de Biologie, (Biologie Animale) **41**: 87–93.

——, 1997. Tanaidacea. In: M. Guțu (ed.), Results of the zoological expedition organized by "Grigore Antipa" museum in the Indonesian archipelago (1991). 1. Peracarida (Crustacea). Travaux du Muséum d'Histoire Naturelle "Grigore Antipa", **38**: 259–327.

——, 1998a. Importance of the pereopodal microstructures for the identification of some apseudomorphan species and genera (Crustacea, Tanaidacea). Travaux du Muséum d'Histoire Naturelle "Grigore Antipa", **40**: 137–152.

——, 1998b. New data on the genus *Parapseudes* G. O. Sars, 1882 (Crustacea, Tanaidacea) and the description of the species *P. trispinosus* sp. nov. from Indonesia. Travaux du Muséum d'Histoire Naturelle "Grigore Antipa", **40**: 165–177.

——, 2002a. Contributions to the knowledge of the genus *Apseudes* Leach, 1814. (Crustacea: Tanaidacea, Apseudomorpha) from the Mediterrane basin and North African Atlantic. Travaux du Muséum d'Histoire Naturelle "Grigore Antipa", **44**: 19–39.

——, 2002b. New apseudid and metapseudid taxa (Crustacea: Tanaidacea) from the Gulf of Mexico and new diagnoses of some genera. Travaux du Muséum d'Histoire Naturelle "Grigore Antipa", **44**: 41–68.

——, 2004. Regarding the position of the fossil superfamily Cretitanaoidea in the Tanaidacean systematics (Crustacea: Peracarida). Travaux du Muséum d'Histoire Naturelle "Grigore Antipa", **46**: 25–28.

Guțu, M. & O. Gomez, 1976. *Pagurapseudes guitarti* new species of Tanaidacea (Crustacea) from the Caribbean Sea. Travaux du Musée d'Histoire Naturelle "Grigore Antipa", **17**: 85–91.

GUŢU, M. & R. W. HEARD, 2002. A new genus and four new species of parapseudid and sphyrapid apseudomorphans (Crustacea: Tanaidacea) from the Caribbean Sea and the Gulf of Mexico. Travaux du Muséum National d'Histoire Naturelle "Grigore Antipa", **44**: 69–92.

GUŢU, M. & T. M. ILIFFE, 1989a. *Apseudes orghidani*, a new species of Tanaidacea (Crustacea) from an anchialine cave on Bermuda. Travaux du Muséum d'Histoire Naturelle "Grigore Antipa", **30**: 161–167.

—— & ——, 1989b. Description of two new species of Tanaidacea (Crustacea) from the marine water caves of the Palau Islands (Pacific Ocean). Travaux du Muséum d'Histoire Naturelle "Grigore Antipa", **30**: 169–180.

—— & ——, 1997. *Pugiodactylus agarthus*, a new species of Apseudomorpha (Crustacea, Tanaidacea) from a marine cave in the south Pacific. Travaux du Muséum d'Histoire Naturelle "Grigore Antipa", **39**: 31–44.

—— & ——, 1998. *Sphyrapoides bicornis*, new genus and new species from a submarine escarpment in the Bahamas and some remarks on Sphyrapinae (Crustacea, Tanaidacea). Travaux du Muséum d'Histoire Naturelle "Grigore Antipa", **40**: 211–225.

—— & ——, 2001. *Grallatotanais antipai*, a new genus and species of the family Leptocheliidae Lang, 1963 [sic] from a marine cave in the Bahamas (Crustacea: Tanaidacea: Tanaidomorpha). Travaux du Muséum d'Histoire Naturelle "Grigore Antipa", **43**: 93–100.

GUŢU, M. & J. SIEG, 1999. Ordre des Tanaïdacés. In: J. FOREST (ed.), Traité de Zoologie **7**(3)(A), Crustacés Péricarides. Mémoires de l'Institut Océanographique, Monaco, **19**: 353–389.

HAEDRICH, R. L., G. T. LOWE & P. T. POLLONI, 1975. Zonation and faunal composition of epibenthic populations on the continental slope of New England. Journal of Marine Research, **33**: 191–212.

HANNIBAL, J. T., 2003. Deepwater Tanaidacea (Crustacea: Neotanaidae) from the Ladinina (Middle Triassic) of Hungary. Abstract # 22–2. Geological Society of America, 37th Annual Meeting (24–25 March 2003), Kansas City, Kansas.

HANSEN, B., 1967. The taxonomy and zoogeography of the deep-sea holothurians in their evolutionary aspects. Studies in Tropical Oceanography, **5**: 480–501.

——, 1975. Systematics and biology of the deep-sea holothurians. Galathea Report, **13**: 1–262.

HANSEN, H. J., 1895. Isopoden, Cumaceen und Stomatopoden der Plankton Expedition. II. Ordung: Tanaidacea. Ergebnissen der Plankton Expedition, **2**(G.C.): 49–50.

——, 1913. Crustacea, Malacostraca, II. Danish Ingolf Expedition, **3**(3): 1–127.

HANSKNECHT, T. & R. W. HEARD, 2001. Tanaidacea (Crustacea: Peracarida) of the Gulf of Mexico. IX. Geographical occurrence of *Apseudes olimpiae* Guţu, 1986 with a review of previous records for the genus *Apseudes* in the Gulf. Gulf and Caribbean Research, **13**: 67–73.

HANSKNECHT, T., R. W. HEARD & R. BAMBER, 2002. *Tanapseudes gutui*, a new species of apseudomorphan Tanaidacea (Crustacea: Peracarida) from the Caribbean Sea and the taxonomic status of the family Tanapseudidae Băcescu, 1978. Gulf and Caribbean Research, **14**: 67–74.

HANSKNECHT, T., R. W. HEARD & J. W. MARTIN, 2001. *Saltipedis navassensis*, a new species of apseudomorph tanaidacean (Crustacea: Peracarida: Parapseudidae) from the tropical northwestern Atlantic. Zootaxa, **18**: 1–11.

HARGER, O., 1879. Notes on New England Isopoda. Proceeding of the U. S. National Museum, **2**: 157–165.

HARTMAN, O. & K. FAUCHALD, 1971. Deep water benthic polychaetous annelids off New England to Bermuda and other North–Atlantic areas. Part II. Allan Hancock Monographs of Marine Biology, **6**: 1–327.

HARTNEY, K. B. & L. TUMYAN, 1998. Temporal changes in diet and foraging habitat of California killifish (*Fundulus parvipinnis*) in Marina del Rey, California. Bulletin of the Southern California Academy of Sciences, **97**(1): 1–8.

HASSACK, E. & D. M. HOLDICH, 1987. The tubicolous habit amongst the Tanaidacea (Crustacea, Peracarida) with particular reference to deep-sea species. Zoologica Scripta, **16**(3): 223–233.

HEARD, R. W., 1982. Guide to common tidal marsh invertebrates of the northeastern Gulf of Mexico: 1–82. (Mississippi–Alabama Sea Grant Consortium, Publ. No. MASGP–79–004; Ocean Springs, Mississippi).

HEARD, R. W., T. HANSKNECHT & K. LARSEN, 2004. An illustrated identification guide to Florida tanaidaceans (Crustacea, Peracarida) occurring in depths of less than 200 meters: 1–92. (State of Florida, Department of Environmental Protection, Tallahassee, Florida).

HESSLER, R. R., 1970. The Desmosomatidae (Isopoda, Asellota) of the Gay Head–Bermuda Transect. Bulletin of the Scripps Institution of Oceanography, **15**: 1–185.

——, 1982. The structural morphology of the walking mechanisms of the eumalacostracan Crustacea. Philosophical Transactions of the Royal Society of London, **296**: 245–298.

HETTLER, W. F., JR., 1989. Nekton use of regularly–flooded saltmarsh cordgrass habitat in North Carolina, USA. Marine Ecology Progress Series, **56**: 111–118.

HICKEY, C. N. & M. L. MARTIN, 1995. Relative sensitivity of five benthic invertebrate species to reference toxicants and resin–acid contaminated sediment. Environment Toxicology and Chemistry, **14**(8): 1401–1409.

HIGHSMITH, R. C., 1982. Induced settlements and metamorphoses of sand dollar (*Dendraster excentricus*) larvae in predator free sites: adult sand dollar beds. Ecology, **63**: 329–337.

——, 1983. Sex reversal and fighting behaviour: coevolved phenomena in a tanaid crustacean. Ecology, **64**: 719–726.

——, 1985. Floating and algal rafting as potential dispersal mechanisms in brooding invertebrates. Marine Ecology Progress Series, **25**: 169–179.

HOBSON, E. & J. R. CHESS, 1976. Trophic interactions among fishes and zooplankton near the shore at Santa Catalina Island, California. Fishery Bulletin, U. S., **74**: 567–598.

—— & ——, 1978. Trophic relationships among fishes and plankton in the lagoon at Enewatak Atoll, Marshall Islands. Fishery Bulletin, U. S., **76**: 133–153.

HODSON, R. G., J. O. HACKMAN & C. R. BENNET, 1981. Food habits of young spots in nursery areas of the Cape Fear River estuary North Carolina. Transactions of the American Fisheries Society, **110**: 495–501.

HOESE, H. D., 1972. Seasonal and spatial settling of fouling organisms in Mobile Bay and eastern Mississippi Sound, Alabama. Alabama Marine Research Bulletin, **8**: 9–17.

HOFMANN, E. L., J. E. CAPELLA, R. M. ROSS & L. B. QUETIN, 1992. Models for the early life history of *Euphausia superba*. Part I. Time and temperature dependence during the descent–ascent cycle. Deep-sea Research, **39**: 1177–1200.

HOLDICH, D. M. & G. J. BIRD, 1985. A preliminary report on 'dikonophoran' tanaidaceans (Crustacea). In: L. LAUBIER & C. MONNIOT, Peuplements profonds du Golfe de Gascogne. Centre National de Tri d'Océanographie Biologique: 441–447. (Centob Ifremer, Brest).

—— & ——, 1986. Tanaidacea (Crustacea) from sublittoral waters of west Scotland, including the description of two new genera. Journal of Natural History, London, **20**: 79–100.

HOLDICH, D. M. & J. A. JONES, 1983a. British tanaids. Synopses of the British Fauna, (n. s.) **27**: 1–95. (Cambridge University Press, Cambridge).

—— & ——, 1983b. The distribution and ecology of British shallow–water tanaid crustaceans (Peracarida, Tanaidacea). Journal of Natural History, London, **17**: 157–183.

HOWARD, A. D., 1952. Molluscan shells occupied by tanaids. Nautilus, **65**: 74–75.

HUH, S. H. & S. N. KWAK, 1998. Feeding habits of *Pseudoblennius cottoides*. Journal of the Korean Fisheries Society, **31**(1): 37–44.

HUYS, R., 1991. Tantulocarida (Crustacea: Maxillopoda) a new taxon from temporary meiobenthos. Pubblicazioni della Stazione Zoologica di Napoli, (1, Marine Ecology) **12**(1): 1–34.

INTERNATIONAL COMMISION ON ZOOLOGICAL NOMENCLATURE, 1985. International code of zoological nomenclature, (3rd ed.): i–xx, 1–388. (International Trust for Zoological Nomenclature, London).

— —, 1999. International code of zoological nomenclature, (4th ed.): i–xxix, 1–306. (International Trust for Zoological Nomenclature, London).

JAMIESON, B. G. M., 1990. Ultrastructure and phylogeny of crustacean spermatozoa. Memoirs of the Queensland Museum, **31**: 109–142.

JAŻDŻEWSKI, K., 1969. Biology of two hermaphroditic Crustacea, *Cyathura carinata* (Krøyer) (Isopoda) and *Heterotanais oerstedi* [sic] (Krøyer) (Tanaidacea) in waters of the Polish Baltic Sea. Zoologica Poloniae, **19**: 5–25.

JOHNSON, S. B. & Y. G. ATTRAMADAL, 1982a. A functional–morphological model of *Tanais cavolinii* Milne-Edwards (Crustacea, Tanaidacea) adapted to a tubicolous life–strategy. Sarsia, **67**: 29–42.

— — & — —, 1982b. Reproductive behaviour and larval development of *Tanais cavolinii* Milne-Edwards (Crustacea, Tanaidacea). Marine Biology, Berlin, **71**: 11–16.

JONES, M. M., 1991. Marine organisms transported in ballast water: a review of the Australian scientific position. Bureau of Rural Resources, **11**: 1–45.

JONES, N. S. & H. L. SANDERS, 1972. Distribution of Cumacea in the deep Atlantic. Deep-sea Research, **19**: 737–745.

JUMARS, P. A. & K. FAUCHALD, 1977. Between–community contrast in successful polychaete feeding strategies. In: B. C. COULL (ed.), Ecology of marine benthos: 1–20. (Columbia University Press, New York).

JUMARS, P. A., L. M. MAYER, J. W. DEMMNING, J. A. BAROSS & R. A. WHEATCROFT, 1990. Deep-sea deposit–feeding strategies suggested by environmental and feeding constraints. Philosophical Transactions of the Royal Society of London, (A) **331**: 85–101.

KILLINGLEY, J. S. & M. A. REX, 1985. Mode of larval development in some deep-sea gastropods indicated by oxygen–18 values of their carbonate shells. Deep-sea Research, **32**. 809–818.

KINNE, O., 1954. Eidonomie, Anatomie und Lebenszyklus von *Sphaeroma hookeri* Leach (Isopoda). Kieler Meersforschungen, **10**(1): 100–120.

KJENNERUD, J., 1950. Ecological observations on *Idothea* [sic] *neglecta* G. O. Sars. Universitetet i Bergen, Årbok 1950, (Naturvitenskapelig Rekke) **7**: 1–47.

KNEIB, R. T., 1984. Patterns in the utilisation of the intertidal salt marsh by larvae and juveniles of *Fundulus heteroclitus* (Linneaus) and *Fundulus luciae* (Baird). Journal of Experimental Marine Biology and Ecology, **83**: 41–51.

— —, 1985. Predation and disturbance by grass shrimp *Palaemonetes pugio* Holthuis, in soft–substratum benthic invertebrate assemblages. Journal of Experimental Marine Biology and Ecology, **93**: 91–102.

— —, 1986. The role of *Fundulus heteroclitus* in saltmarsh trophic dynamics. American Zoology, **26**: 259–269.

— —, 1988. Testing for indirect effects of predation in an intertidal soft–bottom community. Ecology, **69**: 1795–1805.

— —, 1992. Population dynamics of the tanaid *Hargeria rapax* (Crustacea: Peracarida) in a tidal marsh. Marine Biology, Berlin, **113**: 437–445.

KNIGHT, J., K. LARSEN & R. W. HEARD, 2003.Tanaidacea (Crustacea: Peracarida) of the Gulf of Mexico, XI: a new deep-sea tanaidomorphan genus. Proceedings of the Biological Society of Washington, **116**(2): 497–502.

KNOWLES, L. L. & S. BELL, 1998. The influence of habitat structure in faunal–habitat associations in Tampa Bay seagrass system, Florida. Bulletin of Marine Science, **62**(3): 781–794.

KNOWLTON, N., 1993. Sibling species in the sea. Annual Review of Ecology and Systematics, **24**: 189–216.

KNOWLTON, N. & J. B. JACKSON, 1993. Inbreeding and outbreeding in marine invertebrates. In: N. W. THORNHILL, The natural history of inbreeding and outbreeding: 200–249. (Chicago University Press, Chicago, Illinois).

KOCATAS, A., 1977. The species of Peracarida and Eucarida (Crustacea, Malacostraca) from Bafa Lake, Turkey. Fen Fakueltesi Dergisi, **1977**: 311–318.

KRASNOW, L. D., 1992. Particle selection for tube building and the stabilisation of sediment by *Leptochelia dubia* (Crustacea, Tanaidacea) interactions with bacteria and microalgae: 1–372. (Unpubl. Ph.D. Thesis, Oregon State University, Corvallis, Oregon).

KRASNOW, L. D. & G. L. TAGHON, 1997. Rate of tube building and sediment size selection during tube construction by the tanaid crustacean *Leptochelia dubia*. Estuaries, **20**(3): 534–546.

KRØYER, H., 1842. Nye arter af slægten *Tanais*. Naturhistorisk Tidskrift, **4**: 167–188.

KUDINOVA-PASTERNAK, R. K., 1965. Deep-sea Tanaidacea from the Bougainville Trench of the Pacific. Crustaceana, **8**(1): 75–91.

——, 1966a. Tanaidacea Crustacea of the Pacific ultra–abyssals. Zoologicheskii Zhurnal, **45**: 518–535.

——, 1966b. On a new abyssal tanaidacean from the Pacific *Arthrura andriashevi* n. gen., n. sp. Crustaceana, **12**: 257–260.

——, 1968. Tanaidacea. In: L. A. ZENKEVITCH, The Pacific Ocean. Biology of the Pacific Ocean, **2**: 72–80. (Nauka, Moscow).

——, 1969. A case of extramarsupial development of eggs in *Typhlotanais magnificus* living in the tube. Zoologicheskii Zhurnal, **48**: 1737–1738.

——, 1970. Tanaidacea of the Kurile-Kamchatka Trench. Trudy Instituta Okeanologii, Akademiya Nauk SSSR, **86**: 341–380.

——, 1972. Notes about the tanaidacean fauna (Crustacea, Malacostraca) of the Kirmadec Trench. Complex Research of the Nature of the Ocean, Publications of Moscow University, **3**: 257–258.

——, 1973. Tanaidacea (Crustacea, Malacostraca) collected on the R/V Vitjas in regions of the Aleutian Trench and Alaska. Trudy Instituta Okeanologii, Akademiya Nauk SSSR, **91**: 141–168.

——, 1975. Tanaidacea (Crustacea, Malacostraca) from the Atlantic sector of the Antarctic and Subantarctic. Trudy Instituta Okeanologii, Akademiya Nauk SSSR, **103**: 194–228.

——, 1976. Additional notes on the fauna of Tanaidacea (Crustacea, Malacostraca) of the deep water trenches. Trudy Instituta Okeanologii, Akademiya Nauk SSSR, **99**: 115–125.

——, 1978. Deep-sea Tanaidacea (Crustacea, Malacostraca) collected in the Caribbean Sea and Puerto–Rico Trench during the 16-th cruise of R/V "Akademic Kurchatov" and the resemblance of the Caribbean region and the Pacific. Trudy Instituta Okeanologii, Akademiya Nauk SSSR, **113**: 79–198.

——, 1981. Tanaidacea. In: A. P. KUSNJEZOV & A. N. MIRONOV, Benthos of the submarine mountain Markus Necker and adjacent Pacific regions: 94–112. (Shirshov Institute of Oceanology, Academy of Science, Moscow).

— —, 1983. The abyssal Tanaidacea (Crustacea) of Iberian and West–European hollows of the Atlantic Ocean. Zoologicheskii Zhurnal, **62**(8): 1170–1176.

— —, 1984. Tanaidacea (Crustacea, Malacostraca) of the Sea of Japan. Zoologicheskii Zhurnal, **63**: 828–837.

— —, 1986. Abyssal Tanaidacea (Crustacea, Malacostraca) from the south–western part of the Indian Ocean. Suborder Tanaidomorpha. Zoologicheskii Zhurnal, **65**: 66–75.

— —, 1991. Trophic groups of Tanaidacea (Crustacea, Peracarida). Zoologicheskii Zhurnal, **70**(3): 30–37.

— —, 1993. Tanaidacea (Crustacea, Malacostraca) collected in the 43 cruise of the R/V "Dimitri Mendeleev" in the south–western Atlantic and the Weddell Sea. Trudy Instituta Okeanologii, Akademiya Nauk SSSR, **127**: 134–145.

KUDINOVA-PASTERNAK, R. K. & F. A. PASTERNAK, 1978. Deep-sea Tanaidacea (Crustacea, Malacostraca) collected in the Caribbean Sea and Puerto–Rico Trench during the 16th cruise of R/V "Akademic Kurchatov" and the resemblance between fauna of deep-sea Tanaidacea of the Caribbean region and the Pacific. Trudy Instituta Okeanologii, Akademiya Nauk SSSR, **113**: 178–197.

— — & — —, 1981. Tanaidacea collected by the Sovjet Antarctic Expedition during the years 1955–1958 and the correlation of the ranges of the Tanaidacea obtained in the Southern Ocean and their bathymetrical distribution. Trudy Instituta Okeanologii, Akademiya Nauk SSSR, **115**: 108–125.

KUMÉ, M. & K. DAN, 1968. Invertebrate embryology: 1–605. (National Library of Medicine, Washington, D.C.). [Translated from the Japanese.]

KUSSAKIN, O. G. & L. A. TZAREVA, 1974. A new genus of Tanaidacea from the Antarctic. Zoologicheskii Zhurnal, **53**(1): 125–128.

LANG, K., 1953a. *Apseudes hermaphroditicus* sp. nov. a hermaphroditic tanaid from the Antarctic. Arkiv för Zoologi, (2) **4**: 341–350.

— —, 1953b. The postmarsupial development of the Tanaidacea. Arkiv för Zoologi, (2) **4**: 409–422.

— —, 1956a. The Kalliapseudidae, a new family of Tanaidacea. In: K. G. WINGSTRAND, Bertil Hanström; zoological papers in honour of his 65 birthday November 29th 1956: 205–225. (Zoologisk Institut Særtryk, Lund).

— —, 1956b. Tanaidacea aus Brasilien, gesammelt von Prof. Dr. A. Remane und Dr. S. Gerlach. Kieler Meeresforschungen, **12**: 249–260.

— —, 1956c. Neotanaidae nov. fam. with some remarks on the phylogeny of Tanaidacea. Arkiv för Zoologi, (2) **9**(21): 469–475.

— —, 1958. Protogynie bei zwei Tanaidaceen–Arten. Arkiv för Zoologi, (2) **11**(32): 535–542.

— —, 1960. The genus *Oosaccus* Richardson and the brood pouch of some tanaids. Arkiv för Zoologi, (2) **13**(2): 77–81.

— —, 1966. Taxonomische und phylogenetische Unterschungen über die Tanaidaceen, 2. Die Gattung *Parapseudes* G. O. Sars. Arkiv för Zoologi, (2) **18**(24): 549–566.

— —, 1967. Taxonomische und phylogenetische Unterschungen über die Tanaidaceen, 3. Der Umfang der Familien Tanaidae Sars, Lang [sic] und Paratanaidae Lang, nebst Bemerkungen über den taxonomischen Wert der Mandibeln und Maxillulae. Dazu eine taxonomische–monographische Darstellung der Gattung *Tanaopsis* Sars. Arkiv för Zoologi, (2) **19**: 343–368.

— —, 1968. Deep-sea Tanaidacea. Galathea Report, **9**: 23–209.

— —, 1970a. Taxonomische und phylogenetische Untersuchungen über die Tanaidaceen, 4. Aufteilung der Apseudidae in vier Familien nebst Aufstellung von zwei Gattungen und einer neuen Art der Familie Leiopidae. Arkiv för Zoologi, **22**(16): 595–626.

——, 1970b. Taxonomische und phylogenetische Untersuchungen über die Tanaidaceen, 5. Die Gattung *Typhlotanais* G. O. Sars, 1882, nebst Beschreibung einer neuen Art dieser Gattung. Dazu eine Berichtigung der Dornenzahl des Enditen der Maxillulae bei *T. peculiaris* Lang, 1968. Arkiv för Zoologi, **23**(2): 267–291.

——, 1971a. Die Gattungen *Agathotanais* Hansen and *Paragathotanais* n. gen. (Tanaidacea). Crustaceana, **21**: 57–71.

——, 1971b. Taxonomische und phylogenetische Untersuchungen über die Tanaidaceen, 6. Revision der Gattung *Paranarthrura* Hansen, 1913 und Aufstellung von zwei neuen Familien, vier neuen Gattungen und zwei neuen Arten. Arkiv för Zoologi, (2) **23**: 361–401.

——, 1971c. Taxonomische und phylogenetische Untersuchungen über die Tanaidaceen, 7. Revision der Gattung *Strongylura* G. O. Sars, 1882 nebst Beschreibung einer neuen Art dieser Gattung. Arkiv för Zoologi, **23**(7): 403–415.

——, 1972. *Siphonolabrum mirabile* n. gen., sp. nov. (Tanaidacea). Crustaceana, (Supplement) **3**: 214–220.

——, 1973. Taxonomische und phylogenetische Untersuchungen über die Tanaidaceen (Crustacea), 8. Die Gattungen *Leptochelia* Dana, *Paratanais* Dana, *Heterotanais* G. O. Sars und *Nototanais*. Dazu einige Bemerkungen über die Monokophora und ein Nachtrag. Zoologica Scripta, **2**(5–6): 197–229.

LARSEN, K., 1998. Caprellidea (Crustacea; Amphipoda) from the Faroe Islands waters, with a key to the North–Atlantic species. Fródskaparrit, Faroes Journal of Arts and Sciences, **46**: 81–90.

——, 1999a. Deep-sea tanaidaceans (Crustacea: Peracarida) from the Albatross cruises 1885–86, with keys to the suborder Neotanaidomorpha. Journal of Natural History, London, **33**(8): 1107–1132.

——, 1999b. Pacific Tanaidacea (Crustacea) revision of the genus *Agathotanais* with description of three new species. Records of the Australian Museum, **51**: 99–112.

——, 1999c. A new species of the deep-sea genus *Carpoapseudes* (Crustacea: Tanaidacea) from southwestern Atlantic. Zoosystema, **21**(4): 647–659.

——, 2000a. Tanaidacea (Crustacea; Peracarida) biology, polymorphism, distribution and systematics: 1–478. (Unpubl. Ph.D. Thesis, Macquarie University, Sydney).

——, 2000b. Revision of the genus *Collettea* (Crustacea: Tanaidacea). Invertebrate Taxonomy, **14**: 681–693.

——, 2001. Morphological and molecular investigation of polymorphism and cryptic species in tanaid crustaceans: implications for tanaid systematics and biodiversity estimates. Zoological Journal of the Linnean Society, London, **131**: 353–379.

——, 2002a. Tanaidacea (Crustacea: Peracarida) of the Gulf of Mexico X: the question of being male. Gulf and Caribbean Research, **14**: 53–66.

——, 2002b. *Curtipleon heterochelum*, a new species of Tanaidacea (Crustacea: Peracarida) from the Andaman Sea. Species Diversity, **7**(2): 145–154.

——, 2002 [and onwards]. Tanaidacea: Families. Version 1: 2 October 2002. http://crustacea.net'

——, 2003a. Proposed new standardized terminology for tanaidacean Crustacea. Journal of Crustacean Biology, **23**(3): 644–661.

——, 2003b. The Tanaidacean fauna (Peracarida) from a deep-sea cold seep in the Gulf of Mexico. Journal of Crustacean Biology, **23**(4): 777–794.

——, (in press). Tanaidacean fauna from geologically active habitats, I: the hydrothermal vent system of the Juan de Fuca Ridge, Escabana Trough and Gorda Ridge, east Pacific. Zootaxa.

LARSEN, K. & M. BŁAŻEWICZ-PASZKOWYCZ, 2003. A new species of *Neotanais* Beddard (Crustacea: Tanaidacea) from the Subantarctic, off the Falkland Islands. Zootaxa, **339**: 1–11.

LARSEN, K. & T. HANSKNECHT, 2003. Three new species of the genus *Neotanais* Beddard. Journal of Natural History, London, **37**(23): 2787–2806.

— — & — —, 2004. A new genus and species of freshwater tanaidacean *Pseudohalmyrapseudes aquadulcis* (Apseudomorpha; Parapseudidae) from Northern Territory, Australia. Journal of Crustacean Biology, **24**(4):

LARSEN, K. & R. W. HEARD, 2001. A new tanaidacean subfamily, Bathytanaidinae (Tanaidacea; Paratanaidae) from the Australian continental shelf and slope. Zootaxa, **19**: 1–22.

— — & — —, 2002. Two new deep-sea tanaidacean genera *Isopodidus* and *Cetiopyge* (Crustacea: Peracarida) from the Gulf of Mexico. Proceedings of the Biological Society of Washington, **115**(2): 403–411.

— — & — —, 2004a. Revision of the genus *Tanaella* (Crustacea: Tanaidacea). Journal of Natural History, London, **38**(5): 549–579.

— — & — —, 2004b. Tanaidacea (Crustacea: Peracarida) of the Gulf of Mexico, XV. A new species of *Protanaissus*. Gulf and Caribbean Research, **16**: 43–48.

LARSEN, K. & H. RAYMENT, 2002. New species of *Leptochelia* (Crustacea: Tanaidacea) from the Andaman Sea, north-eastern Indian Ocean. Phuket Marine Biological Center, (Special Publication) **23**(1): 17–31.

LARSEN, K. & G. D. F. WILSON, 1998. Tanaidomorphan systematics — is it obsolete? Journal of Crustacean Biology, **18**(2): 346–362.

— — & — —, 2002. Tanaidacean phylogeny. The first step: the superfamily Paratanaidoidea. Journal of Zoological Systematics and Evolutionary Research, **40**(4): 205–222.

LARWOOD, H. J. C., 1954. Crustacea Tanaidacea and Isopoda from the Suez Canal. The Annals and Magazine of Natural History, (7) **12**: 561–577.

LAUBITZ, D. R., 1972. The Caprellidae (Crustacea, Amphipoda) of the Atlantic and Arctic Canada. Publications in Biological Oceanography, **4**: 1–82.

LAUERMAN, L. M. L. & R. S. KAUFMANN, 1998. Deep-sea epibenthic echinoderms and a temporally varying food supply: results from a one year time series in the N.E. Pacific. Deep Sea Research, (2, Topical Studies in Oceanography) **45**(4–5): 817–842.

LAUTERBACH, K. E., 1970. Der Cephalothorax von *Tanais cavolinii* Milne-Edwards (Crustacea–Malacostraca), ein Beitrag zur vergleichenden Anatomie und Phylogenie der Tanaidacea. Zoologische Jahrbücher, (Anatomie und Ontogenie der Tiere) **87**: 94–204.

LEBER, K. M., III, 1983. Feeding ecology of decapod crustaceans and the influence of vegetation on the foraging success in a subtropical seagrass meadow: 1–166. (Unpubl. Ph.D. Thesis, Florida State University, Tallahassee, Florida).

LEITE, F. P. P., 1995. Seasonal and spatial distribution of *Kalliapseudes schubarti* Mañe-Garzon (Tanaidacea, Crustacea), in the Araca region, São Sebastião. Arquivos de Biologia e Tecnologia, **38**(2): 605–618.

LEVINGS, C. D. & F. RAFI, 1978. *Tanais stanfordi* Richardson, 1901 (Crustacea, Tanaidacea) from the Fraser River estuary, British Columbia. Syesis, **11**: 51–53.

LILLJEBORG, W., 1864. Bidrag till Kännedomen om de inom Sverige och Norrige förekommande Crustaceer af Isopodernas underordning och Tanaidernas familj. Uppsala Universitet Årsskrifter, **5**: 1–32.

LIM, S. T. A. & C. G. ALEXANDER, 1986. Reproductive behaviour of the caprellid amphipod, *Caprella scaura typica* Mayer, 1890. Marine Behaviour and Physiology, **12**: 217–230.

LINCOLN, R. J., 1979. British marine Amphipoda: Gammaridea: 1–658. (British Museum of Natural History, London).

LINCOLN, R. J., G. A. BOXSHALL & P. F. CLARK, 1982. A dictionary of ecology, evolution and systematics: 1–446. (Cambridge University Press, Cambridge).

LIVINGSTON, R. J., P. S. SHERIDAN, M. G. MCLANE, F. G. LEWIS, III & G. G. KOBYLINSKI, 1977. The biota of the Apalachicola Bay system: functional relationships. Florida Marine Resources Publications, **26**: 75–100.

LOBEL, P. B., S. P. BELKHODE, S. E. JACKSON & H. P. LONGERICH, 1990. Recent taxonomic discoveries concerning the mussel *Mytilus*: implications for biomonitoring. Archives of Environmental Contamination and Toxicology, **19**: 508–512.

LUTZ, R. A., D. JABLONSKI & R. D. TURNER, 1984. Larval development and dispersal at deep sea hydrothermal vents. Science, New York, **226**: 1452–1454.

MADSEN, F. J., 1961. On the zoogeography and origin of the abyssal fauna, in view of the knowledge of the Porcellanasteridae. Galathea Report, **4**: 177–218.

MAKKAVEEVA, E. B., 1968. Vidovoj sostov I raspredelenie Tanaidovych i ravnogogich Rakov v pribrenznom rajonie o Kuba. In: Issledovanije Central'no–Amerikaniskich Morej, **2**: 99–104.

MANTON, S. M., 1978. The Arthropoda, habits, functional morphology and evolution: 1–527. (Oxford University Press, London).

MARSHALL, N. B., 1979. Developments in deep sea biology: 1–566. (Blandford Press, Poole, Dorset).

MASUNARI, S., 1983. Postmarsupial development and population dynamics of *Leptochelia savignyi* (Krøyer, 1842) (Tanaidacea). Crustaceana, **44**: 151–162.

MAUCHLINE, J., 1980. The biology of mysids and euphausiids. Advances in Marine Biology, **18**: 1–681.

MAYER, M. A., 1985. Ecology of juvenile white shrimp, *Penaeus setiferus* Linnaeus, in a salt marsh habitat: 1–211. (Unpubl. M.Sc. Thesis, Georgia Institute of Technology, Atlanta, Georgia).

MAYR, E., 1971. Principles of systematic zoology: 1–568. (McGraw-Hill, New York).

MCCAIN, J. C., 1968. The Caprellidae (Crustacea: Amphipoda) of the western North Atlantic. United States National Museum Bulletin, **278**: 1–147.

MCSWEENY, E. S., 1982. A new *Pagurapseudes* (Crustacea: Tanaidacea) from southern Florida. Bulletin of Marine Science, **32**(2): 455–466.

MCWILLIAM, P. S., P. F. SALE & D. T. ANDERSON, 1981. Seasonal changes in resident zooplankton sampled by emergence traps in One Tree Lagoon, Great Barrier Reef. Journal of Experimental Marine Biology and Ecology, **52**: 185–203.

MENDOZA, J. A., 1982. Some aspects of the autecology of *Leptochelia dubia* (Krøyer, 1842) (Tanaidacea). Crustaceana, **43**: 225–240.

MENGE, B. A., 1976. Organisation of the New England rocky intertidal community: role of predation, competition and environmental heterogeneity. Ecological Monographs, **46**: 355–393.

MENZIES, R. J., 1953. The apseudid Chelifera of the eastern tropical and north temperate Pacific Ocean. Bulletin of the Museum of Comparative Zoology at Harvard College, **107**: 443–496.

MENZIES, R. J., R. Y. GEORGE & G. T. ROWE, 1973. Abyssal environment and ecology of the world oceans: 1–488. (J. Wiley and Sons, New York).

MEREDITH, W. H. & W. A. LOTRICH, 1979. Production dynamics of a tidal creek population of *Fundulus heteroclitus* (Linnaeus). Estuarine, Coastal and Marine Science, **8**: 99–118.

MESSING, C. G., 1977. *Neotanais persephone*, a new species of hadal Tanaidacea (Crustacea: Peracarida). Bulletin of Marine Science, **27**(3): 511–519.

— —, 1980. *Pagurapseudes* (Crustacea. Tanaidacea) in southeast Florida: functional morphology, post–marsupial development, ecology and shell use: 1–242. (Unpubl. Ph.D. Thesis, University of Miami, Coral Gables, Florida).

— —, 1983. Postmarsupial development and growth of *Pagurapsudes largoensis* McSweeny (Crustacea, Tanaidacea). Journal of Crustacean Biology, **3**(3): 380–408.

MEYER, G. & R. W. HEARD, 1989. Tanaidacea (Crustacea: Peracarida) of the Gulf of Mexico. VII. *Atlantapseudes lindae* sp. nov. (Apseudidae) from the continental slope of the northern Gulf of Mexico. Gulf Research Reports, **8**: 67–105.

MIYADI, D., 1938. Ecological studies on marine relics and landlocked animals in inland waters of Nippon. Philippine Journal of Science, **65**: 239–249.

MODLIN, R. F. & P. A. HARRIS, 1989. Observations on the natural history and experiments on the reproductive strategy of *Hargeria rapax* (Tanaidacea). Journal of Crustacean Biology, **9**(4): 578–586.

MOERS-MESSMER, W., 1936. Das Marsupium der Amphipoden und Tanaidaceen: 1–37. (Unpubl. Ph.D. Thesis, Humboldt University, Berlin).

MONOD, T., 1925. Tanaidacés et Isopodes aquatiques de l'Afrique occidentale et septentrionale. (1re partie: Tanaidacea, Anthuridea, Valvifera). Bulletin de la Société des Sciences Naturelles du Maroc, **3**(1): 61–77.

MOORE, H. F., 1894. *Tanais robustus*, a new species of Anisopoda. Proceedings of the Academy of Natural Sciences, Philadelphia, **46**: 90–94.

MORTON, J., 1954. The crevice fauna of the upper intertidal zone at Wembury. Journal of the Marine Biological Association of the United Kingdom, **33**: 187–224.

MÜLLER, H. G., 1992. *Tanzanapseudes polynesiensis* sp. nov. first species of Tanzanapseudinae from the Pacific (Crustacea: Tanaidacea: Metapseudidae). Cahiers de Biologie Marine, **33**: 191–108.

MUUS, B. J., 1967. The fauna of Danish estuaries and lagoons. Distribution and ecology of dominating species in the shallow reaches of the mesohaline zone. Meddelelser fra Danmarks Fiskeri– og Havundersøgelser, **5**(1): 1–315.

MYERS, A. C., 1977. Sediment processing in a marine subtidal sandy bottom community: I. Physical aspects. Journal of Marine Research, **35**: 609–632.

NELSON, W. G., 1981. Experimental studies of decapod and fish predation on seagrass macrobenthos. Marine Ecology Progress Series, **5**: 141–149.

NIERSTRASZ, H. F., 1913. Die Isopoden der Siboga–Expedition, I. Isopoda Chelifera. Siboga Expeditie Mon., **32**(a): 1–56.

ODUM, W. E. & E. J. HEALD, 1972. Trophic analysis of an estuarine mangrove community. Bulletin of Marine Science, **22**: 671–738.

OGLE, J. T., 1977. Tanaidacea from the Gulf of Mexico. A preliminary summary. Abstract. Journal of the Mississippi Academy of Sciences, **22**: 105.

OGLE, J. T., R. W. HEARD & J. SIEG, 1982. Tanaidacea (Crustacea: Peracarida) of the Gulf of Mexico. II. Introduction and an annotated bibliography of Tanaidacea previously reported from the Gulf of Mexico. Gulf Research Reports, **7**: 101–104.

OLIVER, J. S. & P. N. SLATTERY, 1985. Effects of crustacean predators on species composition and population structure of soft bodied infauna from McMurdo Sound, Antarctica. Ophelia, **24**: 155–175.

OVERSTREET, R. M. & R. W. HEARD, 1978. Food of the Atlantic croaker, *Micropogonias undulatus*, from Mississippi Sound and the Gulf of Mexico. Gulf Research Reports, **6**: 145–152.

OWEN, R. K. & A. T. FORBES, 1997. Salinity, floods and the infaunal macrobenthic community of the St. Lucia Estuary, KwaZulu–Natal, South Africa. South African Journal of Aquatic Sciences, **23**(1): 14–30.

PALUMBI, S. R., 1992. Marine speciation on a small planet. Trends in Ecology and Evolution, **7**: 114–118.

PALUMBI, S. R. & J. BENZIES, 1991. Large mitochondrial DNA differences between morphologically similar penaeid shrimp. Molecular Marine Biology and Biotechnology, **1**(1): 27–34.

PALUMBI, S. R. & A. C. WILSON, 1990. Mitochondrial DNA diversity in sea urchins *Strongylocentrotus purpuratus* and *S. droebachiensis*. Evolution, **44**: 403–415.

PECK, S. B., 1993. Diversity and zoogeography of the non–oceanic Crustacea of the Galapagos Islands, Ecuador (excluding terrestrial Isopoda). Canadian Journal of Zoology, **72**: 54–69.

PEQUEGNAT, W. E., B. J. GALLAWAY & L. H. PEQUEGNAT, 1990. Aspects of the ecology of the deep–water fauna of the Gulf of Mexico. American Zoologist, **30**: 45–64.

PETERSON, C. H., 1977. Competitive organisation of the soft bottom macrobenthic communities of the southern California lagoons. Marine Biology, Berlin, **43**: 343–359.

PIRES, A. M. S., 1980. Ecological studies on intertidal and infralittoral Brazilian Tanaidacea (Crustacea) (Peracarida). Studies on Neotropical Fauna and Environment, **15**: 141–153.

PLATELL, M. E., I. C. POTTER & K. R. CLARKE, 1998. Do the habitats, mouth morphology and diets of the mullids *Upeneichthys stotti* and *U. lineatus* in coastal waters of south–western Australia differ? Journal of Fish Biology, **52**(2): 398–418.

POORE, G. C. B., J. JUST & B. F. COHEN, 1994. Composition and diversity of Crustacea Isopoda of the southeastern Australian continental slope. Deep-sea Research, **41**: 677–693.

PRESTON, N. P. & P. J. DOHERTY, 1994. Cross–shelf patterns in the community structure of coral–dwelling Crustacea in the central region of the Great Barrier Reef. II. Cryptofauna. Marine Ecology Progress Series, **104**: 27–38.

RAPOPORT, E. H., 1982. Areography: geographical strategies of species. (Pergamon Press, Oxford).

REGER, J. F., 1966. A comparative study on the fine structure of developing spermatozoa in the isopod, *Oniscus asellus* and the amphipod, *Orchestoidea* sp. Zeitschrift für Zellforschung und Mikroskopische Anatomie, **75**: 579–590.

REIDENAUER, J. A. & D. THISTLE, 1985. The tanaid fauna from a region of the deep North Atlantic where near–bottom current velocities are high. Oceanologia Acta, **8**(3): 355–360.

REIMER, A. A., 1975. Description of a *Tetraclita stalactifera panamensis* community on a rocky, intertidal Pacific shore of Panama. Marine Biology, Berlin, **35**: 225–238.

REX, M. A., 1977. Zonation in deep-sea gastropods: the importance of biological interactions to rates of zonation. Proceedings of the European Marine Biology Symposium, **11**: 521–530.

— —, 1981. Community structures in the deep-sea benthos. Annual review of Ecology and Systematics, **12**: 331–353.

REX, M. A., R. J. ETTER & C. T. STUART, 1997. Large-scale patterns of species diversity in the deep-sea benthos. In: R. J. A. ORMOND, J. D. GAGE & M. A. ANGEL, Marine biodiversity: causes and consequences: 94–121. (Cambridge University Press, Cambridge).

REX, M. A. & A. WAREN, 1982. Planktotrophic development in deep-sea prosobranch snails from the western North Atlantic. Deep-sea Research, **29**: 171–184.

RHOADS, D. L., 1974. Organism–sediment relations on the muddy sea floor. Oceanography and Marine Biology, an Annual Review, **12**: 263–300.

RICE, A. L., M. H. THURSTON & A. L. NEW, 1990. Dense aggregations of a hexactinellid sponge, *Pheronema carpenteri*, in the Porcupine Seabight (northeast Atlantic Ocean) and possible causes. Progress in Oceanography, **24**: 179–196.

RICHARDS, I., 1969. Tanaidacea (Crustacea: Peracarida) of San Juan Island. Zoological Report, **533**(b): 1–18. (Friday Harbour Laboratories, University of Washington, Friday Harbour, Washington).

RICHARDSON, H., 1901. Papers from the Hopkins Stanford Galapagos Expedition, 1898–1899, VI. The isopods. Proceedings of the Washington Academy of Science, **3**: 565–568.

——, 1903. The marine and terrestrial isopods of the Bermudas, with descriptions of two new genera and species. Transactions of the Connecticut Academy of Arts and Sciences, **11**(1): 277–310.

ROUBAULT, A., 1937. Dimorphisme et croissance chez un Tanaïdacé. Travaux de la Station Biologique de Roscoff, **15**: 135–152.

ROWE, G. T. & R. J. MENZIES, 1969. Zonation of large benthic invertebrates in the deep-sea off the Carolinas. Deep-sea Research, **16**: 531–537.

ROWE, G. T., P. T. POLLONI & R. L. HEADRICH, 1982. The deep-sea macrobenthos on the continental margin of the northwest Atlantic Ocean. Deep-sea Research, **29**: 257–278.

ROZAS, L. P. & L. M. LASALLE, 1990. A comparison of the diets of Gulf killifish, *Fundulus heteroclitus* Baird and Girard, entering and leaving a Mississippi brackish marsh. Estuaries, (B) **13**: 332–336.

SALVAT, B., 1967. La macrofaune carcinologique endogée des sédiments meubles intertidaux (Tanaidacés, Isopodes et Amphipodes), ethologie, bionomie et cycle biologique. Mémoires du Muséum National d'Histoire Naturelle, Paris, (A) **45**: 1–275.

SANDERS, H. L., 1969. Benthic marine diversity and the stability–time hypothesis. Brookhaven Symposium in Biology, **22**: 71–81.

——, 1977. Evolutionary ecology and the deep-sea benthos. Academy of Natural Sciences, Philadelphia, Pennsylvania, (Special Publications) **22**: 223–243.

SANDERS, H. L. & J. F. GRASSLE, 1971. The interactions of diversity, distribution and mode of reproduction among major groupings of the deep-sea benthos. In: M. UDA, The world oceans. Proceedings of the Joint Oceanography Assembly: 260–262. (Society for Promotion of Science, Tokyo).

SANDERS, H. L. & R. R. HESSLER, 1969. Ecology of the deep-sea benthos. Science, New York, **163**: 1419–1424.

SARMA, A. L. N., 1979. Faunal associations of littoral sponges in and around Balugaon in Chilka Lake (Lagoon). Journal of the Bombay Natural History Society, **76**(1): 192–194.

SARS, G. O., 1869. Undersøgelser over Christianiafjordens Dybvandsfauna, anstillede paa en i Sommeren 1868 foretagen zoologisk Reise. Nyt Magazin for Naturvidenskaberne, **16**: 305–362.

——, 1882. Revision af Gruppen: Isopoda Chelifera med Characteristik af nye herhen hørende Arter og Slægter. Archiv for Matematik og Naturvidenskab, **7**: 1–54.

——, 1896. Isopoda. Parts I & II. Apseudidae, Tanaidae. An Account of the Crustacea of Norway with Short Descriptions and Figures of All the Species, **2**: 1–40.

SCHELTEMA, R., 1977. Dispersal of marine invertebrate organisms – paleobiogeographic and biostratigraphic implications. In: E. G. KAUFFMAN & J. E. HAZELS, Concepts and methods of biostratigraphy, **13**: 73–108. (Dowden, Hutchinson and Ross, Inc., Academic Press, Stroudsburg, New York).

——, 1978. On the relationship between dispersal of veliger larvae and the evolution of marine prosobranch gastropods. In: B. BATTAGLIA & J. A. BEARDMORE, Marine organisms: genetics, ecology and evolution: 303–322. (Plenum Press, New York).

SCHMIDT, A., 2001. The Tanaidacea fauna of the Beagle Channel and its relationship to the fauna of the Antarctic continental shelf. Reports on Polar Research, **333**: 1–113. [For the year 1999.]

SCHMIDT, A. & A. BRANDT, 2001a. The Tanaidacea fauna of the Beagle Channel (southern Chile) and its relationship to the fauna of the Antarctic continental shelf. Antarctic Sciences, **13**(4): 420–429.

—— & ——, 2001b. Diversity of Subantarctic Tanaidacea (Crustacea, Malacostraca) in and off the Beagle Channel. Polish Polar Research, **22**(3–4): 213–226.

SCHMIDT, A., V. SIEGLE & A. BRANDT, 2002. Postembryonic development of *Apseudes heroa* and *Allotanais hirsutus* (Tanaidacea, Crustacea) in Magellanic and Subantarctic waters. Antarctic Sciences, **14**(3): 201–211.

SCHRAM, F. R., J. SIEG & E. MALZAHN, 1986. Fossil Tanaidacea. Transactions of the San Diego Society of Natural History, **21**(8): 127–144.

SEGERSTRÅLE, S. G., 1962. Investigations on Baltic populations of the bivalve *Macoma balthica* (L.) Part II. What are the reasons for the periodic failure of recruitment and the scarcity of *Macoma* in the deeper waters of the inner Baltic? Commentationes Biologicae, **24**: 1–26.

SEXTON, E. W., 1924. The moulting and growth–stages of *Gammarus,* with descriptions of the normals and intersexes of *G. chevreuxi*. Journal of the Marine Biological Association of the United Kingdom, (new series) **13**: 340–401.

SHERIDAN, P., 1997. Benthos of adjacent mangrove, seagrass and non–vegetated habitats in Rookery Bay, Florida, U.S.A. Estuarine, Coastal and Shelf Science, **44**(4): 455–469.

SHICK, J. M., M. P. LESSER, W. C. DUNLAP, W. R. STOHAJ, B. E. CHALKER & J. W. WON, 1995. Transmission of *Pseudoterranova decipiens* (Nematoda: Ascaridoidea) via benthic macrofauna to sympatric flatfishes (*Hippoglossoides platessoides*, *Pleuronectes ferrugineus*, *P. americanus*) on Sable Island Bank, Canada. Marine Biology, Berlin, **122**(1): 129–135.

SHIINO, S. M., 1937. On *Apseudes nipponicus* sp. nov. (Crustacea Tanaidacea). Annotationes Zoologicae Japonenses, **16**: 53–62.

— —, 1951. On two new species of the family Apseudidae found at Seto. Report of Faculty of Fisheries, Prefectural University of Mie, **1**(1): 11–25.

— —, 1965. Tanaidacea from the Bismarck Archipelago. Videnskabelige Meddelelser fra Dansk Naturhistorisk Forening, (Noona Dan Papers, 17) **128**: 177–203.

— —, 1966. On *Kalliapseudes* (*Kalliapseudes*) *tomiokaenis* sp. nov. (Crustacea: Tanaidacea) from Japanese Waters. Report of Faculty of Fisheries, Prefectural University of Mie, **5**(3): 473–488.

— —, 1970. Paratanaidae collected in Chile Bay, Greenwich Island by the XXII Chilean Expedition, with an *Apseudes* from Porvenir Point, Tierra del Fuego Island. Instituto Antarctico de Chile, Santiago, (Serie Cientifico) **1**(2): 77–122.

SHILLAKER, R. O. & P. G. MOORE, 1987. The biology of brooding in the amphipods *Lembos websteri* Bate and *Corophium bonnellii* Milne-Edwards. Journal of Experimental Marine Biology and Ecology, **110**: 113–132.

SHOUKR, F. A., M. H. MONA & H. EL-SHEIKH, 1991. Identification of marine fouling Crustacea in the Suez Canal, with a note on distribution and migration. Pakistan Journal of Zoology, **23**(1): 79–83.

SIEBENALLER, J. F., 1978. Genetic variability in deep sea fishes of the genus *Sebastolobus*, Scorpaenidae. In: B. BATTAGLIA & J. BEARDMORE, Marine organisms: genetics, ecology and evolution: 95–122. (Plenum Press, New York).

SIEBENALLER, J. F. & G. N. SOMERO, 1978. Pressure adaptive differences in lactate dehydrogenases of congeneric fishes living at different depths. Science, New York, **201**: 255–257.

SIEBOLD, C. T., 1836. Über die Spermatozoen der Crustaceen, Insecten, Gastropoden und anderer wirbellosen Thiere. Archiv für Anatomie, Physiologie und Wissenschaftliche Medizin, **1836**: 13–53.

SIEG, J., 1972. Untersuchungen über Tanaidaceen. Kieler Meeresforschungen, **28**(2): 232–236.

— —, 1976. Zum natürlichen System der Dikonophora Lang (Crustacea, Tanaidacea). Zeitschrift für Zoologische Systematik und Evolutionsforschung, **14**: 177–198.

— —, 1977. Taxonomische Monographie der Familie Pseudotanaidae (Crustacea, Tanaidacea). Mitteilungen aus dem Zoologischen Museum Berlin, **53**(1): 1–109.

——, 1978. Bemerkungen zur Möglichkeit der Bestimmung der Weibchen bei den Dikonophora und der Entwicklung der Tanaidacea. Zoologischer Anzeiger, **200**: 233–241.
——, 1979. Erstnachweis von *Expina typica* Lang für den Atlantischen Ozean (Tanaidacea). Crustaceana, **36**(2): 189–190.
——, 1980a. Taxonomische Monographie der Tanaidae Dana, 1849 (Crustacea: Tanaidacea). Abhandlungen der Senckenbergischen Naturforschenden Gesellschaft, **537**: 1–267.
——, 1980b. Sind die Dikonophora eine polyphyletische Gruppe? Zoologischer Anzeiger, **205**: 401–416.
——, 1982a. Über ein "connecting link" in der Phylogenie der Tanaidomorpha (Tanaidacea). Crustaceana, **43**(1): 65–77.
——, 1982b. Anmerkungen zur Gattung *Tannaissus* Norman & Scott, 1906 (Crustacea: Tanaidacea: Nototanaidae). Senckenbergiana Biologica, **63**: 113–135.
——, 1982c. Tanaidacea. In: S. H. HURLBERT & A. VILLALOBOS-FIGUEROA, Aquatic biota of Mexico, Central America and the West Indies: 203–205. (State University of California at San Diego, San Diego, California).
——, 1983a. Tanaidacea. In: H.-E. GRUNER & L. B. HOLTHUIS, Crustaceorum catalogus, **6**: 1–552. (Dr. W. Junk Publishers, The Hague).
——, 1983b. Evolution of Tanaidacea. In: F. R. SCHRAM (ed.), Crustacean phylogeny. Crustacean Issues, **1**: 229–256. (A. A. Balkema, Rotterdam).
——, 1983c. Tanaidomorpha (Crustacea: Tanaidacea) from the Ross Sea, Antarctica. Journal of the Royal Society of New Zealand, **13**(4): 395–418.
——, 1984. Neuere Erkenntnisse zum natürlichen System der Tanaidacea. Eine phylogenetische Studie. Zoologica, Stuttgart, **136**: 1–132.
——, 1986a. Crustacea Tanaidacea of the Antarctic and Subantarctic. 1. On material collected at Tierra del Fuego, Isla de los Estados and the west coast of the Antarctic peninsula. Biology of the Antarctic Seas, **45**: 1–180.
——, 1986b. Distribution of the Tanaidacea: synopsis of the known data and suggestions on possible distribution patterns. In: R. H. GORE & K. L. HECK (eds.), Crustacean biogeography. Crustacean Issues, **4**: 165–194. (A. A. Balkema, Rotterdam).
——, 1988. Das phylogenetische System der Tanaidacea und die Frage nach Alter und Herkunft der Crustaceenfauna des antarktischen Festlandsockels. Zeitschrift für Zoologische Systematik und Evolutionsforschung, **26**(5): 363–379.
SIEG, J. & M. DOJIRI, 1989. Remarks on *Araphura* Bird & Holdich (Crustacea, Tanaidacea) and allied genera, including descriptions of three new species. Zoologica Scripta, **18**: 115–137.
SIEG, J. & R. W. HEARD, 1985. Tanaidacea (Crustacea: Peracarida) of the Gulf of Mexico. IV. On *Nototanioides trifurcatus* gen. nov. sp. nov. with a key to the genera of the Nototanaidae. Gulf Research Report, **8**(1): 51–62.
—— & ——, 1988. Tanaidacea (Crustacea: Peracarida) of the Gulf of Mexico. V. The family Pseudotanaidae from less than 200 m, with the description of *Pseudotanais mexikolpos* sp. nov. and a key to the known genera and species of the world. Proceedings of the Biological Society of Washington, **101**: 39–59.
—— & ——, 1989. Tanaidacea (Crustacea: Peracarida) of the Gulf of Mexico. VI. On the genus *Mesotanais* Dollfus, 1897 with descriptions of 2 new species, *M. longisetosus* and *M. vadicola*. Gulf Research Report, **8**: 73–95.
SIEG, J., R. W. HEARD & J. T. OGLE, 1982. Tanaidacea (Crustacea: Peracarida) of the Gulf of Mexico. II. The occurrence of *Halmyrapseudes bahamensis* Băcescu and Guţu, 1974 (Apseudidae) in the eastern Gulf with redescription and ecological notes. Gulf Research Report, **7**(2): 105–113.

SIEG, J. & H. ZIBROWIUS, 1988. Association of a tube inhabiting tanaidacean, *Bifida scleractinicola* gen. nov., sp. nov., with bathyal scleractinians off New Caledonia (Crustacea Tanaidacea – Cnidaria Scleractinia). Mésogée, **48**: 189–199.

SIEWING, R., 1953. Morphologische Untersuchungen an Tanaidaceen und Lophogastriden. Zeitschrift für Wissenschaft, (Zoologie) **157**: 333–426.

——, 1957. Anatomie und histologie von *Thermosbaena mirabilis*. Ein beitrag zur phylogenie den reihe Pancarida (Thermosbaenacea). Abhandlung der Akademie den Wissenschaft und Literatur, Mainz, (Mathematische und Naturwissenschaftliche Klasse) **1957**(7): 195–270.

SILVA-BRUM, I. N., 1978. *Paraleiopus macrochelis* g. n., sp. n. do litoral sudeste do Brasil (Tanaidacea, Monokonophora). Revista Brasileira de Biologia, **38**(3): 639–645.

SLATKIN, M., 1985. Gene flow in natural populations. Annual Review of Ecology and Systematics, **16**: 393–430.

SMITH, C. R. & R. R. HESSLER, 1987. Colonisation and succession in the deep-sea ecosystems. Trends in Ecology and Evolution, **2**(12): 359–363.

SNELGROVE, P. V. R. & C. A. BUTMAN, 1994. Animal–sediment relationships revisited: cause versus effect. Oceanography and Marine Biology, an Annual Review, **32**: 111–177.

SOKOLOVA, M. N., 1959. On the distribution of deep–water bottom animals in relation to their feeding habits and the character of sedimentation. Deep-sea Research, **6**: 1–4.

——, 1972. Trophic structure of deep-sea macrobenthos. Marine Biology, Berlin, **16**: 1–12.

STEBBING, T. R. R., 1896. On the isopod genus *Leptochelia*. Annals and Magazine of Natural History, (6) **17**: 156–265.

——, 1900. On Crustacea brought by Dr. Willey from the South Seas. In: A. WILLEY, Zoological results based on material from New Britain, New Guinea, Royalty Islands and elsewhere collected during 1895, 1896 and 1897, **5**: 605–690. (Cambridge University Press, Cambridge).

STEELE, D. H., 1967. The life cycle of the marine amphipod *Stegocephalus inflatus* Krøyer in the northwest Atlantic. Canadian Journal of Zoology, **45**: 623–628.

STEPHENSEN, K., 1913. Grønlands krebsdyr og pycnogonider (Conspectus Crustaceorum et Pycnogoridorum Groenlandiae). Meddelelser om Grønland, **22**: 1–479.

——, 1915. Isopoda, Tanaidacea, Cumacea, Amphipoda (excl. Hyperiidea). Report on the Danish Oceanographic Expeditions, 1908–10, to the Mediterranean and Adjacent Seas, (Biology) **2**(D)(1): 1–53.

STEVENS, G. C., 1989. The latitudinal gradient in geographical range: how so many species coexist in the tropics. American Naturalist, **133**: 240–256.

——, 1992. The elevational gradient in altitudinal range: an extension of Rapoport's rule to altitude. American Naturalist, **140**: 893–911.

STONER, A. W., 1986. Cohabitation on algal habitat islands by two hermaphroditic Tanaidacea (Crustacea: Peracarida). Journal of Crustacean Biology, **6**(4): 719–728.

SVAVARSSON, J. & B. DAVIDSDOTTIR, 1995. *Cibicides* spp. (Protozoa, Foraminifera) as epizoites on the Arctic antenna–brooding *Arcturus baffini* (Crustacea, Isopoda, Valvifera). Polar Biology, **15**: 569–574.

SVAVARSSON, J., G. GUDMUNDSSON & T. BRATTEGARD, 1993. Feeding by asellote isopods (Crustacea) on Foraminifera (Protozoa) in the deep sea. Deep-sea Research, **40**: 1225–1239.

THIEL, M., 1999. Duration of extended parental care in marine amphipods. Journal of Crustacean Biology, **19**(1): 60–71.

THIEL, M., S. SAMPSON & L. WATLING, 1997. Extended parental care in two endobenthic amphipods. Journal of Natural History, London, **31**: 713–725.

THISTLE, D., J. Y. YINGST & K. FAUCHALD, 1985. A deep-sea benthic community exposed to strong near–bottom currents on the Scotland Rise (western Atlantic). Marine Geology, **66**: 91–112.

THOMSON, G. M., 1880. New species of Crustacea from New Zealand. Annals and Magazine of Natural History, (5) **6**: 1–6.

THORSON, G., 1966. Some factors influencing the recruitment and establishment of marine benthic communities. Netherlands Journal of Sea Research, **3**: 267–293.

THURSTON, M. H., D. S. M. BILLETT & E. HASSACK, 1987. An association between *Expina typica* Lang (Tanaidacea) and deep-sea holothurians. Journal of the Marine Biological Association of the United Kingdom, **67**: 11–15.

TROCH, M. DE, J. MEES & E. WAKWABI, 1998. Diets of abundant fishes from beach seine catches in seagrass beds of a tropical bay (Gazi Bay, Kenya). Belgian Journal of Zoology, **128**(2): 135–154.

VALENTIN, C., 1978. Zur Biozoenotik von Benthos Crustaceen: ein Beitrag zur Biologie und Autökologie der Cumaceen aus dem Küstenbereich der Insel Ischia (Golf von Neapel/Italien): 1–138. (Unpubl. Thesis, Christian–Albrechts Universität, Kiel).

VALENTINE, J. W. & D. JABLONSKI, 1983. Speciation in the shallow sea: general patterns and biogeographic controls. In: R. W. SIMS, J. H. PRICE & P. E. S. WHALLEY (eds.), Evolution, time and space: the emergence of the biosphere: 201–226. (Academic Press, London).

VALIELA, I., J. E. WRIGHT, J. M. TEAL & S. B. VOLKMAN, 1977. Growth, production and energy transformation in the saltmarsh killifish *Fundulus herteroclitus*. Marine Biology, Berlin, **40**: 135–144.

VISKUP, B. A. & R. W. HEARD, 1989. Tanaidacea (Crustacea: Peracarida) of the Gulf of Mexico. VII. *Pseudosphyrapus siegi* sp. nov. (Sphyrapidae) from the continental slope of the northern Gulf of Mexico. Gulf Research Reports, **8**: 107–115.

WALLACE, N. A. 1919. The Isopoda of the Bay of Fundy. University of Toronto Studies, **18**: 8–10.

WARÉN, A. & F. CARROZZA, 1994. *Arculus sykesi* (Chaster), a leptonacean bivalve living on a tanaid crustacean in the Gulf of Genova. Bolletino Malacologico, **29**(9–12): 303–306.

WIEDENBURG, S., U. NOLTE & N. L. WUERDIG, 1997. Macrobenthos of a coastal lake in southern Brazil. Archiv für Hydrobiologie, **140**(4): 533–548.

WILSON, G. D. F., 1983. Systematics of a species complex in the deep-sea genus *Eurycope*, with a revision of six previously described species (Crustacea, Isopoda, Eurycopidae). Bulletin of Scripps Institute of Oceanography, **24**: 1–64.

——, 1987. Crustacean communities of the Manganese Nodule Province. Report for the National Oceanic and Atmospheric Administration Office of Ocean and Coastal Resource Management (La Jolla, California): 1–35.

WILSON, G. D. F. & R. R. HESSLER, 1987. Speciation in the deep sea. Annual Review of Ecology and Systematics, **18**: 185–207.

WILSON, W. H., JR., 1980. A laboratory investigation of the effect of a terebellid polychaete on the survivorship of a nereid polychaete larva. Journal of Experimental Marine Biology and Ecology, **46**: 73–80.

——, 1981. Sediment–mediated interactions in a densely populated infaunal assemblage: the effect of the polychaete *Abarenicola pacifica*. Journal of Marine Research, **39**: 735–748.

WOLFF, T., 1956a. Crustacea Tanaidacea from depths exceeding 6000 m. Galathea Report, **2**: 187–241.

——, 1956b. Six new abyssal species of *Neotanais* (Crustacea, Tanaidacea). Videnskabelige Meddelelser fra Dansk Naturhistorisk Forening i København, **118**: 41–52.

— —, 1962. The systematics and biology of bathyal Isopoda Asellota. Galathea Report, **6**: 7–320.
— —, 1977. Diversity and composition of deep-sea benthos. Nature, London, **267**: 780–785.
WOODIN, S. A., 1974. Polychaete abundance patterns in a marine soft-sediment environment: the importance of biological interactions. Ecological Monographs, **44**: 171–187.
— —, 1976. Adult-larvae interactions in dense infaunal assemblages: patterns of abundance. Journal of Marine Research, **34**: 25–41.

APPENDIX I

Station list for the DGoMB Program, with the R/V "Gyre"

Cruise (#)	Station	Repli-cate (#)	Date (2000–2002)	Time	Depth (m)	Locality	
						Latitude N	Longitude W
1	WC 5	1	4 May 00	15:25	400	27°46'2838"	91°45'7293"
1	WC 5	2	4 May 00	17:23	320	27°47'1150"	91°45'8502"
1	WC 5	3	4 May 00	18:57	348	27°46'5547"	91°45'9407"
1	WC 5	4	4 May 00	20:18	330	27°46'906"	91°45'9578"
1	WC 5	5	4 May 00	22:06	325	27°46'9893"	91°45'8800"
1	WC12	1	5 May 00	13:01	1168	27°19'3945"	91°33'3486"
1	WC12	2	5 May 00	14:32	1125	27°19'1120"	91°33'2314"
1	WC12	3	5 May 00	15:49	1175	27°19'4320"	91°33'1625"
1	WC12	4	5 May 00	17:15	1169	27°19'1851"	91°33'0650"
1	WC12	5	5 May 00	18:55	1194	27°19'6198"	91°33'0871"
1	B 1	1	6 May 00	07:25	2260	27°12'2050"	91°24'1077"
1	B 1	2	6 May 00	08:57	2255	27°12'1374"	91°24'1806"
1	B 1	3	6 May 00	12:07	2253	27°12'1704"	91°24'3638"
1	B 1	4	6 May 00	13:48	2254	27°12'0086"	91°24'3648"
1	B 1	5	6 May 00	15:14	2254	27°12'1525"	91°24'3131"
1	NB 2	1	7 May 00	08:43	1530	27°08'0243"	91°59'9584"
1	NB 2	2	7 May 00	10:44	1530	27°08'2040"	91°59'9207"
1	NB 2	3	7 May 00	12:02	1530	27°07'9710"	92°00'0170"
1	NB 2	4	7 May 00	13:28	1530	27°08'1027"	91°59'9891"
1	NB 2	5	7 May 00	15:45	1530	27°08'0900"	92°00'0041"
1	NB 3	1	8 May 00	08:55	1875	26°33'4820"	91°49'3530"
1	NB 3	2	8 May 00	08:41	1875	26°33'3912"	91°49'4653"
1	NB 3	3	8 May 00	10:10	1875	26°32'9725"	91°49'1853"
1	NB 3	4	8 May 00	11:31	1875	26°32'3039"	91°49'4705"
1	NB 3	5	8 May 00	12:48	1875	26°32'3039"	91°49'5132"
1	NB 5	1	9 May 00	04:14	2070	26°15'1137"	91°12'6102"
1	NB 5	2	9 May 00	06:08	2060	26°15'0855"	91°12'7524"
1	NB 5	3	9 May 00	10:53	2060	26°14'8501"	91°12'7500"
1	NB 5	4	9 May 00	12:24	2060	26°15'2157"	91°12'6639"
1	NB 5	5	9 May 00	13:57	2065	26°14'7240"	91°12'5945"
1	B 3	1	10 May 00	09:07	2650	26°09'8667"	91°44'1060"
1	B 3	2	10 May 00	10:55	2580	26°09'7884"	91°43'9954"
1	B 3	3	10 May 00	12:41	2650	26°09'9750"	91°43'9772"
1	B 3	4	10 May 00	14:42	2610	26°09'9760"	91°44'1386"

1	B 3	5	10 May 00	16:37	2600	26°09'8542"	91°44'0725"
1	NB 4	1	11 May 00	09:01	2030	26°15'2711"	92°23'6978"
1	NB 4	2	11 May 00	10:51	2050	26°14'9693"	92°23'4731"
1	NB 4	3	11 May 00	13:37	2020	26°15'0096"	92°23'7449"
1	NB 4	4	11 May 00	15:14	2050	26°14'9712"	92°23'5933"
1	NB 4	5	11 May 00	16:43	2060	26°14'8050"	92°23'3372"
1	B 2	1	12 May 00	06:50	2630	26°33'0774"	92°12'9990"
1	B 2	2	12 May 00	08:57	2630	26°33'3750"	92°13'4716"
1	B 2	3	12 May 00	10:55	2625	26°32'9491"	92°13'1357"
1	B 2	4	12 May 00	12:47	2625	26°32'9158"	92°13'3885"
1	B 2	5	12 May 00	14:32	2635	26°33'0007"	92°12'9049"
1	W 1	1	13 May 00	20:42	379	27°34'7791"	93°32'8573"
1	W 1	2	13 May 00	21:20	389	27°34'8829"	93°33'0319"
1	W 1	3	13 May 00	22:17	420	27°34'5256"	93°33'0286"
1	W 1	4	13 May 00	23:09	420	27°34'5441"	93°33'0153"
1	W 1	5	13 May 00	23:54	415	27°34'6299"	93°33'0603"
1	W 2	1	14 May 00	08:26	625	27°24'8008"	93°20'2579"
1	W 2	2	14 May 00	09:18	625	27°24'7019"	93°20'3849"
1	W 2	3	14 May 00	10:41	625	27°24'8517"	93°20'4010"
1	W 2	4	14 May 00	11:30	625	27°24'8855"	93°20'4223"
1	W 2	5	14 May 00	12:14	625	27°24'8356"	93°20'4198"
1	W 3	1	14 May 00	20:21	860	27°10'3711"	93°19'3081"
1	W 3	2	14 May 00	21:26	860	27°10'3840"	93°19'3315"
1	W 3	3	14 May 00	22:29	875	27°10'4482"	93°19'4034"
1	W 3	4	14 May 00	23:40	860	27°10'4739"	93°19'2691"
1	W 3	5	15 May 00	00:27	859	27°10'3438"	93°19'3976"
1	W 4	1	15 May 00	11:42	1420	26°43'9027"	93°19'1708"
1	W 4	2	15 May 00	13:24	1460	26°43'8500"	93°19'1843"
1	W 4	3	15 May 00	14:43	1460	26°43'8088"	93°19'2061"
1	W 4	4	15 May 00	15:53	1460	26°43'7675"	93°19'2500"
1	W 4	5	15 May 00	16:59	1460	26°43'8494"	93°19'1836"
1	W 5	1	16 May 00	03:23	2755	26°15'9670"	93°21'6800"
1	W 5	2	16 May 00	07:47	2740	26°16'5781"	93°21'7309"
1	W 5	3	16 May 00	10:04	2750	26°16'1395"	93°21'7915"
1	W 5	4	16 May 00	12:05	2745	26°16'1871"	93°22'0669"
1	W 5	5	16 May 00	14:50	2775	26°16'0663"	93°19'9634"
1	W 6	1	17 May 00	11:48	3140	26°00'1707"	93°19'2166"
1	W 6	2	17 May 00	13:53	3145	25°59'7526"	93°18'8003"
1	W 6	3	17 May 00	15:42	3150	25°59'7601"	93°18'5516"
1	W 6	4	17 May 00	17:44	3145	25°59'8101"	93°18'4085"
1	W 6	5	17 May 00	19:00	3150	25°59'1728"	93°18'6528"
1	RW 6	1	18 May 00	12:00	3015	25°59'9222"	94°29'7608"

1	RW 6	2	18 May 00	16:43	3015	26°00'142"	94°29'381"
1	RW 6	3	18 May 00	19:07	3000	25°59'9982"	94°29'5629"
1	RW 6	4	18 May 00	20:59	3000	25°59'8684"	94°29'9071"
1	RW 6	5	18 May 00	22:59	3010	25°59'8382"	94°29'7347"
1	AC 1	1	19 May 00	21:20	2505	26°23'4995"	94°33'5773"
1	AC 1	2	19 May 00	01:19	2450	26°23'2813"	94°33'2633"
1	AC 1	3	19 May 00	05:14	2480	26°23'5394"	94°33'4048"
1	AC 1	4	19 May 00	09:44	2440	26°23'1654"	94°33'2954"
1	AC 1	5	19 May 00	11:48	2519	26°23'6140"	94°34'3849"
1	RW 5	1	21 May 00	00:12	1620	26°30'2819"	95°00'1866"
1	RW 5	2	21 May 00	01:36	1620	26°30'0261"	95°00'1315"
1	RW 5	3	21 May 00	02:48	1620	26°30'0206"	95°00'0512"
1	RW 5	4	21 May 00	04:05	1620	26°30'0206"	95°00'0025"
1	RW 5	5	21 May 00	05:25	1620	26°30'4516"	94°59'8033"
1	RW 4	1	21 May 00	15:14	1570	26°44'8861"	95°14'7680"
1	RW 4	2	21 May 00	16:19	1580	26°44'9468"	95°14'6826"
1	RW 4	3	21 May 00	17:34	1575	26°44'9707"	95°14'9825"
1	RW 4	4	21 May 00	18:59	1570	26°45'1621"	95°15'0741"
1	RW 4	5	21 May 00	19:59	1575	26°45'0852"	95°15'0105"
1	RW 3	1	22 May 00	03:50	1320	27°00'2983"	95°30'1920"
1	RW 3	2	22 May 00	05:03	1325	27°00'2956"	95°30'0541"
1	RW 3	3	22 May 00	06:19	1340	27°00'1881"	95°29'7639"
1	RW 3	4	22 May 00	07:27	1330	27°00'2706"	95°29'8642"
1	RW 3	5	22 May 00	08:45	1335	27°00'5014"	95°29'5417"
1	RW 2	1	22 May 00	16:06	950	27°15'2435"	95°44'6186"
1	RW 2	2	22 May 00	17:07	950	27°15'2852"	95°44'6402"
1	RW 2	3	22 May 00	18:05	950	27°14'9757"	95°44'8287"
1	RW 2	4	22 May 00	19:04	950	27°15'0531"	95°45'1451"
1	RW 2	5	22 May 00	19:54	950	27°15'2416"	95°44'8084"
1	RW 1	1	23 May 00	05:00	213	27°30'0242"	96°00'1437"
1	RW 1	2	23 May 00	05:39	213	27°29'9333"	96°00'2164"
1	RW 1	3	23 May 00	06:12	212	27°30'0733"	95°59'9273"
1	RW 1	4	23 May 00	06:58	212	27°30'0550"	96°00'1154"
1	RW 1	5	23 May 00	07:40	213	27°30'0085"	96°00'1708"
1	C 1	1	30 May 00	05:54	334	28°03'6046"	90°14'9329"
1	C 1	2	30 May 00	06:50	336	28°03'5783"	90°14'9483"
1	C 1	3	30 May 00	07:42	336	28°03'5629"	90°14'9494"
1	C 1	4	30 May 00	08:30	336	28°03'5830"	90°14'9610"
1	C 1	5	30 May 00	09:15	336	28°03'4286"	90°15'3698"
1	C 7	1	30 May 00	18:53	1080	27°43'6967"	89°58'7782"
1	C 7	2	30 May 00	20:04	1070	27°43'9713"	89°58'6211"
1	C 7	3	30 May 00	21:05	1066	27°43'8913"	89°59'0119"

1	C 7	4	30 May 00	22:04	1070	27°43'8089"	89°59'1213"
1	C 7	5	30 May 00	23:03	1072	27°43'8262"	89°58'9220"
1	C 4	1	31 May 00	10:43	1455	27°27'5640"	89°47'1391"
1	C 4	2	31 May 00	12:42	1452	27°27'6103"	89°46'7720"
1	C 4	3	31 May 00	13:55	1463	27°27'1450"	89°46'5588"
1	C 4	4	31 May 00	15:04	1476	27°27'0322"	89°45'7167"
1	C 4	5	31 May 00	16:16	1470	27°27'2854"	89°45'7098"
1	C14	1	1 Jun 00	08:33	2487	26°55'7737"	89°34'2830"
1	C14	2	1 Jun 00	11:16	2487	26°55'7970"	89°34'2245"
1	C14	3	1 Jun 00	13:31	2495	26°55'7900"	89°33'8676"
1	C14	4	1 Jun 00	15:25	2487	26°55'7932"	89°34'2542"
1	C14	5	1 Jun 00	17:19	2478	26°56'2946"	89°34'3503"
1	C12	1	2 Jun 00	15:04	2922	26°22'7651"	89°14'4849"
1	C12	2	2 Jun 00	17:09	2920	26°22'9752"	89°14'4854"
1	C12	3	2 Jun 00	19:15	2924	26°22'4978"	89°14'5836"
1	C12	4	2 Jun 00	21:30	2918	26°22'4015"	89°14'2552"
1	C12	5	3 Jun 00	00:08	2920	26°22'7838"	89°14'4179"
1	MT 5	1	3 Jun 00	21:49	2275	27°19'9308"	88°40'0690"
1	MT 5	2	3 Jun 00	23:44	2290	27°19'5819"	88°40'1733"
1	MT 5	3	4 Jun 00	01:59	2267	27°20'0759"	88°39'7337"
1	MT 5	4*	4 Jun 00	08:44	2263	27°20'1910"	88°39'5716"
1	MT 5	5	4 Jun 00	05:25	2280	27°19'9703"	88°39'3639"
1	MT 6	1	4 Jun 00	23:25	2745	27°00'0067"	87°59'8697"
1	MT 6	2	5 Jun 00	01:28	2750	27°00'0892"	87°59'2938"
1	MT 6	3	5 Jun 00	03:33	2743	26°59'7904"	87°59'9247"
1	MT 6	4	5 Jun 00	05:41	2745	26°59'9689"	87°59'7695"
1	MT 6	5	5 Jun 00	07:35	2743	27°00'0989"	87°59'9478"
1	S39	1	6 Jun 00	09:23	3007	27°29'9500"	87°00'0849"
1	S39	2	6 Jun 00	11:20	3000	27°29'6965"	86°59'8982"
1	S39	3	6 Jun 00	13:19	3000	27°29'2661"	87°00'0122"
1	S39	4	6 Jun 00	15:40	3000	27°29'2518"	86°59'8236"
1	S39	5	6 Jun 00	17:38	2998	27°29'0205"	86°59'9889"
1	S40	1	7 Jun 00	13:26	2975	27°50'3354"	86°45'1549"
1	S40	2	7 Jun 00	15:07	2975	27°50'1707"	86°45'0821"
1	S40	3	7 Jun 00	16:51	2972	27°50'2944"	86°45'0489"
1	S40	4	7 Jun 00	18:53	2978	27°49'9965"	86°45'2125"
1	S40	5	7 Jun 00	21:07	2894	27°50'3686"	86°45'0849"
1	S41	1	8 Jun 00	19:02	2974	28°00'8463"	86°34'5587"
1	S41	2	8 Jun 00	21:36	2974	28°01'4243"	86°34'2753"
1	S41	3	9 Jun 00	00:17	2978	28°00'7328"	86°34'1930"
1	S41	4	9 Jun 00	02:37	2983	28°00'6970"	86°33'9732"
1	S41	5	9 Jun 00	05:10	2976	28°00'8185"	86°34'4009"

1	S42	1	9 Jun 00	23:20	767	28°15'1557"	86°25'0663"
1	S42	2	10 Jun 00	00:11	768	28°15'1070"	86°25'0209"
1	S42	3	10 Jun 00	01:01	763	28°15'2344"	86°24'7995"
1	S42	4	10 Jun 00	01:57	766	28°15'1049"	86°24'8363"
1	S42	5	10 Jun 00	02:52	772	28°15'0602"	86°25'1562"
1	S43	1	10 Jun 00	12:55	366	28°30'1055"	86°04'9983"
1	S43	2	10 Jun 00	13:31	360	28°30'1434"	86°04'8562"
1	S43	3	10 Jun 00	14:19	362	28°30'0613"	86°04'8672"
1	S43	4	10 Jun 00	16:33	357	28°30'1982"	86°04'5881"
1	S43	5	10 Jun 00	17:13	358	28°30'1766"	86°04'6074"
1	S44	1	11 Jun 00	01:05	215	28°45'0140"	85°44'9642"
1	S44	2	11 Jun 00	01:51	213	28°45'0188"	85°44'9499"
1	S44	3	11 Jun 00	02:31	212	28°45'0632"	85°44'8807"
1	S44	4	11 Jun 00	03:09	212	28°45'0093"	85°44'8686"
1	S44	5	11 Jun 00	03:49	212	28°44'9996"	85°44'8622"
1	S35	1	11 Jun 00	18:47	658	29°20'0500"	87°03'3758"
1	S35	2	11 Jun 00	19:32	667	29°19'9897"	87°02'9021"
1	S35	3	11 Jun 00	20:25	668	29°19'9337"	87°03'0166"
1	S35	4	11 Jun 00	21:17	658	29°19'9877"	87°02'7818"
1	S35	5	11 Jun 00	22:08	666	29°20'1091"	87°02'7818"
1	S36	1	12 Jun 00	10:13	1825	28°55'1647"	87°40'2232"
1	S36	2	12 Jun 00	11:36	1832	28°55'0080"	87°40'0627"
1	S36	3	12 Jun 00	12:55	1826	28°55'1634"	87°40'2154"
1	S36	4	12 Jun 00	13:59	1827	28°55'1792"	87°40'5743"
1	S36	5	12 Jun 00	15:08	1830	28°55'1108"	87°40'3290"
1	S37	1	13 Jun 00	06:23	2388	28°33'4054"	87°45'7357"
1	S37	2	13 Jun 00	08:00	2382	28°33'4292"	87°45'6441"
1	S37	3	13 Jun 00	09:17	2387	28°33'2345"	87°45'8373"
1	S37	4	13 Jun 00	10:35	2382	28°33'2196"	87°45'7768"
1	S37	5	13 Jun 00	11:49	2381	28°33'2176"	87°46'0109"
1	S38	1a	14 Jun 00	07:16	2090	28°17'0006"	87°14'6903"
1	S38	1c	14 Jun 00	11:24	2635	28°16'4749"	87°19'9043"
1	S38	2	14 Jun 00	13:23	2642	28°16'3144"	87°19'5165"
1	S38	3	14 Jun 00	15:14	2627	28°16'7846"	87°20'1025"
1	S38	4	14 Jun 00	18:40	2630	28°16'7579"	87°20'2446"
1	S38	5	14 Jun 00	20:37	2632	28°16'7968"	87°19'6555"
1	MT4	1	15 Jun 00	22:18	1401	27°49'6198"	89°09'9526"
1	MT4	2	15 Jun 00	23:22	1401	27°49'7018"	89°09'8829"
1	MT4	3	16 Jun 00	00:14	1401	27°49'6810"	89°10'0723"
1	MT4	4	16 Jun 00	01:32	1402	27°50'0087"	89°09'9487"
1	MT4	5	16 Jun 00	02:38	1402	27°49'6563"	89°09'9687"
1	MT 3	1	16 Jun 00	12:31	983	28°13'2246"	89°29'7679"

1	MT 3	2	16 Jun 00	13:29	987	28°13'1533"	89°29'7860"
1	MT 3	3	16 Jun 00	14:22	990	28°13'1426"	89°29'5065"
1	MT 3	4	16 Jun 00	15:14	988	28°13'0561"	89°29'6289"
1	MT 3	5	16 Jun 00	16:16	985	28°13'2906"	89°29'6427"
1	MT 2	1	17 Jun 00	01:56	676	28°27'0646"	89°40'3563"
1	MT 2	2	17 Jun 00	02:45	677	28°27'0726"	89°40'2206"
1	MT 2	3	17 Jun 00	03:30	676	28°27'0327"	89°40'3636"
1	MT 2	4	17 Jun 00	04:15	677	28°27'0168"	89°40'4000"
1	MT 2	5	17 Jun 00	04:54	680	28°26'8755"	89°40'3167"
1	MT 1	1	17 Jun 00	14:07	480	28°32'5129"	89°49'7324"
1	MT 1	2	17 Jun 00	14:48	482	28°32'3703"	89°49'7338"
1	MT 1	3	17 Jun 00	15:25	482	28°32'4636"	89°49'6283"
1	MT 1	4	17 Jun 00	16:07	481	28°32'4352"	89°49'7257"
1	MT 1	5	17 Jun 00	16:52	481	28°32'4666"	89°49'5011"
2	MT1	1	2 Jun 01	22:53	487	28°32'2850"	89°49'6638"
2	MT1	2	2 Jun 01	23:56	490	28°32'1101"	89°49'5365"
2	MT1	3	3 Jun 01	00:52	485	28°32'3296"	89°49'8205"
2	MT1	4	3 Jun 01	01:50	480	28°32'4628"	89°49'8489"
2	MT1	5	3 Jun 01	02:47	478	28°32'5111"	89°49'7656"
2	MT3	1	4 Jun 01	04:50	980	28°13'4727"	89°30'7579"
2	MT3	2	4 Jun 01	07:09	982	28°13'4659"	89°30'3941"
2	MT3	3	4 Jun 01	09:24	984	28°13'3579"	89°30'3455"
2	MT3	4	4 Jun 01	10:58	984	28°13'2498"	89°30'3220"
2	MT3	5	4 Jun 01	12:22	985	28°13'2624"	89°30'5530"
2	HiPro	1	5 Jun 01	14:51	1574	28°33'0955"	88°34'7428"
2	HiPro	2	5 Jun 01	16:30	1567	28°33'2054"	88°34'6205"
2	HiPro	3	5 Jun 01	19:41	1562	28°33'1285"	88°34'2763"
2	HiPro	4	5 Jun 01	21:38	1578	28°33'0216"	88°34'0368"
2	HiPro	5	5 Jun 01	23:28	1565	28°33'1855"	88°34'3232"
2	S42	1	7 Jun 01	08:13	773	28°15'2223"	86°25'7206"
2	S42	2	7 Jun 01	09:52	773	28°15'2676"	86°25'7309"
2	S42	3	7 Jun 01	10:56	775	28°15'1132"	86°25'6429"
2	S42	4	7 Jun 01	12:08	774	28°15'2364"	86°25'6948"
2	S42	5	7 Jun 01	16:22	774	28°14'9967"	86°25'4675"
2	S41	1	8 Jun 01	11:08	2982	27°59'0570"	86°34'1492"
2	S41	2	8 Jun 01	13:18	2990	27°57'6364"	86°34'9326"
2	S41	3	8 Jun 01	15:17	3015	27°56'5569"	86°35'0140"
2	S41	4	8 Jun 01	21:52	2995	27°57'6441"	86°34'7163"
2	S36	1	10 Jun 01	03:17	1849	28°54'7195"	87°40'7687"
2	S36	2	10 Jun 01	04:59	1850	28°54'5855"	87°40'8908"
2	S36	3	10 Jun 01	06:53	1825	28°55'1485"	87°40'9469"
2	S36	4	10 Jun 01	08:22	1850	28°54'5677"	87°40'6751"

2	S36	5	10 Jun 01	09:46	1840	28°54'9072"	87°40'5773"
2	MT6	1	13 Jun 01	16:06	2740	26°59'4407"	88°00'8396"
2	MT6	2	13 Jun 01	18:23	2733	27°00'2038"	88°00'8728"
2	MT6	3	13 Jun 01	20:16	2742	26°59'1498"	88°00'6761"
2	MT6	4	13 Jun 01	22:32	2737	26°59'6643"	88°00'6912"
2	MT6	5	14 Jun 01	00:27	2740	26°59'2898"	88°00'8510"
2	GKF	1	15 Jun 01	01:07	2452	26°55'6309"	90°12'6372"
2	GKF	2	15 Jun 01	03:38	2473	26°55'2967"	90°13'2297"
2	GKF	3	15 Jun 01	06:48	2477	26°55'1654"	90°14'3360"
2	GKF	4	15 Jun 01	10:57	2468	26°56'1511"	90°13'9245"
2	GKF	5	15 Jun 01	15:22	2455	26°55'3965"	90°12'2895"
2	GKT	1	16 Jun 01	01:03	1880	27°00'5379"	90°17'9100"
2	GKT	2	16 Jun 01	02:40	1884	27°00'5193"	90°17'9343"
2	C7	1	16 Jun 01	13:44	1057	27°44'1073"	89°58'8717"
2	C7	2	16 Jun 01	14:59	1045	27°44'1045"	89°59'0927"
2	C7	3	16 Jun 01	16:00	1072	27°43'7961"	89°58'9058"
2	C7	4	16 Jun 01	17:00	1077	27°43'8876"	89°58'3715"
2	C7	5	16 Jun 01	18:04	1076	27°43'7424"	89°58'6605"
2	Bush Hill	1	18 Jun 01	08:20	546	27°47'8832"	91°28'2238"
2	Bush Hill	2	18 Jun 01	09:24	548	27°47'8935"	91°28'2050"
2	Bush Hill	3	18 Jun 01	10:19	540	27°47'9416"	91°28'1321"
2	Bush Hill	4	18 Jun 01	11:06	545	27°47'9088"	91°28'1539"
2	Bush Hill	5	18 Jun 01	11:53	533	27°47'9889"	91°28'1670"
3	S1	BC 3	3 Jun 02	21:12	3528	24°59'7900"	92°00'5180"
3	S1	BC 9	5 Jun 02	06:00	3523	25°00'3660"	92°00'1245"
3	S2	BC14	6 Jun 02	14:20	3732	23°28'7808"	92°00'4808"
3	S2	BC16	7 Jun 02	00:09	3732	23°29'3388"	92°00'3906"
3	S3	4	6 Aug 02	11:43	3675	24°45'1431"	90°45'2636"
3	S4	BC26	10 Jun 02	20:26	3410	24°14'5230"	85°29'0792"
3	S4	BC27	10 Jun 02	22:33	3410	24°14'6470"	85°28'8971"
3	S4	BC28	11 Jun 02	00:52	3410	24°14'9211"	85°29'3372"
3	S4	BC31	11 Jun 02	14:46	3405	24°15'3919"	85°28'7323"
3	S5	BC33	13 Jun 02	00:21	3312	25°29'4724"	88°15'7796"
3	S5	BC37	13 Jun 02	10:43	3316	25°29'5313"	88°16'2453"

TAXONOMIC INDEX

A[*pseudes*]. *acutifrons* 23, 49
A[*pseudes*]. *africanus* 49
A[*pseudes*]. *agulhensis* 49
A[*gathotanais*]. *ahyongi* 103
A[*pseudes*]. *annabensis* 49
A[*pseudes*]. *apocryphus* 49
A[*pseudes*]. *arguinensis* 49
A[*pseudes*]. *armatus* 49
A[*pseudes*]. *babelmandebensis* 49
A[*pseudes*]. *bacescui* 49
A[*pseudes*]. *bermudensis* 49
A[*raphuroides*]. *bombus* 232, 239
A[*pseudes*]. *bowmani* 49
A[*raphura*]. *breviaria* 219, 231
A[*raphura*]. *brevimanus* 219, 224, 231
A[*gathotanais*]. *brevis* 103
A[*raphuroides*]. *brevispina* 219, 232, 240
A[*pseudes*]. *browni* 49
A[*pseudes*]. *bruneinigma* 49
A[*pseudes*]. *cedroensis* 49
A[*pseudes*]. *chilkensis* 24, 49
A[*pseudes*]. *cooperi* 49
A[*pseudes*]. *crozetensis* 49
A[*raphura*]. *curticauda* 219, 231
A[*raphura*]. *cuspirostris* 219, 231
A[*tlantapseudes*]. *diversus* 52, 53, 54
A[*narthruropsis*]. *edentula* 142, 146, 147
A[*pseudes*]. *elisae* 49
A[*raphura*]. *elongata* 219, 224, 231
A[*pseudes*]. *espinosus* 49
A[*raphura*]. *extensa* 219, 224, 231
A[*raphura*]. *filiformis* 219, 224, 231
A[*pseudes*]. *galapagensis* 49
A[*narthruropsis*]. *galatheae* 142, 146, 147
A[*pseudes*]. *galeroni* 49
A[*gathotanais*]. *ghilarovi* 103
A[*pseudes*]. *graciloides* 49
A[*pseudes*]. *graziellae* 49
A[*pseudes*]. *grossimanus* 23, 49
A[*gathotanais*]. *hanseni* 103, 104
A[*pseudes*]. *hastifrons* 49
A[*pseudes*]. *heroae* 30, 49
A[*raphura*]. *higginsi* 219, 224, 231
A[*llotanais*]. *hirsutus* 30
A[*pseudes*]. *holthuisi* 49
A[*gathotanais*]. *ingolfi* 103, 104
A[*pseudes*]. *intermedius* 23, 49
A[*raphura*]. *joubinensis* 219, 224, 231
A[*pseudes*]. *lagenirostris* 49
A[*narthruropsis*]. *langi* 142, 146, 147
A[*pseudes*]. *latreillii* 49
A[*pseudes*]. *latus* 49
A[*pseudes*]. *levis* 50
A[*tlantapseudes*]. *lindae* 52, 53
A[*narthruropsis*]. *longa* 142, 146, 147
A[*gathotanais*]. *manganicus* 103
A[*pseudes*]. *mediterraneus* 50
A[*pseudes*]. *meridionalis* 50
A[*pseudes*]. *minimus* 50
A[*rmaturatanais*]. *minipodus* 244
A[*pseudes*]. *minutus* 50
A[*pseudes*]. *misarai* 50
A[*pseudes*]. *mussauensis* 50
A[*pseudes*]. *nagae* 50
A[*pseudes*]. *nhatrangensis* 50
A[*tlantapseudes*]. *nigrichela* 52, 53
A[*pseudes*]. *nipponicus* 24
A[*pseudes*]. *olimpiae* 50, 51
A[*pseudes*]. *orghidani* 50
A[*pseudes*]. *ostroumovi* 50
A[*raphura*]. *parabrevimanus* 219, 224, 231
A[*raphuroides*]. *parabreviremis* 232, 240
A[*pseudes*]. *paragracilis* 50
A[*pseudes*]. *portnovensis* 50
A[*raphura*]. *rectifrons* 219, 224, 231
A[*pseudes*]. *rikiteanus* 50
A[*pseudes*]. *robustus erythraeicus* 50
A[*pseudes*]. *robustus robustus* 50
A[*pseudes*]. *rotundifrons* 50
A[*pseudes*]. *sarsi* 23
A[*raphuroides*]. *septentrionalis* 232, 240
A[*pseudes*]. *setiferus* 50
A[*pseudes*]. *seurati* 50
A[*pseudes*]. *sicilianus* 50
A[*pseudes*]. *siegi* 50
A[*narthruropsis*]. *similis* 142, 146, 147
A[*pseudes*]. *spectabilis* 23, 24, 50
A[*gathotanais*]. *spinipoda* 103

A[*raphura*]. *spinithenari* 219, 224, 231
A[*pseudes*]. *spinosus* 24, 51, 52
A[*pseudes*]. *spinosus albidus* 50
A[*pseudes*]. *spinosus spinosus* 50
A[*gathotanais*]. *splendidus* 103
A[*pseudes*]. *talpa* 50
A[*pseudes*]. *tenuimanus* 50
A[*pseudes*]. *tenuipes* 50
A[*pseudes*]. *tenuis* 50
A[*pseudes*]. *tridens* 50
A[*pseudes*]. *tropicalis* 50
A[*pseudes*]. *tropicus* 50
A[*pseudes*]. *uncidigitatus* 50
A[*pseudes*]. *unicus* 50
A[*pseudes*]. *vicinus* 50
A[*pseudes*]. *vitjazi* 50
Agathotanaidae 10, 14, 15, **102**, 150
Agathotanaididae 102
Agathotanais 7, 15, 26, 102, **103**, 104, 107, 121, 261, 293, 298, 301
Agathotanais hanseni **104**, 105, 106, 313
Agathotanais ingolfi 103, 291
Allodaposia 242
Allotanais hirsutus 30, 41
Alotanais 96
Anarthrura 141, 150, 242, 257, 270, 298
Anarthrurella 141
Anarthruridae 14, 102, **141**, 150, 187
Anarthruropsis 1, 8, **141**, 142, 147, 301
Anarthruropsis edentula **142**, 143, 145, 317
Anarthruropsis galatheae 142
Anarthruropsis longa 147
Anasphyrapus 78
Androgнathia 242
Androtanais 242
Anisopechys 141
Anisopoda 44
Anthracocarididae 45
Anthracocaridomorpha 2, 45
Anthracocaris 45
Apsendes 48
Apsendidæ 46
Apseude 48
Apseudella typica 62
Apseudes 42, 46, 47, **48**, 49, 51, 52, 298, 300, 301
Apseudes algicola 94
Apseudes bowmani 54
Apseudes hermaphroditicus 23
Apseudes heroae 30, 41
Apseudes latifrons 94
Apseudes latreillii 24, 41
Apseudes olimpiae **50**, 306
Apseudes propinquus 3
Apseudes sibogae 76
Apseudes spectabilis 291, 337
Apseudes spinosa 51
Apseudes cf. *spinosus* **51**, 306
Apseudes spinosus 51
Apseudes talpa 29, 30, 51
Apseudiadae 46
Apseudidæ 46
Apseudidae 5, **46**, 47, 49, 52, 53, 77, 92
Apseudidae sp. A 53
Apseudidea 46
Apseudides 46
Apseudii 46
Apseudinae **47**, 48
Apseudomorpha 2, 5, 6, 8, 10, 11, 12, 16, 18, 19, 20, **23**, 24, 25, 29, 34, 35, 42, 44, 45, **46**, 300
Apseudopsis 48
Araphura 1, **218**, 219, 230, 231, 232, 300, 301
Araphura breviaria 224
Araphura curticauda **220**, 221, 222, 224, 327
Araphura cuspirostris 224
Araphura extensa **224**, 328
Araphura higginsi **225**, 328
Araphura spinithenari **225**, 226, 228, 230, 329
Araphuroides 1, 218, 219, **231**, 239, 301
Araphuroides bombus **232**, 233, 234, 236, 237, 238, 329
Araphuroides breviremis 232
Armatognathia 242
Armaturatanais 1, 242, **243**, 301
Armaturatanais minipodus 243, **244**, 245, 246, 248, 249, 250, 331
Arthrura 11, 217, 218, 270
Ascumnella 93, 300
Atlantapseudes 1, 46, 47, **52**, 53, 54, 300, 301
Atlantapseudes lindae **53**, 307
Atlantapseudes nigrichela 52

B[*athyleptochelia*]. *oculata* 184
Bathyleptochelia **183**, 184, 301
Bathyleptochelia oculata **184**, 322

Bathytanais 5, 21
Bathytanaissus 201
Bifida 242
Bilobatus 47
Bombus 232
Bunakenia 47, 300

C[*ryptocope*]. *abbreviata* 213
C[*hauliopleona*]. *amdrupii* 252
C[*ollettea*]. *antarctica* 157
C[*hauliopleona*]. *armata* 252
C[*ollettea*]. *arnaudi* 157
C[*ollettea*]. *cylindrata* 157, 159
C[*ollettea*]. *cylindratoides* 157
C[*hauliopleona*]. *dentata* 252
C[*ollettea*]. *elongata* 157
C[*hauliopleona*]. *faini* 252
C[*hauliopleona*]. *hastata* 252
C[*ollettea*]. *humbolti* 157
C[*ollettea*]. *inflatus* 258
C[*ollettea*]. *larviformis* 157
C[*ollettea*]. *longipedia* 157
C[*arpoapseudes*]. *longissimus* 62
C[*ollettea*]. *minima* 157
C[*etiopyge*]. *mira* 155
C[*arpoapseudes*]. *oculicornutus* 62
C[*ollettea*]. *pegmata* 157
C[*audalonga*]. *quatropleon* 150, 155
C[*ollettea*]. *rotundicauda* 157
C[*ollettea*]. *subtilis* 157
C[*ollettea*]. *vermiformis* 157
C[*ryptocope*]. *vitjazi* 213
C[*ollettea*]. *wilsoni* 157, 159
Calozidion 300
Cancer (*Gammarus*) *talpa* 48
Carlclausus 45
Carololangia 96, 301
Carpoapseudes 47, 52, 63, 65, 298, 301
Carpoapseudes kudinovae 61
Carpoapseudes serratispinosus 61
Caudalonga 1, 148, **149**, 155, 301
Caudalonga quatropleon **150**, 151, 152, 154, 318
Cerambycidae 63
Cetiopyge 148, 149, **155**, 301
Cetiopyge mira 155, **156**, 318
Chaetozone setosa 296
Chauliopleona 1, 242, **251**, 301
Chauliopleona dentata 252
Chauliopleona faini **252**, 254, 255, 331
Cirratodactylus floridensis 24
Coalecerotanais 10, 242, **257**, 258, 259, 301
Coalecerotanais inflatus 257, **258**, 332
Coleoptera 63
Collettea 26, 121, 148, 149, 155, **156**, 157, 159, 160, 293, 298, 301
Collettea sp. 159
Collettea sp. A **157**, 158, 159, 319
Collettea spp. 159
Collettea cylindrata 291
Collettea cylindratoides 37
Collettea elongata **157**, 159, 288, 319
Collettea longipedia 155
Collettea minima 291
Colletteidae 14, 15, **147**, 148, 149, 150
Cretitanaidae 45
Cretitanais 45
Cristotanais 148
Crurispina 1, 258, 259
Crurispina insolituchelia 259
Crustacea 42, 299
Cryptocope 211, **212**, 302
Cryptocopinae **212**
Cryptocopoides 212
Cryptocopoides arcticus 291, 337
Ctenapseudes 93
Curtichelia 212
Curtipleon 5, 12, 15, 34, 257
Cycloapseudes 300

D[*actyloprion*]. *anakanthes* 54
Dactyloprion 46, 47, **54**, 302
Dactyloprion anakanthes **54**, 307
Dalapseudes 93
Dalapseudes pedispinis 94
Dikonophora 44
Dimorphognathia 242
Discapseudes 93, 300
Discapseudes surinamensis 41

E[*xpina*]. *typica* 34, 260
Eucryptocaris 45
Euphée 48
Eupheus 48
Expina 155, 242, **259**, 302
Expina typica 33, **260**, 290, 291, 297, 333, 337

F[*ilitanais*]. *curticaudus* 160, 165
F[*ilitanais*]. *filiformis* 160, 165
F[*ilitanais*]. *moskalevi* 160, 165
F[*ilitanais*]. *rebainsi* 160, 165
F[*ilitanais*]. *vulgaris* 160
Fageapseudes 47
Familae indeterminatae **242**
Filiform species 157, 165, 181, **288**
Filitanais 1, 148, 149, **159**, 160, 165, 302
Filitanais curticaudus 160
Filitanais filiformis **160**, 162, 163, 288, 320
Filitanais moskalevi 159, 160
Filitanais vulgaris 160, 165
Foraminifera 30
Francapseudes 80

Gigantapseudidae 14
Glabroapseudes larseni 61
Gollumudes 47
Gutuapseudes 47

H[*eterotanais*]. *oerstedii* 23, 24, 40
Hainanius 48
Halmyrapseudes 93, 300
Haplocope 31, 148, 188
Hargeria 187, 300
Hargeria rapax 21, 292
Herpotanais 15, 96
Heterotanaidinae **183**
Heterotanais 16, 23, 24, 34, 183, 298, 300
Heterotanais antarcticus 23
Heterotanais oerstedii 17, 20, 21, 22, 23, 24, 36, 38, 39, 41
Heterotanaoides 182, 183
Hexapleomera 300
Hoplomachus 48
Hoplomachus propinquus 3

I[*sopodidus*]. *janum* 166
I[*ungentitanais*]. *primitivus* 213
I[*nsociabilitanais*]. *ventrospinatus* 261
Imitapseudes 34
Insociabilitanais 1, 208, 242, 243, **260**, 302
Insociabilitanais ventrospinatus 208, 261, **262**, 263, 264, 265, 266, 333
Intermedichelia 186, 187
Isopoda 44
Isopoda aberrantia 44
Isopoda chelifera 44
Isopoda vagantia 44
Isopodidus 148, 149, 150, **165**, 166, 243, 257, 261, 302
Isopodidus janum 165, **166**, 320
Ithyomus 141
Iungentitanais 211, 212, **213**, 300, 302
Iungentitanais primitivus 212, **213**, 326

Jurapseudes 45
Jurapseudidae 45

K[*udinopasternakia*]. *bispinosa* 91, 92
K[*udinopasternakia*]. *dispar* 91, 92
K[*udinopasternakia*]. *larisae* 91, 92
K[*udinopasternakia*]. *siegi* 91, 92
Kalliapseudes 35, 298, 300
Kalliapseudidae 10, 17, 19, 29, 33, 35, 91
Kudinopasternakia 1, 46, 80, **91**, 92, 302
Kudinopasternakia bispinosa **91**, 310
Kudinopasternakia siegi **92**, 310

L[*eviapseudes*]. *aberrans* 62, 65
L[*eptognathiella*]. *abyssi* 167, 174, 181
L[*eptognathia*]. *abyssorum* 188
L[*eptognathia*]. *acanthifera* 188
L[*eptognathia*]. *angustocephala* 188
L[*eptognathia*]. *arctophylax* 188
L[*eviapseudes*]. *bipartitus* 65, 76
L[*eptognathia*]. *birsteini* 188
L[*eptognathia*]. *brachiata* 188
L[*eptognathia*]. *breviremis* 188
L[*eptognathia*]. *breviremoides* 188
L[*eviapseudes*]. *caecus* 65
L[*eptognathiella*]. *clivicola* 167, 181
L[*eviapseudes*]. *conspicuus* 62, 65
L[*eptognathia*]. *crassa* 188
L[*eviapseudes*]. *demerarae* 65
L[*eptognathia*]. *dentifera* 188
L[*eviapseudes*]. *desbruyeresi* 65
L[*eptognathia*]. *distincta* 188
L[*eviapseudes*]. *drachi* 65
L[*eptochelia*]. *dubia* 37, 38
L[*eptognathia*]. *elegans* 188
L[*eptochelia*]. *forresti* 38
L[*eviapseudes*]. *galatheae* 65
L[*eptognathia*]. *glacialis* 188
L[*eptognathia*]. *glandiceps* 188
L[*eviapseudes*]. *gracilis* 65
L[*eviapseudes*]. *gracillimus* 23, 65

L[eptognathia]. greveae 188
L[eptognathia]. gyreae 188, 200
L[eviapseudes]. hanseni 65
L[eptognathia]. indivisa 188
L[eptognathia]. langi 188
L[eptognathia]. latiremis 188
L[eviapseudes]. leptodactylus 65
L[eptognathia]. linearis 188
L[eptognathia]. lineata 189
L[eptognathia]. longa 189
L[eptognathia]. longimanus 189, 194, 200
L[eviapseudes]. longispina 65
L[eptognathia]. luykeni 189
L[eptognathia]. manca 189
L[eptognathia]. microcephala 189
L[eptognathia]. mironovi 189
L[eptognathiella]. occidentis 167, 181
L[eptognathia]. parabranchiata 189
L[eptognathia]. paramanca 189
L[eptognathia]. parelegans 189
L[eviapseudes]. preamazonica 65
L[eptognathia]. profunda 189
L[eptognathia]. rotundicauda 189, 200
L[eptochelia]. savignyi 21
L[eviapseudes]. segonzaci gasconicus 65
L[eviapseudes]. segonzaci segonzaci 65
L[eviapseudes]. shiinoi 65
L[eviapseudes]. cf. *sibogae* 76
L[eviapseudes]. sibogae 62, 65
L[eptognathiella]. spinicauda 167, 181
L[eptognathiella]. subaequalis 167, 181
L[eptognathia]. tenella 189
L[eviapseudes]. thomassini 65
L[eptognathia]. tuberculata 189
L[eptognathia]. uncinata 189
L[eptognathia]. vanhoeffeni 189
L[eptognathia]. ventralis 189
L[eptognathiella]. vermiformis 167, 181
L[eptognathia]. vinogradovae 189
L[eptognathia]. vitjazi 189
L[eptognathia]. voeringi 189
L[eviapseudes]. weberi 62, 65
L[eviapseudes]. wolffi 62, 65
L[eptognathia]. zenkevitchi 189
L[eviapseudes]. zenkevitchi 65
L[eviapseudes]. zenkevitchioides 65
L[eptognathia]. zezinae 189
Langapseudes 48
Latitanais 212
Leiopidae 63
Leiopinae 63
Leiopus 63, 64
Leiopus longidactylus 64
Leiopus sibogae 76
Leptochelia 25, 34, 42, 186, 187, 298, 300
Leptochelia sp. 20
Leptochelia dubia 21, 36, 38, 291
Leptochelia forresti 36
Leptocheliidae 5, 10, 11, 14, 15, 25, 30, 36, **182**, 183, 185
Leptocheliinae **186**
Leptognathia 1, 25, 31, 36, 43, 166, 167, 187, **188**, 194, 200, 218, 240, 251, 276, 293, 298, 302
Leptognathia sp. G 200
Leptognathia brevimanus 219, 291
Leptognathia breviremis 290, 291
Leptognathia gyreae **189**, 190, 191, 192, 324
Leptognathia longimanus **194**, 195, 196, 198, 200, 324
Leptognathia parabrevimanu 219
Leptognathidae 187
Leptognathiella 1, 31, 148, 149, **166**, 167, 181, 302
Leptognathiella abyssi 166
Leptognathiella occidentis **167**, 169, 170, 171, 172, 321
Leptognathiella vermiformis **174**, 176, 177, 179, 180, 288, 321
Leptognathiidae 5, 10, 14, 15, 26, 39, 148, **187**, 213
leptognathiid genus H 200
Leptognathioides 242
Leptognathiopsis 148
Leviapseudes 1, 5, 12, 14, 15, 25, 31, 46, 47, 63, **64**, 65, 76, 298, 302
Leviapseudes bipartitus 47, 64, **65**, 66, 68, 70, 72, 73, 75, 91, 309
Leviapseudes galatheae 23
Leviapseudes hanseni 64
Leviapseudes shiinoi 62
Leviapseudes cf. *sibogae* **76**
Leviapseudes zenkevitchioides 64
Leviapseudidae 63
Leviapseudinae 48, 53, **63**, 65
Libanius 148, 149, 302
Longiflagrum 93

M[*esotanais*]. *dubius* 185
M[*esotanais*]. *elongatus* 185
M[*eromonakantha*]. *gilrowei* 203
M[*esotanais*]. *longisetosus* 185
M[*eromonakantha*]. *macrocephala* 203
M[*esotanais*]. *vadicola* 185
M[*irandotanais*]. *vorax* 34, 41
Macrinella 148
Macrolabrum 12, 15
Malacostraca 2
Mendamanus 48
Meromonakantha 1, 31, **202**, 203, 209, 302
Meromonakantha gilrowei **203**, 204, 205, 207, 325
Meromonakantha macrocephala 291
Mesotanais 1, 182, 183, **184**, 185, 186, 302
Mesotanais dubius 185, 186
Mesotanais elongatus 186
Mesotanais longisetosus **185**, 186, 323
Mesotanais vadicola **186**, 323
Metagathotanais 102, 150, 258
Metagathotanais insulcatus 257
Metapseudes 300
Metatanais 201
Mimicarhaphura 242
Mirandotanais 148, 155
Mirandotanais vorax 34, 41, 291, 297
Monokonophora 44
Monstrotanais 242
Mortensenia 213
Mortensenia primitiva 213
Mystriocentrus 214

N[*eotanais*]. *affinis* 98
N[*eotanais*]. *americanus* 97, 98
N[*eotanais*]. *antarcticus* 98
N[*eotanais*]. *armiger* 98
N[*eotanais*]. *bacescui* 98
N[*eotanais*]. *barfoedi* 98
N[*eotanais*]. *calcarulus* 98
N[*eotanais*]. *curvimanus* 98
N[*eotanais*]. *dinotomer* 98
N[*eotanais*]. *gardineri* 15, 98
N[*eotanais*]. *giganteus* 98
N[*eotanais*]. *hadalis* 98
N[*eotanais*]. *hamatus* 98
N[*eotanais*]. *hastiger* 98
N[*eotanais*]. *hessleri* 98
N[*eotanais*]. *insignis* 98
N[*eotanais*]. *krappschickelae* 98
N[*eotanais*]. *kurchatovi* 98
N[*eotanais*]. *laevispinosus* 98
N[*eotanais*]. *magnificus* 98
N[*eotanais*]. *mesostenoceps* 98
N[*eotanais*]. *micromopher* 23, 97, 98
N[*eotanais*]. *minimus* 98
N[*eotanais*]. *noelitaiti* 98
N[*eotanais*]. *peculiaris* 98
N[*eotanais*]. *persephone* 98
N[*eotanais*]. *pfaffi* 98
N[*eotanais*]. *pfaffioides* 98
N[*eotanais*]. *robustus* 98
N[*esotanais*]. *rugula* 35
N[*eotanais*]. *sandersi* 98
N[*eotanais*]. *triangulocephalus* 98
N[*eotanais*]. *tricarinatus* 98
N[*ototanoides*]. *trifurcatus* 268
N[*eotanais*]. *tuberculatus* 98
N[*eotanais*]. *vemae* 98
N[*eotanais*]. *wolffi* 98
N[*eotanais*]. *zenkevitchi* 98
Nematotanais 148
Neotanaidae **96**
Neotanaidomorpha 2, 6, 7, 8, 10, 11, 12, 14, 15, 16, 17, 18, 19, 20, **23**, 26, 29, 35, 42, 44, 45, **96**, 294
Neotanais 23, 45, **96**, 97, 98, 298, 302
Neotanais affinis 291, 337
Neotanais americanus 97
Neotanais armiger **98**, 311
Neotanais curvimanus **99**, 312
Neotanais hadalis 291
Neotanais micromopher 20, 23
Neotanais minimus **99**, 312
Nesotanais 14, 201
Nesotanais rugula 35
Nototanaidae 11, 14, 26, **201**
Nototanaididae 201
Nototanais 201
Nototanoides 242, **268**, 300, 302
Nototanoides trifurcatus **268**, 334

Ophthalmapseudes 45

P[*aragathotanais*]. sp. A 118
P[*arapseudotanais*]. *abyssalis* 215
P[*seudotanais*]. *abyssi* 216
P[*seudotanais*]. *affinis* 216

P[*aranarthrura*]. *angolensis* 121, 138
P[*seudosphyrapus*]. *anomalus* 80, 81
P[*aranarthrura*]. *bacescui* 121
P[*seudotanais*]. *baresnauti* 216
P[*aranarthrura*]. *bispinosa* 121, 129, 138
P[*aranarthrura*]. *borealis* 121, 140
P[*seudotanais*]. *californiensis* 216
P[*ectinapseudes*]. *carolinensis* 55, 62
P[*arafilitanais*]. *caudatus* 269
P[*seudosphyrapus*]. *centobi* 81
P[*seudotanais*]. *colonus* 216
P[*seudotanais*]. *corollatus* 216
P[*aranarthrura*]. *crassa* 121, 129, 138
P[*seudotanais*]. *denticulatus* 216
P[*seudotanais*]. *falcicula* 216
P[*seudotanais*]. *forcipatus* 216
P[*aranarthrura*]. *fortispina* 121, 132, 140
P[*arapseudes*]. *francispori* 94
P[*seudotanais*]. *gaussi* 216
P[*seudotanais*]. *gerlachi* 216
P[*aragathotanais*]. *gracilis* 107, 118
P[*seudotanais*]. *guillei* 216
P[*seudosphyrapus*]. *gutui* 81
P[*seudoarthrura*]. *heardi* 270
P[*arapseudes*]. *inermis* 94
P[*aranarthrura*]. *insignis* 121, 130, 138, 140, 210
P[*aranarthrura*]. *intermedia* 121, 129, 140
P[*seudotanais*]. *jonesi* 216
P[*aranarthrura*]. *kurchatovi* 121, 140
P[*agurapseudes*]. *largoensis* 24
P[*arapseudes*]. *latifrons* 94
P[*seudotanais*]. *lilljeborgi* 216
P[*seudotanais*]. *longipes* 216
P[*seudotanais*]. *longisetosus* 216
P[*seudotanais*]. *longispinus* 216
P[*aranarthrura*]. *lusitanus* 121, 138
P[*aragathotanais*]. *macrocephalus* 107, 118
P[*seudotanais*]. *macrochelis* 216
P[*ectinapseudes*]. *magnus* 55, 61, 62
P[*seudotanais*]. *makrothrix* 216
P[*seudotanais*]. *malayensis* 216
P[*seudotanais*]. *mediterraneus* 216
P[*aragathotanais*]. *medius* 107, 120
P[*aranarthrura*]. *meridionalis* 121, 140
P[*ectinapseudes*]. *mesteri* 55, 61, 62
P[*arafilitanais*]. *mexicanus* 269
P[*seudotanais*]. *mexikolpos* 216, 217
P[*seudotanais*]. *mortenseni* 216
P[*aragathotanais*]. *nanus* 108, 118
P[*seudotanais*]. *nordenskioldi* 216
P[*seudotanais*]. *oculatus* 216
P[*arapseudes*]. *pedispinis* 24
P[*aragathotanais*]. *robustus* 108, 120
P[*seudotanais*]. *scalpellum* 216
P[*seudosphyrapus*]. *serratus* 81
P[*seudotanais*]. *siegi* 216
P[*arafilitanais*]. *similis* 269
P[*aranarthrura*]. *similis* 121, 140
P[*seudotanais*]. *similis* 216
P[*seudotanais*]. *spatula* 216
P[*seudotanais*]. *spicatus* 216
P[*aragathotanais*]. *spinosus* 108, 118
P[*aranarthrura*]. *subtilis* 121, 140
P[*aranarthrura*]. *tenuimanus* 121, 129, 140
P[*aranarthrura*]. *tridens* 121, 140
P[*arapseudes*]. *trispinosus* 94
P[*aragathotanais*]. *typicus* 108, 118, 120
P[*aranarthrura*]. *undulata* 121, 140
P[*seudotanais*]. *unicus* 216
P[*aranarthrura*]. *vitjazi* 121, 140
P[*seudotanais*]. *vitjazi* 216
P[*seudosphyrapus*]. *vladimiri* 81
P[*seudotanais*]. *vulsella* 216
P[*aranarthrura*]. *zevinae* 121, 140
Pagurapseudes 12, 24, 37, 42, 301
Pagurapseudes largoensis 24, 41
Pakistanapseudes 93, 301
Paradoxapseudes 48
Parafilitanais 26, 121, 242, 243, **269**, 293, 302
Parafilitanais caudatus 269
Parafilitanais mexicana 269
Parafilitanais mexicanus **269**, 270, 334
Paragathotanais 1, 26, 102, **104**, 107, 118, 121, 293, 302
Paragathotanais sp. A **118**, 119, 315
Paragathotanais medius **108**, 313
Paragathotanais spinosus **108**, 109, 111, 314
Paragathotanais typicus 1, 107, **113**, 114, 115, 117, 118, 314
Paraiungentitanais 212
Paraleptognathia 242
Paraleptognathia gracilis 291, 337
Paraleptognathia longiremis 28, 291
Paranarthrura 1, 14, 26, 31, 41, 102, **120**, 121, 138, 232, 293, 298, 302
Paranarthrura sp. 32, 33

Paranarthrura abbreviata 140
Paranarthrura bacescui 140
Paranarthrura bispinosa **121**, 122, 124, 125, 127, 128, 129, 315
Paranarthrura cf. *insignis* **129**, 316
Paranarthrura insignis 120, 129, 130, 210, 291
Paranarthrura tenuimanus **130**, 131, 133, 134, 135, 136, 138, 316
Paranarthrura cf. *undulata* **138**, 139, 317
Paranarthrura undulata 138
Paranarthrurella 102, 270
Parapseudes 46, 47, **93**, 94, 301, 302
Parapseudes algicola 94
Parapseudes goodei 94
Parapseudes grubei 94
Parapseudes cf. *latifrons* **94**, 311
Parapseudes latifrons 23, 24, 94
Parapseudes latinfrons 94
Parapseudes neglectus 94
Parapseudes pedispinis 94
Parapseudes similis 94
Parapseudes spongicola 94
Parapseudidae 14, **92**, 93
Parapseudotanais 211, 212, **214**, 302
Parapseudotanais abyssalis 215
Paratanaidae 5, 10, 14, 26, 30, 39, 182, 211
Paratanais 5, 21, 25, 43, 215, 298, 301
Paratanais spinotandus 155
Paratanaoidea 5, 9, 10, 11, 12, 24, 45, **101**
Paratyphlotanais 201, 203, 208, 261, 302
Pectinapseudes 1, 46, 48, **55**, 55, 62, 302
Pectinapseudes carolinensis 55
Pectinapseudes magnus 1, **55**, 56, 58, 60, 62, 308
Pectinapseudes mesteri **62**, 308
Peracarida 2, 20, 290, 292
Peraeospinosus 201, 203, 293
Protanais 21
Protanaissus 201, 301
Psammokallioapseudes 301
Pseudoapseudes 93
Pseudoarthrura 1, 242, 243, **270**, 302
Pseudoarthrura heardi 270, **271**, 272, 273, 274, 335
Pseudobathytanais 31
Pseudoleptochelia 187, 301
Pseudoleptochelia antarctica 21, 23
Pseudoleptognathia 148
Pseudomacrinella 242
Pseudonototanais 187, 301
Pseudoparatanais 242
Pseudosphyrapinae 79
Pseudosphyrapodinae 77, **79**
Pseudosphyrapus 46, 77, 79, **80**, 90, 91, 302
Pseudosphyrapus sp. A **81**, 82, 83, 85, 87, 88, 90, 309
Pseudosphyrapus larisae 91
Pseudosphyrapus siegi 92
Pseudotanaidae 11, 24, **211**, 213
Pseudotanainae **214**
Pseudotanais 211, 212, 214, **215**, 293, 301, 302
Pseudotanais abyssi 291
Pseudotanais mexikolpos **216**, 326
Pseudotanais mortenseni 214, **217**, 327
Pseudozeuxidae 11
Pseudozeuxo 257
Pugiodactylinae 48

R[*obustochelia*]. *angusticephala* 276, 281
R[*obustochelia*]. *longa* 276, 281
R[*obustochelia*]. *robusta* 276, 281
R[*obustochelia*]. *solida* 276, 281
Rhœa 48
Rhœa spinosa 51
Rhoé 48
Rhoés 48
Rhoë 48
Rhoëa 48
Rhoëa latifrons 94
Rhoe 48
Rhoea 48
Rhoea latifrons 94
Rhoea spinosa 51
Robustochelia 1, 242, 243, **276**, 281, 302
Robustochelia longa 276, 281
Robustochelia solida **276**, 278, 279, 335
Robustognathia 242

S[*phyrapoides*]. *bicornis* 78, 79
S[*tenotanais*]. *crassiseta* 282, 288
S[*tenotanais*]. *hamicauda* 282, 288
S[*pinitanaopsis*]. *insolituchelia* 259
S[*tenotanais*]. *macrodactylus* 282, 288
S[*phyrapoides*]. *tuberculifrons* 78, 79
Safaritanais 242
Saksisopoder 44

Salemia 242
Saltipedis 93, 301
Scaphopoda 37
Scoloura 242
Sinelobus 25, 298, 301
Sinelobus stanfordi 28, 290, 291, 337
Siphonolabrum 8, 10, 141, 142, 242
Sphyrapidae 77
Sphyrapinae 77
Sphyrapodidae **77**
Sphyrapodinae 8, 77, **78**
Sphyrapoides 1, 46, **78**, 79, 301, 302
Sphyrapoides bicornis 78, **79**
Sphyrapus 77, 78
Sphyrapus serratus 80
Spinitanaopsis 1, 10, 242, 243, **258**, 259, 301
Spinitanaopsis insolituchelia **259**, 332
Spinosapseudes 48
Stenotanais 1, 242, 243, **281**, 282, 288, 302
Stenotanais crassiseta 282
Stenotanais macrodactylus **282**, 283, 285, 286, 336
Strongylura 156, 157
Strongylura antarctica 260
Strongylura cylindrata 156
Strongylurella 188
Subulella 148
Synanarthrura 141
Synapseudes 12, 298, 301

T[yphlotanais]. aequiremis 209
T[yphlotanais]. angularis 209
T[yphlotanais]. assimilis 209
T[yphlotanais]. brachyurus 209
T[yphlotanais]. brevicornis 209
T[anais]. cavolinii 33, 39
T[yphlotanais]. compactus 209
T[yphlotanais]. cornutus 209
T[yphlotanais]. crassus 209
T[yphlotanais]. dubius 209
T[anais]. dulongii 40
T[yphlotanais]. elegans 209
T[anaella]. eltaninae 241
T[yphlotanais]. eximius 209
T[yphlotanais]. filatovae 209
T[yphlotanais]. finmarchicus 209
T[anaella]. forcifera 241
T[yphlotanais]. gracilipes 209
T[yphlotanais]. grandis 209
T[yphlotanais]. greenwichensis 209
T[yphlotanais]. gruzovi 209
T[yphlotanais]. inaequipes 209
T[yphlotanais]. inermis 209
T[yphlotanais]. irregularis 209
T[yphlotanais]. kerguelenensis 209
T[anaella]. kimi 241
T[yphlotanais]. kussakini 209
T[yphlotanais]. longicephala 209
T[yphlotanais]. longimanus 209
T[yphlotanais]. longisetosus 209
T[yphlotanais]. longus 209
T[yphlotanais]. magnificus 41, 209
T[yphlotanais]. magnus 209
T[yphlotanais]. mananensis 209
T[anaella]. mclellandi 241
T[yphlotanais]. messinensis 209
T[yphlotanais]. microcheles 209
T[yphlotanais]. mucronatus 210
T[anaella]. paraforcifera 241
T[yphlotanais]. parangularis 210
T[yphlotanais]. parvus 210
T[yphlotanais]. peculiaris 210
T[yphlotanais]. penicillatus 210
T[yphlotanais]. plebejus 210
T[yphlotanais]. plicatus 210
T[yphlotanais]. proctagon 210
T[anaella]. profunda 241
T[yphlotanais]. profundus 210
T[anaella]. prolixcauda 241
T[anaella]. propinqua 241
T[anaella]. propinquus 241
T[yphlotanais]. pulcher 210
T[umidochelia]. randyi 182
T[yphlotanais]. rectus 210
T[yphlotanais]. richardi 210
T[anaella]. rotundicephala 241
T[yphlotanais]. rotundirostris 210
T[yphlotanais]. sandersi 210
T[yphlotanais]. setosus 210
T[yphlotanais]. simplex 210
T[yphlotanais]. solidus 210
T[yphlotanais]. spinicauda 210
T[yphlotanais]. spinipes 210
T[yphlotanais]. spiniventris 210
T[yphlotanais]. tenuicornis 210
T[yphlotanais]. tenuimanus 210
T[yphlotanais]. trispinosus 210
T[anaella]. tuberculata 240, 241

T[anaella]. unguicillata 241
T[anaella]. unisetosa 241
T[yphlotanais]. variabilis 210
T[yphlotanais]. williamsae 210
Tanabnormia 242
Tanaella 26, 121, 218, 232, 239, **240**, 251, 302
Tanaella mclellandi **241**, 330
Tanaella prolixcauda **241**, 330
Tanaella unguicillata 240
Tanaellidae 14, 15, 148, 150, **217**, 218
Tanaidacea 1, 2, 3, 6, 8, 11, 16, 20, 24, 26, 27, 31, **44**, 45, 95, 101, 291, 293, 295, 296, 298, 299, 300, 301
Tanaidae 16, 24, 25, 36, 37, 44, 211
Tanaidomorpha 2, 5, 6, 7, 9, 10, 11, 12, 14, 15, 16, 18, 19, 20, **21**, 23, 24, 25, 29, 35, 36, 38, 39, 42, 44, 45, **101**, 290, 295
Tanais 16, 34, 40, 43, 188, 202, 209, 215
Tanais abbreviatus 212
Tanais breviremis 188
Tanais cavolinii 6, 19
Tanais dulongii 6, 19, 21, 33, 38, 39, 40, 41, 292
Tanais forcipatus 215
Tanaissus 201, 301
Tanaissus lilljeborgi 291
Tanaoidea 5, 9, 10, 11, 12, 15, 18, 45, **101**
Tanaopsis 242, 259, 302
Tanapseudes 301
Tanapseudes gutui 35
Teleotanais 201, 301
Thorkelius 141
Trichapseudes 93
Tuberapseudes 48
Tumidochelia 148, 149, **181**, 302
Tumidochelia randyi **182**, 322
Typhlapseudes 48, 301
Typhlotanaidae 201
Typhlotanais 31, 201, 203, **209**, 240, 302
Typhlotanais magnificus 41
Typhlotanais cf. *trispinosus* **210**, 325
Typhlotanais trispinosus 210
Typhlotanaoides 201, 202, 203
Typhlotanoides 202

Venusticrus 96, 302

Whitelegiidae 46

Zeuxo 301